Geophysical Monograph Series

Including

IUGG Volumes
Maurice Ewing Volumes
Mineral Physics Volumes

GEOPHYSICAL MONOGRAPH SERIES

Geophysical Monograph Volumes

1 **Antarctica in the International Geophysical Year** *A. P. Crary, L. M. Gould, E. O. Hulburt, Hugh Odishaw, and Waldo E. Smith (Eds.)*
2 **Geophysics and the IGY** *Hugh Odishaw and Stanley Ruttenberg (Eds.)*
3 **Atmospheric Chemistry of Chlorine and Sulfur Compounds** *James P. Lodge, Jr. (Ed.)*
4 **Contemporary Geodesy** *Charles A. Whitten and Kenneth H. Drummond (Eds.)*
5 **Physics of Precipitation** *Helmut Weickmann (Ed.)*
6 **The Crust of the Pacific Basin** *Gordon A. Macdonald and Hisashi Kuno (Eds.)*
7 **Antarctica Research: The Matthew Fontaine Maury Memorial Symposium** *H. Wexler, M. J. Rubin, and J. E. Caskey, Jr. (Eds.)*
8 **Terrestrial Heat Flow** *William H. K. Lee (Ed.)*
9 **Gravity Anomalies: Unsurveyed Areas** *Hyman Orlin (Ed.)*
10 **The Earth Beneath the Continents: A Volume of Geophysical Studies in Honor of Merle A. Tuve** *John S. Steinhart and T. Jefferson Smith (Eds.)*
11 **Isotope Techniques in the Hydrologic Cycle** *Glenn E. Stout (Ed.)*
12 **The Crust and Upper Mantle of the Pacific Area** *Leon Knopoff, Charles L. Drake, and Pembroke J. Hart (Eds.)*
13 **The Earth's Crust and Upper Mantle** *Pembroke J. Hart (Ed.)*
14 **The Structure and Physical Properties of the Earth's Crust** *John G. Heacock (Ed.)*
15 **The Use of Artificial Satellites for Geodesy** *Soren W. Henricksen, Armando Mancini, and Bernard H. Chovitz (Eds.)*
16 **Flow and Fracture of Rocks** *H. C. Heard, I. Y. Borg, N. L. Carter, and C. B. Raleigh (Eds.)*
17 **Man-Made Lakes: Their Problems and Environmental Effects** *William C. Ackermann, Gilbert F. White, and E. B. Worthington (Eds.)*
18 **The Upper Atmosphere in Motion: A Selection of Papers With Annotation** *C. O. Hines and Colleagues*
19 **The Geophysics of the Pacific Ocean Basin and Its Margin: A Volume in Honor of George P. Woollard** *George H. Sutton, Murli H. Manghnani, and Ralph Moberly (Eds.)*
20 **The Earth's Crust: Its Nature and Physical Properties** *John G. Heacock (Ed.)*
21 **Quantitative Modeling of Magnetospheric Processes** *W. P. Olson (Ed.)*
22 **Derivation, Meaning, and Use of Geomagnetic Indices** *P. N. Mayaud*
23 **The Tectonic and Geologic Evolution of Southeast Asian Seas and Islands** *Dennis E. Hayes (Ed.)*
24 **Mechanical Behavior of Crustal Rocks: The Handin Volume** *N. L. Carter, M. Friedman, J. M. Logan, and D. W. Stearns (Eds.)*
25 **Physics of Auroral Arc Formation** *S.-I. Akasofu and J. R. Kan (Eds.)*
26 **Heterogeneous Atmospheric Chemistry** *David R. Schryer (Ed.)*
27 **The Tectonic and Geologic Evolution of Southeast Asian Seas and Islands: Part 2** *Dennis E. Hayes (Ed.)*
28 **Magnetospheric Currents** *Thomas A. Potemra (Ed.)*
29 **Climate Processes and Climate Sensitivity (Maurice Ewing Volume 5)** *James E. Hansen and Taro Takahashi (Eds.)*
30 **Magnetic Reconnection in Space and Laboratory Plasmas** *Edward W. Hones, Jr. (Ed.)*
31 **Point Defects in Minerals (Mineral Physics Volume 1)** *Robert N. Schock (Ed.)*
32 **The Carbon Cycle and Atmospheric CO_2: Natural Variations Archean to Present** *E. T. Sundquist and W. S. Broecker (Eds.)*
33 **Greenland Ice Core: Geophysics, Geochemistry, and the Environment** *C. C. Langway, Jr., H. Oeschger, and W. Dansgaard (Eds.)*
34 **Collisionless Shocks in the Heliosphere: A Tutorial Review** *Robert G. Stone and Bruce T. Tsurutani (Eds.)*
35 **Collisionless Shocks in the Heliosphere: Reviews of Current Research** *Bruce T. Tsurutani and Robert G. Stone (Eds.)*
36 **Mineral and Rock Deformation: Laboratory Studies—The Paterson Volume** *B. E. Hobbs and H. C. Heard (Eds.)*
37 **Earthquake Source Mechanics (Maurice Ewing Volume 6)** *Shamita Das, John Boatwright, and Christopher H. Scholz (Eds.)*
38 **Ion Acceleration in the Magnetosphere and Ionosphere** *Tom Chang (Ed.)*
39 **High Pressure Research in Mineral Physics (Mineral Physics Volume 2)** *Murli H. Manghnani and Yasuhiko Syono (Eds.)*
40 **Gondwana Six: Structure, Tectonics, and Geophysics** *Gary D. McKenzie (Ed.)*
41 **Gondwana Six: Stratigraphy, Sedimentology, and Paleontoloty** *Garry D. McKenzie (Ed.)*
42 **Flow and Transport Through Unsaturated Fractured Rock** *Daniel D. Evans and Thomas J. Nicholson (Eds.)*

43 Seamounts, Islands, and Atolls *Barbara H. Keating, Patricia Fryer, Rodey Batiza, and George W. Boehlert (Eds.)*

44 Modeling Magnetospheric Plasma *T. E. Moore, J. H. Waite, Jr. (Eds.)*

45 Perovskite: A Structure of Great Interest to Geophysics and Materials Science *Alexandra Navrotsky and Donald J. Weidner (Eds.)*

46 Structure and Dynamics of Earth's Deep Interior (IUGG Volume 1) *D. E. Smylie and Raymond Hide (Eds.)*

47 Hydrological Regimes and Their Subsurface Thermal Effects (IUGG Volume 2) *Alan E. Beck, Grant Garvin and Lajos Stegena (Eds.)*

48 Origin and Evolution of Sedimentary Basins and Their Energy and Mineral Resources (IUGG Volume 3) *Raymond A. Price (Ed.)*

49 Slow Deformation and Transmission of Stress in the Earth (IUGG Volume 4) *Steven C. Cohen and Petr Vaníček (Eds.)*

50 Deep Structure and Past Kinematics of Accreted Terranes (IUGG Volume 5) *John W. Hillhouse (Ed.)*

51 Properties and Processes of Earth's Lower Crust (IUGG Volume 6) *Robert F. Merev, Stephan Mueller and David M. Fountain (Eds.)*

52 Understanding Climate Change (IUGG Volume 7) *Andre L. Berger, Robert E. Dickinson and J. Kidson (Eds.)*

Maurice Ewing Volumes

1 Island Arcs, Deep Sea Trenches, and Back-Arc Basins *Manik Talwani and Walter C. Pitman III (Eds.)*

2 Deep Drilling Results in the Atlantic Ocean: Ocean Crust *Manik Talwani, Christopher G. Harrison, and Dennis E. Hayes (Eds.)*

3 Deep Drilling Results in the Atlantic Ocean: Continental Margins and Paleoenvironment *Manik Talwani, William Hay, and William B. F. Ryan (Eds.)*

4 Earthquake Prediction—An International Review *David W. Simpson and Paul G. Richards (Eds.)*

5 Climate Processes and Climate Sensitivity *James E. Hansen and Taro Takahashi (Eds.)*

6 Earthquake Source Mechanics *Shamita Das, John Boatwright, and Christopher H. Scholz (Eds.)*

IUGG Volumes

1 Structure and Dynamics of Earth's Deep Interior *D. E. Smylie and Raymond Hide (Eds.)*

2 Hydrological Regimes and Their Subsurface Thermal Effects *Alan E. Beck, Grant Garvin and Lajos Stegena (Eds.)*

3 Origin and Evolution of Sedimentary Basins and Their Energy and Mineral Resources *Raymond A. Price (Ed.)*

4 Slow Deformation and Transmission of Stress in the Earth *Steven C. Cohen and Petr Vaníček (Eds.)*

5 Deep Structure and Past Kinematics of Accreted Terranes *John W. Hillhouse (Ed.)*

6 Properties and Processes of Earth's Lower Crust *Robert F. Merev, Stephan Mueller and David M. Fountain (Eds.)*

7 Understanding Climate Change *Andre L. Berger, Robert E. Dickinson and J. Kidson (Eds.)*

Mineral Physics Volumes

1 Point Defects in Minerals *Robert N. Schock (Ed.)*

2 High Pressure Research in Mineral Physics *Murli H. Manghnani and Yasuhiko Syono (Eds.)*

Geophysical Monograph 53

T.M

Plasma Waves and Instabilities at Comets and in Magnetospheres

Bruce T. Tsurutani
Hiroshi Oya
Editors

American Geophysical Union

Associate Editors

Library of Congress Cataloging-in-Publication Data

Plasma waves and instabilities at comets and in magnetospheres.

(Geophysical monograph ; 53)
Invited papers at the Chapman conference held in Sendai/Mt. Zao, Japan, Oct. 12–16, 1987
1. Plasma waves—Congresses. 2. Magnetospheres—Congresses. 3. Comets—Congresses. I. Tsurutani, Bruce T. II. Oya, Hiroshi, 1926– . III. American Geophysical Union. IV. Series
QC718.5.W3P53 1989 538'.766 89-6900
ISBN 0-87590-073-9

Printed in the United States of America.

This work is dedicated to the late Frederick L. Scarf, a pioneer in the study of plasma waves and instabilities in space. Dr. Scarf was a leading experimentalist and theorist in the field. Fred also loved to stimulate good work in others, and as busy as he was, he always found time to help graduate students. We hope this book carries a bit of Dr. Scarf's spirit and enthusiasm forward to the next generation of space plasma physicists.

CONTENTS

PREFACE

The following chapters were presented as a tutorial and as invited papers at the Chapman Conference held in Sendai/Mt. Zao, Japan, October 12–16, 1987. The authors were chosen both for their expertise in their subject matter and also for their willingness to address an audience of graduate students and newcomers to the field. We hope they have succeeded. The contributed papers presented at the conference were published in a special section of the January 1989 issue of the Journal of Geophysical Research, Space Physics.

Bruce T. Tsurutani
Hiroshi Oya

Notation

A	anisotropy
α	angle between $\vec{V}_{SW}$ and $\vec{B}_0$ or pitch angle
B	magnetic field, G or nT
β	plasma beta, equal to $8\pi NkT/B^2$
C	wave velocity: C_A, C_s
c	velocity of light, equal to 3×10^{10} cm/s
Γ	growth rate
γ	ratio of specific heats
δ	deviation, $\delta\mathbf{B}$
E	dc electric field, statvolts/cm
e	electron charge, equal to 4.8×10^{-10} statcoulomb
f	distribution function
F	function
θ	latitude angle or angle between two vectors, θ_{kB}
i	ion
J	current density
j	current
k	wave propagation vector
$k(T)$	Boltzmann's constant, equal to 1.38×10^{-16} erg/deg
K	Kelvin
l	length
λ	wavelength
M	Mach number
m	mass (m_e, electron mass; m_p, proton mass)
n	normal (n_s, shock normal direction)
N	number density, particles/cm^3
P	pressure, dyne/cm^2
π	3.14159
q	charge
ρ	mass density, g/cm^3
S	sound: C_S
s	shock: v_S
Σ	summation
σ	conductivity, s^{-1}
T	temperature, °K
t	time, s
τ	time constant
ϕ	azimuth angle
Φ	potential
V	volts
v	velocity, cm/s
x,y,z	Cartesian cordinates
ω	frequency, rad/s
Ω	frequency constant (Ω_p, proton gyrofrequency; Ω_{LHR})
Π	plasma frequency (Π_e, electron plasma frequency)

Subscripts

1	upstream (B_1)
2	downstream (v_2)
e	electron (Ω_e, electron gyrofrequency)
p	proton
0	ambient (B_0, ambient magnetic field)
A	Alfvén ($C_A=(\gamma P/\rho)^{1/2}$; M_A, sonic Mach number)
S	sound ($C_S=(\gamma P/\rho)^{1/2}$; M_S, sonic Mach number)
f	fast mode (C_f, M_f)
s	shock (v_s, shock velocity)
n	normal component
t	tangential
α	alpha particle
$\perp$, $\parallel$	w/ respect to $\vec{B}_0$
max	γ_{max}, k_{max}

SPECTRA OF PLASMA TURBULENCE, PARTICLE ACCELERATION AND HEATING BY PLASMA WAVES IN THE INTERACTING PLASMA

A. A. Galeev

Space Research Institute of USSR Academy of Sciences,
117810 GSP-7, Moscow, USSR

Abstract. The present state of weak and strong plasma turbulence and plasma heating by waves are reviewed, with emphasis on the exact solutions of the wave kinetic equations, such as the power-law Kolmogorov spectra and streamers in $\vec{k}$-space. The strong plasma turbulence generated by the modulational instability of linearly excited waves and their collapse is represented by strong Langmuir turbulence. Because of stabilizing Landau damping effects on the weak plasma instabilities most of the excited waves interact only with suprathermal particles and generate the suprathermal tails of the particle velocity distribution. This happens also in the process of collisionless energy transfer from one species of plasma to another, and plays an important role in the interaction of plasmas in planetary magnetospheres and comets. This is reviewed here.

Introduction

In collisionless magnetospheric plasmas, wave-particle interactions play an important role in the heating and acceleration of plasma particles and in the collisionless energy transfer from one species of plasma to another. If this interaction is due to the Cherenkov resonance of particles with waves, then the quasilinear diffusion of particles in the velocity space is necessarily accompanied by the diffusion in energy. Whether this diffusion in energy will result in plasma heating or particle acceleration depends critically on the spectral energy distribution of excited waves, i.e., on the spectrum of turbulence. Heating of plasma species takes place when waves fall into Cherenkov resonance with the main body of the velocity distribution of the corresponding species. The classical example of this is the Joule heating of plasma due to the ion sound anomalous resistivity [Galeev and Sagdeev, 1984]. But much more often, weak linear plasma instabilities result in the excitation of waves with phase velocities higher than the thermal velocity of the given plasma species. Also, in the case of the developed plasma turbulence the wave energy cascading in $\vec{k}$-space can lead to a wave energy flux from a region of high phase velocity where waves are excited linearly to a region of phase velocities of the order of the particle thermal velocity. In both of these cases a non-Maxwellian tail of suprathermal particles is generated; i.e., acceleration of particles takes place instead of plasma heating. In this paper we will consider in detail two examples of this kind. The first one concerns the acceleration of suprathermal electrons by strong Langmuir turbulence. The second example illustrates the energy transfer from the ion to the electron component of a plasma via oblique Langmuir waves. This process is of specific importance for magnetospheric physics. For example, the energy for heating the Earth's plasmasphere and the middle latitude stable red arcs can come only from the energetic ring current plasma interacting with the plasmasphere [Cornwall et al., 1972]. Similar phenomena can take place in the Jovian magnetosphere, where the energetic magnetospheric particles interact with the Io plasma torus [Thorne, 1983]. But the most basic problem of this kind is the interaction of the magnetized plasma flow with the neutral gas when the avalanche gas ionization starts with the flow velocity higher than the critical one [Alfvén, 1954]. The necessary condition for such anomalous ionization is the collisionless transfer of a fraction of the plasma kinetic energy to electrons ionizing the gas. The efficiency of this energy transfer defines the value of the critical velocity.

It is clear that to calculate the plasma heating or the efficiency of the energy transfer we first should find the spectrum of turbulence. We can do this with the help of current theories of weak and strong plasma turbulence. The theory of weak plasma turbulence is based on the assumption that the turbulent plasma can be described as a gas of weakly interacting charged particles and plasma waves (plasmons, phonons, helicons, etc.) [Sagdeev and Galeev, 1969; Kadomtsev, 1966]. This assumption is valid as a rule when the energy density of the waves is sufficiently smaller than the thermal or magnetic energy density. In contrast to the hydro-dynamical turbulence, here we can use the expansion in the small ratio of these energies to obtain a set of well-defined equations to describe the weak plasma turbulence. Therefore the spectra of turbulence could be obtained in the form of exact solutions of these equations (see review of Zakharov, [1984] for details). In this paper we will limit ourselves to the discussion of the physics of energy cascading towards smaller scales in a system of resonantly interacting plasma waves and show how to obtain the power-law Kolmogorov spectra from simple dimensional estimates.

We should note here that the low ratio of the wave and plasma energy densities does not guarantee the validity of the weak turbulence theory. One of the best known cases of this kind is Langmuir wave turbulence. Here the collapse of Langmuir waves [Zakharov, 1972] results in a strong interaction of waves in the collapsing plasma cavity. Since the wave energy again cascades towards smaller scales, we can construct the Kolmogorov spectrum of the Langmuir turbulence (see review by Shapiro and Shevchenko [1984] for details).

Finally, in a plasma the wave-particle interaction not only provides the source and sink of waves but also strongly influences the nonlinear interaction of waves. In the case when the three-wave resonant interaction is impossible the spectrum of turbulence is formed as a result of the induced scattering of waves by plasma particles. In this particular case the energy cascading sometimes can be described as the differential energy transfer towards larger scales in an approximation of isotropic turbulence [Kadomtsev, 1966]. But in general such turbulence is very anisotropic and takes the form of streamers in the wave vector space. While discussing all three above mentioned types of turbulence we will pay special attention to the spectra of turbulence in the absorption region, since the wave energy dissipated here goes into plasma heating and particle acceleration.

Kolmogorov Spectra of Weak Plasma Turbulence

Wave Kinetic Equation

To the lowest (second) order in the ratio of the wave energy to the plasma energy the wave kinetic equation can be written in the form

$$\frac{\partial n_{\vec{k}}}{\partial t} = 2\left[\left(\gamma^{i}_{\vec{k}} + \gamma^{e}_{\vec{k}}\right) + \int d\vec{k}_1\, T(\vec{k},\vec{k}_1)\, n_{\vec{k}_1}\right] n_{\vec{k}} + \\ + 4\pi \int d\vec{k}_1\, d\vec{k}_2 |V(\vec{k},\vec{k}_1,\vec{k}_2)|^2 \, (n_{\vec{k}_1} n_{\vec{k}_2} - \\ - \mathrm{Sign}\, \omega_{\vec{k}_2} \omega_{\vec{k}} n_{\vec{k}} n_{\vec{k}_1} - \mathrm{Sign}\, \omega_{\vec{k}_1} \omega_{\vec{k}}\, n_{\vec{k}_1} n_{\vec{k}_2}) \cdot \\ \cdot \delta\left(\omega_{\vec{k}} - \omega_{\vec{k}_1} - \omega_{\vec{k}_2}\right) \delta(\vec{k} - \vec{k}_1 - \vec{k}_2) \tag{1}$$

where $W_{\vec{k}} \equiv n_{\vec{k}} |\omega_{\vec{k}}|$ is the spectral energy density of waves with the frequency $\omega_{\vec{k}}$ and wave vector $\vec{k}$ that is assumed here to be positive for simplicity. The first term on the right hand side describes the wave growth (damping) with the rate $\gamma^j_{\vec{k}}$ due to the interaction of the wave with particles of the j-species. The second term represents the induced scattering of waves by plasma particles. Accordingly, the kernel $T(\vec{k}, \vec{k}_1)$ of the integral operator here contains a delta function reflecting the fact that the total energy of the plasma and of the waves is conserved in this process. Finally, the last term describes the three-wave resonant interaction. Here the energy and momentum conservation laws are represented by two delta functions in an explicit form. The technique to calculate the kernel $T(\vec{k}, \vec{k}_1)$ and the matrix element $V(\vec{k}, \vec{k}_1, \vec{k}_2)$ of the three-wave interaction is given in textbooks on weak plasma turbulence (see, for instance, the book by Sagdeev and Galeev [1969]).

Kolmogorov Spectra of Isotropic Turbulence

The source and sink of the plasma wave represented in equation (1) by the linear wave growth are often separated in the $\vec{k}$-space (Fig. 1). When waves are excited in the long wave range and absorbed in the short wave range, we can expect that a wave energy cascade from large to small scales will be established in a way similar to the Kolmogorov turbulence in fluids. In contrast to the fluid turbulence, here we have equation (1) whose solution can provide the spectrum of turbulence. The first exact solutions of this equation were obtained by Zakharov [1964] for the case when the three-wave resonant interaction is the dominant process. Let us note here that the three-wave interaction within one branch of plasma oscillations is possible only for waves with a dispersion such that the energy and momentum conservation equations

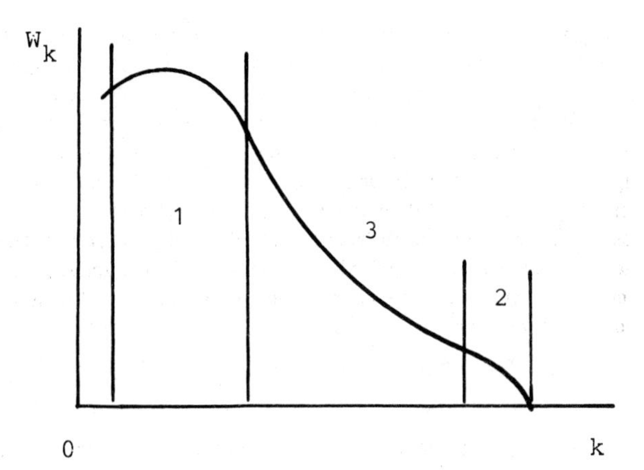

Fig. 1. The wave spectral energy distribution in the case of Kolmogorov turbulence. The source and absorption region and the inertial range are marked by numbers 1, 2, 3, respectively.

$$\omega_{\vec{k}} = \omega_{\vec{k}_1} + \omega_{\vec{k}_2}; \quad \vec{k} = \vec{k}_1 + \vec{k}_2 \tag{2}$$

can be satisfied. This condition imposes severe restriction on the wave dispersion curves. For example, for the dispersion curves drawn schematically in Fig. 2 the three-wave interaction is possible for the branches 3, 5, and 6. However, the conservation laws (2) could be easily satisfied if the three interacting waves are from different branches.

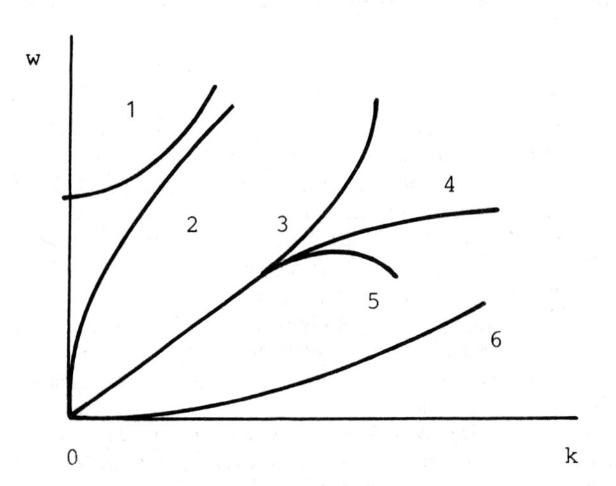

Fig. 2. Different types of wave dispersion curves.

The exact solution of equation (1) for the inertial range of stationary turbulence, where we can neglect all terms except the last one, was obtained under even more specific conditions when the wave frequency and the matrix element of the three-wave interaction are functions of the wave vectors, namely homogeneous

$$\omega(\varepsilon\vec{k}) = \varepsilon^{\alpha}\, \omega(\vec{k}), \quad V(\varepsilon\vec{k}, \varepsilon\vec{k}_1, \varepsilon\vec{k}_2) = \varepsilon^{\beta}\, V(\vec{k}, \vec{k}_1, \vec{k}_2) \tag{3}$$

and the plasma is uniform and isotropic. Here $\alpha > 1$ in order to satisfy the conservation laws (2) within one branch of waves.

Two exact solutions of equation (1) were found in this case. One solution $n_{\vec{k}} = T/\omega_{\vec{k}}$ obviously corresponds to the Rayleigh-Jeans distribution with the temperature T. Here we are interested in the second solution that describes the constant energy flux in the $\vec{k}$-space from the source region to the sink region (Fig. 1). We obtain it here using dimensional estimates, referring those who are interested in the exact solutions to the excellent review by Zakharov [1984].

Assuming the locality of turbulence, i.e., that only waves with wave vectors of comparable magnitude $|\vec{k}| \sim |\vec{k}_1| \sim |\vec{k}_2|$ interact most strongly, we can write the energy flux P in the inertial range of turbulence in the form

$$P \cong \frac{\omega_k n_k k^s}{\tau_{NL}} \tag{4}$$

where s is the dimensionality of the $\vec{k}$-space and τ_{NL} is the characteristic time of the energy transport in the $\vec{k}$-space due to the three-wave resonant interaction. From equation (1), using the locality condition, we obtain

$$\tau_{NL}^{-1} \cong V(\vec{k}, \vec{k}_1, \vec{k}-\vec{k}_1)^{\,2}\, |\vec{k}|-|\vec{k}_1| \cdot \frac{n_k k^s}{\omega_k} \tag{5}$$

Combining equations (4) and (5) and taking into account relations (3), we obtain the power law index of the turbulence spectrum:

$$n_k\, \omega_k\, k^{s-1} \propto P^{1/2} k^{\alpha-\beta-1} \tag{6}$$

Let us note here that this expression can be used only if the assumption of the locality of plasma turbulence is confirmed by a direct check of the convergence of the integrals in equation (1).

There are no straightforward applications of exact solutions of this kind to space plasmas. For example, in the finite-β solar wind plasma the fast and slow magnetosonic waves are strongly damped, and interaction of Alfvén waves among themselves is impossible because the spin conservation law is violated in this process [Galeev and Oraevskii, 1962]. Moreover, the matrix element of Alfvén waves in interaction with fast magnetosonic waves, which are slowly damped in the moderately low-β region of the solar wind, is strongly anisotropic [Galeev and Karpman, 1963]:

$$V(\vec{k}_f, \vec{k}_A', \vec{k}_A'') = \sqrt{\frac{\pi\omega_f\, \omega_A'\, \omega_A''}{B_o^2}} \quad \frac{k_{\perp f}}{k_f} \tag{7}$$

where the indices f and A refer to the fast magnetosonic wave and the Alfvén wave, respectively. However, assuming that the MHD turbulence resulting from this interaction will be isotropic and taking into account that relations (3) are satisfied in this case, we can use dimensional analysis to find the spectrum of isotropic magnetic field fluctuations with an accuracy within a factor of the order of unity:

$$k^2\, |\vec{B}_{\vec{k}}|^2 \cong (P/v_A)^{1/2}\, B_o^2 k^{-3/2} \tag{8}$$

The power law index here is very close to that observed in interplanetary space [Intrilligator, 1981]. It was also found by a similar dimensional analysis by McIvor [1977]. We see that the energy-containing region here is in the long wave range. So this spectrum is the true Kolmogorov spectrum. It is easy to see from equation (6) that this statement is valid whenever $\beta > \alpha$ in relations (3).

Kolmogorov Spectra of Anisotropic Turbulence

Kolmogorov spectra can also be found for an axially symmetric plasma turbulence. Let us consider here the case when the wave dispersion relation and the matrix element of the three wave interaction are axially symmetric with respect to the magnetic field (compare with relations (3)):

$$\begin{gathered}\omega(\varepsilon k_{\parallel}, \mu\vec{k}_{\perp}) = \varepsilon^a \mu^b\, \omega(k_{\parallel}, \vec{k}_{\perp}) \\ V\left(\varepsilon k_{\parallel}, \varepsilon k_{\parallel}', \varepsilon k_{\parallel}'', \mu\vec{k}_{\perp}, \mu\vec{k}_{\perp}', \mu\vec{k}_{\perp}''\right) = \varepsilon^u \mu^v \\ V\left(k_{\parallel}, k_{\parallel}', k_{\parallel}'', \vec{k}_{\perp}, \vec{k}_{\perp}', \vec{k}_{\perp}''\right)\end{gathered} \tag{9}$$

where $k_{\parallel}$ and $\vec{k}_{\perp}$ are the wave vector components along and across the magnetic field $\vec{B}_o$, respectively. Again using dimensional analysis, we can construct the energy flux P and the momentum flux $Q_{\parallel}$ along the magnetic field as

$$P = |V\left(\vec{k}, \vec{k}', \vec{k}-\vec{k}'\right)|^2 \quad \left(n_{\vec{k}}\, k_{\perp}^2 k_{\parallel}\right)^2 \tag{10}$$

$$Q_{\parallel} = |V\left(\vec{k}, \vec{k}', \vec{k}-\vec{k}'\right)|^2 \quad \left(n_{\vec{k}}\, k_{\perp}^2 k_{\parallel}\right)^2 \frac{k_{\parallel}}{\omega_k} \tag{11}$$

As a result we find two spectra of turbulence corresponding to the conservation of the energy flux:

$$W_{\vec{k}} = n_{\vec{k}}\, \omega_{\vec{k}} \propto \frac{P^{1/2}}{k_{\parallel}^{1+u-a}\, k_{\perp}^{2+v-b}} \tag{12}$$

and the momentum flux:

$$W_k \propto \frac{Q_{\parallel}^{1/2}}{k_{\parallel}^{3(1-a)/2+u}\, k_{\perp}^{2+v-3b/2}} \tag{13}$$

It is very tempting to apply this analysis to oblique Langmuir wave turbulence in a magnetized plasma, since it plays an important role in numerous magnetospheric plasma physics applications (loss-cone instability of the ring current, super-critical shocks, critical velocity ionization phenomena, etc.). The relations (9) are satisfied for these waves, since in the low-β plasma the wave dispersion relation and the matrix element of their interaction can be represented in the form (for $k_{\parallel} \ll k_{\perp}$):

$$\omega_k = \omega_c\, k_{\parallel}/k;\; V(\vec{k}, \vec{k}', \vec{k}-\vec{k}')\Big|_{|\vec{k}'| \cong |\vec{k}|} \cong \sqrt{\frac{\omega_k^3}{n_e m_e\, (\omega_{\vec{k}}/k_{\perp})^2}} \tag{14}$$

The matrix element here is given by an order of magnitude estimate taken from exact calculations [Galeev, 1965]. Comparing expressions (9) and (14), we find

$$a \cong 1; \quad b \cong -1; \quad u = 1/2; \quad v \cong 1/2 \tag{15}$$

Therefore, according to equation (12), the integral over the spectral energy density $W_{\vec{k}} \propto k_{\parallel}^{-1/2} k_{\perp}^{-1/2}$ diverges with $k_{\parallel}$ so that $|k_{\parallel}| \sim k_{\perp}$ in the energy-containing region. Moreover, the locality of the turbulence is violated here, and the approach described above fails.

Strong Plasma Turbulence

It is obvious that for sufficiently large amplitudes of interacting waves the weak turbulence approach becomes incorrect. Actually, we have used two assumptions while deriving the wave kinetic equation (1):

1) the waves interact weakly in the sense that

$$1/\omega\tau_{NL} << 1 \tag{16}$$

2) under this condition a stochastization of interacting-wave phases takes place.

In the case of MHD waves, condition (16) is violated according to equations (5) and (7) only when the wave magnetic field is comparable to the ambient magnetic field. Therefore this case falls into the class of strong hydrodynamical turbulence, which is beyond the scope of this paper.

The situation changes for oblique Langmuir waves in the low-β magnetized plasma briefly considered in the previous section. Combining equations (5) and (14), we rewrite the condition in the form

$$\frac{1}{\omega_k\tau_{NL}} \cong \frac{W_{\vec{k}}/k_{\parallel}k_{\perp}^2}{n_e m_e (\omega_{\vec{k}}/k_{\perp})^2} \leq 1 \tag{17}$$

where $W_{\vec{k}} = (\omega_{pe}^2/\omega_{ce}^2)\, |\vec{E}_{\vec{k}}|^2/4\pi$ is the spectral energy density of the waves expressed through the spectral energy density of the wave electric field $\vec{E}_{\vec{k}}$. The phase velocity region along a field line occupied by these two waves is larger than the electron thermal velocity. Thus for $|k_{\parallel}| < k_{\perp}$ the weak turbulence approach fails because the inequality (17) fails when the wave energy density is much smaller than the thermal energy density of the plasma.

To describe the strong plasma turbulence when condition (17) is violated, we rewrite the latter in the form of a condition for the wave profile turning over (wave breaking):

$$u_{\perp} \cong (c/B)\left(|\vec{E}_{\vec{k}}|^2\, k_{\parallel}k_{\perp}^2\right)^{1/2} > \omega_k/k_{\perp} \tag{18}$$

where $u_{\perp}$ is the electric drift velocity. As in the case of gravity waves on a fluid surface [see Zakharov, 1984, and references therein] the spectrum in the phase space region of strong turbulence does not depend on the energy flux, which is no longer conserved due to wave dissipation caused by wave breaking:

$$|\vec{E}_{\vec{k}}|^2\, k_{\parallel}k_{\perp}^2 \cong B^2(\omega_k/k_{\perp}c)^2 \tag{19}$$

This spectrum is found under the assumption that the strength of the wave interaction is limited by the condition (17). Here the spectral energy density of waves diverges with $k_{\parallel}$ as in the weak turbulence case. However, the notion of locality of turbulence no longer makes sense here, since a jump to very short scales becomes impossible. Therefore we can describe the strong turbulence spectrum by expression (19) with $|k_{\parallel}| \cong k_{\perp}$.

Spectrum of Strong Langmuir Wave Turbulence in the Inertial Range

The interaction of Langmuir waves in the lowest order of expansion in the ratio of the wave energy density to the plasma thermal energy density can be described by the induced scattering of these waves by ion sound waves in a nonisothermal plasma or by plasma ions in an isothermal plasma. For both of these processes the Langmuir wave turbulence can be considered to be weak under the condition [Galeev and Sagdeev, 1983]

$$\frac{1}{\omega\tau_{NL}} \cong \frac{\omega_{pe}|\vec{E}_{\vec{k}}|^2\, k^3}{4\pi n_e T_e}\,\frac{1}{\Delta\omega} < 1 \tag{20}$$

(compare with condition (17)), where ω_{pe} is the electron plasma frequency, λ_D the Debye length, and $\Delta\omega = 3\omega_{pe}k\Delta k\lambda_D^2$ the frequency difference of the interacting Langmuir waves, which is of the order of the ion sound frequency or the resonant beat wave frequency $\sim kv_{Ti}$. When this condition is violated, plasma cavities filled by trapped Langmuir waves are formed as a result of the modulational instability. It is easy to see that these cavities are unstable against collapse (see Fig. 3). Indeed, in a shrinking cavity the high-frequency pressure of trapped Langmuir waves increases. The pressure in turn pushes the plasma out of the cavity, making the latter deeper, so that the plasma waves fall to the bottom, where the cavity is narrower. Then the process repeats itself. As a result, the plasma cavity collapses, thereby making the wavelength of the trapped waves shorter. When the wavelength becomes so short that the Langmuir waves fall into Cherenkov resonance with thermal electrons, the Langmuir waves will damp. In the absence of high-frequency wave pressure the plasma cavity decays into outgoing sound waves. The whole picture of the Langmuir wave evolution in the collapsing cavity can be described as an energy cascade towards shorter scales. If we take into account the fact that numerous cavities are created in the plasma volume due to the modulational instability of an initially uniform Langmuir wave turbulence and that each of these cavities with equal probability could be in any particular stage of collapse, we can characterize this plasma state as that of strong Langmuir wave turbulence. Let $N_k dk$ be the number density of cavity size ℓ and wave energy ε_k in the range $(k^{-1}, (k + dk)^{-1})$. If we assume that the cavities are continually formed as a result of the modulational instability of Langmuir waves, then the flux of cavities in k-space will be constant:

$$N_k\, dk/dt(k) = \text{const} \tag{21}$$

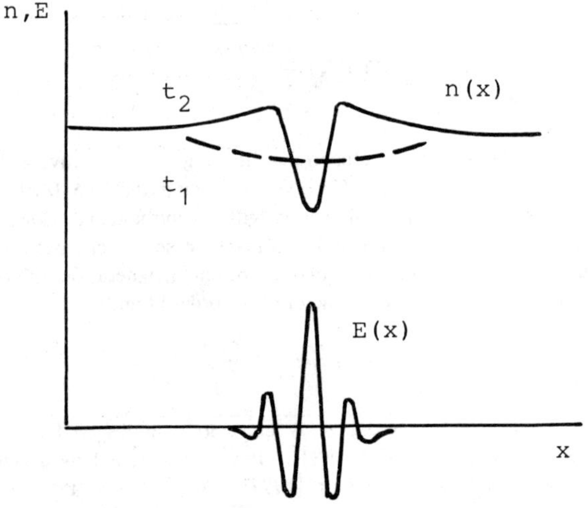

Fig. 3. Evolution of spatial profiles of the plasma density and the Langmuir wave electric field in the collapsing plasma cavity ($t_2 > t_1$).

where dt(k) is the time during which the size of a cavity is reduced from the value $\ell = k^{-1}$ to the value $\ell' = (k+dk)^{-1}$. It can be shown [Zakharov, 1983] that the wave energy in this collapsing cavity is conserved, i.e.,

$$\varepsilon_k = \int |E|^2 d^3\vec{r} = \text{const} \tag{22}$$

We will assume here that the wave energies in different cavities are approximately of the same order of magnitude. Then the wave spectral energy density can be written as

$$k^2 W_k = \frac{2\pi k^2 |E_k|^2}{4\pi} = \varepsilon_k N_k \tag{23}$$

As a result, the wave energy flux in k-space takes the form

$$P = \frac{2\pi k^2 |E_{\vec{k}}|^2}{4\pi} \frac{dk}{dt(k)} \tag{24}$$

To find the value of dt(k)/dk we shall use dimensional analysis of the equations describing the Langmuir wave collapse. As we have already discussed, the dynamics of the cavity is controlled by the wave pressure and described by the equation

$$\frac{\partial^2 \delta n}{\partial t^2} = \nabla^2 \left(\delta n\, T_e + \frac{|E|^2}{8\pi} \right) / m_i \tag{25}$$

In the case of supersonic collapse we can neglect the variation of the thermal plasma pressure in the cavity. Taking into account that the wavelength of a trapped wave is defined by the depth δn of the density cavity, i.e.,

$$k^2 \lambda_D^2 \cong -\delta n / n_o \tag{26}$$

we can roughly estimate the wave pressure gradient in equation (25) and obtain the time behavior of wave pressure in the cavity:

$$|E|^2 \propto (t_o - t)^{-2} \tag{27}$$

Combining (22) and (27), we find

$$\ell \cong k^{-1} \propto (t_o - t)^{2/3} \text{ or } dt/dk \propto k^{-5/2} \tag{28}$$

Using this result and assuming the wave energy flux (24) to be constant, we obtain the spectrum of strong Langmuir wave turbulence [Galeev et al., 1975]:

$$k^2 |E_k|^2 \propto k^{-5/2} \tag{29}$$

Langmuir Wave Turbulence in the Absorption Region and Acceleration of Suprathermal Electrons

While the wavelength of Langmuir waves in the collapsing cavities is shortening, their phase velocity decreases, and finally the waves fall into Cherenkov resonance with tail electrons. The spectrum of the turbulence in the absorption region of the phase space (see Fig. 4) has to be found selfconsistently by the solution of the system of equations consisting of the quasilinear equation for the suprathermal electron distribution and of the equation for the spectral energy density of waves. Assuming that both the electron distribution and the wave spectrum are isotropic, we write these equations in the form [*Galeev et al.*, 1975]

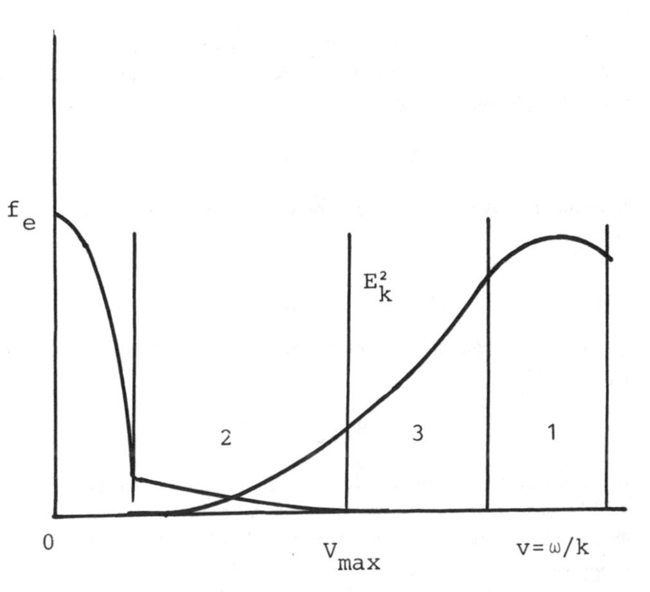

Fig. 4. The electron velocity distribution and the Langmuir wave spectral energy density as a function of the phase velocity. The arrows show the direction of the wave energy flux from the pump region 1 to the absorption region 3 and the electron flux from the Maxwellian distribution to the suprathermal tail.

$$\frac{\partial f}{\partial t} = \frac{1}{v^2}\frac{\partial}{\partial v}\left[\frac{4\pi e^2 \omega_{pe}^2}{m_e^2} \int_{\omega_{pe}/v}^{\infty} \frac{dk\, k^2 W_k}{k^2\sqrt{k^2 v^2 - \omega_{pe}^2}} \frac{\partial f}{\partial v} \right] \tag{30}$$

$$k^2 \frac{\partial W_k}{\partial t} + \frac{\partial}{\partial k}\left(k^2 W_k \frac{dk}{dt} \right) = \frac{\omega_{pe}^4}{n_e k^2} \int_{\omega_{pe}/k}^{\infty} \frac{2\pi v^2 dv}{\sqrt{k^2 v^2 - \omega_{pe}^2}} \frac{\partial f}{v \partial v} k^2 W_k \tag{31}$$

where $k^2 W_k = k^2 |E_k|^2/2$ is the spectral energy density of Langmuir waves in the range (k, k+dk).

We are looking for a stationary solution of these equations in the form of power law functions of the particle velocity and wave vector. We also assume that the wave collapse is self-similar in the absorption range as well as in the inertial range. Then in the stationary case we obtain from the quasilinear equation (30) that the spherically symmetrical outward flux of suprathermal electrons in the velocity space (the term in square brackets) is constant:

$$k^2 W_k\, f\left(\frac{\omega_{pe}}{k}\right) = \text{const} \tag{32}$$

Similarly, we reduce equation (31) to the form

$$\frac{1}{k}\frac{dk}{dt} \propto \frac{1}{k^3} f\left(\frac{\omega_{pe}}{k}\right) \tag{33}$$

As was noted by Pelletier [1982], we cannot use here the self-similarity law of collapse (29), since it was obtained using the assumption that the wave energy in the collapsing cavities is conserved. To find a new law, we assume that the number flux of collapsing cavities in $\vec{k}$-space is still conserved (see equation (21)). Also, we can show that the time behaviour of the electric field in the cavity is still described by relation (27). Then the energy (22) of the damped wave in the collapsing cavity can be represented in the form

$$\varepsilon_k \equiv |E|^2 k^{-3} \propto t(k)^{-2} k^{-3} \tag{34}$$

Substituting this value into equation (23) and using equation (21), we find

$$k^2 W_k \propto \frac{1}{k^3 t^2(k)} \frac{dt(k)}{dk} \tag{35}$$

On the other hand, from relations (32) and (33) we obtain a second equation for W_k and $t(k)$:

$$k^2 W_k \propto 1/k^2 \, dk/dt \tag{36}$$

It is easy to see that the cavities now collapse more slowly ($dk/dt \propto k^{3/2}$) because of the wave dissipation on this scale ($k^2 W_k \propto k^{-7/2}$). We also obtain scaling for the velocity distribution of suprathamal electrons:

$$f(v) \propto v^{-7/2} \tag{37}$$

The Langmuir wave absorption by suprathermal electrons in the collapsing cavities could play an important role in beam plasma discharge, where the energy transfer from beam electrons to ionizing plasma electrons is the crucial process [Galeev et al., 1976; Papadopoulos, 1982].

Spectra of Weak Turbulence Formed by the Wave-Particle-Wave Interaction

Streamer Type Spectra. If in the lowest order in the wave energy the three wave interaction is forbidden by the conservation laws or wave-polarization specifics, then the wave kinetic equation takes the following form:

$$\frac{1}{2}\frac{\partial n(\vec{k})}{\partial t} = \Gamma(\vec{k})\, n(\vec{k}) \equiv \left\{ \sum_j \gamma^j_{QL}(k) + \gamma_{NL}(k) \right\} n(\vec{k}) \tag{38}$$

where γ^j_{QL} is the quasilinear growth (damping) rate and γ_{NL} the damping (growth) rate of waves due to their induced scattering by plasma particles. Both of these rates functionally depend on the spectral energy distribution of waves; for example, γ_{NL} is given explicitly by the second term in the square brackets on the right-hand side of equation (1).

Following Breizman et al. [1973], let us consider now which stationary solutions of equation (38) may exist. It is obvious that a stable solution must have the form

$$\begin{aligned} \Gamma(\vec{k}) &= 0 \quad \text{for } n(\vec{k}) \neq 0 \\ \Gamma(\vec{k}) &< 0 \quad \text{for } n(\vec{k}) = 0 \end{aligned} \tag{39}$$

In order to satisfy the second of these conditions one should require that the three-dimensional surface $\gamma = \gamma^j_{QL}(k)$ should only touch the surface $\gamma = \gamma_{NL}(k)$ from below along a two-dimensional surface or even a line. Such surfaces (lines) at which the spectral distribution of waves is concentrated were named [Breizman et al., 1973] "streamers" (two-dimensional or one-dimensional, respectively). The above described requirements for the functions $\gamma^j_{QL}(\vec{k})$ and $\gamma_{NL}(\vec{k})$ can be written in a formal mathematical way. For example, in the axially symmetric case the stationary solution of equation (38) can be written in the form

$$n(k,x) = \sum_j n_j(k)\, \delta[x - x_j(k)] \tag{40}$$

with the requirement that

$$\Gamma(k, x_j(k)) \equiv 0 \tag{41}$$

$$\frac{d}{dx}\Gamma(k,x)\Big|_{x=x_j(k)} = 0 \tag{42}$$

$$\frac{d^2}{dx^2}\Gamma(k,x)\Big|_{x=x_j(k)} < 0 \tag{43}$$

Differential Energy Transfer and Power-Law Spectra

The analog of induced scattering of waves by particles of an unmagnetized plasma species in classical electrodynamics is Cherenkov absorption (emission) of a beat wave ($\omega_{\vec{k}} \pm \omega_{\vec{k}_1}, \vec{k} \pm \vec{k}_1$) by a particle with velocity $\vec{v}$. The condition for Cherenkov resonance in this case

$$\omega_{\vec{k}} \pm \omega_{\vec{k}_1} = (\vec{k} \pm \vec{k}_1) \cdot \vec{v} \tag{44}$$

enters the expression for the kernel $T(\vec{k}, \vec{k}_1)$ of the integral equation (38) as an argument of the delta function and guarantees energy conservation in the wave-induced scattering process. Since the number of waves is also conserved, the kernel $T(\vec{k}, \vec{k}_1)$ must be antisymmetric in its arguments.

In a cold plasma, where the thermal velocity of particles is smaller than the phase velocity of the beat wave ($\omega - \omega_1 \gg (\vec{k} - \vec{k}) \cdot \vec{v}$), the above specified properties of $T(\vec{k}, \vec{k}_1)$ permit one to simplify equation (38) by transforming it into an integro-differential equation. To illustrate this, we again limit ourselves to the axially symmetric case. Introducing the new variables ($\omega(k,x)$, x) instead of the old ones (k, x), we find that by using condition (44) we can perform the integration over ω_1 in (38). But because of the antisymmetry of $T(\omega, x; \omega_1, x_1)$ in ω, the lowest-order term in the expansion over the small parameter $kv_T/\omega \ll 1$ vanishes. In the next order, equation (38) takes the integrodifferential form

$$\frac{\partial n(\omega,x)}{\partial t} = 2\left\{\gamma(\omega,x) + \int \frac{\partial}{\partial \omega_1}\left[\tilde{T}(\omega,x;\omega_1,x_1)\, n(\omega_1,\Sigma_1)\right]\Big|_{\omega_1=\omega} dx_1\right\} \cdot n(\omega,x) \tag{45}$$

where $T(\vec{k}, \vec{k}_1) = T(\omega, x; \omega_1, x_1)\,(d/d\omega)\delta(\omega - \omega_1)$ for the case of wave scattering by the unmagnetized plasma species (see, for example, Sagdeev and Galeev [1969]. In general, equation (45) describes the differential number of waves transferred along the streamers $x = x_j(\omega)$ towards the lower frequencies [*Breizman et al.*, 1973].

However, in some particular cases, equation (45) could have different types of solutions. One example is the current-driven long-wave ion sound turbulence. Here the waves are scattered by ions, and the corresponding kinetic equation for the waves takes the form [see Galeev and Sagdeev, 1984, and references therein]

$$\frac{\partial n(k,x)}{\partial t} = 2\{\gamma_{oe}(k)\,\gamma_{QL}(x) - k^m \frac{\partial}{\partial k} k^{m+s-1} \int T(x,x')n(k,x')dx'\} \cdot n(k,x) \tag{46}$$

where s = 3 is the dimensionality of the $\vec{k}$-space and m = 2 for the particular case of long-wave current-driven ion sound turbulence. Here $\gamma_{QL}(x)$ is the quasilinear growth rate normalized to the Landau damping rate $\gamma_{oe}(k)$ in a plasma without electron currents. Bychenkov and Silin [1982] have solved this equation in the stationary case and shown that the well-known Kadomtsev-Petviashvili spectrum of ion sound turbulence $n(k)\omega_s(k) \propto k^{-3}$ [Kadomtsev, 1966] still holds in the quasilinear regime described earlier by Korablev and Rudakov [1966]. Let us note that this spectrum looks similar to the Kolmogorov type spectrum, but it differs strongly in that the wave energy flows towards longer waves, where it is absorbed by ions due to their viscosity.

Let us note also that in the one-dimensional case the equation is reduced to a differential one. An interesting example of this kind is the induced scattering of Alfvén waves by ions in the solar wind or near comets. Assuming that these waves are propagating preferentially along magnetic field lines (both forward and backward), we obtain the equation in the form (46) with T = const, s = 1, m = 1. Since the long Alfvén waves are only weakly damped, here we put $\gamma = 0$ and obtain the spectrum of turbulence in the inertial range, where the flux of the number of waves (not energy) is conserved:

$$|B_k|^2/4\pi \equiv n(k)\,k\,V_A \propto \frac{1}{k} \tag{47}$$

Again, the wave energy flows towards longer waves, and the integral over the spectral energy density of the waves diverges logarithmically. It seems that this nonlinear process is important neither in the solar wind nor near the comets [Galeev et al., 1987].

Spectra of Turbulence and Electron Acceleration in Interacting Plasmas

We have already mentioned that the collisionless energy transfer from moderately energetic ions to cold plasma electrons plays an important role in the acceleration of suprathermal electrons by collisionless shocks [Papadopoulos, 1981; Vaisberg et al., 1983], in the heating of the Earth's plasmasphere [Cornwall et al., 1971] and of the Io torus [Thorne, 1983] by energetic magnetospheric ions, and finally in electron heating at the expense of the kinetic energy of neutral gas moving across the magnetic field through the plasma (so-called critical velocity ionization [Alfvén, 1954]).

The interaction of energetic magnetospheric ions with the cold Earth plasmasphere and the Io torus was considered with the assumption that the free ion energy is transferred by low-frequency whistler waves to the main body of the electron distribution; i.e., electron heating takes place. In shocks and gas-plasma interaction the energy is transferred by the oblique Langmuir [Formisano et al., 1982] or whistler [Papadopoulos, 1981] waves in the lower-hybrid range of frequencies. In the shocks the whistler waves are excited by a strong ($n_b \leq n_o$) beam of reflected ions, and nonlinear effects play an important role. It was shown [Musher et al., 1986] that the streamer type plasma wave spectrum is formed in this case, and the wave energy flows along the streamers towards the lower-frequency region where the wave condensate is formed. The modulational instability of this condensate followed by the lower-hybrid wave collapse provides an energy flux towards shorter scales, where it is absorbed mainly by electrons.

In this review we consider in detail only the case of the gas-plasma interaction where the plasma turbulence is weak and find an exact solution to the problem.

Collisionless Energy Transfer from Ions to Electrons in a Gas Ionized by Plasma Flow

In this section we consider in detail the key element of the critical velocity ionization phenomenon, which is energy transfer via plasma waves. We assume here that the ionization of a gas by electron impact in a rarefied space medium is so slow that the plasma turbulence resulting from the excitation of plasma waves is weak and adequately described by the quasilinear plasma theory. We shall show later that the condition for ionized atoms to be magnetized is satisfied whenever the weak plasma turbulence approximation is valid. Therefore in the case under consideration each atom ionized by the plasma flow starts to move along a cyclotron orbit with a velocity $\vec{V}$ equal to that of the gas in the plasma system and simultaneously drifts with the plasma. Since the temperature of the gas is usually negligibly small in comparison with the kinetic energy of its motion through the plasma, the velocity distribution function of newly ionized atoms in the coordinate system moving with the plasma has the form

$$f_{oi} = (n_o/\pi)\,\delta\left(v_\perp^2 - V^2\right)\delta(v_\parallel) \tag{48}$$

where n_o is the density of ionized atoms and $v_\perp$ and $v_\parallel$ are their velocities perpendicular and parallel to the magnetic field, respectively. In addition, we have assumed that the motion of the gas relative to the plasma is exactly perpendicular to the magnetic field, so that ionized atoms do not move along the field lines. It is well known [Rosenbluth and Post, 1965] that a plasma with an ion velocity distribution of the form (48) is unstable with respect to the excitation of electrostatic plasma oscillations with frequencies ω and wave vectors $\vec{k}$ in the ranges defined by the following inequalities:

$$\omega_{ce} >> \omega >> \omega_{ci},\ Jm\omega >> \omega_{ci}$$
$$k_\perp\rho_i >> 1 >> k_\perp\rho_e,\ k_\parallel << k_\perp \tag{49}$$

Here $\rho_j = v_{Tj}/\omega_{cj}$ is the Larmor radius of particles of the j-th species with thermal velocity $v_{Tj} = \sqrt{2T_j/m_j}$ and temperatuare T_j. Under these conditions one can neglect the magnetic field influence on the ion motion and use the drift approximation for electrons. Then the dispersion equation for the plasma waves in the frequency range (8) takes the form

$$\varepsilon(\omega,k) \equiv 1 + \frac{\omega_{pe}^2}{\omega_{ce}^2} - \frac{\omega_{pp}^2}{\omega^2} - \frac{\omega_{pe}^2}{\omega^2}\frac{k_\parallel^2}{k^2} - \frac{4\pi e_i^2}{m_i k^2} \int d\vec{v}\,\frac{\partial f_i}{v_\perp \partial v_\perp}\left(1 - \frac{\omega}{\sqrt{\omega^2 - k_\perp^2 v_\perp^2}}\right) - \frac{i\pi\omega_{pe}^2}{k^2 n_e}\int dv_\parallel\, k_\parallel \cdot \frac{\partial f_{Te}}{\partial v_\parallel}\,\delta(\omega - k_\parallel v_\parallel) = 0 \tag{50}$$

where $\omega_{pj}^2 n_j / = 4\pi e_j^2 n_j/m_j$ is the plasma frequency for the j-th species with density n_j, and $f_i(v)$ and $f_{Te}(v)$ the velocity distribution functions of the ionized atoms and suprathermal electrons, respectively. Here we have assumed that the plasma is cold and used an expansion in the ratios of the thermal particle velocities to the phase velocity of waves ($k_\perp v_{Tp}/\omega << 1$, $k_\| v_{Te}/\omega << 1$) in order to calculate the contribution of plasma ions (labelled by the index p) and thermal electrons to the dispersion equation (50). In the quasilinear approximation, when nonlinear effects are neglected, the wave growth is described by equation (50):

$$\frac{\partial}{\partial t}|\phi_k|^2 = (\partial\varepsilon/\partial\omega_k)^{-1}\left\{\frac{4\,e_i^2\,\omega_k}{m_i k^2}\int dv\,\frac{\partial f_i}{v_\perp \partial v_\perp}\left(k_\perp^2 v_\perp^2 - \omega_k^2\right)^{-1/2} + \frac{4\pi e^2}{k^2 m_e}\int_{-\infty}^{+\infty} d v_\| \, k_\| \frac{\partial f_{Te}}{\partial v_\|}\pi\delta(\omega_k - k_\| v_\|)\right\}|\phi_{\vec{k}}|^2 \tag{51}$$

where $k^2|\phi_k|^2/8\pi$ is the spectral energy density of the electric field oscillations in plasma waves. The first term in equation (51) describes the generation of waves by a beam of ionized atoms with growth rate γ_k^i, and the second term accounts for the electron Landau damping with growth rate γ_k^e.

A complete set of quasilinear equations besides equation (51) for the plasma waves includes also the equation for the velocity distributions of ionized atoms and suprathermal electrons:

$$\frac{\partial f_i}{\partial t} = \frac{\partial}{v_\perp \partial v_\perp}\int\frac{d\vec{k}}{(2\pi)^3}\,\frac{\omega_k e^2|\phi_k|^2}{m_i^2\sqrt{k_\perp^2 v_\perp^2 - \omega_k^2}}\,\frac{\partial f_i}{v_\perp \partial v_\perp} + \pi^{-1}\frac{d n_i}{dt}\,\delta(v^2 - V^2)\,\delta(v) \tag{52}$$

$$\frac{\partial f_{Te}}{\partial t} = \frac{\partial}{\partial v}\frac{e^2}{m_e^2}\int\frac{d\vec{k}}{(2\pi)^3}\,k_\|^2|\phi_{\vec{k}}|^2\,\pi\delta(\omega_k - k_\| v_\|)\frac{\partial f_{Te}}{\partial v_\|} + \frac{\partial}{\partial v_\|}\sigma_{ion}\,N\!\left(\frac{e\phi_I}{m_e}\right) f_{Te} \tag{53}$$

The first of these equations was derived earlier to describe the loss-cone instability of the plasma in the mirror machines [Galeev, 1965]. Here we included in it the source of the ionized atoms with the velocity distribution in the form (48). The production of ionized atoms is described by

$$\frac{d n_i}{dt} = \nu_{ion}\, n_{Te} \tag{54}$$

where

$$\nu_{ion} = N\int_{-\infty}^{+\infty}\sigma_{ion}(v_\|)\,|v_\||f_{Te}(v_\|, t)\,d v_\|/n_{Te}(t)$$

is the rate of ionizational energy losses of electrons, and N is the neutral gas density.

Equation (54) describes the quasilinear diffusion of suprathermal electrons in velocity space in the drift approximation.

With a continuous ionization of the gas by electron impact we can expect that a final state will be reached that is characterized by a stationary shape of the velocity distributions of ionized atoms and suprathermal electrons whose densities, however, are growing according to equation (54). Then the ionized atom distribution can be represented in the form

$$f_i(v,t) = \left[n_i(t)/\pi V^2\right]\psi(v^2/V^2)\,\delta(v) \tag{55}$$

Moreover, after a sufficiently long interaction of the gas with the plasma flow, the quasilinear diffusion of electrons in velocity space under the action of excited waves will result in the formation of a long suprathermal tail in the electron distribution. We can expect the number of suprathermal electrons to finally reach a value such that the electron Landau damping of waves will balance wave generation by the beam of ionized atoms. If the suprathermal tail in the electron distribution extends through the whole range of possible phase velocities of waves, then all the energy released in the course of the quasilinear relaxation of the velocity distribution of ionized atoms will be transferred to electrons. Thus the fraction of transferred energy is controlled in fact by the quasilinear diffusion of ionized atoms in the velocity space described by equation (52). However, to solve this equation we have to discuss first the spectrum of the excited plasma turbulence. As follows from expression (51) for the growth rate γ_k^i, the latter reaches its maximum for the particular value of the wave phase velocity V_{ph} depending on the degree of relaxation of the ionized atom beam (see below). Therefore it is reasonable to assume that the spectrum of plasma turbulence has the form of a streamer in $\vec{k}$-space [Galeev and Khabibrakhmanov, 1985]:

$$k^2|\phi_k|^2 = E^2(k_\perp)\,\delta\!\left(k_\| \pm k_\perp^2 V_{ph}/\omega_{ce}\right) \tag{56}$$

where for clarity we used the explicit expression for the wave frequency in a cold plasma,

$$\omega_k = \omega_{ce}\,|k_\||/k \tag{57}$$

in the limit of oblique propagation $k_\|/k >> (m_e/m_i)^2$. The balance of the growth and damping of waves along this streamer can be reached only for the specific shape of the velocity distribution of suprathermal electrons defined by equation (49):

$$\frac{\pi k_\|}{|k_\||}\frac{\partial f_{Te}}{\partial v_\|}\Big|_{v_\| = \omega/k_\|} = \frac{2m_e n_i}{m_i}\,w_*^{1/2}\int_{w_*}^{\infty} dw\,\frac{d\psi(w)}{dw}\,[w - w_*]^{1/2} \tag{58}$$

where $w = v_\perp^2/V^2$ and $w_* = V_{ph}^2/V^2$. Since the right-hand side is independent of the longitudinal phase velocity, we obtain

$$f_{Te}(v_\|, t) = \left[n_{Te}(t)/V_{oe}^2\right]\,(V_{oe} - |v_\||) \tag{59}$$

The shape of the velocity distribution of ionized atoms can be found from equation (52) which, with the help of equation (55), takes the simple form

$$\frac{16a^2}{25}\,\frac{d}{dw}(w - w_*)^{-1/2}\frac{d\psi}{dw} - \psi = -\,\delta(w-1) \tag{60}$$

where we introduce the parameter a characterizing the ratio of the quasilinear relaxation rate and the ionization rate:

$$a^2 = \int\frac{d\vec{k}}{(2\pi)^3}\,\frac{4\omega_k}{kV}\,\frac{e^2|\psi_{\vec{k}}|^2}{m_i^2 V^4}\,\frac{16\nu_{ion}\,n_{Te}}{25 n_i} \tag{61}$$

The general solution of equation (19) has the form [*Formisano et al.*, 1982]

$$\psi(w,a) = \frac{1}{1-w_*}\,\psi_o\left(\frac{w - w_*}{1 - w_*}, a_*\right)\theta(w - w_*) \tag{62}$$

where

$$\psi_o(w,a) = \frac{5w^{3/4}}{4a^2}\{I_{3/5}(\frac{w^{5/4}}{a})K_{3/5}\theta\left(\frac{1}{a}\right)(1-w)$$

$$+K_{3/5}\left(\frac{w^{3/5}}{a}\right)I_{3/5}\left(\frac{1}{a}\right)\theta(w-1)+\frac{2}{\Gamma(2/5)\Gamma(3/5)}\cdot \tag{63}$$

$$\cdot K_{3/5}\left(\frac{w5/4}{a}\right)K_{3/5}\left(\frac{1}{a}\right)\}\theta(w)$$

Here $\Gamma(x)$ is the gamma function, and $\theta(w)=1$ for $w>0$ and $\theta(w) = 0$ for $w < 0$.

The fraction η of the kinetic energy of ionized atoms transferred to electrons as a result of the quasilinear relaxation is calculated as the difference betwen the initial and final energies:

$$\eta(a) = 1 - \int_0^\infty w\psi(w,a)\,dw = \frac{16a^2}{25}\int_{w_*(a)}^\infty dw\frac{d\psi}{dw}(w-w_*)^{-1/2} \tag{64}$$

A second expression for η is obtained here with the help of equation (60) and can be rewritten in the obvious form

$$\eta = \int\frac{dk}{(2\pi)^3}2\gamma_k^i\frac{\partial\omega_e}{\partial\omega}\frac{k^2|\phi_k|^2}{8\pi}\Big/\nu_{\rm ion}n_{Te}m_iV^2/2 \tag{65}$$

Formisano et al. [1982] calculated this value as a function of the parameter for the limit of small phase velocity of excited waves, i.e., for $w_*\rightarrow 0$:

$$\eta_o(a) = 1-\int_0^\infty w\psi_o(w,a)\,dw \tag{66}$$

We have mentioned that the phase velocity itself is a function of the parameter a, calculated from the condition that the growth rate γ_k^i peaks at the particular value of w_*. Using expressions (64) and (66), we can rewrite the wave balance equation in the stationary case as

$$\frac{\partial}{\partial t}|\phi_k|^2 = \frac{2\omega_{LH}^2}{k^2V^2}\omega_k\left(\bar{\gamma}_i - \frac{m_iV^2}{m_eV_{oe}^2}\frac{\pi n_{Te}}{2n_i}\right)|\phi_k|^2 = 0 \tag{67}$$

where $\omega_{LH} = \sqrt{|\omega_{ci}\,\omega_{ce}|}$ is the lower hybrid frequency, and

$$\bar{\gamma}_i = w_*^{1/2}(1-w*)\frac{25}{16a^2}\eta_o\left(\frac{a}{(1-w_*)^{5/4}}\right) \tag{68}$$

Here we have assumed that in the narrow energy range $0 < w < w_* << 1$ the velocity distribution function can be made flat, i.e., $\psi(0) = \psi(w_*,a)$ (see Fig. 5) without changing much the total energy content of the ionized atoms. The phase velocity w_* of excited waves is found from the condition $d\gamma_i/dw_* = 0$. Both quantities w_* and γ_i as well as the efficiency of the energy transfer $\eta(a) = (1-w_*)\eta_o(a/(1-w_*^{5/4})$ are plotted in Figure 6 as functions of the parameter a.

From the calculated shape of the velocity distribution of ionized atoms (Fig. 5) we see that the ionized atoms diffuse in velocity space to energies both higher and lower than the initial energy. This is because a wave with a given phase velocity interacts with all ionized atoms having perpendicular velocities higher than the wave phase velocity, resulting in their diffusion. The fast cyclotron rotation guarantees the isotropy of the perpendicular velocity distribution. Let us note that the positive contribution to the growth rate comes from particles in the region of a positive derivative of the velocity distribution. Since the diffusion in velocity space results in a decrease of this derivative, the growth rate of the waves also drops. The width of the distribution function increases with the intensity of the waves characterized by the dimensionless parameter a (see equation (61)), and the growth rate decreases correspondingly until it drops to zero at a = 0.6 (see Fig. 6).

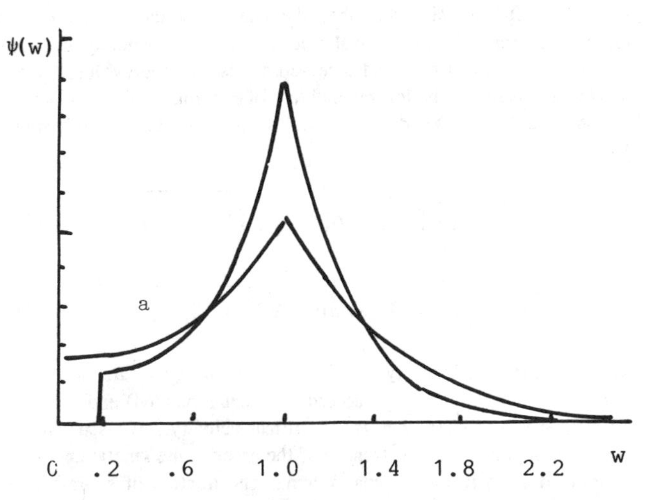

Fig. 5. The normalized distribution functions of ionized atoms versus perpendicular velocities in the case of the complete quasilinear stabilization of the instability (a) and in the case of maximum efficiency of energy transfer to electrons.

In the absence of suprathermal electrons this corresponds to the only existing stationary solution of the wave equation (51). In this case the energy is not transferred to electrons, i.e., $\eta(0.6) = 0$, but is simply redistributed among the ionized atoms.

As the number of suprathermal electrons increases, the stationary energy density of waves characterized by the parameter a decreases, and simultaneously, the efficiency of the energy transfer to electrons increases, reaching its maximum $\eta_{max} = 0.018$ for a = 0.38.

The density n_{Te} and the maximum velocity V_{oe} of suprathermal electrons can be found from the wave energy balance equation (67) and the equation for the particle energy balance resulting from the quasilinear equations (51)-(53):

$$\frac{dn_{Te}}{dt}m_eV_{oe}^2/12 = \nu_{\rm ion}n_{Te}\left[\eta(a)m_iV^2/2 - e\,\phi_I\right] \tag{69}$$

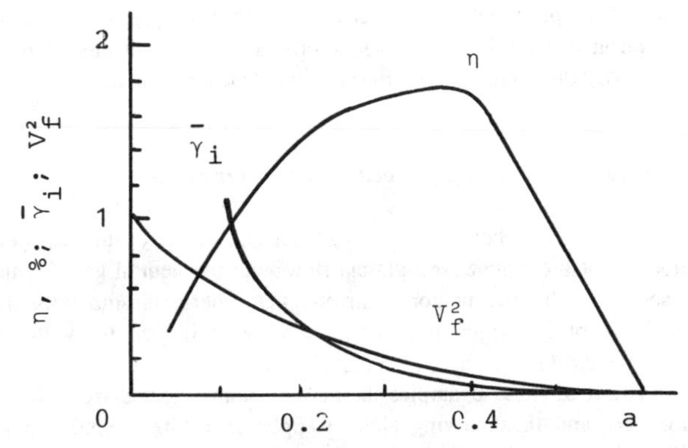

Fig. 6. The dependence of the energy transfer efficiency η in percent, the squared phase velocity $V_f = \omega_k/kV$, and the normalized growth rate $\bar{\gamma}_i$ of the most unstable wave on the parameter a, i.e., on the ratio of the quasilinear diffusion rate to the ionization rate.

Here the left-hand side describes the change of the energy of the growing number of suprathermal electrons with the velocity distribution (59), and the right-hand side represents the energy lost by the ionized atoms minus the ionizational loss of electrons.

From equations (67) and (69) we find [*Galeev and Khabibrakhmanov*, 1983]

$$\frac{n_{Te}}{n_i} = \frac{5\eta(a)}{2a}\,[w_*(a)]^{1/4}\sqrt{\frac{3}{\pi}\left(1 - V_c^2/V^2\right)} \tag{70}$$

$$\frac{m_e V_{oe}^2}{m_i V^2} = \frac{4a}{5}\,[w_*(a)]^{1/4}\sqrt{3\pi\left(1-V_c^2/V^2\right)} \tag{71}$$

where the efficiency of the energy transfer is $\eta(a) = (1-w_*)\,\eta_o\,(a/(1-w_*)^{5/4})$ according to equations (64) and (66) (see Fig. 6), and $V_c = \sqrt{2e\,\phi_I/\eta m_i}$ is the critical velocity. We see that the energy of suprathermal electrons is of the order of the kinetic energy of an ionized atom in the plasma system. The fraction of suprathermal electrons happens to be of the order of the efficiency of the energy transfer.

Finally, the spectral energy density of plasma waves is found from the kinetic equation (53) for the electrons. Taking into account that the shape of the velocity distribution of suprathermal electrons is stationary (see equation (59) and that their number density increases proportionally to the number density of ionized atoms (see equations (70) and (73)), we find from equations (53) and (56),

$$\frac{k^2|\phi_{\vec{k}}|^2}{4\pi} = \left(\frac{e\phi_I}{m_e c^2}\right)B^2\,\{\sigma_{\rm ion}V_{oe}N + 6\nu_{\rm ion}\left(V^2/V_c^2-1\right)$$
$$\left(1-\frac{\omega_{ce}}{k_\perp V_{oe}}\right)\}\left(1-\frac{\omega_{ce}}{k_\perp V_{oe}}\right)\frac{1}{k_\perp^3 V}\,\delta\left(k_\parallel \pm \frac{k_\perp^2 w_*^{1/2} V}{\omega_{ce}}\right). \tag{72}$$

Thus we have found the dependence of the density and energy of suprathermal electrons on the velocity of the plasma flow through the gas and therefore have solved the problem of the anomalous ionization of the gas by the plasma flow. However, we should point out here that the parameter a was left as a free parameter of the problem. This means that the quasistationary ionization of the gas can be provided by a different number of suprathermal electrons, which is controlled by the process of the pulling of thermal electrons into the suprathermal tail of the electron distribution. This latter process was not discussed here, since we still do not have a rigorous analytical description of it.

Observations of the Critical Velocity Ionization in Space

We briefly discuss here only the two most spectacular examples of the interaction of the magnetized plasma flow with the neutral gas: 1) the interaction of the Jovian corotating magnetospheric plasma with the neutral gas of Io origin and 2) the solar wind interaction with the extended neutral atmosphere of comet Halley.

In the first of these examples the neutral atoms ejected from the Io atmosphere and thus moving along Kepplerian orbits around Jupiter with a velocity of 17 km/s form, as a result of their ionization, a beam of ions corotating with the planet with a velocity of 75 km/s at the Io orbit, in the same direction as Io itself. Thus the relative velocity of the gas through the plasma is equal to V = 58 km/s. Since the expected neutral gas density is $N \cong 20\ \mathrm{cm}^{-3}$ and the main components of this gas are sulfur and oxygen, the above velocity exceeds the critical velocity in the limit of a rarefied gas at and beyond Io's orbit. For the measured magnetic field $B \cong 1.9\ 10^{-2}$ G, the condition for the applicability of the quasilinear theory is well satisfied. Though for the typical parameters of the Io plasma torus ($n_e \cong 1500\ \mathrm{cm}^{-3}$, $T_e \cong 8$ ev) the main contribution to the ionization of the neutral gas comes from the Maxwellian tail electrons, the consequences of the development of the beam instability of the newly ionized atoms are well observed (for details see Galeev and Khabibrakhmanov [1983, 1988]). The latter can be formulated as follows:

1. Electric field oscillations in the lower hybrid range of frequencies have been detected [Coroniti et al., 1980]. Moreover, if one calculates the frequency distribution of the spectral energy density of the electric field oscillations in the spacecraft system, then this expression describes both the shape and the absolute value of the observed wave spectrum (Fig. 7).
2. The energy (~1 keV), the density ($n_{Te} \simeq 0.01\ n_e$), and the shape of the velocity distribution of the observed suprathermal electrons [Scudder et al., 1981] agree with the theoretical estimates of Galeev and Khabibrakhmanov [1988, section 2.1].
3. Bursts of the electric field oscillations at the frequency $f \cong 0.5\ f_{ce}$ have been detected [Coroniti, et al., 1980] and could be the result of whistler wave generation by the anisotropic suprathermal electrons [Galeev and Khabibrakhmanov, 1988].

However, the absolute value of the intensity of the registered oscillations happens to be orders of magnitude lower than the theoretically predicted values for the largest possible anisotropy of the suprathermal electron tail. This can be explained by good confinement of these electrons by the self-consistent electric fields in a plasma whose ions are confined in the Io torus by centrifugal forces. Therefore, we expect that the losses of these electrons along the magnetic field are very low and, as a result of this, the asymmetry of the suprathermal electron tail

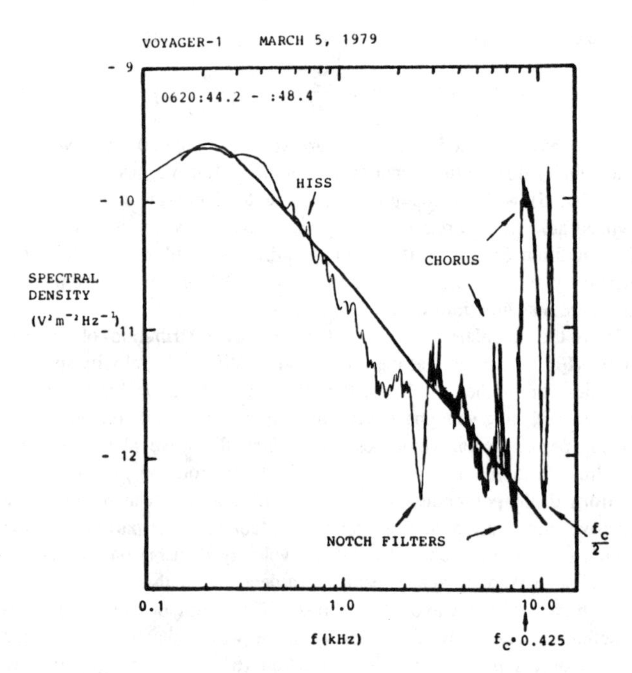

Fig. 7. The measured [*Coroniti et al.*, 1980] and theoretical (smooth thick curve) spectra of the electric field oscillations in the Io torus.

is orders of magnitude lower than that used to calculate the whistler wave intensity.

These data have led us [Galeev and Khabibrakhmanov, 1982] to the conclusion that the suprathermal electrons are generated in the process of the collective interaction of the newly ionized atoms with the corotating magnetospheric plasma and that although their contribution to the gas ionization is small, the presence of the multiplying ionized atoms in the Io torus and the fast variation of the cold electron temperature could be explained only by the influence of the suprathermal electrons. An analysis of the energy balance of suprathermal electrons with an accounting or the observed intensity of the lower hybrid waves [Barbosa et al., 1985] has confirmed the first of these conclusions.

Effects of the collective interaction of newly ionized atoms and molecules of the cometary atmosphere with the solar wind loaded by cometary ions were observed during the flyby of spacecraft Vega-1, Vega-2, and Giotto through the coma of comet Halley. Here, as in the Io torus, the number density of suprathermal electrons was so low [d'Uston et al., 1986] that ionization by electron impact in the inner coma is negligible compared to photoionization by the solar UV radiation. However, we have analyzed the plasma density measurements of cometary ions by the wide-angle electrostatic analyzer Plasmag-1 and the integral intensity of plasma waves in the lower-hybrid range of the frequencies from 1 Hz to 32 Hz ($f_{LH} \cong 8$ Hz in the magnetic field B_o = 50 nT) measured by the low-frequency plasma wave analyzer aboard Vega-2 [Savin et al., 1986, Fig. 8]. This analysis showed that the plasma density build-up towards the inner coma is an intermittent process: the slow density growth due to photoionization is interrupted by shorter periods of faster growth. At the beginning of a faster growth period the intensity of plasma waves peaks. Therefore it is natural to assume that this faster density growth could be explained by the suprathermal electrons generated by the excited lower-hybrid

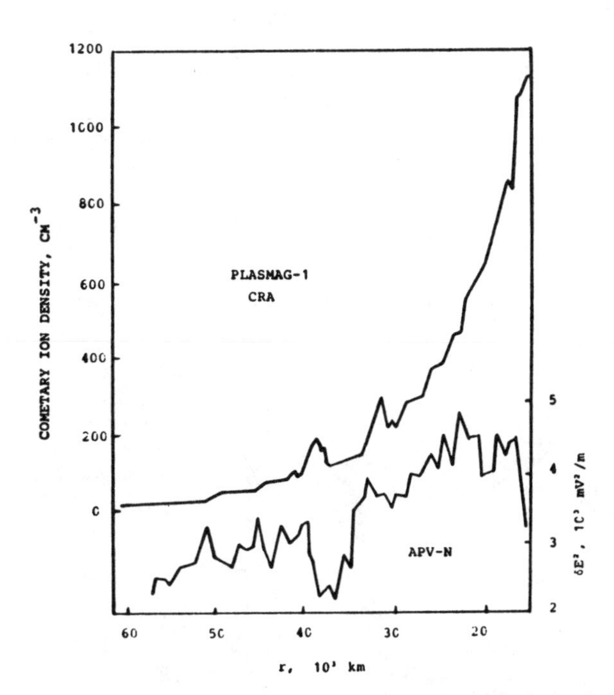

Fig. 8. Profiles of the heavy ion density (top curve) and the integral wave intensity (lower curve) measured aboard Vega-2 [from *Savin et al.*, 1986].

plasma waves. These regions of denser plasma are viewed by ground observers as a receding envelope structure.

References

Alfvén, H., On the origin of the solar system, Oxford, University Press, Oxford, 1954.

Barbosa, D. D., F. V. Coroniti, W. S. Kurth, and F. L. Scarf, Detection of lower-hybrid noise in the Io plasma torus and plasma heating rates, *Astrophys. J.*, *289*, 392, 1985.

Bychenkov, V.Yu., and V. P. Silin, The theory of ion-sound instability in a plasma with an electric field, *Zhurn. Eksper. Teor. Fiz.*, *82*, 1886, 1982.

Breizman, B. N., V. E. Zakharov, and S. L. Musher, Kinematics of stimulated scattering of Langmuir waves by plasma ions, *Zhurn. Eksper. Teor. Fiz.*, *64*, 1297, 1973.

Coroniti, F. V., F. L. Scarf, C. F. Kennel, W. S. Kurth, and D. A. Gurnett, Detection of Jovian whistler mode chorus: Implication for the Io plasma torus aurora, *Geophys. Res. Lett.*, *7*, 45, 1980.

Cornwall, J., F. V. Coroniti, and R. M. Thorne, Turbulent loss of ring current protons, *J. Geophys. Res.*, *75*, 4699, 1970.

Formisano, V., A. A. Galeev, and R. Z. Sagdeev, The role of the critical ionization velocity phenomena in the production of inner coma cometary plasma, *Planet. Space Sci.*, *30*, 491-497, 1982.

Galeev, A. A., and V. N. Oraevskii, On the instability of Alfvén waves, *Soviet Phys. "Doklady,"* *7*, 988, 1962.

Galeev, A. A., and V. N. Karpman, Turbulent theory of a weakly nonequilibrium rarefied plasma and the structure of shock waves, *Soviet Phys. JETP*, *17*, 403, 1963.

Galeev, A. A., Ion escape from a magnetic mirror trap due to development of instability connected with the "loss cone," *Soviet Phys. JETP*, *23*, 672, 1965 (in Russian).

Galeev, A. A., R. Z. Sagdeev, Yu. S. Sigov, V. D. Shapiro, and V. I. Shevchenko, Nonlinear theory of modulational instability of Langmuir waves, *Sov. J. Plasma Phys.*, *1*, 5, 1975.

Galeev, A. A., E. M. Mishin, R. Z. Sagdeev, V. D. Shapiro, and V. I. Shevchenko, Nearby rocket discharge following electron beam injection in the ionosphere, *Soviet Phys. Doklady*, *21*, 641, 1976.

Galeev, A. A., and I. Kh. Khabibrakhmanov, Origin and energetics of the Io plasma torus, *Adv. Space Res.*, *3*, No. 3, 71, 1983.

Galeev, A. A., and R. Z. Sagdeev, Theory of weakly turbulent plasma, in *Basic Plasma Physics I*, edited by A. A. Galeev and R. N. Sudan, 677, North-Holland, Amsterdam, 1983.

Galeev, A. A., and R. Z. Sagdeev, Current instabilities and anomalous resistivity of plasma, Basic Plasma Physics II, 271, 1984.

Galeev, A. A., and I. Kh. Khabibrakhmanov, On the nature of plasma waves in the Io torus, *Pis'ma Astron. Zhurn.*, *11*, 292, 1985.

Galeev, A. A., A. N. Polyudov, R. Z. Sagdeev, K. Sego, V. D. Shapiro, V. I. Shevchenko, MHD turbulence in solar wind interacting with comet, *Zhurn. Eksper. Teor. Fiz.*, *92*, 2090, 1987.

Galeev, A. A., and I. Kh. Khabibrakhmanov, The critical ionization velocity phenomena in astrophysics, in *Reviews of Science and Technology: Space Research, 27*, edited by R. Z. Sagdeev, 56, VINITI, Moscow, 1988.

Intrilligator, D. S., Solar wind turbulence and fluctuations, in *Solar Wind Four*, edited by H. Rosenbauer, 359, Report No. MPAE-W-100-81-31, 1981.

Kadomtsev, B. B., *Plasma Turbulence*, Academic Press, London, 1966.

Korablev, L. V., and L. I. Rudakov, Quasilinear theory of current instability in plasma, *Zhurn. Eksper. Teor. Fiz.*, *50*, 220, 1966.

Kuznetsov, E. A., On turbulence of ion sound in a plasma located in a magnetic field, *Zhurn. Eksper. Teor. Fiz., 62,* 584, 1972.

McIvor, I., The inertial range of weak magnetohydrodynamic turbulence in the interstellar medium, *M.N.R. Astr. Soc., 178,* 85, 1977.

Musher, S. L., A. M. Rubenchik, and I. Ya. Shapiro, Nonlinear effects on the propagation of an ion beam across a magnetic field, *Zhurn. Eksper. Teor. Fiz., 90,* 890, 1986.

Papadopoulos, K., Theory of beam plasma discharge, in *Artificial Particle Beams in Space, Plasma Studies,* edited by B. Grandal, 505, 1981.

Papadopoulos, K., Electron acceleration in magnetosonic shock fronts, in *Plasma Astrophysics,* edited by T. D. Guyenne and G. Levy, 313, ESA SP-161, 1981.

Pelletier, G., Generation of a high-energy tail by strong Langmuir turbulence in a plasma, *Phys. Rev. Lett., 49,* 782, 1982.

Sagdeev, R. Z., and A. A. Galeev, *Nonlinear Plasma Theory,* Benjamin, New York, 1969.

Savin, S., G. Avanesova, M. Balikhin, S. Klimov, A. Sokolov, P. Oberc, D. Orlowskii, and Z. Krawczyk, ELF waves in the plasma regions near the comet, 20th ESLAB Symp., ESA SP-250, *3,* 433, 1986.

Scudder, J. D., E. C. Sittler, Jr., H. I. Bridge, *J. Geophys. Res., 86,* 8157, 1981.

Shapiro, V. D., and V. I. Shevchenko, Strong turbulence of plasma oscillations, in *Basic Plasma Physics II,* edited by A. A. Galeev and R. N. Sudan, 123, North-Holland, Amsterdam, 1984.

Thorne, R. M., and J. Moses, Electromagnetic ion cyclotron instability in the multi-ion Jovian magnetosphere, *Geophys. Res. Lett., 10,* 631, 1983.

d'Uston, C., H. Reme, J. A. Sauvaud, et al., Description of the main boundaries seen by the Giotto electron experiment inside the comet Halley—solar wind interaction regions, in *Exploration of Halley's Comet,* edited by M. Grewing, F. Praderie, R. Reinhard, Springer-Verlag, Berlin, 137, 1987.

Vaisberg, O. L., A. A. Galeev, G. N. Zastenker, S. I. Klimov, M. N. Nozdrachev, R. Z. Sagdeev, A. Yu. Sokolov, V. D. Shapiro, Electron acceleration at the front of strong collisionless shock, *Zhurn. Eksper. Teor. Fiz., 85,* 1232, 1983.

Zakharov, V. E., Collapse of Langmuir waves, *Zhurn. Eksper. Teor. Fiz., 62,* 1745, 1972.

Zakharov, V. E., Kolmogorov spectra in weak turbulence problems, in *Basic Plasma Physics II,* edited by A. A. Galeev and R. N. Sudan, 3, North-Holland, 1984.

ULTRA-LOW FREQUENCY WAVES AT COMETS

Martin A. Lee

Institute for the Study of Earth, Oceans and Space, University of New Hampshire, Durham, NH 03824

Abstract. This paper is a review of the observations and theory of ultra-low frequency (ULF) waves or turbulence (spacecraft frequencies less than 1 Hz) upstream of comets with an emphasis on the region upstream of the cometary bow wave. It attempts to be comprehensive in its coverage of the literature through Spring, 1988. First the general morphology of the cometary foreshock with its pickup ion and ULF wave distributions is described briefly. The observations of the ULF waves at comet Giacobini-Zinner by the ICE spacecraft and at comet Halley by the Giotto, Vega and Sakigake spacecraft are then reviewed. The major spectral feature at $\sim$ 0.01 Hz is shown to arise from cyclotron resonance of the waves with water-group pickup ions. The stability analysis is presented for hydromagnetic and whistler waves propagating parallel to the ambient magnetic field in the presence of narrow and broad pickup ion ring distributions. The growth rate of the right-hand circularly polarized sunward propagating hydromagnetic wave is sufficient to account for the spectral feature at $\sim$ 0.01 Hz. Instabilities which could account for the 0.1 - 1 Hz wave packets observed at comet Giacobini-Zinner are reviewed but none appears to be promising. The quasilinear theory of the pickup ion - ULF wave interaction is shown to account rather well for ULF wave intensities near the cometary bow waves of the two comets. Finally work on the nonlinear evolution of the $\sim$ 0.01 Hz ULF waves is reviewed and found to provide a plausible origin for the 0.1 - 1 Hz wave packets.

1. Introduction

One of the most beautiful examples of the plasma wave-particle interaction found in space has presented itself in the regions upstream of comets Giacobini-Zinner (G-Z) and Halley. Atoms and molecules (predominantly water group) are evaporated from the cometary surface when the surface warms in the vicinity of the Sun.

Following dissociation reactions in the cometary coma, the particles leave the coma with speeds ranging from $\sim$ 1 km/s (O, OH, H_2O) to $\sim$ 10 km/s (H) relative to the comet and form an expanding gas halo subject to photoionization by solar radiation or ionization by charge exchange with the solar wind with a timescale of $\sim 10^6$ s at 1 AU. The freshly ionized atoms and molecules are "picked up" by the solar wind and gyrate about the solar wind magnetic field. In the solar wind frame the pickup ions initially stream toward the Sun with the solar wind speed, V_{SW} (actually the solar wind speed relative to the comet), and pitch angle, $\alpha = \cos^{-1} [(\underset{\sim}{V}_{SW} \cdot \underset{\sim}{B})/|\underset{\sim}{V}_{SW}||\underset{\sim}{B}|]$, where $\underset{\sim}{B}$ is the local magnetic field averaged over the first gyration. In the frame of the comet their guiding center velocity is initially approximately equal to $\underset{\sim}{B}_o \times (\underset{\sim}{V}_{SW} \times \underset{\sim}{B}_o)/B_o^2$, where $\underset{\sim}{B}_o$ is the solar wind magnetic field averaged over the cometary foreshock.

The configuration upstream of comet G-Z is shown schematically in Figure 1, with the observed bow wave indicated by a solid heavy curve at a distance of about 10^5 km from the comet. The light curves indicate field lines of $\underset{\sim}{B}$. The vector is the solar wind velocity which defines the local α. Indicated by a solid curve is the motion of a water-group pickup ion (the trajectory actually lies in the plane normal to $\underset{\sim}{B}$). Superposed on the fieldlines of $\underset{\sim}{B}$ are waves propagating parallel to $\underset{\sim}{B}$ in the solar wind frame with wavelength equal to the distance a water-group pickup ion moves along $\underset{\sim}{B}$ in the solar wind frame in one gyroperiod. The arrows on the waves indicate the expected direction of wave propagation in the solar wind frame (see Section 5). Figure 1 is drawn approximately to scale. Noting that the neutral gas cloud density and the ion pickup rate vary as $\sim r^{-2}$, where r is radial distance from the comet, it is clear that the configuration near the bow wave is not spatially homogeneous on the scale of a pickup ion gyroradius or the designated wavelength, nor are the ion distributions gyrotropic. These complications should be kept in mind when applying theory for spatially homogeneous systems to the wave-particle

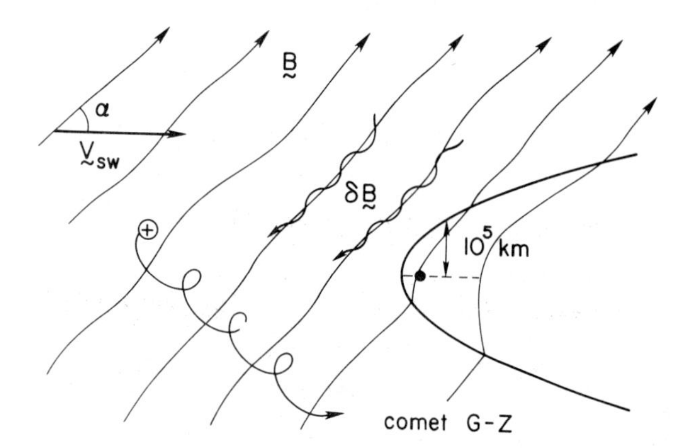

Fig. 1. Schematic diagram of the foreshock of comet Giacobini-Zinner. The dark solid curve is the cometary bow wave; light curves are magnetic fieldlines. Also indicated is the solar wind velocity, $\underset{\sim}{V}_{SW}$, the angle α between $\underset{\sim}{V}_{SW}$ and the local magnetic field $\underset{\sim}{B}$, a water-group pickup ion trajectory, and magnetic field lines of $\underset{\sim}{B} + \delta\underset{\sim}{B}$ associated with upstream ULF turbulence. The bow wave, ion trajectory and waves are drawn approximately to scale.

interaction. They should be less important upstream of comet Halley whose bow wave has a standoff distance of $\sim 10^6$ km.

Two other features of the pickup process are apparent in Figure 1. The indicated variations in the direction of $\underset{\sim}{B}$ imply that in any volume of solar wind pickup ions with a range of pitch angles are present, although those picked up most recently dominate the pickup ion distribution. This range is increased in the presence of enhanced turbulence due both to the increased range of α and to the increased scattering of the pickup ions in pitch angle. It is also clear that unless $\langle\alpha\rangle = 0°$ or $90°$ the pickup ion spatial distribution is asymmetric about the axis through the comet parallel to $\underset{\sim}{V}_{SW}$. As indicated by the ion trajectory the pickup ions are initially skewed to one side of the comet. It is possible that this asymmetry could account in part for asymmetries observed at both comets Halley and G-Z between inbound and outbound spacecraft trajectories.

Since the pickup ions in the solar wind frame have the solar wind speed and are approximately gyrotropic away from the bow wave where the ion gyroradius is much less than r, their phase-space distribution is conveniently displayed as a function of $\mu = \cos\alpha$, where $\mu < 0$ indicates streaming sunwards in the solar wind frame. Figure 2 depicts a characteristic pickup ion "ring" distribution $f(\mu)$ with $\langle\mu\rangle \cong -0.5$. The spread in μ has been estimated by computing $|\delta\underset{\sim}{B}|_{RMS}$ based on the integrated fluctuation power during typical solar wind conditions at scales $\lesssim 10^6$ km. The result is $|\delta\underset{\sim}{B}|_{RMS} \cong 0.2$ B_o. Taking the fluctuations to be equally distributed in the 3 spatial directions, we find $\Delta\mu \sim 0.1$, as shown in Figure 2.

As we shall discuss in Section 5, there is plenty of free energy in the pickup ion distribution shown in Figure 2 to excite waves in the ULF frequency range (< 1 Hz in the spacecraft frame) as shown schematically in Figure 1. The ULF waves grow and enhance the pitch-angle scattering of the pickup-ion ring distribution toward isotropy. A small ion anisotropy is insured by the continual pickup of freshly ionized particles, although the enhanced turbulence near the bow wave may eliminate much of the anisotropy within one gyration [Wu et al., 1986; Price and Wu, 1987]. The isotropization of the pickup ions erases the spatial anisotropy imposed on the upstream region by the pickup ion trajectories as shown in Figure 1, and the ions are then convected on average with the solar wind. The pickup ions mass-load the solar wind, prepare the wind to bypass the near-comet region, and determine the location and structure of the bow wave [Galeev, 1986; Galeev et al., 1986a; Sagdeev et al., 1987]. In turn the pickup ions are accelerated at the bow wave [Amata and Formisano, 1985], in the compression associated with mass-loading, and via second-order Fermi acceleration by the enhanced ULF turbulence [Isenberg, 1987; Ip and Axford, 1986; Terasawa, 1988] or possibly enhanced electrostatic turbulence [Buti and Lakhina, 1987]. The "clutch" providing the interaction between the ionized cometary gas and solar wind and determining the structure of the upstream region is a combination of the average magnetic field and the ULF waves, which dominate the enhanced turbulent wave field around the comet.

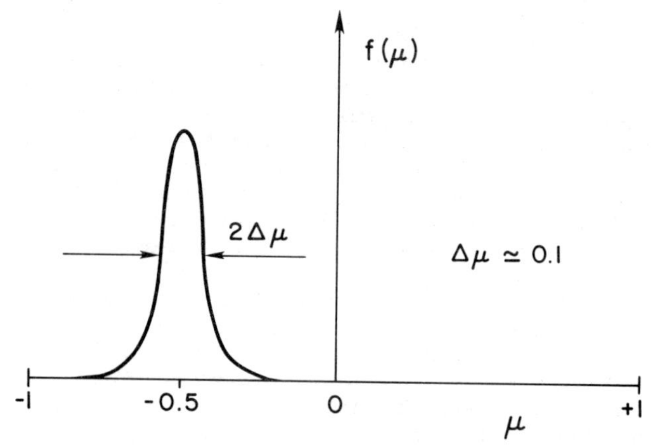

Fig. 2. Schematic diagram of a pickup ion distribution function in the frame of the solar wind as a function of the cosine of ion pitch angle, μ. The domain $\mu < 0$ indicates sunward streaming. Representative values of $\langle\mu\rangle$ and $\Delta\mu$ are given.

In Sections 2 and 3 we describe the observed characteristics of ULF waves at comets G-Z and Halley. In Section 4 we discuss the cyclotron resonance condition and in Section 5 we review theoretical work on the instabilities contributing to ULF wave excitation. In Section 6 we discuss the quasilinear evolution of the ULF wave spectrum. Finally, in Section 7 we address nonlinear evolution of the ULF waves and assess theoretical attempts to account for the observed ULF waveforms.

In closing this introduction we note that the region upstream of comets containing pickup ions and ULF waves is similar to the so-called "foreshock" region upstream of interplanetary shocks and planetary bow shocks, and we shall use the term (perhaps indiscriminately) to describe the region upstream of the cometary bow wave. The foreshock at interplanetary shocks and planetary bow shocks is populated by ions which are sufficiently energetic to escape the shock upstream in spite of their being constrained to move along the average upstream magnetic field. These ions may form energetic diffuse distributions or, at Earth's bow shock, field-aligned beams or gyrating beams. The gyrating beam distributions are not dissimilar to pickup ion rings once they have become approximately gyrotropic further upstream of the shock. In both foreshocks and cometary upstream regions the ions excite via similar instabilities ULF waves which are convected back to the shock or cometary bow wave where they are compressed to large amplitudes in the downstream region. There are, however, important differences. The pickup ions are predominantly very massive, they are also created on magnetic field lines not connected to the shock or bow wave, and they do not depend on a strong shock for their initial injection into the upstream region. Thus cometary upstream regions do not exhibit the rather abrupt ion "foreshock" boundary found upstream of Earth's bow shock. The large mass of water-group ions increases ULF wavelengths and increases timescales for ion transport, so that the ions upstream of comets are less evolved from their initial distributions than their proton foreshock counterparts. Finally the bow wave at comets is generally weak and not responsible for producing many ions in the upstream region. The wave-particle foreshock at Earth is treated in depth in a special issue of JGR (86, June 1, 1981). Articles by Hoppe et al. [1981] and Greenstadt and Baum [1986] are of particular interest for the structure of the ULF wave foreshock at Earth.

2. ULF Waves Upstream of Comet Giacobini-Zinner

ICE observations of the average magnetic field and ULF fluctuations at comet G-Z are described by Smith et al. [1986], Jones et al. [1986], Tsurutani and Smith [1986a,b] and Tsurutani et al. [1987b; 1988a,b]. An overview of the magnetic field observations is given in Figure 3, taken from Tsurutani and Smith [1986a], in which the field strength, elevation angle out of the ecliptic, and azimuthal angle measured from the spacecraft - Sun axis are presented for the 6 hours about closest approach. The bow wave and magnetic tail lobes are apparent and indicated in the trace of the field strength. The average field has an azimuth close to the Parker spiral at 1 AU (∿135°), particularly on the inbound leg, but with an inclination of 30° - 45° out of the ecliptic.

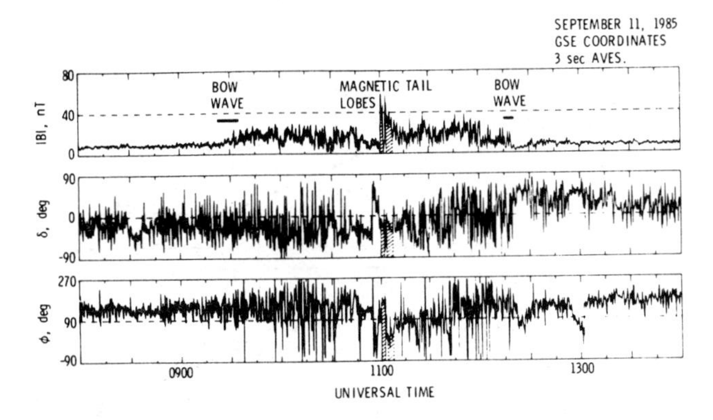

Fig. 3. An overview of the magnetic field near comet G-Z (Figure 1 from Tsurutani and Smith [1986a]), including field strength, elevation angle out of the ecliptic, and azimuthal angle measured from the spacecraft - Sun axis. ULF turbulence is apparent during the entire interval.

It is clear from Figure 3 that virtually the entire interval shown is characterized by very large amplitude fluctuations. Upstream of the bow wave the fluctuations are predominantly in the direction of $\mathbf{B}$; downstream of the bow wave large fluctuations in $|\mathbf{B}|$ occur also. It is even possible in this figure to pick out a dominant period of about 2 minutes. The enhanced turbulence extends to a distance ∿ 10^6 km from the comet: the transverse fluctuation level upstream of the bow wave decreases by a factor of 10 over a distance of ∿ 4.5×10^5 km (∿ 2×10^5 km) on the inbound (outbound) leg [Tsurutani and Smith, 1986a].

Figure 4 (taken from Tsurutani and Smith [1986a]) presents the power spectra of the three components of the magnetic field parallel (X) and transverse (Y, Z) to the average field taken during the first 25 minutes of the interval shown in Figure 3. In comparison with levels in "active" solar wind , the wave power levels are enhanced. The levels for "active" solar wind indicated in Figure 4 reflect a correction of the original figure by the authors. The spectra

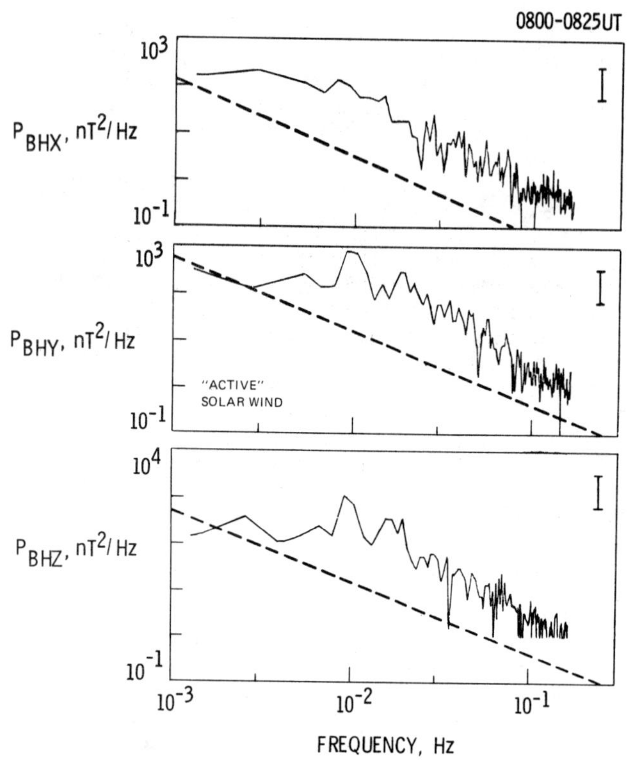

Fig. 4. The wave power spectral density along (BHX) and orthogonal to the average field direction computed during the first 25 minutes of the interval shown in Figure 3 (Figure 3 from Tsurutani and Smith [1986a]). Spectral peaks in the transverse components of the magnetic field are evident near 0.01 Hz. The indicated power levels for "active" solar wind reflect a correction of the original figure by the authors.

exhibit a peak at ∿ 0.01 Hz with an approximately Kolmogoroff decrease at higher frequencies. Within a bandwidth of 0.03 Hz about 0.01 Hz, the total power in the fluctuating field just upstream and downstream of the bow wave is approximately equal to that in the ambient magnetic field. The ULF fluctuations with dominant frequency of ∿ 0.01 Hz are also observed in the solar wind electron number density and the direction of the solar wind electron bulk flow velocity [Gosling et al. 1986; Tsurutani et al., 1987b].

The detailed waveforms apparent in the magnetic field fluctuations have been described in detail by Tsurutani and Smith [1986b] and Tsurutani et al. [1987b]. At larger distances from the comet, > 2-3 x 10^5 km, the approximately 100s-period waves tend to be left-hand elliptically polarized in the spacecraft frame. Closer to the bow wave the waves tend to exhibit linear polarization, an example of which, ∿ 2.5 x 10^5 km from the comet, is shown in Figure 5 [Tsurutani and Smith, 1986b]. The magnetic field is given in principal axis components with maximum (minimum) variation in the 1(3) direction. The similarity of the B_3 and $|\tilde{B}|$ traces implies that the wave propagates roughly parallel to the direction of the average field. Closer to the comet, ∿ 1-3 x 10^5 km, the waveforms are more compressional and tend to exhibit a left-polarized partial rotation of the magnetic field associated with each wave crest where $|\tilde{B}|$ is maximum. This waveform gives the impression that a left-polarized wave has been deformed to place most of the helicity at the wave crest [Tsurutani et al., 1987b; see also Section 7]. Still closer to the bow wave and within it, wave compression increases and the partial rotation is apparently extended into several rotations to form a wave packet composed of 1-7 second period oscillations associated with the crest of the underlying 100s-period wave, as shown for one example in Figure 6. Here the magnetic field components are in solar-ecliptic coordinates. The wave packets are observed to be left-hand circularly polarized, often to have large amplitude, and to exhibit minimum variance direction along the average field. A beautiful example is shown in Figure 7. The packets consist of from 1 to 20 oscillations, which decrease linearly in amplitude with distance upstream of the 100s-period wave crest, and exhibit a positive

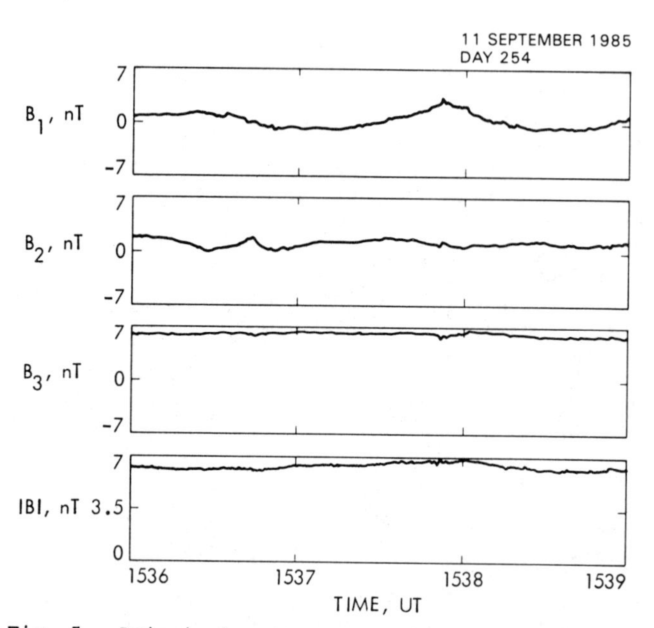

Fig. 5. Principal axis components for a long period (∿100s) wave (Figure 1 from Tsurutani and Smith [1986b]). The wave is approximately linearly polarized and propagates approximately along the average magnetic field.

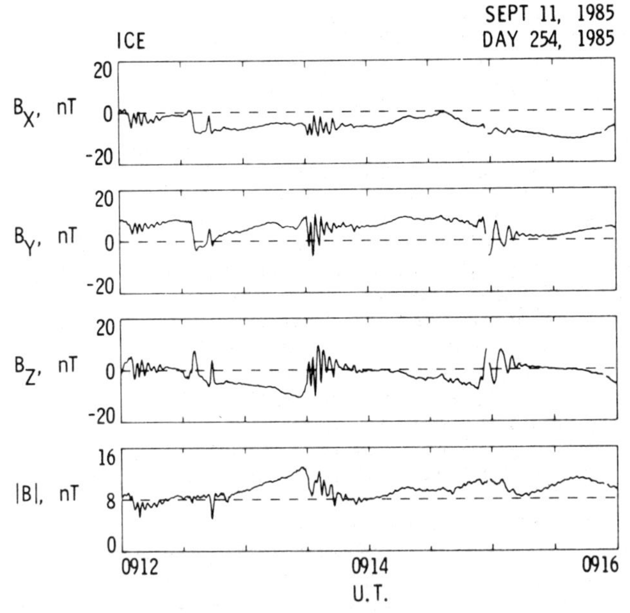

Fig. 6. Approximately two periods of the long period linearly polarized waves in solar-ecliptic coordinates showing the wave packets associated with the crests of the long period waves (Figure 2 from Tsurutani and Smith [1986b]).

correlation between the number of oscillations and wave frequency (Tsurutani et al., preprint entitled "Properties of whistler mode wave packets at the leading edge of steepened magnetosonic waves: Comet Giacobini-Zinner", 1988).

During those periods further from the comet ($3.5 - 7 \times 10^5$ km) when $\langle\alpha\rangle \cong 90°$, ULF wave activity is minimal. The 100s-period waves are absent. The only obvious signatures are 6-10s oscillations which occur sporadically as pulses with one, or occasionally a few, oscillations [Tsurutani et al., 1988a,b].

Pickup ions associated with the ULF waves have been observed at comet G-Z and have been inferred to be water-group ions predominantly [Sanderson et al., 1986; Ipavich et al., 1986; Gloeckler et al., 1986]. The radial dependence of their number density is consistent with theoretical expectations and a cometary water-group atom or molecule production rate of $\sim 2.6 \times 10^{28}$ s^{-1} assuming spherical symmetry. The pickup ion density at the bow wave is ~ 0.02 cm^{-3} [Gloeckler et al., 1986]. The form of their velocity distribution is difficult to determine precisely. Nevertheless a good correlation of $\sin^2\alpha$ with ion intensity at distances $\gtrsim 2 \times 10^5$ km is consistent with a ring distribution as shown in Figure 2 [Sanderson et al., 1986]. The existence and equivalence of the ion "rest frame" velocity and the solar wind velocity at distances $\lesssim 2 \times 10^5$ km is consistent with nearly isotropic distributions [Gloeckler et al., 1986]. Thus pitch-angle scattering begins to become important approximately 2×10^5 km from comet G-Z.

3. ULF Waves Upstream of Comet Halley

Enhanced ULF turbulence was also observed upstream of comet Halley by the Giotto, Vega and Sakigake spacecraft [Neubauer et al., 1986; Acuña et al., 1986; Johnstone et al., 1986, 1987a,b; Galeev, 1986; Galeev et al., 1986b; Riedler et al., 1986; Glassmeier et al., 1986, 1987, 1988; Saito et al., 1986; Yumoto et al., 1986a,b]. Due to the larger gas production rate at comet Halley than at comet G-Z and the correspondingly greater shock stand-off distance ($\sim 10^6$ km), the turbulence was observed at greater distances, but was of somewhat lower amplitude upstream of the bow wave than at comet G-Z. Observations by Giotto were complicated by multiple crossings of the interplanetary current sheet.

An example of a magnetic field fluctuation power spectrum upstream of the bow wave $\sim 2 \times 10^6$ km from the comet is shown in Figure 8 for the field component parallel to the average field (B_z) and a transverse component (B_x) (Glassmeier, private communication, 1987). The fluctuations are predominantly transverse and exhibit a peak near 0.01 Hz. Similar spectra have been recorded upstream of the bow wave by Acuña et al. [1986], Johnstone et al. [1986] and

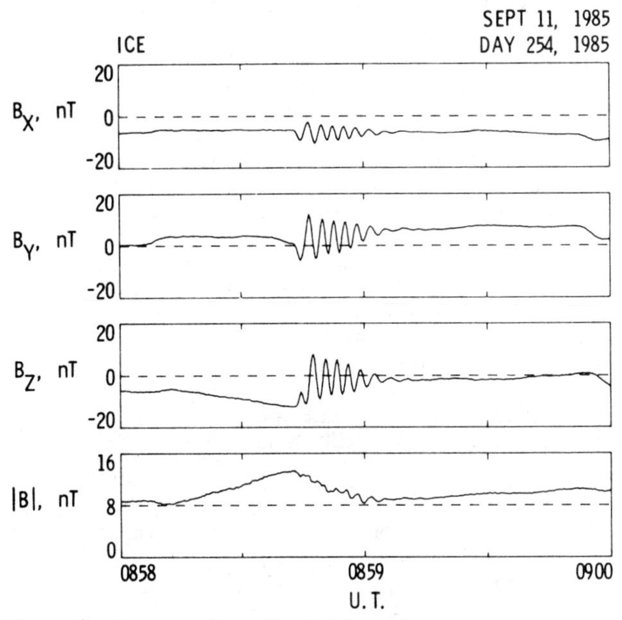

Fig. 7. An example of a higher frequency wave packet in solar-ecliptic coordinates (Figure 3 from Tsurutani and Smith [1986b]).

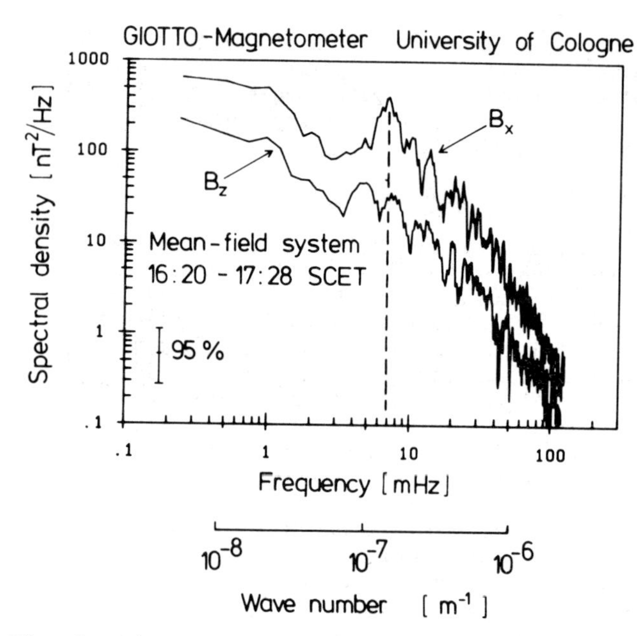

Fig. 8. The wave power spectral density for field components parallel (B_z) and perpendicular (B_x) to the average field measured by Giotto upstream of comet Halley (Glassmeier, private communication, 1987). A spectral peak near 0.01 Hz is apparent in the transverse component.

Glassmeier et al. [1986, 1988], and downstream by Glassmeier et al. [1986, 1987].

The spacecraft Sakigake observed 150-380s period waves at a distance of 7 x 10^6 km from the comet [Yumoto et al., 1986a,b]. The waveforms were observed to be approximately linearly polarized (but often left-hand elliptically polarized in the spacecraft frame) and to propagate nearly parallel to the average field [Yumoto et al., 1986a,b]. Closer to the comet the waveforms upstream of the bow wave tended to be linearly polarized, but wavefront steepening and the associated 1-7s period wave packets observed at comet G-Z have not yet been identified [Acuña et al., 1986]. During one particular period the wave was linearly polarized with magnetic field perturbation in the ecliptic. Johnstone et al. [1986] found a strong Alfvénic correlation between fluctuations in the magnetic field, $\delta \mathbf{B}$, and the proton bulk velocity, $\delta \mathbf{V}$, during some time intervals upstream of the bow wave (i.e., $B_o \delta \mathbf{V} = \pm C_A \delta \mathbf{B}$, where C_A is the Alfvén speed). At a distance of 28 x 10^6 km from comet Halley, the ICE spacecraft observed no cometary ULF turbulence [Tsurutani et al., 1987a].

Glassmeier et al. [1988] distinguishes the ULF turbulence observed by Giotto on the inbound leg between quasiperpendicular ($\langle\alpha\rangle \sim 90°$) and quasiparallel ($\langle\alpha\rangle \sim 30°$) geometries. As observed upstream of comet G-Z, the 100s-period waves are only clearly present in quasiparallel geometries. There, correlation studies between magnetic field and flow velocity show that the fluctuations are Alfvénic and propagate sunward. Second and higher harmonics of the fundamental frequency near 0.01 Hz have also been detected; these also have minimum variance directions roughly parallel to the magnetic field. In the quasiperpendicular geometry the only clear ULF signature is a shoulder near 0.03 Hz, although there may be a broad maximum near 0.001 Hz. The shoulder near 0.03 Hz is left-hand circularly polarized in the spacecraft frame and propagates nearly parallel to the ambient field; Glassmeier et al. [1988] argue that the Doppler shift is small so that the 0.03 Hz waves are also left polarized in the frame of the solar wind.

The ULF turbulence observed downstream of the bow wave was highly turbulent [Glassmeier et al., 1986, 1987]; the spectral peak at $\sim$ 0.01 Hz is not distinct, but a slight shoulder near 0.01 Hz (a flatter spectrum at lower frequencies and a spectrum steeper than Kolmogoroff at higher frequencies) provides evidence for maximum wave enhancement near 0.01 Hz.

The water-group and proton pickup ions have been observed at comet Halley [Terasawa et al., 1986; Mukai et al., 1986; Neugebauer et al., 1986; Neugebauer and Neubauer, 1988]. It is difficult to extract the radial dependence of the ion anisotropy from the published data, but the distributions for both protons and water-group ions are consistent with partially filled shells of speed V_{SW} with approximate ring distributions at $\gtrsim 3 \times 10^6$ km and nearly isotropic distributions downstream of the bow wave. The radial dependence of the pickup ion number density seems to be in accord with expectations with a water-group pickup ion density of $\sim$ 0.05 - 0.1 cm^{-3} [Mukai et al., 1986] and a pickup proton density of $\sim$ 0.09 cm^{-3} [Neugebauer and Neubauer, 1988] just upstream of the bow wave .

4. Cyclotron Resonance

Since the observed ULF turbulence at both comets G-Z and Halley is transverse and characterized by a spectral peak near 0.01 Hz and since water-group ions dominate the pickup ions, it is natural to investigate whether the water-group ions and waves are in cyclotron resonance. We do not expect the nonresonant firehose instability to contribute substantially to wave excitation since (1) the observed spectral feature at 0.01 Hz is well-defined, (2) at comet G-Z the pressure in water-group pickup ions, NmV_{SW}^2, only attains that in the average field $\sim 2 \times 10^5$ km from the nucleus where we expect pitch-angle scattering to produce near isotropy, and (3) also at comet G-Z the magnetic field is not nearly radial (see Figure 3).

The cyclotron resonance condition for wave propagation nearly parallel to the average field

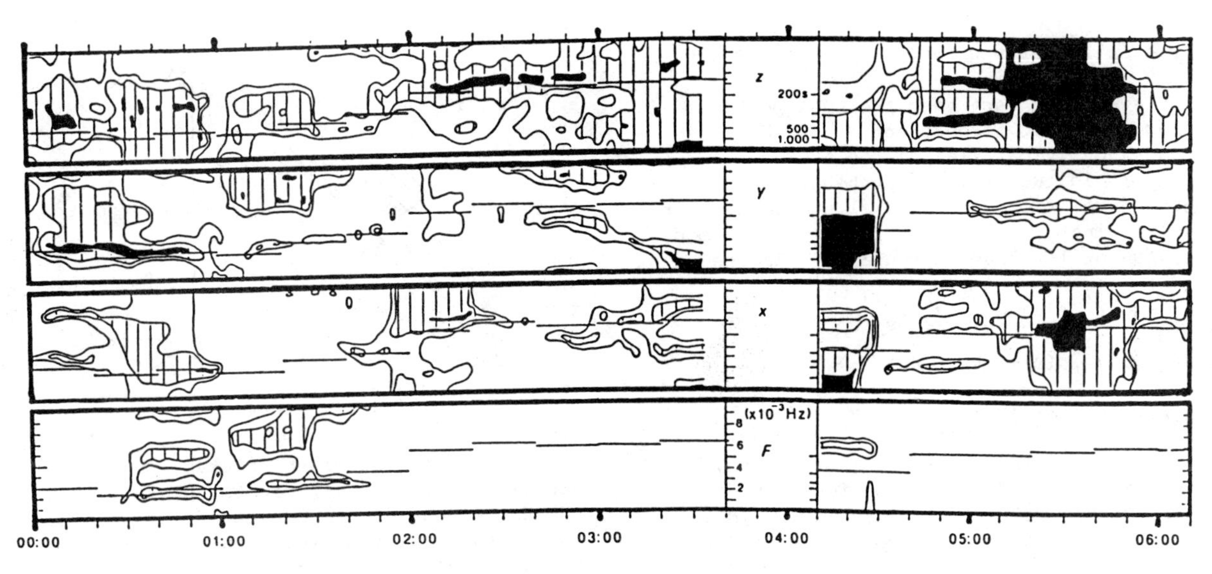

Fig. 9. Dynamic spectrograms for the X,Y and Z components of the magnetic field in solar-ecliptic coordinates and the field magnitude (F) as measured during the closest approach of Sakigake to comet Halley (Figure 2 from Yumoto et al. [1986b]). The horizontal bars give the O^+ cyclotron frequency based on 20-minute averages of the magnetic field strength.

is $\omega - kv_z + \Omega = 0$, where in the solar wind frame ω and k are wave frequency and wavenumber, v_z is the component of ion velocity parallel to $\underset{\sim}{B}$ and oriented away from the Sun, and Ω is the ion cyclotron frequency. For an unscattered pickup ion $v_z \cong - V_{SW} |\cos\alpha|$, so that the resonance condition becomes $\omega + \underset{\sim}{k} \cdot \underset{\sim}{V}_{SW} + \Omega \cong 0$. The combination, $\omega + \underset{\sim}{k} \cdot \underset{\sim}{V}_{SW}$, is simply the wave frequency in the comet frame, ω_c. The resonance condition is then $\omega_c \cong -\Omega$. Thus an unscattered O^+ pickup ion will resonate in an 8nT magnetic field (characteristic of the solar wind magnetic field at comet G-Z) with a wave of approximately 7.6×10^{-3} Hz or a period of 130 s, consistent with the observed spectral peak near 0.01 Hz. The minus sign in the resonance condition is important. Since the ions are positively charged, $\omega_c/B < 0$, which implies left-hand circular polarization in the comet (approximately spacecraft) frame, consistent with the observed polarization of the 100s-period waves far upstream of both comets.

A convincing verification of the relation $\omega_c \cong -\Omega$ is presented in Figure 9, taken from Yumoto et al. [1986b]. It shows a dynamic spectrogram of the three components plus the magnitude of the magnetic field taken over a 6-hour period near the closest approach of Sakigake to comet Halley ($\sim 7 \times 10^6$ km). Also shown with horizontal bars is the O^+ cyclotron frequency based on 20-minute averages of the magnetic field strength. It is clear that the power enhancements near 0.01 Hz correlate very well with the local cyclotron frequency. There is a slight tendency for the cyclotron frequency to lie below the spectral peak. It is unlikely that the frequency shift is due to the difference between comet and spacecraft frames. The shift is, however, quite reasonable, as we shall see in the next section, since the dominant unstable waves resonate with the inner flank of the pickup ion distribution shown in Figure 2 ($<\mu> < \mu < 0$), which resonates with waves of somewhat higher frequency. A similar shift in frequency is noted by Glassmeier et al. [1988]: the spectral peak during one time period upstream of comet Halley is at 7×10^{-3} Hz, whereas the water-group ion cyclotron frequency yields 5.3×10^{-3} Hz. Furthermore, the 130s cyclotron period estimated above at comet G-Z is about the upper limit of the observed range of wave periods of 75s - 135s [Tsurutani and Smith, 1986b].

The Sakigake measurements are ideal to test the resonance condition. Closer to the comet, pickup ions are scattered in pitch angle and resonate with a range of frequencies. Nonlinear effects should also deplete the spectral peak, cascading power to higher and lower frequencies, as the waves grow in amplitude. Finally wave compression at the bow wave may cause a frequency shift in the wave spectrum.

The correspondence of the 0.01 Hz peak with the water-group ion cyclotron frequency has been noted in many papers [e.g. Tsurutani and Smith, 1986b; Yumoto et al., 1986a,b; Saito et al., 1986; Neubauer et al., 1986; Acuña et al., 1986; Galeev, 1986].

5. Pickup-Ion Driven Instability of ULF Waves

In view of the fact that the 100s-period waves are observed to be transverse and propagate nearly parallel to the average magnetic field, we start with the transverse dispersion relation for parallel propagation in the solar wind frame, for which the normal modes are right- or left-hand circularly polarized:

$$0 = \omega^2 - c^2k^2 + 4\pi^2\omega \sum_s q_s^2 m_s^{-1}$$
$$\cdot \int dv_z\, dv_\perp\, v_\perp^2\, (\omega - kv_z + \Omega_s)^{-1}$$
$$\cdot [\partial f_s/\partial v_\perp + (k/\omega)(v_\perp \partial f_s/\partial v_z - v_z \partial f_s/\partial v_\perp)] \quad (1)$$

where $\mathrm{Im}(\omega) > 0$ (instability). Here $f_s(v_z,v_\perp)$ is the phase-space distribution function of nonrelativistic particle species s with $\int d^3v\, f_s = N_s$, the number density of species s, and $\underset{\sim}{B} = B\underset{\sim}{\hat{e}}_z$. Also q_s, m_s and Ω_s ($=q_sB/m_sc$) are, respectively, the charge, mass and cyclotron frequency of species s. The wave polarization in the solar wind frame is right if $\omega/B > 0$ and left if $\omega/B < 0$.

The v_z-integration in equation (1) for the pickup ions contains the product of two peaked functions: $(\omega - kv_z + \Omega_i)^{-1}$ and Of_i, where subscript i indicates pickup ions and O indicates the operator in square brackets. If $\gamma < k\Delta v_z$, where $\omega = \omega_r + i\gamma$ and Δv_z is the width of the pickup ion distribution, then $(\omega - kv_z + \Omega_i)^{-1}$ is the narrower function, which may then be replaced by $P(\omega - kv_z + \Omega_i)^{-1} - i\pi\, \delta(\omega - kv_z + \Omega_i)$. Assuming for simplicity that the pickup ions do not modify ω_r ($\gamma << \omega_r$) and the solar wind protons and electrons do not resonate with the wave, and taking the hydromagnetic limit $\omega_r \lesssim \Omega_p$ (where p indicates protons), we obtain $\omega_r^2 = k^2C_A^2$, where C_A is the Alfvén speed, and

$$\gamma = 2\pi^3 \frac{C_A^2}{c^2} \frac{q_i^2}{m_i} \frac{1}{|k|} \cdot \int_{|\Omega_i/k|}^{\infty} dv' [(v')^2 - \frac{\Omega_i^2}{k^2}] \frac{k}{\omega_r} \frac{\partial f_i}{\partial \mu'} \Bigg|_{\mu' = \Omega_i/(kv')} \quad (2)$$

where $\mu' = v_z'/v'$ and a prime denotes velocity components in the wave frame moving with velocity $(\omega_r/k)\, \hat{e}_z$. This class of instabilities was first investigated in detail by Sagdeev and Shafranov [1960]. Actually ω_r is probably modified by the pickup ions near the bow wave (firehose instability would be an extreme modification; e.g. Gary et al. [1984]), but these effects are neglected here and are probably not essential.

Instability requires $(k/\omega_r)\, \partial f_i/\partial \mu'|_{res} > 0$, where subscript "res" implies evaluation at $\mu' = \Omega_i/(kv')$, which is the cyclotron resonance condition. A cut through the pickup-ion ring distribution in the $v_x - v_z$ plane is shown in Figure 10 for an intermediate value of $<\alpha>$. The curves give the contours of the ion distribution. The dashed circular arcs denote curves of constant v' in the two wave frames moving with $\pm C_A \hat{e}_z$. The two domains of instability are indicated by hatched regions where $|\partial f_i/\partial \mu'|$ is large, along the arcs corresponding to the unstable wave propagation direction. Thus, waves propagating in the $+\hat{e}_z$ direction feed on ions in the domain where $\partial f_i/\partial \mu' > 0$ and waves propagating in the $-\hat{e}_z$ direction feed on ions in the domain where $\partial f_i/\partial \mu' < 0$. Since in both domains $\mu' < 0$, also $B/k < 0$. Thus, the unstable waves with $\omega_r/k > 0$ are left polarized and those with $\omega_r/k < 0$ are right polarized. However, since both sets of waves are resonant with ions with $\mu' < 0$, they have the same sense of helicity. In the spacecraft frame with $\underset{\sim}{V}_{SW} \cdot \hat{e}_z > C_A$, both sets would therefore be left polarized. In Figure 11 pickup-ion ring distributions as functions of μ ($\cong \mu'$ with $V_{SW} >> C_A$) are shown schematically for $<\alpha> = 0°$, 60° and 90°. Domains of large $|\partial f_i/\partial \mu'|$ are denoted in each by a dashed line under which the propagation direction and solar wind polarization of the corresponding unstable waves are given. It is clear that both right and left polarized waves can be excited but unless $<\alpha> \sim 90°$ both are left polarized in the spacecraft frame. The computed maximum growth rates are approximately

$$\gamma \sim |\Omega_i| \frac{C_A|v_z|}{(\Delta v_z)^2} \frac{\frac{1}{2} N_i m_i v_\perp^2}{B_o^2/8\pi} \quad (3)$$

where we have assumed $v_\perp^2 >> v_z \Delta v_z$.

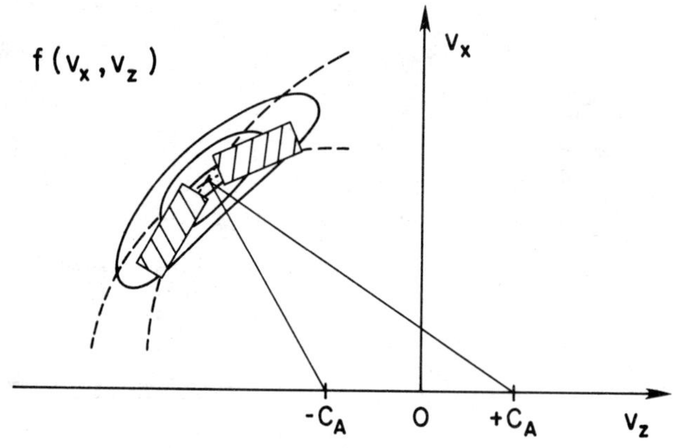

Fig. 10. Contours of the pickup ion distribution, $f(v_x,v_z)$, in the plane $v_y = 0$. The dashed curves are curves of constant speed in the two frames moving with $\pm C_A$ along the ambient field. Hatched areas are domains responsible for maximum instability drawn parallel to the dashed curve corresponding to the unstable wave propagation direction for that domain.

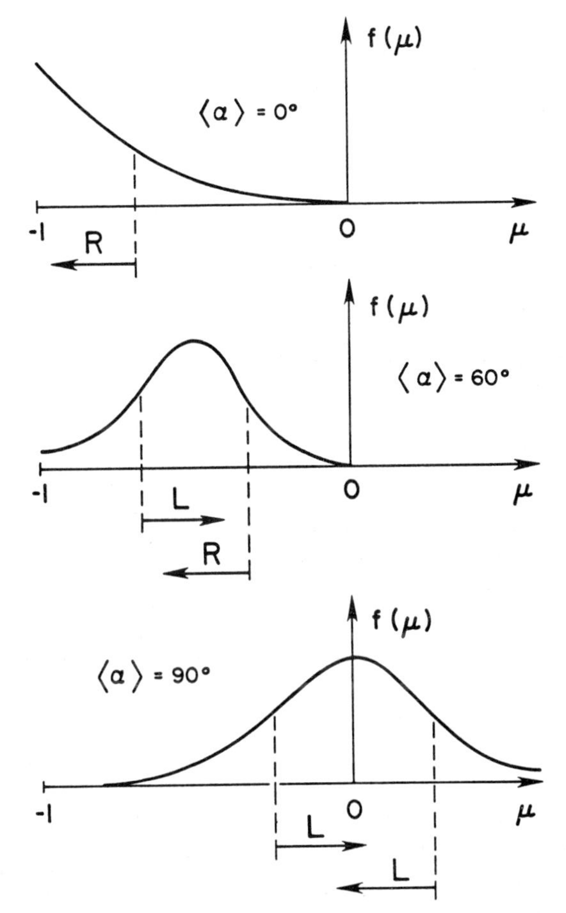

Fig. 11. Schematic diagrams of pickup ion distribution functions in the solar wind frame for $\langle\alpha\rangle$ = 0°, 60° and 90°. The most unstable domains of μ are denoted by vertical dashed lines under which are shown the solar wind polarization and propagation direction of the unstable waves.

Incidentally, the condition for instability, $(\omega_r/k)\ \partial f_i/\partial\mu'|_{res} > 0$, has a simple interpretation. In the frame of the transverse wave particles are scattered in pitch angle only and $|\partial f_i/\partial\mu'|$ is reduced. If $(\omega_r/k)\ \partial f_i/\partial\mu'|_{res} > 0$, then the scattered resonant ions lose energy on average in the frame of the solar wind. That energy must be transferred to the waves.

Far upstream of the comet before pickup-ion pitch-angle scattering, the ion distribution may be very sharp in v since initially all ions have $v = V_{SW}$ apart from the initial speeds of the atoms and molecules relative to the comet. Then the growth rate may satisfy $\gamma > k\Delta v_z$, or more explicitly $\gamma > kv\Delta\mu'$, so that ∂f_i may be a narrower function of μ' than $(\omega - kv_z + \Omega_i)^{-1}$. The distribution f_i may then be assumed proportional to $\delta(v_z - v_{zo})\ \delta(v_\perp - v_{\perp o})$; further assuming that the solar wind electrons and protons are cold, or nonresonant, dispersion relation (1) reduces to a quartic equation

$$0 = \bar{\omega}^4 + 2\,\bar{\omega}^3(kv_{zo} - \Omega_i) + \bar{\omega}^2[(kv_{zo} - \Omega_i)^2 - k^2C_A^2] + \Pi_i^2\,C_A^2c^{-2}[\bar{\omega}\,\Omega_i^{-1}(\bar{\omega} - \Omega_i)^2 - \tfrac{1}{2}\,k^2v_{\perp o}^2] \qquad (4)$$

for $\omega \ll \Omega_p$, where $\bar{\omega} = \omega - kv_{zo} + \Omega_i$ and $\Pi_i^2 = 4\pi q_i^2N_i/m_i$. The unstable roots of equation (4) have been investigated by Wu and Davidson [1972], Wu and Hartle [1974], Winske et al. [1985] and Lee and Ip [1987].

Since we expect $\Pi_i^2\Omega_i^{-2}C_A^2c^{-2}$ [$\cong (N_i/N_p)(m_i/m_p)$] to be small, the unstable roots can be shown to satisfy $\omega - kv_{zo} + \Omega_i \cong 0$, that is they are cyclotron resonant with the narrow ring. For any given k there is at most one unstable root of equation (4). The two most unstable roots occur for the two values of k satisfying $\omega_r \cong kv_{zo} - \Omega_i \cong \pm\,kC_A$, which are Alfvén waves with growth rates approximately

$$\gamma \sim |\Omega_i| \left(\frac{C_A}{|v_{zo}|}\ \frac{\tfrac{1}{2}\,N_im_iv_\perp^2}{B_o^2/8\pi} \right)^{1/3} \qquad (5)$$

The left (right) polarized unstable wave propagating in the $\hat{e}_z$ direction ($-\hat{e}_z$ direction) is resonant with a v_z slightly less (greater) than v_{zo}. They are then analogous to the two modes shown in Figure 11(b). Indeed, if Δv_z in equation (3) is replaced by $\gamma/|k|$ (thus extending that result to the limit of its validity) and $|k| = |\Omega_i/v_z|$ then growth rate (5) is obtained.

If $|kv_{zo} - \Omega_i| > |kC_A|$ all roots of equation (4) are stable. If $|kv_{zo} - \Omega_i| < |kC_A|$, the unstable root has a phase speed less than C_A, a growth rate

$$\gamma \sim |\Omega_i| \left(\frac{\tfrac{1}{2}\,N_im_iv_\perp^2}{B_o^2/8\pi} \right)^{1/2} \qquad (6)$$

and only exists by virtue of the ring. This unstable root, first discussed by Wu and Davidson [1972], is less unstable than the Alfvén waves [equation (5)] and not likely to play an essential role at comets.

To these parallel-propagating resonant instabilities should be added the nonresonant firehose instability, which as mentioned in Section 4 is unlikely to be of importance. These and related instabilities have also been discussed in the cometary context by Brinca and Tsurutani [1987a,b; 1988a,c], Galeev et al. [1986a], Sagdeev et al. [1986], Sharma and Patel [1986], Thorne and Tsurutani [1987], Winske and

Gary [1986], Gary and Madland [1988], Rogers et al. [1985], Verheest [1987], and Lakhina and Verheest [1988]. A general treatment is also presented by Gary et al. [1984].

The right-hand polarized (R) wave propagating sunwards can account for many of the observed characteristics of the 100s-period waves. For $\langle\alpha\rangle$ not close to 90° the left-hand polarized wave should be suppressed by the quasilinear evolution of the pickup ion distribution as described in the next section [for $\langle\alpha\rangle = 0$ it is not excited in the first place; see Figure 11(a)]. Since the right-hand wave resonates with the inner flank of the ion ring distribution ($\langle\mu\rangle < \mu < 0$) its spacecraft frequency is somewhat larger than the ion gyrofrequency, as observed for the frequencies of the peak wave enhancement in Figure 9. The spacecraft polarization of the wave, dominated by Doppler shift, is left-hand polarized as observed far from comets G-Z and Halley. The predicted growth rate maximizes for parallel propagation, consistent with observed directions of minimum variance. The propagation direction agrees with that inferred at comet Halley from the correlation between velocity and magnetic field fluctuations [Glassmeier et al., 1988]. Furthermore, inside the bow wave the instability may be weaker or no longer operative since the pickup ions have a smaller speed in the solar wind frame and the high level of turbulence would isotropize the ions in their first gyration [Wu et al., 1986; Price and Wu, 1987]. The resulting ion distribution (neglecting those just picked up) could cause the waves to decay, possibly as evident near closest approach to comet G-Z shown in Figure 3.

The wave growth rate taken from equation (3) and evaluated for characteristic parameters far upstream of the bow wave [$B_o = 5$ nT, $N_i = 10^{-3}$ cm^{-3}, $m_i = 18\ m_p$, $V_{SW} = 400$ km/s, $\langle\alpha\rangle = 45°$, and somewhat arbitrarily $C_A|v_z| = (\Delta v_z)^2$] yields $\gamma \sim 0.1\ |\Omega_i|$ or a growth scalelength, $V_{SW}\ \gamma^{-1}$, of $\sim 10^5$ km. This growth rate is sufficient in principle to account for the observed 100s-period wave intensities in the region upstream of the cometary bow wave, a conclusion which is supported using quasilinear theory in the next section.

However, the R mode cannot account for the linearly polarized waveforms observed closer to the comets. The L and R modes cannot combine to produce a linearly polarized wave since they have different frequencies and growth rates, and are both L polarized in the spacecraft frame. The most likely explanation appears to be wave propagation oblique to the magnetic field [Gary and Winske, 1986; Brinca and Tsurutani, 1987a,b, 1988c; T. Hada, unpublished manuscript, 1987]. The transition of hydromagnetic waves from circular polarization on-axis to linear polarization occurs when the angle θ between $\underset{\sim}{k}$ and $\underset{\sim}{B}$ satisfies $\sin^4\theta\ \Omega_p^2 \cong 4\omega_r^2 \cos^2\theta$ [Stix, 1962, p.43]. For representative parameters ($V_{SW} \cdot \hat{e}_z \cong 5\ C_A$) we find $\theta \cong 8°$ for the 100s-period waves, a small angle probably consistent with the observed approximate alignment of the direction of minimum variance and $\underset{\sim}{B}$. Direct excitation of the off-axis linear waves is unlikely since γ decreases with the transition to linear polarization, and all planes of polarization would be equally unstable for a gyrotropic ion distribution. A possible further difficulty is that the transition angle θ increases in the presence of a dense ring distribution with $N_i/N_p \gtrsim 10^{-3}$ [Gary and Winske, 1986; T. Hada, unpublished manuscript, 1987]. A more likely possibility is that near the cometary bow wave the wavelength λ is comparable to the foreshock scalelength r so that excited waves must be viewed as having a component of their wavevector normal to $\underset{\sim}{B}$ [Gary and Winske, 1986] and/or parallel-propagating waves will be refracted off-axis through the transition angle as suggested by Hada et al. [1987] in the context of Earth's foreshock. An apparent difficulty with this explanation is that the effect should be greater at comet G-Z where $\lambda \sim r$ than at comet Halley where $\lambda << r$ upstream of the bow wave, but linearly polarized waveforms have been observed at both. Also a prediction of the plane of polarization based on refraction has not been made. Another possible explanation of the linear polarization is as a direct result of nonlinear evolution [Tsurutani et al., 1987b; Omidi and Winske, 1988; Hoshino, 1987] and will be described in Section 7.

Whistler waves are also unstable in the presence of a pickup ion ring [Wu and Davidson, 1972; Goldstein and Wong, 1987; Thorne and Tsurutani, 1987; Brinca and Tsurutani, 1988b; Wong and Goldstein, 1988; Wu et al., 1988]. The growth rate of parallel-propagating whistler waves is also proportional to $\partial f_i/\partial\mu'|_{res}$ so that they can also feed on the steep gradients of the ion ring distribution. For whistler waves ($\omega >> \Omega_i$) the cyclotron resonance condition becomes $\omega_r \cong kv_z$, where v_z specifies the domain where $|\partial f_i/\partial\mu'|$ is large. Using the whistler dispersion relation, $\omega_r = -\ c^2k^2\Omega_e/\Pi_e^2$, where Π_e is the electron plasma frequency, the frequency and wavenumber of the unstable whistler are $\omega_r = -\ v_z^2\ c^{-2}\Pi_e^2/\Omega_e$ and $k = -\ v_z\ c^{-2}\ \Pi_e^2/\Omega_e$. In the frame of the comet the frequency of the unstable whistler is then

$$\omega_c = \omega_r + \underset{\sim}{k} \cdot \underset{\sim}{V}_{SW} = -\ \Pi_e^2 v_z c^{-2} \Omega_e^{-1}\ (v_z + V_{SW}\ |\cos\alpha|) \qquad (7)$$

Note that if the pickup ion distribution is a

sharp ring, $v_z \cong - V_{SW} |\cos\alpha|$ so that $\omega_c \cong 0$; more accurately, by the arguments presented in Section 4, $\omega_c = -\Omega_i << \omega$. However, with $v_z + V_{SW} |\cos\alpha| = \Delta v_z$, rewriting terms yields $\omega_c = \Omega_p v_z \Delta v_z C_A^{-2}$. With $|\underset{\sim}{B}| = 8$ nT, this expression yields a wave period $T \cong 8 C_A^2 (v_z \Delta v_z)^{-1}$ s. Since reasonable values of v_z and Δv_z could yield T in the range 1-7s, Goldstein and Wong [1987] have suggested that the 1-7s wave packets observed at comet G-Z arise from this instability. Since it can also be shown for whistlers that the growth rate is controlled by $\gamma \propto (\omega_r/k)\, \partial f_i/\partial\mu'|_{res}$, these waves would also feed on the inner flank of the pickup ion distribution. Thus for instability $\Delta v_z > 0$, which implies $\omega_c/B < 0$ or left-hand spacecraft-frame (equivalent to comet-frame) polarization as observed.

This explanation of the 1-7s period wave packets, however, has drawbacks: Their absence further upstream of comet G-Z, their association with the crests of the underlying 100s-period waves, and their absence at comet Halley (which exhibited upstream wave levels somewhat lower than at comet G-Z) would rather favor an origin based on the nonlinear evolution of the 100s-period waves as described in Section 7. The pickup ions are nearly isotropic near the bow wave, a configuration which should reduce the growth rate unless the waves are unstable to those ions picked up within the previous gyroperiod [Wong and Goldstein, 1988; Wu et al., 1988]. Finally, Winske et al. [1985] and Kojima et al. (preprint entitled "Nonlinear evolution of high frequency R-mode waves excited by water group ions near comets: computer experiments", 1988) find based on simulations that the water-group pickup-ion excited whistlers saturate rapidly and do not grow to the large amplitudes of the wave packets.

Another possible explanation for the 1-7s wave packets is excitation by pickup hydrogen rings. This explanation would readily account for the observed polarization and field-aligned minimum variance direction of most packets. The predicted frequency in the comet or spacecraft frame, $\omega_c = - \Omega_p$, yields a period of 8 seconds if $|\underset{\sim}{B}| = 8$nT. This period is markedly greater than those normally observed. This explanation would also require the existence of the wave packets at comet Halley and far from comet G-Z, since neutral hydrogen escapes to greater distances, neither of which is observed. The association of the wave packets with the crests of the 100s-period waves could be interpreted as trapping of the packets by the variation of the Alfvén speed due to the underlying 100s-period wave. However, in any case it would appear that the observed frequency is not consistent with excitation by the pickup hydrogen ring.

The most likely origin of the 1-7s wave packets is the nonlinear evolution of the 100s-period waves, which we shall discuss in Section 7.

6. Quasilinear Evolution of ULF Waves and Pickup Ions

The water-group pickup ions are effectively scattered in pitch angle by the enhanced intensity of cyclotron resonant waves due to the instability just described. A distribution of pickup ions will evolve schematically as shown in Figure 12 for 4 consecutive times: t_1 (soon after pickup) $< t_2 < t_3 < t_4 \to \infty$. The time-asymptotic state is a "shell" distribution. Fresh pickup ions are continually produced in a given volume of solar wind but these too are more rapidly isotropized in the turbulence present. It is apparent in Figure 12 that pitch-angle scattering suppresses the excitation of the left-hand polarized wave with $\omega_r/k > 0$, reduces the growth rate of the right-hand mode (due to a given group of pickup ions), increases the frequency range of the unstable right-hand waves, and even produces left-hand waves with $\omega_r/k < 0$ resonant with ions with $v_z > 0$.

This coupled evolution of the pickup ions and ULF waves is readily described by quasilinear theory under the assumption that the pickup ions are nearly isotropic and wave propagation is predominantly parallel to the ambient field [Galeev et al., 1986a,b; Sagdeev et al., 1986, 1987; Galeev, 1986; Rogers et al., 1985]. The model outlined by Galeev et al. [1986a] is most complete and addresses mirroring of pickup ions through $\mu = v_z/v = 0$ by waves with frequencies below the resonant frequency, mass-loading and deceleration of the solar wind, modification of the bow wave location due to ion isotropization by the ULF waves, and the nonlinear process of induced wave scattering. It should be noted, however, that the water

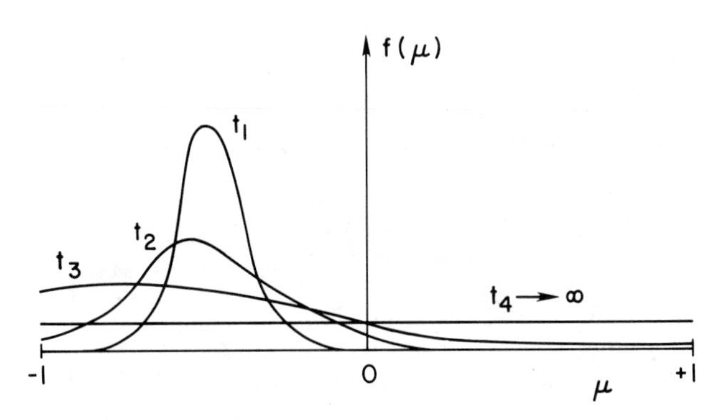

Fig. 12. Schematic diagram of the distribution function of a given group of pickup ions undergoing pitch-angle scattering soon after pickup (t_1), at very long times ($t_4 \to \infty$), and at two intermediate times (t_2 and t_3).

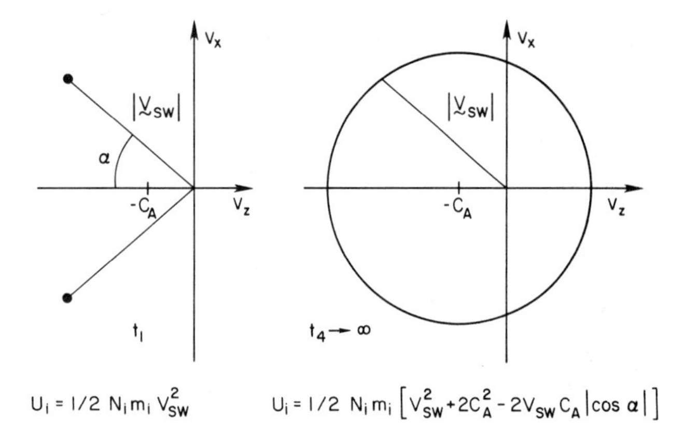

Fig. 13. A schematic diagram of the pickup ion distribution in the plane $v_y = 0$ just after pickup (t_1) and at long times ($t_4 \to \infty$) after pitch-angle scattering in the wave frame. The ion energy densities of the two configurations in the solar wind frame are given.

group pickup ions are only nearly isotropic just upstream of the bow wave. Farther upstream the ion distribution is approximately a ring and the predictions of quasilinear theory for nearly isotropic distributions are not rigorously valid.

One simple prediction of quasilinear theory can be checked immediately. The ULF waves are expected and inferred from observations (see also Section 7) to be predominantly sunward propagating along the ambient magnetic field. As the pickup ions scatter from $\mu \cong -|\cos\alpha|$ to near isotropy in the wave frame, they lose energy density in the solar wind frame by an amount $\Delta U_i = -N_i m_i C_A (V_{SW}|\cos\alpha| - C_A)$ as shown in Figure 13. In this frame the bulk kinetic energy density of the solar wind does not change; the energy is therefore transferred to the ULF waves, half of which appears as enhanced power in the fluctuating magnetic field (assuming no nonlinear coupling of the waves to the solar wind plasma and neglecting wave amplification due to deceleration and compression of the solar wind):

$$\langle \delta \underset{\sim}{B} \cdot \delta \underset{\sim}{B} \rangle B_o^{-2} =$$

$$(N_i/N_p)(m_i/m_p)[(V_{SW}/C_A)|\cos\alpha| - 1] \qquad (8)$$

This prediction of quasilinear theory has also been discussed by Lee and Ip [1987, p. 11046] in application to interstellar pickup ions in the solar wind.

Just upstream of the bow wave of comet G-Z the water-group pickup ions are observed to be nearly isotropic with $N_i \cong 2 \times 10^{-2}$ cm^{-3}. With $m_i = 17\ m_p$, $N_p \cong 6$ cm^{-3}, $V_{SW} \cong 400$ km/s, $B_o \cong$ 8nT, and $|\cos\alpha| \cong 2^{-\frac{1}{2}}$ we obtain $\langle \delta \underset{\sim}{B} \cdot \delta \underset{\sim}{B} \rangle B_o^{-2} \cong$ 0.17. For a linearly polarized waveform this implies a wave amplitude $|\Delta B| B_o^{-1} \cong 0.6$, in quantitative agreement with observations [Tsurutani and Smith, 1986a]. A similar comparison between quasilinear theory, including wave saturation by induced wave scattering on solar wind protons, and observations by Vega-1 at comet Halley yields qualitative agreement as a function of distance from the nucleus [Galeev, 1986; Galeev et al., 1986b]. The close agreement at comet G-Z may indicate that nonlinear saturation effects are not important upstream of the bow wave at comet G-Z. In a similar vein, Johnstone et al. [1987a] compare the energy density in the ULF bulk velocity fluctuations observed on Giotto with the water-group pickup ion energy density inferred from the deceleration of the solar wind and find results in qualitative agreement with quasilinear theory [equation (8)].

Neugebauer and Neubauer [1988] have compared the pickup proton density, N'_p (distinct from the solar wind proton density N_p), measured on Giotto with the magnetic field variance δB^2 integrated over the frequency band, 0.016 - 14 Hz, which includes the proton (excludes the water group) cyclotron-resonant frequency. At distances greater than 2×10^6 km from comet Halley they find that $\delta \underset{\sim}{B}^2$ is linearly proportional to $\frac{1}{2} N'_p m_p V_{SW}^2$, as expected from quasilinear theory [equation (8) with C_A/V_{SW} approximately constant] and simulations [Gary et al., 1986]. The constant of proportionality is roughly consistent with expectations [i.e., equation (8)], although the large observed proton anisotropy undermines comparison with the simplest predictions of quasilinear theory.

7. Nonlinear Evolution of the ULF Waves

Since the power in magnetic fluctuations near the bow waves of both comets was observed to be comparable with that in the ambient field, quasilinear theory, while useful, is fundamentally suspect, and a growing body of work deals with the nonlinear evolution of the pickup ions and ULF waves.

Several papers have addressed the wave-particle interaction using one-dimensional hybrid computer simulations which can treat large-amplitude fluctuations [Omidi and Winske, 1986, 1988; Wu et al., 1986; Gary et al., 1986, 1988; Price and Wu, 1987; Gaffey et al., 1988]. The simulations are qualitatively in accord with quasilinear theory: initial growth of the right-hand mode, pitch-angle scattering of the ions to form a shell in velocity space, and energy diffusion on a longer timescale. Equation (8) apparently provides a reasonable estimate of the saturation wave power before substantial ion energy diffusion takes place, although there are indications that simulations yield a higher saturation level [Winske and Gary, 1986; Winske and Leroy, 1984]. The simulations show that ions are readily scattered through 90° pitch angle. They also show that at high levels of turbulence ($|\delta B| \sim B_o$) the isotropization occurs immediately within a gyroperiod because the distribution of magnetic field orientations is approximately isotropic. Similar results have been obtained in a recent two-dimensional simulation by Galeev et al. [1987].

Goldstein et al. [1987] have investigated the turbulent evolution of an initially coherent wave mode at ≅ 0.01 Hz using an incompressible - MHD 2-dimensional spectral-method code. Within a few eddy-turnover times they obtain an inertial range spectrum at higher frequencies with Kolmogoroff slope which is in accord with the observed spectra. However, other quasilinear and nonlinear effects not included in this code may also play a role in determining the spectral slope: (1) quasilinear wave excitation also occurs at higher frequencies as ions scatter through 90° pitch angle, (2) a nonlinear cascade to lower frequencies due to induced wave scattering on solar wind protons [Galeev et al., 1986a], and (3) nonlinear effects associated with compressibility, which we now discuss.

It is apparent from the discussion at the end of Section 5 that two major observed features of the ULF waves, the linear polarization of the 100s-period waves near the cometary bow wave and the 1-7s wave packets associated with the crests of the 100s-period waves, are not readily explained by linear instabilities, particularly the wave packets. Three recent papers have considered these observed waveforms as the natural result of nonlinear evolution of the 100s-period waves: Tsurutani et al. [1987b] and Omidi and Winske [1988] consider ULF waves in the region upstream of the cometary bow wave and Hada et al. [1987] the related problem of ULF waves in Earth's foreshock.

Tsurutani et al. [1987b] first notes that the large-amplitude 100s-period ULF waves near the comet G-Z bow wave are compressive with electron density fluctuations well correlated with variations in $|B|^2$. Thus nonlinear steepening of the initially excited right-hand circularly polarized nearly parallel-propagating wave should occur. The steepening should transfer some of the 360° rotation per wavelength in δB to a rapid variation at the crest of the 100s-period wave which could account for the partial rotations often observed. Tsurutani et al. [1987b] note that the portion of the wave trailing the crest could appear as a linearly polarized wave. The rapid rotation may contain sufficiently high frequency components to create a whistler wave packet with higher phase speed which precedes the crest of the underlying

PROPAGATION TOWARDS SUN

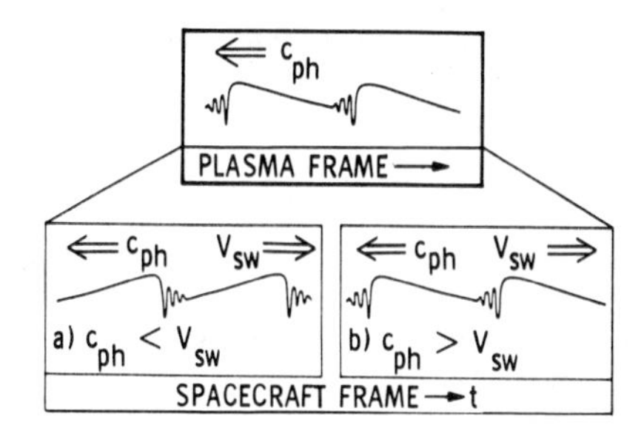

PROPAGATION AWAY FROM SUN

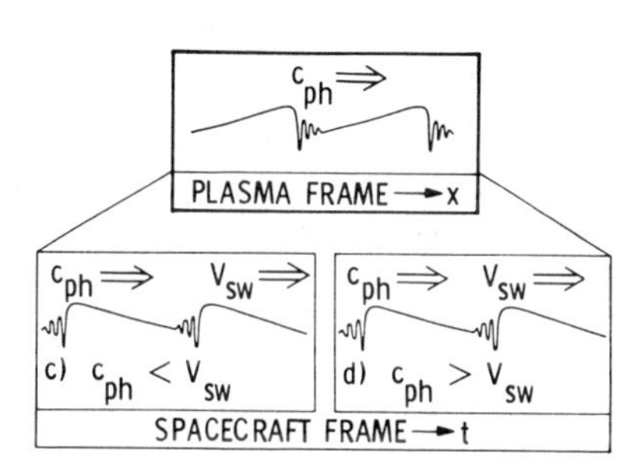

Fig. 14. A schematic diagram (Figure 4 from Tsurutani et al. [1987b]) of the two possible propagation directions of a steepened wave in the solar wind frame and its predicted waveform in the spacecraft frame if $C_A > V_{SW}$ and if $C_A < V_{SW}$. Only combination (a) is consistent with the observations as shown, for example, in Figure 6.

100s-period wave and accounts for the observed 1-7s wave packets. This scenario is strongly supported by the observed temporal sequence apparent in Figure 6: first trailing portion followed by crest and wave packet. As shown in Figure 14 taken from Tsurutani et al. [1987b] this sequence is only consistent with a steepened wave which propagates towards the Sun in the solar wind frame but is convected outwards with the solar wind. And these are precisely the propagation characteristics expected of the unstable right-hand polarized magnetosonic wave and are consistent with the propagation direction inferred by Glassmeier et al. [1988] and the left-hand polarization observed in the spacecraft frame further from the comet before the waveforms become linearly polarized.

Similarly, Hada et al. [1987] argues based on simulation results that the discrete wave packets observed in Earth's foreshock arise from nonlinear steepening of magnetosonic waves. They point out, however, that nearly parallel-propagating right-hand polarized magnetosonic waves are not sufficiently compressive to produce the steepening. Rather, similar to the suggestion of Gary and Winske [1986], they argue that the waves are excited preferentially on-axis but refract off-axis in the inhomogeneous foreshock to produce elliptically or nearly linearly polarized waves which are sufficiently compressive to steepen and produce the discrete wave packets by dispersion. In application to cometary foreshocks, as discussed in Section 5, this scenario provides a natural explanation for the observed linearly polarized waveforms near the cometary bow wave where spatial inhomogeneity is important. It is, however, noteworthy that the linearly polarized waveforms apparently occur as frequently at comet Halley with spatial scalelengths an order of magnitude larger than at comet G-Z.

Omidi and Winske [1988] have also studied the nonlinear evolution of the 100s-period waves using a 1-D hybrid code with an angle of 30° between wavevectors and the ambient magnetic field. Their simulation shows the growth of the waves, their steepening, and the evolution of linear waveforms and the discrete wave packets. They emphasize that the linear waveform is a direct consequence of steepening rather than off-axis propagation, and that the packets are not standing with respect to the 100s-period wave crest but rather evolving continually through dispersion. The linear rather than exponential decay of the wave packets (Tsurutani et al., preprint entitled "Properties of whistler mode wave packets at the leading edge of steepened magnetosonic waves: Comet Giacobini-Zinner," 1988) may be a direct manifestation of dispersion rather than damping.

Taken together these three papers offer convincing evidence that the linear waveforms and wave packets originate from nonlinear steepening of obliquely propagating magnetosonic waves. The simulation results displayed in Omidi and Winske [1988] are almost indistinguishable from the comet G-Z observations. The precise origin of the oblique propagation is unclear, but based on observations a small angle between propagation direction and ambient field suffices and is probably hard to avoid. However, although the simulations provide sound evidence, they cannot replace the understanding provided by analytical work, which remains to be done. It should finally be noted that two observed features of the wave packets remain puzzling: (1) they are not observed at comet Halley in spite of large 100s-period wave amplitudes and (2) the Fourier spectra of the packets often seem to show the clear dominance of a single frequency (apparent in Figure 7), which would appear to be characteristic of an instability rather than nonlinear steepening. A recent paper by Kaya, Matsumoto and Tsurutani (preprint entitled "Test particle simulation study of whistler wave packets observed near comet Giacobini-Zinner", 1988) shows with simulations that water-group pickup ions can be trapped in the formed whistler wave packet and contribute to its growth.

8. Summary

The regions containing pickup ions and ULF waves discovered upstream of comets Giacobini-Zinner and Halley by spacecraft ICE, Vega, Giotto, Sakigake and Suisei provide an excellent opportunity for detailed investigation of the plasma wave-particle interaction in space initiated by an unstable pickup ion distribution. The large gyroradii, large mass and high densities of the water-group pickup ions produce dramatic effects in comparison with, for example, the more subtle effects associated with interstellar ion pickup in the inner heliosphere.

The coupling of the freshly ionized ions to the solar wind is partially accomplished within a gyroperiod by the solar wind magnetic field. But complete ion pickup (ion isotropization) occurs via excitation of ULF hydromagnetic waves which are cyclotron resonant with the pickup ions and unstable to their initial ring distribution. The observed general features of this interaction are well accounted for qualitatively by linear and quasilinear theory (i.e. the dominant ULF wave frequency, wave propagation direction, wave helicity far from the comet, radial evolution of total wave power, minimum variance directions, pickup ion ring distributions far from the comet, ion isotropization and eventual energization closer to the comet). However, the observed detailed features of the ULF waves (their specific waveforms, associated higher frequency wave packets, power spectra, and detailed wave spatial dependence) need more work. Although the basic theoretical ideas governing off-axis propagation, nonlinear evolution and the influence of spatial inhomogeneity may be in place, further work is required on the fundamental properties of large-amplitude waves [e.g. Kennel et al., 1988; Hoshino, 1987] and detailed realistic modeling of the cometary foreshock interaction region.

Acknowledgments. The author is grateful to Alec Galeev, Peter Gary, Karl-Heinz Glassmeier, Mel Goldstein, Wing Ip, Bruce Tsurutani, Ching Wu and especially Tohru Hada and Toshio Terasawa for invaluable discussions. He is also very thankful to Bruce Tsurutani and Hiroshi Oya for thorough reviews of the original typescript and for the excellent organization of the Chapman Conference on Plasma Waves and Instabilities in Magnetospheres and at Comets, at which this paper was presented. This work was supported, in part, by NSF Grant ATM-8513363, NASA Grant

NAG 5-728, and NASA Solar Terrestrial Theory Program Grant NAGW-76.

References

Acuña, M.H., K.H. Glassmeier, L.F. Burlaga, F.M. Neubauer, and N.F. Ness, Upstream waves of cometary origin detected by the GIOTTO magnetic field experiment, Proc. 20th ESLAB Symposium, Heidelberg, 3, 447, 1986.

Amata, E., and V. Formisano, Energization of positive ions in the cometary foreshock region, Planet. Space Sci., 33, 1243, 1985.

Brinca, A.L., and B.T. Tsurutani, On the polarization, compression and non-oscillatory behavior of hydromagnetic waves associated with pickup ions, Geophys. Res. Lett., 14, 495, 1987a.

Brinca, A.L., and B.T. Tsurutani, Unusual characteristics of electromagnetic waves excited by cometary newborn ions with large perpendicular energies, Astron. Astrophys., 187, 311, 1987b.

Brinca, A.L., and B.T. Tsurutani, Survey of low-frequency electromagnetic waves stimulated by two coexisting newborn ion species, J. Geophys. Res., 93, 48, 1988a.

Brinca, A.L. and B.T. Tsurutani, Temperature effects on the pickup process of water-group and hydrogen ions: extensions of "A theory for low-frequency waves observed at Comet Giacobini-Zinner" by M.L. Goldstein and H.K. Wong, J. Geophys. Res., 93, 243, 1988b.

Brinca, A.L., and B.T. Tsurutani, The oblique behavior of low-frequency electromagnetic waves excited by newborn cometary ions, J. Geophys. Res., in press, 1988c.

Buti, B., and G.S. Lakhina, Stochastic acceleration of cometary ions by lower hybrid waves, Geophys. Res. Lett., 14, 107, 1987.

Gaffey, J.D., Jr., D. Winske, and C.S. Wu, Time scales for formation and spreading of velocity shells of pickup ions in the solar wind, J. Geophys. Res., 93, 5470, 1988.

Galeev, A.A., Theory and observations of solar wind/cometary plasma interaction processes, Proc. 20th ESLAB Symposium, Heidelberg, 1, 3, 1986.

Galeev, A.A., R.Z. Sagdeev, V.D. Shapiro, V.I. Shevchenko, and K. Szego, Mass loading and MHD turbulence in the solar wind/comet interaction region, Proc. Varenna-Abastumani International School & Workshop on Plasma Astrophysics, Sukhumi, USSR, p. 307, 1986a.

Galeev, A.A., B.E. Gribov, T. Gombosi, K.I. Gringauz, S.I. Klimov, P. Oberz, A.P. Remizov, W. Riedler, R.Z. Sagdeev, S.P. Savin, A. Yu. Sokolov, V.D. Shapiro, V.I. Shevchenko, K. Szego, M.I. Verigin, and Ye. G. Yeroshenko, Position and structure of the comet Halley bow shock: Vega-1 and Vega-2 measurements, Geophys. Res. Lett., 13, 841, 1986b.

Galeev, A.A., A.S. Lipatov, and R.Z. Sagdeev, Two-dimensional numerical simulation of the relaxation of cometary ions and MHD turbulence in the flow of the solar wind around a cometary atmosphere, Sov. J. Plasma Phys., 13, 323, 1987.

Gary, S.P., and C.D. Madland, Electromagnetic ion instabilities in a cometary environment, J. Geophys. Res., 93, 235, 1988.

Gary, S.P. and D. Winske, Linearly polarized magnetic fluctuations at comet Giacobini-Zinner, J. Geophys. Res., 91, 13699, 1986.

Gary, S.P., C.W. Smith, M.A. Lee, M.L. Goldstein, and D.W. Forslund, Electromagnetic ion beam instabilities, Phys. Fluids, 27, 1852, 1984.

Gary, S.P., S. Hinata, C.D. Madland, and D. Winske, The development of shell-like distributions from newborn cometary ions, Geophys. Res. Lett., 13, 1364, 1986.

Gary, S.P., C.D. Madland, N. Omidi, and D. Winske, Computer simulations of two-ion pickup instabilities in a cometary environment, J. Geophys. Res., in press, 1988.

Glassmeier, K.H., F.M. Neubauer, M.H. Acuña, and F. Mariani, Strong hydromagnetic fluctuations in the comet P/Halley magnetosphere observed by the Giotto magnetic field experiment, Proc. 20th ESLAB Symposium, Heidelberg, 3, 167, 1986.

Glassmeier, K.H., F.M. Neubauer, M.H. Acuña, and F. Mariani, Low-frequency magnetic field fluctuations in comet P/Halley's magnetosheath: Giotto observations, Astron. Astrophys., 187, 65, 1987.

Glassmeier, K.H., A.J. Coates, M.H. Acuña, M.L. Goldstein, A.D. Johnstone, F.M. Neubauer, and H. Rême, Spectral characteristics of low-frequency plasma turbulence upstream of Comet P/Halley, J. Geophys. Res., in press, 1988.

Gloeckler, G., D. Hovestadt, F.M. Ipavich, M. Scholer, B. Klecker, and A.B. Galvin, Cometary pick-up ions observed near Giacobini-Zinner, Geophys. Res. Lett., 13, 251, 1986.

Goldstein, M.L., and H.K. Wong, A theory for low frequency waves observed at comet Giacobini-Zinner, J. Geophys.Res., 92, 4695, 1987.

Goldstein, M.L., D.A. Roberts, and W.H. Matthaeus, Numerical simulation of the generation of turbulence from cometary ion pick-up, Geophys. Res. Lett., 14, 860, 1987.

Gosling, J.T., J.R. Asbridge, S.J. Bame, M.F. Thomsen, and R.D. Zwickl, Large amplitude, low frequency plasma fluctuations at comet Giacobini-Zinner, Geophys. Res. Lett., 13, 267, 1986.

Greenstadt, E.W., and L.W. Baum, Earth's compressional foreshock boundary revisited; observations by the ISEE 1 magnetometer, J. Geophys. Res., 91, 9001, 1986.

Hada, T., C.F. Kennel and T. Terasawa, Excitation of compressional waves and the formation of shocklets in the Earth's foreshock, J. Geophys. Res., 92, 4423, 1987.

Hoppe, M.M., C.T. Russell, L.A. Frank, T.E. Eastman, and E.W. Greenstadt, Upstream hydromagnetic waves and their association with backstreaming ion populations: ISEE 1 and 2 observations, J. Geophys. Res., 86, 4471, 1981.

Hoshino, M., Evolution of polarization in localized nonlinear Alfvén waves, Phys. Rev. Lett., 59, 2639, 1987.

Ip, W.-H., and W.I. Axford, The acceleration of particles in the vicinity of comets, Planet. Space Sci., 34, 1061, 1986.

Ipavich, F.M., A.B. Galvin, G. Gloeckler, D. Hovestadt, B. Klecker, and M. Scholer, Comet Giacobini-Zinner: in situ observations of energetic heavy ions, Science, 232, 366, 1986.

Isenberg, P.A., Energy diffusion of pickup ions upstream of comets, J. Geophys. Res., 92, 8795, 1987.

Johnstone, A.D., et al., Waves in the magnetic field and solar wind flow outside the bow shock at comet Halley, Proc. 20th ESLAB Symposium, Heidelberg, 1, 277, 1986.

Johnstone, A.D., et al., Alfvénic turbulence in the solar wind flow during the approach to comet P/Halley, Astron. Astrophys., 187, 25, 1987a.

Johnstone, A.D., et al., Waves in the magnetic field and solar wind flow outside the bow shock at comet P/Halley, Astron. Astrophys., 187, 47, 1987b.

Jones, D.E., E.J. Smith, J.A. Slavin, B.T. Tsurutani, G.L. Siscoe, and D.A. Mendis, the bow wave of comet Giacobini-Zinner: ICE magnetic field observations, Geophys. Res. Lett., 13, 243, 1986.

Kennel, C.F., B. Buti, T. Hada, and R. Pellat, Nonlinear, dispersive, elliptically polarized Alfvén waves, Phys. Fluids, in press, 1988.

Lakhina, G.S., and F. Verheest, Alfvén wave instabilities and ring current during solar wind-comet interaction, Astrophys. Space Sci., 143, 329, 1988.

Lee, M.A., and W.-H. Ip, Hydromagnetic wave excitation by ionized interstellar hydrogen and helium in the solar wind, J. Geophys. Res., 92, 11041, 1987.

Mukai, T., W. Miyake, T. Terasawa, M. Kitayama, and K. Hirao, Ion dynamics and distribution around comet Halley: Suisei observation, Geophys. Res. Lett., 13, 829, 1986.

Neubauer, F.M., et al., First results from the Giotto magnetometer experiment at comet Halley, Nature, 321, 352, 1986.

Neugebauer, M., and F.M. Neubauer, The density of cometary protons upstream of Comet Halley's bow shock, abstract, EOS, 69, 396, 1988.

Neugebauer, M., A.J. Lazarus, K. Altwegg, H. Balsiger, B.E. Goldstein, R. Goldstein, F.M. Neubauer, H. Rosenbauer, R. Schwenn, E.G. Shelley, and E. Ungstrup, The pick-up of cometary protons by the solar wind, Proc. 20th ESLAB Symposium, Heidelburg, 1, 19, 1986.

Omidi, N., and D. Winske, Simulation of the solar wind interaction with the outer regions of the coma, Geophys. Res. Lett., 13, 397, 1986.

Omidi, N., and D. Winske, Subcritical dispersive shock waves upstream of planetary bow shocks and at comet Giacobini-Zinner, Geophys. Res. Lett., in press, 1988.

Price, C.P., and C.S. Wu, The influence of strong hydromagnetic turbulence on newborn cometary ions, Geophys. Res. Lett., 14, 856, 1987.

Riedler, W., et al., Magnetic field observations in comet Halley's coma, Nature, 321, 288, 1986.

Rogers, B., S.P. Gary, and D. Winske, Electromagnetic hot ion beam instabilities: Quasi-linear theory and simulation, J. Geophys. Res., 90, 9494, 1985.

Sagdeev, R.Z., and V.D. Shafranov, The instability of plasma with the anisotropic velocity distribution in magnetic field, JETF, 39, 181, 1960.

Sagdeev, R.Z., V.D. Shapiro, V.I. Shevchenko, and K. Szego, MHD turbulence in the solar wind--comet interaction region, Geophys. Res. Lett., 13, 85, 1986.

Sagdeev, R.Z., V.D. Shapiro, V.I. Shevchenko, and K. Szego, The effect of mass loading outside cometary bow shock for the plasma and wave measurements in the coming cometary missions, J. Geophys. Res., 92, 1131, 1987.

Saito, T., K. Yumoto, K. Hirao, T. Nakagawa and K. Saito, Interaction between comet Halley and the interplanetary magnetic field observed by Sakigake, Nature, 321, 303, 1986.

Sanderson, T.R., K.-P. Wenzel, P. Daly, S.W.H. Cowley, R.J. Hynds, E.J. Smith, S.J. Bame, and R.D. Zwickl, The interaction of heavy ions from comet P/Giacobini-Zinner with the solar wind, Geophys. Res. Lett., 13, 411, 1986.

Sharma, O.P., and V.L. Patel, Low-frequency electromagnetic waves driven by gyrotropic gyrating ion beams, J. Geophys. Res., 91, 1529, 1986.

Smith, E.J., et al., ICE encounter with Giacobini-Zinner: magnetic field observations, Science, 232, 382, 1986.

Stix, T.H., The Theory of Plasma Waves, McGraw-Hill, New York, 1962.

Terasawa, T., Particle scattering and acceleration in a turbulent plasma around comets, this volume, 1988.

Terasawa, T., T. Mukai, W. Miyake, M. Kitayama, and K. Hirao, Detection of cometary pickup ions up to 10^7 km from comet Halley: Suisei observation, Geophys. Res. Lett., 13, 837, 1986.

Thorne, R.M., and B.T. Tsurutani, Resonant interactions between cometary ions and low frequency electromagnetic waves, Planet. Space Sci., 35, 1501, 1987.

Tsurutani, B.T., and E.J. Smith, Strong hydromagnetic turbulence associated with comet Giacobini-Zinner, Geophys. Res. Lett., 13, 259, 1986a.

Tsurutani, B.T., and E.J. Smith, Hydromagnetic waves and instabilities associated with cometary-ion pickup: ICE observations, Geophys. Res. Lett. 13, 263, 1986b.

Tsurutani, B.T., A.L. Brinca, E.J. Smith, R.M. Thorne, F.L. Scarf, J.T. Gosling, and F.M. Ipavich, MHD waves detected by ICE at distances > 28 x 10^6 km from comet Halley: cometary or solar wind origin?, Astron. Astrophys., 187, 97, 1987a.

Tsurutani, B.T., R.M. Thorne, E.J. Smith, J.T. Gosling and H. Matsumoto, Steepened magnetosonic waves at comet Giacobini-Zinner, J. Geophys. Res., 92, 11074, 1987b.

Tsurutani B.T., D.E. Page, E.J. Smith, B.E. Goldstein, A.L. Brinca, R.M. Thorne, H. Matsumoto, I.G. Richardson, and T.R. Sanderson, Low frequency plasma waves and ion pitch-angle scattering at large distances (>3.5 x 10^5 km) from Giacobini-Zinner: IMF α dependences, J. Geophys. Res., in press, 1988a.

Tsurutani, B.T., A.L. Brinca, B. Buti, E.J. Smith, R.M. Thorne, and H. Matsumoto, Magnetic pulses with durations near the local proton cyclotron period: Comet Giacobini-Zinner, J. Geophys. Res., in press, 1988b.

Verheest, F., Alfvén wave plasma turbulence during solar wind-comet interaction, Astrophys. Space Sci., 138, 209, 1987.

Winske, D., and S.P. Gary, Electromagnetic instabilities driven by cool heavy ion beams, J. Geophys. Res., 91, 6825, 1986.

Winske, D., and M.M. Leroy, Diffuse ions produced by electromagnetic ion beam instabilities, J. Geophys. Res., 89, 2673,1984.

Winske, D., C.S. Wu, Y.Y. Li, Z.Z. Mou, and S.Y. Guo, Coupling of newborn ions to the solar wind by electromagnetic instabilities and their interaction with the bow shock, J. Geophys. Res., 90, 2713, 1985.

Wong, H.K., and M.L. Goldstein, Proton beam generation of oblique whistler waves, J. Geophys. Res., 93, 4110, 1988.

Wu, C.S., and R.C. Davidson, Electromagnetic instabilities produced by neutral particle ionization in interplanetary space, J. Geophys. Res., 77, 5399, 1972.

Wu, C.S., and R.E. Hartle, Further remarks on plasma instabilities produced by ions born in the solar wind, J. Geophys. Res., 79, 283, 1974.

Wu, C.S., D. Winske, and J.D. Gaffey, Jr., Rapid pickup of cometary ions due to strong magnetic turbulence, Geophys. Res. Lett., 13, 865, 1986.

Wu, C.S., X.T. He, and C.P. Price, Excitation of whistlers and waves with mixed polarization by newborn cometary ions, J. Geophys. Res., 93, 3949, 1988.

Yumoto, K., T. Saito, and T. Nakagawa, Hydromagnetic waves near O^+ (or H_2O^+) ion cyclotron frequency observed by Sakigake at the closest approach to comet Halley, Geophys. Res. Lett., 13, 825, 1986a.

Yumoto, K., T. Saito, and T. Nakagawa, Long-period HM waves associated with cometary O^+ (or H_2O^+) ions; Sakigake observations, Proc. 20th ESLAB Symposium, Heidelberg, 1, 249, 1986b.

PLASMA WAVE OBSERVATIONS AT COMETS GIACOBINI-ZINNER AND HALLEY

Frederick Scarf

TRW Space and Technology Group, Redondo Beach, California, USA 90278

Abstract. The 1985-86 spacecraft encounters with Comets Giacobini-Zinner and Halley provided direct evidence that the solar wind interaction with comets involves plasma physics phenomena that extend over distance scales that are measured in millions of kilometers. The comet interaction phenomena that are dominated or controlled by plasma instabilities include pickup and energization of newly-born cometary ions at great distances, solar-wind mass loading effects, formation of diffuse low-Mach number bow shocks, development of intense plasma turbulence that provides scattering and heating, and onset of critical ionization. In this review we describe and evaluate the cometary plasma wave measurements in the spectral range extending from the lower hybrid resonance frequency (a few Hertz in the upstream regions for Giacobini-Zinner and Halley) to the electron plasma frequency (a few MegaHertz during the Giacobini-Zinner tail crossing). We also discuss the difficulties of making plasma wave measurements in the presence of large fluxes of dust particles that impact the spacecraft at high speeds.

Introduction

Between December 1984 (when the first AMPTE artificial comet releases were studied), and March 1986 (when six spacecraft encountered Comet Halley at a range of distances), we learned many significant new facts about plasma physics of the comet-solar wind interaction. The top panel in Figure 1 illustrates the very different sizes of the interaction regions, as measured by the nominal bow shock distance. Four of the six spacecraft in the 1985-86 armada to the natural comets were instrumented to measure plasma waves. The International Cometary Explorer (ICE) carried both a plasma wave instrument and a radio wave receiver to a close encounter with Comet Giacobini-Zinner (7800 km downstream) and a distant approach to Comet Halley (28 million km upstream). The VEGA-1 and VEGA-2 spacecraft, which passed about 9000 km in front of Comet Halley, had a pair of plasma wave investigations, and Sakigake, which traversed the Halley upstream region at a distance of 7 million kilometers, also carried a plasma wave receiver. In this report we will focus attention on these particular encounters with Giacobini-Zinner and Halley, although there will be occasional reference to observations associated with the AMPTE releases.

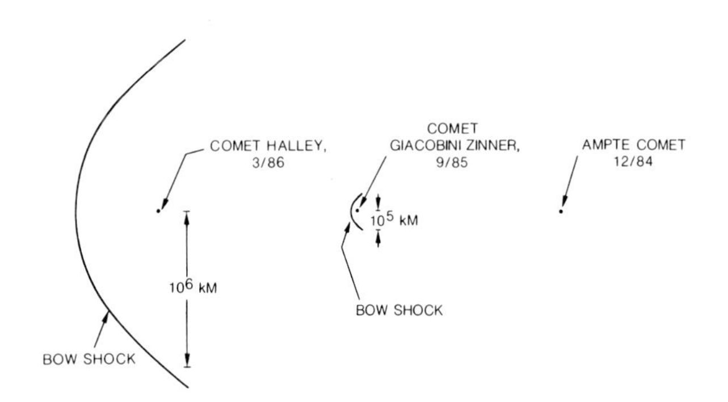

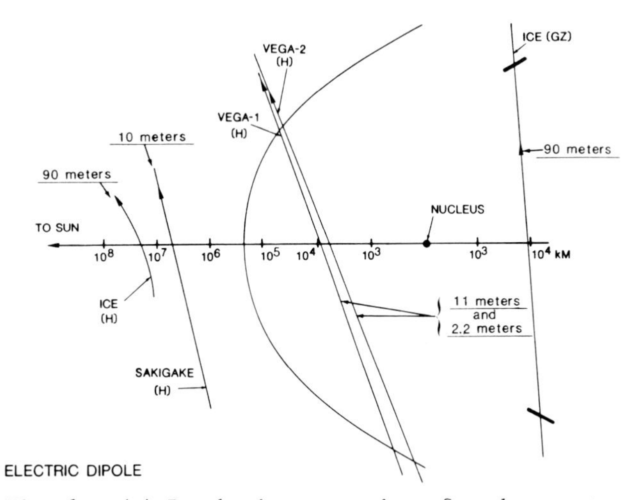

Fig. 1. (a) Rough size comparison for the comet interaction regions probed by flyby spacecraft in the period 1984-1986. (b) Sketch showing the flyby trajectories for the four spacecraft (ICE, VEGA 1,2 and Sakigake) that carried plasma wave instrumentation to Giacobini-Zinner and Halley.

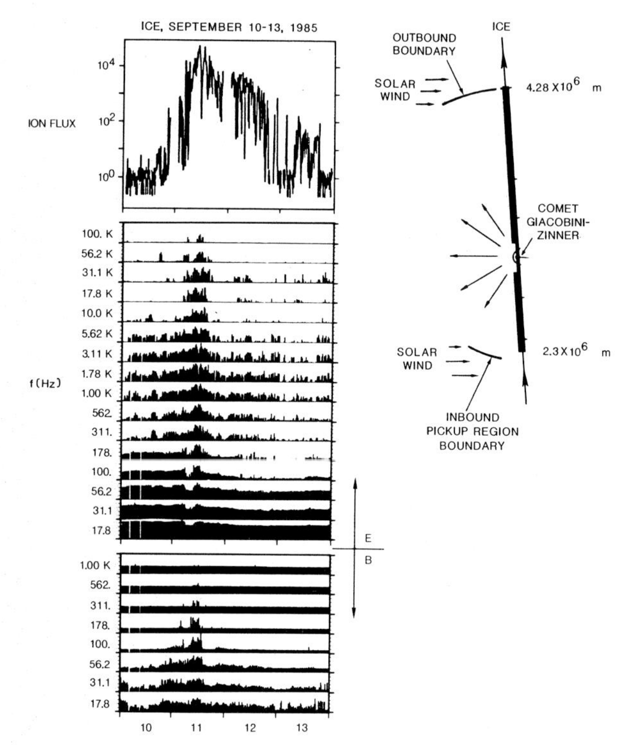

Fig. 2. Overview of the energetic ion and plasma wave observations in a vast region surrounding Comet Giacobini-Zinner. The cometary neutrals generate heavy ions far from the nucleus and these particles are energized in a solar wind pickup process, leading to enhanced wave levels.

The instrument capabilities of the plasma wave investigations varied for many reasons. The lower panel in Figure 1 shows the spacecraft trajectories in comet-centered coordinates, and the effective lengths of the electric antennas are indicated on the sketch (the only high sensitivity magnetic search coil data came from the ICE wave investigation).

The Ion Pick-up Region and Foreshock

The most spectacular result of the ICE encounter with Giacobini-Zinner (G-Z) involved the discovery of a huge region around the comet characterized by the presence of energetic heavy pick-up ions [Hynds et al., 1986; Ipavich et al., 1986] that were detected along with unusually high levels of plasma turbulence [Scarf et al., 1986a, b; Tsurutani and Smith, 1986; Gosling et al., 1986]. The top part of Figure 2 shows that energetic pick-up ions were detected by the Hynds instrument over a distance range extending to several million kilometers from the comet, and the lower panels show that these flux enhancements were found in association with high levels for electromagnetic and electrostatic plasma waves.

For the low frequency turbulence in the distant solar wind, it appears that the pick-up acceleration of the heavy ions generates a ring distribution that excites an ion loss-cone instability with peak wave growth near the lower hybrid frequency [Hartle and Wu, 1973; Sagdeev et al., 1986; see also Galeev, 1986]. Recently, Richardson et al. [1988] examined details of the correlations involving the higher frequency waves, the magnetic field and the ion spectra and Brinca, Tsurutani and Scarf [1988] discussed how these electrostatic bursts could be generated in association with steepened magnetosonic waves.

All of the Halley bound spacecraft with payloads that included magnetometers and energetic ion sensors detected similar profiles of pick-up ions and enhanced low frequency turbulence in a vast region extending to several million kilometers around Halley, and the Sakigake plasma wave investigation clearly detected electrostatic noise bursts in the pick-up region [Oya, 1986].

ICE also passed far in front of Halley (see Figure 1), and during this distant encounter, the energetic ion detector and the plasma wave instrument measured characteristics that appeared to be very similar to those found in the G-Z pick-up region. Figure 3 shows some ICE positive ion fluxes and B-field wave amplitudes observed during the closest approach to Halley, and Figure 4 contains a plot that compares these wave observations

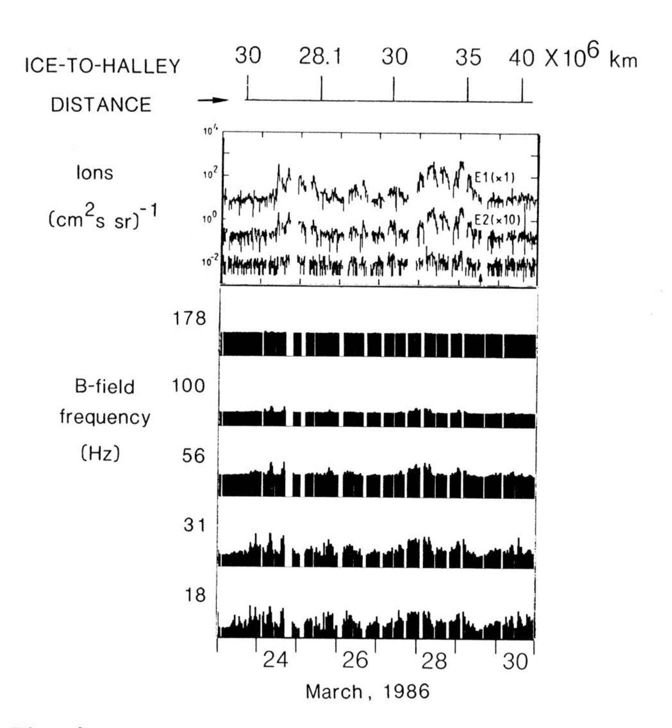

Fig. 3. ICE observations of ion flux enhancements and elevated whistler wave levels during closest approach to Comet Halley.

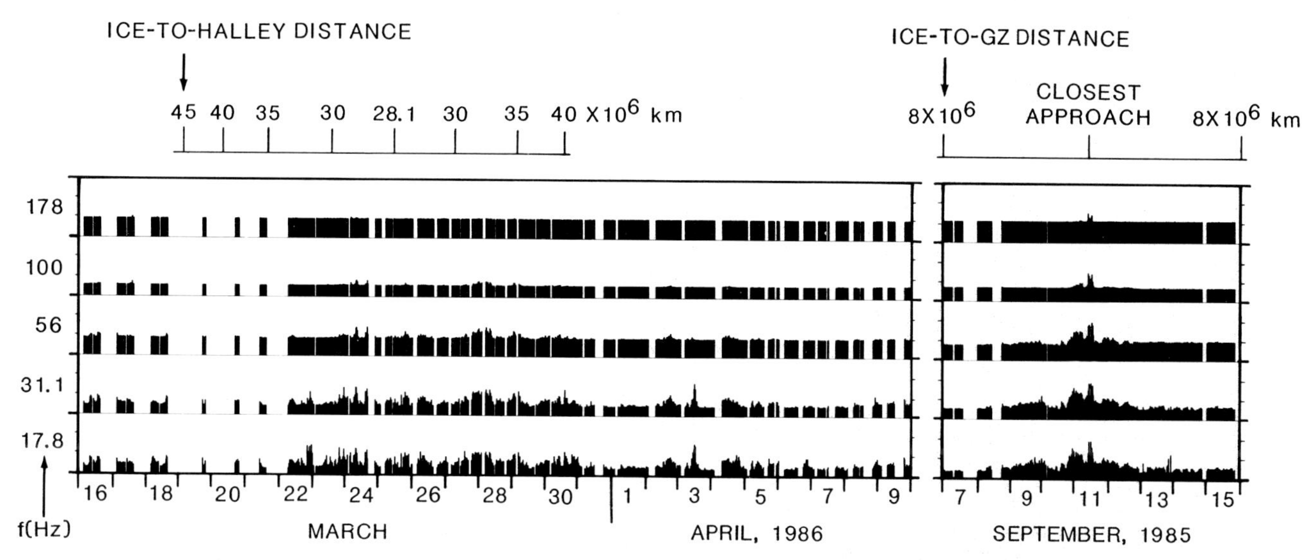

Fig. 4. Comparison of whistler mode wave enhancements detected in the ion pick-up region of Giacobini-Zinner and near closest approach to Halley.

with corresponding measurements from G-Z. Wenzel et al. [1986] and Scarf et al. [1986b] interpreted these March 1986 ICE measurements in terms of passage through the Halley ion pick-up region, but Tsurutani et al. [1986] suggested that these correlated observations simply represented detection of an interplanetary disturbance.

At present, neither one of these interpretations can be rejected conclusively, but there are additional considerations that put significant constraints on the Halley source explanation. Daly [1987] noted that the neutrals from Halley would travel in well-defined solar orbits, and he computed characteristics for a range of trajectories that could yield suitable fluxes of newly-born ions during the ICE closest approach. Daly showed that the lowest ejection speeds would be near 7.5 km/sec, if the neutrals were emitted at perihelion (Feb. 9, 1986); these speeds are higher than the commonly assumed sublimation speeds (near one km/sec), but many chemical reactions can lead to such high ejection velocities, and many of the observed outbursts from the comet had speeds much higher than 7.5 km/sec. It is likely that the determination of the true extent of Halley's pick-up region will remain open until the comet makes its next perihelion passage.

Other upstream plasma and wave phenomena were detected at G-Z and at Halley. As the ICE spacecraft approached Comet Giacobini-Zinner, it moved in and out of a conventional foreshock region that was characterized by: a) magnetic connection to the main comet interaction region, b) enhanced heat flux from the comet, and c) high plasma wave levels (whistlers, electron plasma oscillations, and broadband electrostatic bursts) [see Fuselier et al., 1986 for details]. Some ICE E- and B-field spectra in the foreshock region are shown in the left hand panels for Figure 5.

For the Halley encounter, Figure 1 suggests that Sakigake was actually in an optimum location to detect foreshock phenomena, and Figure 6 compares selected upstream spectra from Sakigake [Oya et al., 1986; Oya, 1986] with the ICE G-Z measurements. These spectra are quite similar in shape,

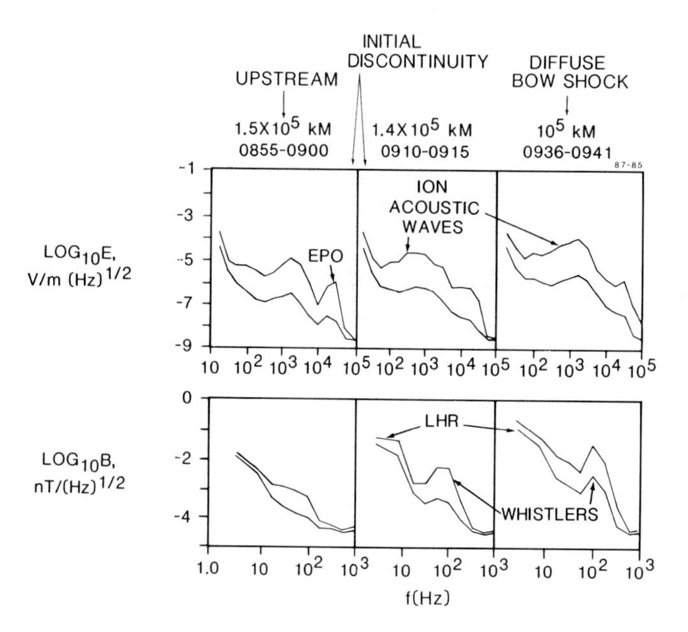

Fig 5. Characteristics E and B wave spectra detected by the ICE plasma wave instrumentation during the approach to Giacobini-Zinner.

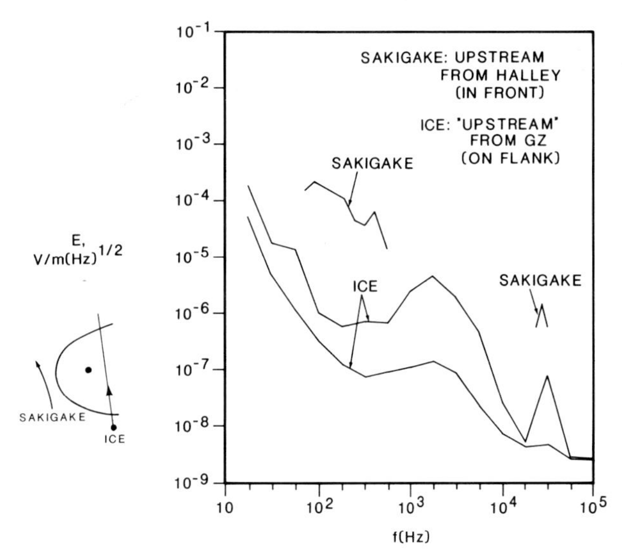

Fig. 6. Sakigake and ICE observations of ion acoustic waves and electron plasma oscillations in the foreshock regions of Halley and G-Z.

but it seems that the wave amplitudes in the Halley foreshock were much more intense than those detected at G-Z.

The Comet Bow Shock Region

Many years ago Wallis [1971] noted that the mass loading of the solar wind by the heavy pick-up ions would lower the Mach number, and he raised the possibility that a well-defined bow shock might not develop at a comet. The initial interpretation of the ICE plasma and field data from Comet Giacobini-Zinner seemed to support this view; Bame et al. [1986] and Smith et al. [1986a] found much disorder but no well defined parameter jumps that would correspond to a conventional bow shock.

However, some of the plasma wave observations from ICE did seem to support the concept of a thin shock [Scarf et al., 1986a; 1986c]. The wave data allowed the investigators to identify a series of discontinuities involving a) detectable levels of ion acoustic waves and electron plasma oscillations on the upstream side, and b) significantly enhanced levels of lower hybrid emissions, ion waves and whistlers on the downstream side. Plate 1 has color-coded E and B frequency-time displays of peak E and B wave amplitudes that clearly show a sequence of discontinuities ending at 0911 on the inbound leg, and a single event of the same type at 1220 on the outbound pass. The display should be compared with the central panels in Figure 5 that show the E and B spectra on the downstream side of the 0911 discontinuity; it is evident from Figure 5 and Plate 1 that these boundary crossings involved great changes in all the relevant plasma turbulence levels. Moreover, although these variations were not accompanied by familiar jumps in the characteristics of the magnetic field or the plasma electrons, Richardson et al. [1986] showed that the heavy ion velocities did shift in these regions more or less as would be expected for a conventional shock crossing. Thus, Scarf et al. [1986c] suggested that in a sense there were two shock crossings at G-Z; one involving the plasma wave discontinuities and the jumps in the heavy ion fluxes, and a second one involving the solar wind plasma and the B-field.

For G-Z it is now generally recognized that the pick-up ions produced so much turbulence and so much mass loading that the shock crossing of the second type was extremely thick and difficult to identify. Smith et al. [1986b] addressed these difficulties. They assumed a thick and diffuse shock and they compared averages of the measured parameters (far upstream and far downstream) with values derived from with Rankine-Hugoniot equations. This analysis verified that at least on the inbound pass, ICE detected a low Mach number quasiperpendicular type of bow shock. For this diffuse shock, we have an extended region of strong coupling and the central panels in Figure 5 give the plasma wave spectra. Coroniti et al. [1986] analyzed the likely wave generation mechanisms in this region, and Kennel et al. [1986] showed that these wave spectra were actually similar to those detected at quasiparallel interplanetary shocks; presumably the very large turbulence levels allow the parallel configuration to develop locally although the average large scale shock was more nearly perpendicular.

In some ways the bow shock crossings at Comet Halley were very different from those detected at G-Z. At Halley, the shock jumps were clearly defined in many of the observations from the plasma probes and magnetometers on Giotto, VEGA and Suisei. However, since VEGA carried the only wave investigations into the Halley shock region, we confine further attention to the VEGA measurements.

The VEGA 1 and 2 plasma probes [Gringauz et al., 1986] and magnetometers [Reidler et al., 1986] detected features resembling bow shocks at a range of about 1.1 to 1.2 million km from the nucleus. These discontinuities were accompanied by significant changes in low frequency ion flux and electric field turbulence levels and Klimov et al., [1986a] interpreted these variations in terms of increased wave activity at the shock. More details of the low frequency VEGA wave observations in the shock region were discussed in a number of additional reports [Galeev et al., 1986a; Galeev, 1986; Klimov et al., 1986b] and Russell et al. [1987] suggested that ultra low frequency magnetic oscillations detected in the post-shock magnetosheath were associated with the mirror instability. In Figure 7 the low frequency VEGA E-field shock spectrum is compared with ICE G-Z observations in the strong coupling region (the bow shock data from the high frequency wave

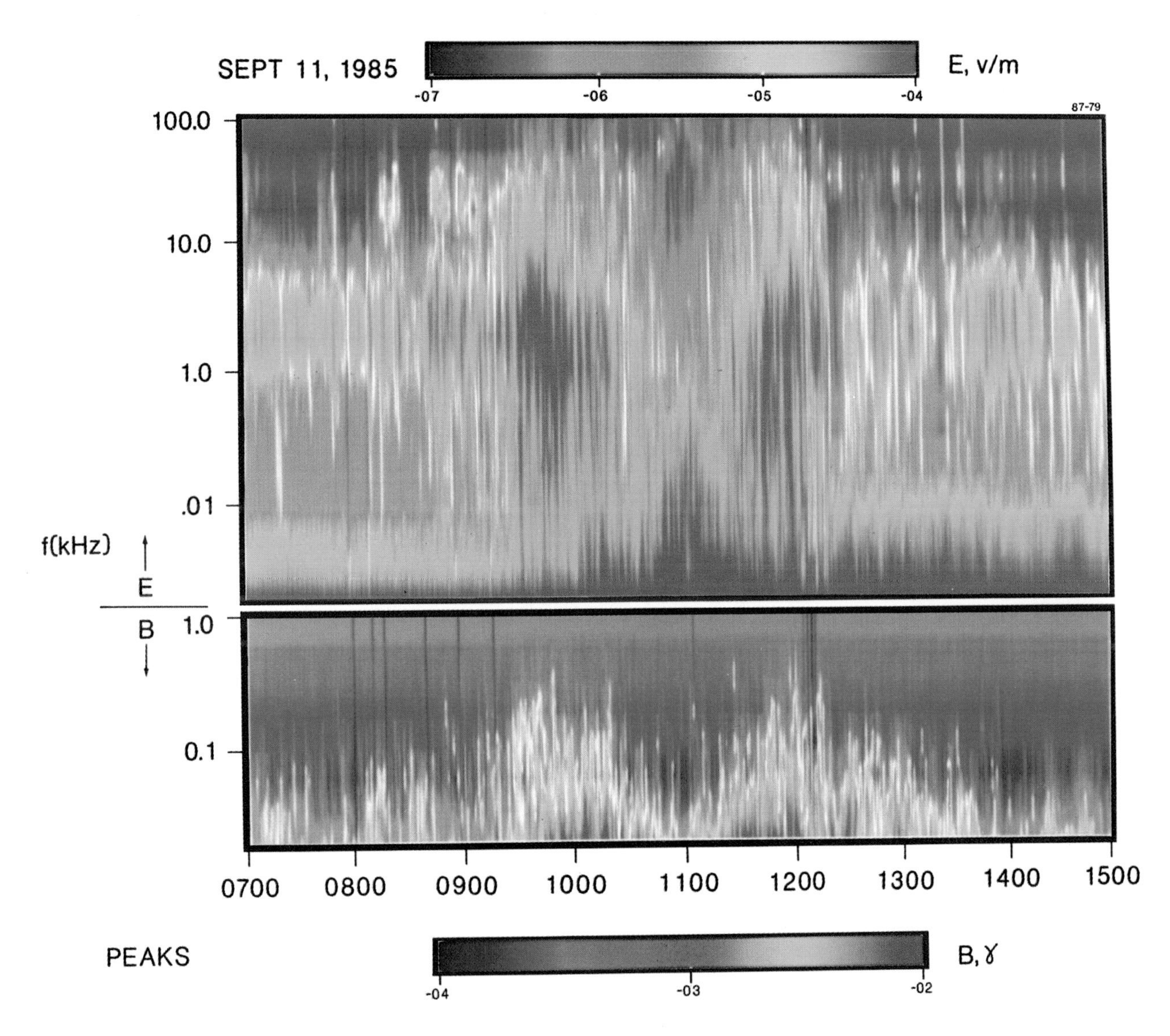

Plate 1. Color-coded E and B frequency-time displays showing variations in peak wave amplitudes during the ICE encounter with Giacobini-Zinner.

instrument on VEGA have not yet been fully processed, but Figure 7 shows that the instrumental sensitivity was quite adequate to detect levels similar to those found at G-Z). It can be seen that at Halley the peak intensity was associated with very low frequency oscillations in the lower hybrid resonance (LHR) range; at G-Z the search coil data also showed a significant LHR enhancement (see Figure 5), but here the whistler and ion acoustic mode turbulence levels were also significant.

It is of interest to compare these measurements at the natural comets with observations of shock-like phenomena detected in connection with the AMPTE artificial comet releases. The size of the AMPTE Barium cloud was always very small in comparison with all relevant plasma scale lengths such as the ion Larmor radius, and thus it was never anticipated that a true shock could form at the solar wind-comet interface. Nevertheless, Gurnett et al. [1986a] showed that as the Barium cloud expanded, they detected extremely intense broadband turbulence in the region where a shock would develop. Figure 8 contrasts the AMPTE, Halley and G-Z spectra in these shock-like regions. It is somewhat surprising that the AMPTE wave levels are so much higher than those detected on VEGA or ICE; perhaps it is because the obstacle

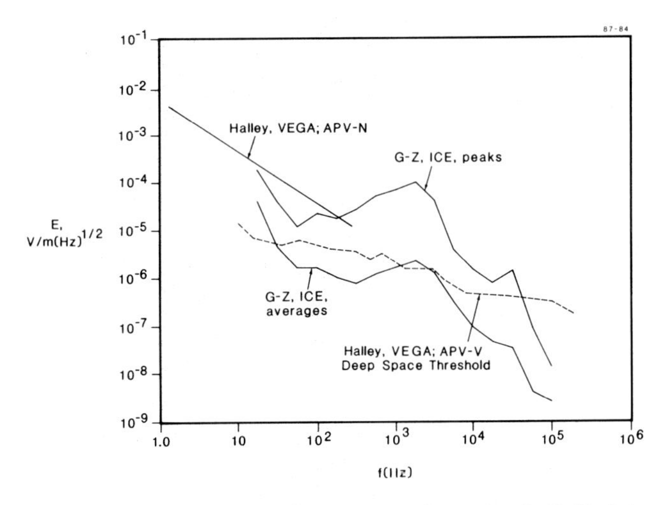

Fig. 7. Comparison of cometary bow shock E-field wave spectra. The VEGA data come from the low frequency instrument; the corresponding observations from the high frequency analyzer are not yet available, but the threshold curve shows that the instrument is sensitive enough to detect peaks comparable to those measured in the G-Z strong coupling region.

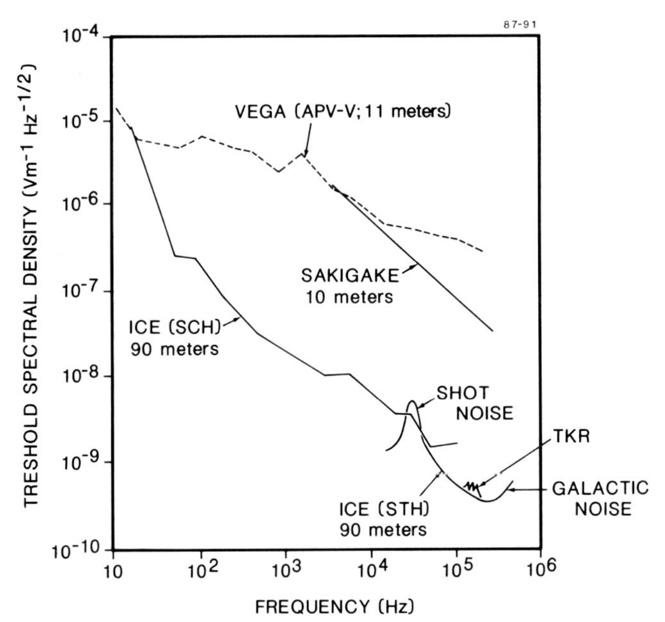

Fig. 9. In-flight threshold levels for wave investigations on ICE, Sakigake and VEGA, together with characteristics high frequency spectra detected by the very sensitive ICE radio astronomy instrument (STH).

is so small in comparison with plasma scale lengths that charge separation effects lead to greatly enhanced turbulence.

Near-Encounter Wave Observations

The lower drawing in Figure 1 shows that the ICE and VEGA encounter trajectories were very different close to the comets, and care must be taken when one attempts to correlate the Halley and G-Z measurements. Grard et al. [1987] carefully examined the ICE and VEGA plasma and field measurements and they concluded that the G-Z tail discontinuity designated as the "magnetopause" [Smith et al., 1986a; Slavin et al., 1986; Scarf, 1986] was physically the same as the "cometopause" boundary detected on the sunward side of Comet Halley [Gringauz et al., 1986].

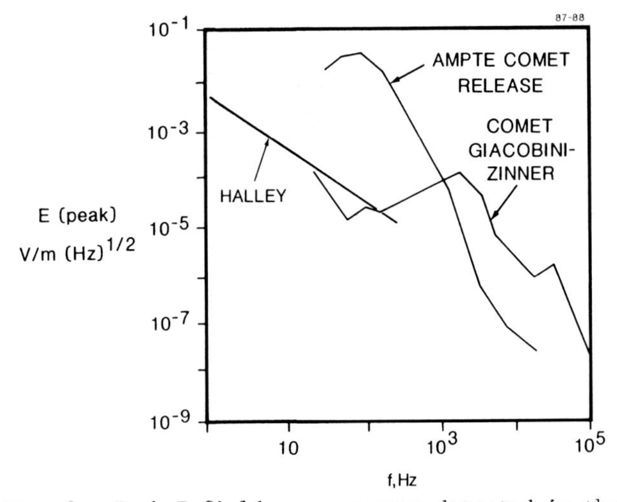

Fig. 8. Peak E-field wave spectra detected in the AMPTE, Halley and G-Z "shock-like" regions.

In the region of this boundary and during the entire tail crossing, the ICE radio astronomy instrument was able to provide unique and very important information on characteristics of the thermal plasma. The basis for these observations is indicated in Figure 9, where the in-flight noise levels for the wave instrument on Sakigake, VEGA and ICE are compared. It can be seen that the ICE radio astronomy instrument (SCH) had exceptional sensitivity in the range above 30 kHz, and this allowed the investigators to detect galactic noise, terrestrial kilometric radiation (TKR) and shot noise. By analyzing the shot noise spectrum, Meyer-Vernet et al. [1986a,b] were able to derive the electron density and temperature profiles for the entire near-encounter region, and some results are shown in the upper panel of Figure 10, [adapted from a drawing first shown by Slavin et al., 1986]. The lower panel in Figure 10 shows the magnetic pressure profile during this part of the ICE encounter with Giacobini-Zinner, and the labels "MP" mark the magnetopause crossings. By comparing the electron and B-field pressure profiles it is easy to see that well-defined lobe regions with high magnetic fields and low plasma densities were separated by a plasma sheet

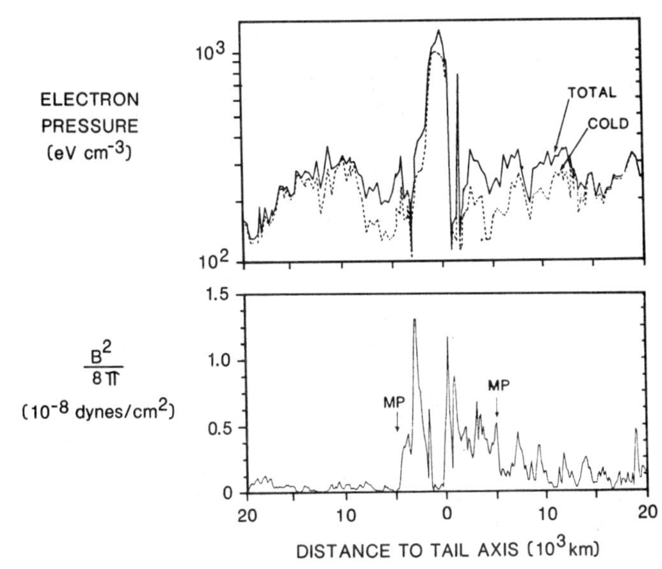

Fig. 10. Particle and field pressure variations detected by the radio astronomy instrument (top) and magnetometer (bottom) during the ICE traversal of the G-Z tail.

with a near-null B-field and a high electron concentration.

Although the particle and field characteristics detected in the G-Z tail region were relatively straightforward, the wave measurements were not so easily interpreted. Figure 11 shows the ICE E-

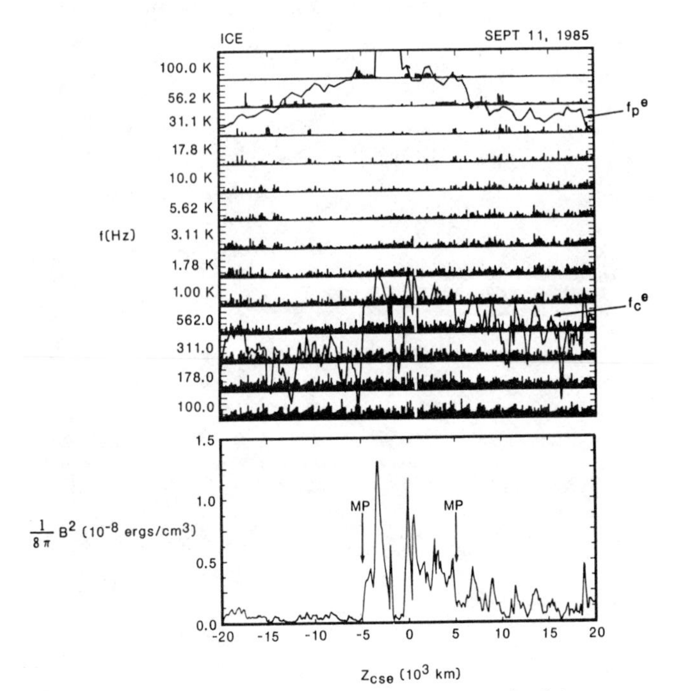

Fig. 11. Variations in plasma wave E-field amplitudes detected on ICE during the G-Z tail traversal.

field observations along with profiles of the electron plasma frequency (derived from the shot noise analysis) and the electron cyclotron frequency. The f_p^e trace does fit very well with enhancements in the highest frequency plasma wave channels, but the low frequency wave level variations are not at all well correlated with changes in f_c^e. It is also evident from Figure 11 that the wave amplitudes did not change in any significant way as ICE traversed the magnetopause, although significant broadband noise bursts were detected beyond these magnetic boundaries.

Some of the near-encounter observations at Halley were similarly difficult to interpret. Figure 12 shows amplitude plots from the two multi-channel analyzers on VEGA 2 [Grard et al., 1986a; Klimov et al., 1986a], and the labels indicate one serious problem that developed here; as the VEGA spacecraft approached the comet nitrogen gas was released to provide cooling for a science instrument, and this generated significant artificial increases in the background noise

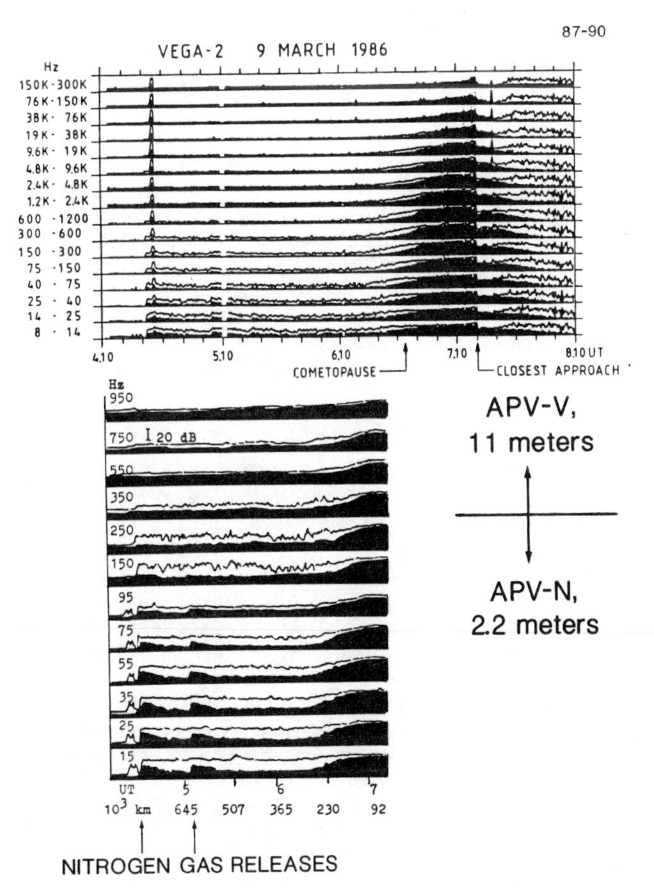

Fig. 12. Variations in plasma wave E-field amplitudes during the VEGA-2 inbound passage. The nitrogen gas releases caused interference as shown, but both wave instruments clearly detected enhanced levels when the cometopause was crossed.

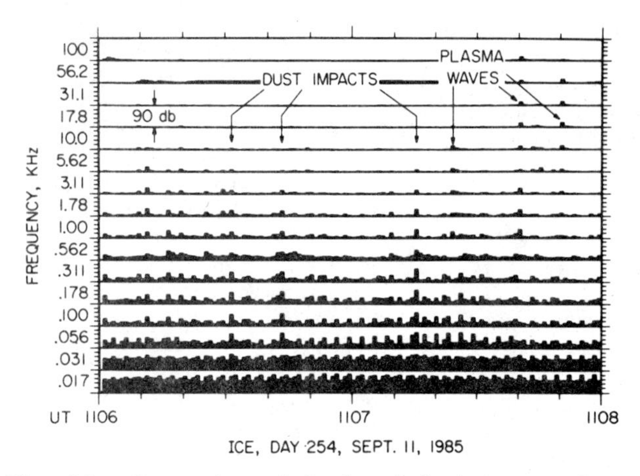

Fig. 13. Detection of isolated dust impacts by the ICE plasma wave instrument.

levels [see comments by Grard et al., 1985; Klimov et al., 1986a; Oberc et al., 1986a]. However, despite this difficulty, both wave investigations detected clear increases in wave activity at the cometopause boundary labeled on Figure 12 [see discussions by Mogilevsky et al., 1986; Grard et al., 1987; Savin et al., 1986].

As the VEGA spacecraft continued on to closest approach, other variations in wave intensity were detected. One of the most interesting interpretations was provided by Galeev et al. [1986b], who suggested that the correlated variations in heavy ion density and low frequency (1-32 Hz) electric field strength detected at a range of about 30,000 km could be explained in terms of processes associated with the onset of the critical ionization phenomenon [see also Galeev, 1986; Savin et al., 1986].

Detection of Dust by Wave Instruments

It was always recognized that it might be very difficult to measure plasma waves during the flybys of the natural comets because at the high encounter speeds (20 km/sec for G-Z to about 80 km/sec for Halley), impacts of neutral gas molecules and dust particles on the spacecraft surfaces would naturally produce transient emissions of sputtered ions and secondary electrons that would generate local impulses in the electric antennas [see, for instance, Grard et al., 1986b, and references therein]. The first opportunity to study this problem came in September 1985 when ICE flew across the dust tail of Giacobini-Zinner. Scarf et al. [1986a] provided an initial qualitative report showing that relatively few dust impacts were detected during this encounter, and Gurnett et al. (1986) then carried out a detailed quantitative analysis. Figure 13, adapted from the paper by Gurnett et al., shows that for G-Z the impact rate was low enough so that it was easy to distinguish dust events from naturally occurring plasma waves.

Figure 14 shows an earth view of the ICE encounter trajectory superimposed on a 1959 groundbased photograph of Giacobini-Zinner (photograph taken by P. Roemer). The trajectory tic marks represent one hour intervals and distance scales of approximately 75,000 km. During the encounter, the ICE plasma wave instrument detected less than 100 strong impacts [Gurnett et al., 1986b], and the number of events measured in each 3.75 minute interval is given by the height of each black bar plotted in the direction perpendicular to the trajectory line. The peak corresponds to 16 impacts in 3.75 minutes, and the rate decreased with an r^{-2} falloff.

It was always expected that the situation with Comet Halley would be very different because groundbased observations showed that the dust flux from Halley is much greater than that from G-Z; moreover, since the spacecraft-Halley encounter speed was approximately four times higher, the relative impact energy on VEGA would be about 16 times greater than the corresponding value for ICE. Indeed, the actual dust collisions with VEGA caused considerable damage as well as confusion about the relative roles of impacts and plasma waves. Several intensive preliminary studies revealed that large numbers of impacts were detected, especially within the cometopause [Grard et al., 1986b; Klimov et al., 1986c; Oberc et al., 1986b; Trotignon et al., 1986], and more detailed studies are now underway. One technique used by

Fig. 14. ICE flyby trajectory and dust impact rate profile superimposed on a ground-based photograph of Comet Giacobini-Zinner.

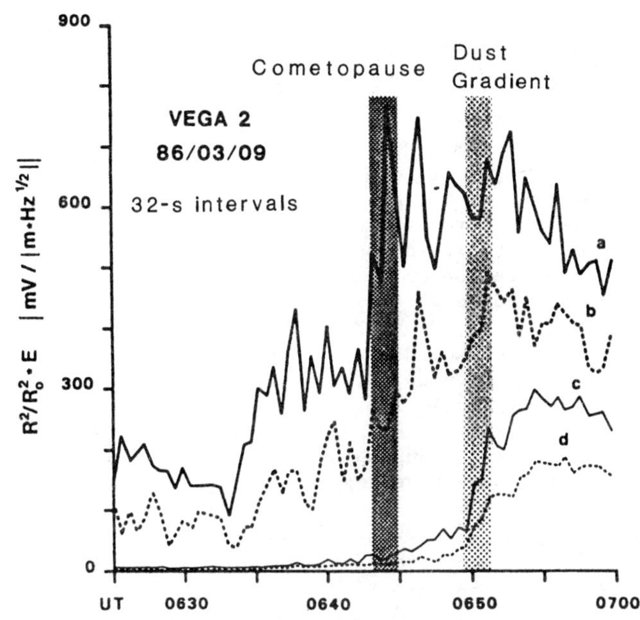

Fig. 15. Normalized plasma wave spectral density profiles measured by VEGA-2 in four frequency intervals: a) 8-14 Hz; b) 14-25 Hz; c) 150-300 Hz; d) 0.6-1.2 kHz.

Grard et al. [1987] to identify dust gradients is illustrated in Figure 15; here the spectral density is normalized with a geometric factor to search for flux changes, and this analysis shows that a distinct change, attributed to dust, was detected after the VEGA 2 cometopause passage. Other studies will include detailed comparisons of VEGA plasma wave observations with measurements from the on-board dust detectors and with results from the langmuir probes.

Conclusions

The measurements from ICE, Sakigake and VEGA, together with the corresponding observations from the AMPTE spacecraft, conclusively illustrate the significant role that plasma physics plays in controlling the solar wind interaction with a comet. We find that a variety of very strong wave-particle interactions develop in the entire region where strong mass loading takes place, and additional plasma instabilities of importance are found in the vast region where newly-born ions are picked up by the solar wind. However, the encounter observations show that the comet-solar wind interaction is an extremely complex one, with many unique features that should provide excellent opportunities for future theoretical studies using analytical techniques and numerical simulations.

Acknowledgments. I am indebted to the large number of cometary plasma physics colleagues whose names appear in the reference list; they have provided me with invaluable information and insights. In addition, I wish to express my special thanks to Terry Averkamp of the University of Iowa for the color processing that was used to make up Plate 1. This research was supported by the ICE Project at NASA Goddard Space Flight Center under Contract NAS5-28703.

References

Bame, S. J., R. C. Anderson, J. R. Asbridge et al., Comet Giacobini-Zinner: plasma description, Science, 232, 356, 1986.

Brinca, A. L., B. T. Tsurutani, and F. L. Scarf, Local generation of electrostatic bursts at Comet Giacobini-Zinner: modulation by steepened magnetosonic waves, submitted to J. Geophys. Res., 1988.

Coroniti, F. V., C. F. Kennel, F. L. Scarf, E. J. Smith, B. T. Tsurutani, S. J. Bame, M. F. Thomsen, R. Hynds, and K.-P. Wenzel, Plasma wave turbulence in the strong coupling region at Comet Giacobini-Zinner, Geophys. Res. Lett., 13, 869, 1986.

Daly, P. W., Can neutral particles from Comet Halley have reached the ICE spacecraft?, Geophys. Res. Lett., 14, 648, 1987.

Fuselier, S. A., W. C. Feldman, S. J. Bame, E. J. Smith, and F. L. Scarf, Heat flux observations and the location of the transition region boundary of Giacobini-Zinner, Geophys. Res. Lett., 13, 247, 1986.

Galeev, A. A., B. E. Gribov, T. Gombosi et al., Position and structure of the Comet Halley bow shock: Vega-1 and Vega-2 measurements, Geophys. Res. Lett., 13, 841, 1986a.

Galeev, A. A., K. I. Gringauz, S. I. Klimov et al., Critical ionization velocity effects in the inner coma of Comet Halley: measurements by Vega-2, Geophys. Res. Lett., 13, 845, 1986b.

Galeev, A. A., Theory and observations of solar wind/cometary plasma interaction processes, ESA SP-250, I, p. 3, 1986.

Gosling, J. T., J. R. Asbridge, S. J. Bame et al., Large amplitude, low frequency plasma fluctuations at Comet Giacobini-Zinner, Geophys. Res. Lett., 13, 267, 1986.

Grard, R., A. Pederson, K. Knott et al., Adv. Space Res., 5, 175, 1985.

Grard, R., A. Pederson, J. G. Trotignon et al., Observations of waves and plasma in the environment of Comet Halley, Nature, 321, 290, 1986a.

Grard, R. J. L., J. A. M. McDonnell, E. Grun, and K. I. Gringauz, Secondary electron emission induced by gas and dust impacts on Giotto, Vega-1 and Vega-2 in the environment of Comet Halley, ESA SP 250, III, 327, 1986b.

Grard, R., F. L. Scarf, J. G. Trotignon, and M. Mogilevsky, A comparison between wave observations performed in the environments of Comets Halley and Giacobini-Zinner, ESA SP 278, 97, 1987.

Gringauz, K. I., T. I. Gombosi, A. P. Rezimov et al., First in situ plasma and neutral gas measurements at Comet Halley, Nature, 321, 282, 1986.

Gurnett, D.A., R. R. Anderson, B. Hausler et al., Plasma waves associated with the AMPTE artificial comet, Geophys. Res. Lett., 12, 851, 1985.

Gurnett, D. A., T. Averkamp, F. L. Scarf, and E. Grun, Dust particles detected near Giacobini-Zinner by the ICE plasma wave instrument, Geophys. Res. Lett., 13, 291, 1986.

Hartle, R. E. and C. S. Wu, Effects of electrostatic instabilities on planetary and interstellar ions in the solar wind, J. Geophys. Res., 78, 5802, 1973.

Hynds, R. J., S. W. H. Cowley, T. R. Sanderson et al., Observations of energetic ions from Comet Giacobini-Zinner, Science, 232, 366, 1986.

Ipavich, F. M., A. B. Galvin, G. Gloeckler et al., Comet Giacobini-Zinner: in situ observations of energetic heavy ions, Science, 232, 366, 1986.

Kennel, C. F., F. V. Coroniti, F. L. Scarf et al., Plasma waves in the shock interaction regions at Comet Giacobini-Zinner, Geophys. Res. Lett., 13, 921, 1986.

Klimov, S., S. Savin, Ya. Aleksevich et al., Extremely-low frequency plasma waves in the environment of Comet Halley, Nature, 321, 292, 1986a.

Klimov, S., A. Galeev, M. Nozdrachev et al., Fine structure of the near-cometary bow shock from plasma wave measurements (APV-N) experiments, ESA SP 250, I, 255, 1986b.

Klimov, S., M. Balikhin, M. Nozdrachev et al., Cometary dust distribution from APV-N data, ESA SP 250, III, 511, 1986c.

Meyer-Vernet, N., P. Couturier, S. Hoang et al., Plasma diagnosis from the thermal noise and limits on dust flux or mass in Comet Giacobini-Zinner, Science, 232, 370, 1986a.

Meyer-Vernet, N., P. Couturier, S. Hoang et al., Physical parameters for hot and cold electron populations in Comet Giacobini-Zinner with the ICE radio experiment, Geophys. Res. Lett., 13, 279, 1986b.

Mogilevsky, M., Y. Mihalov, O. Molchanov et al., Identification of boundaries in the cometary environment from a.c. electric field measurements, ESA SP 250, III, 467, 1986.

Oberc, P., D. Orlowski, P. Wronowski et al., Plasma waves in Halley's inner coma as measured by the APV-N experiment during the Vega mission, ESA SP 250, I, 89, 1986a.

Oberc, P., D. Orlowski, and S. Klimov, Dust impact effects recorded by the APV-N experiment during Comet Halley encounters, ESA SP 250, III, 323, 1986b.

Oya, H., A. Morioka, W. Miyake et al., Discovery of cometary kilometric radiations and plasma waves at Comet Halley, Nature, 321, 307, 1986.

Oya, H., Interaction processes of cometary plasma with the solar wind plasma in regions remote from the nucleus of the Comets Halley and Giacobini-Zinner, ESA SP 250, I, 117, 1986.

Reidler, W., K. Schwingenschuh, Ye G. Yeroshenko et al., Magnetic field observations in Comet Halley's coma, Nature, 321, 288, 1986.

Richardson, I. G., S. W. H. Cowley, R. J. Hynds et al., Three dimensional energetic ion bulk flows at Comet P/Giacobini-Zinner, Geophys. Res. Lett., 13, 415, 1986.

Richardson, I. G., K.-P. Wenzel, S. W. H. Cowley et al., Correlated plasma wave, magnetic field and energetic ion observations in the ion pickup region of Comet Giacobini-Zinner, submitted to J. Geophys. Res., 1988.

Russell, C. T., W. Reidler, K. Schwingenschuh, and Ye Yeroshenko, Mirror instability in the magnetosphere of Comet Halley, Geophys. Res. Lett., 14, 644, 1987.

Sagdeev, R. Z., V. D. Shapiro, V. I. Shevchenko, and K. Szego, MHD turbulence in the solar wind-comet interaction region, Geophys. Res. Lett., 13, 85, 1986.

Savin, S., G. Avanesova, M. Balikhin et al., ELF waves in the plasma regions near the comet, ESA SP 250, III, 433, 1986.

Scarf, F. L., Characteristics of the tail of Comet Giacobini-Zinner, Adv. Space Res., 6, 329, 1986.

Scarf, F. L., F. V. Coroniti, C. F. Kennel, D. A. Gurnett, W.-H. Ip, and E. J. Smith, Plasma wave observations at Comet Giacobini-Zinner, Science, 232, 377, 1986a.

Scarf, F. L., F. V. Coroniti, C. F. Kennel et al., ICE plasma wave measurements in the ion pick-up region of Comet Halley, Geophys. Res. Lett., 13, 857, 1986b.

Scarf, F. L., F. V. Coroniti, C. F. Kennel et al., Observations of cometary plasma wave phenomena, ESA SP 250, I, 163, 1986c.

Smith, E. J., B. T. Tsurutani, J. A. Slavin et al., International Cometary Explorer encounter with Giacobini-Zinner: magnetic field observations, Science, 232, 382, 1986a.

Smith, E. J., J. A. Slavin, S. J. Bame et al., Analysis of the Giacobini-Zinner bow wave, ESA SP 250, III, 461, 1986b.

Slavin, J. A., E. J. Smith et al., The P/Giacobini Zinner magnetotail, ESA SP 250, I, 81, 1986.

Trotignon, J. G., C. Beghin, R. Grard et al., Dust observations of Comet Halley by the Vega plasma-wave analyzer, ESA SP 250, III, 409, 1986.

Tsurutani, B. T., and E. J. Smith, Strong hydromagnetic turbulence associated with Comet Giacobini-Zinner, Geophys. Res. Lett., 13, 259, 1986.

Tsurutani, B. T., A. L. Brinca, E. J. Smith et al., MHD waves detected by ICE at distances $\geq 28 \times 10^6$ km from Comet Halley: cometary or solar wind origin? ESA SP 250, III, 451, 1986.

Wallis, M. K., Shock-free deceleration of the solar wind? Nature, 233, 1971.

Wenzel, K.-P., T. R. Sanderson, I. G. Richardson et al., In-situ observations of cometary pick-up ions $\geq$ 0.2 AU upstream of Comet Halley: ICE observations, Geophys. Res. Lett., 13, 861, 1986.

PARTICLE SCATTERING AND ACCELERATION IN A TURBULENT PLASMA AROUND COMETS

Toshio Terasawa

Geophysical Insitute, Kyoto University, Kyoto 606, Japan
and
Institut fuer extraterrestrische Physik Max-Planck-Institut fuer Physik und Astrophysik, 8046 Garching, FRG

Abstract. Efficient particle acceleration is believed to occur in the region around comets. We first review several possibilities for the acceleration process. For a detailed study, we choose the second-order Fermi acceleration. We shortly summarize the results of quasi-linear theory on this acceleration. The theoretical results are then compared with those of test particle simulations, in which particle (ion) motions are traced in hydromagnetic turbulence of given amplitude and spectra. If the energy density of the turbulent magnetic field is less than ≈10% of that of the background magnetic field, the quasi-linear results fit the simulation results well. It is found that nonlinear effects tend to *enhance* the acceleration rate when the amplitude of turbulence becomes stronger.

1. Introduction

The interaction of the solar wind plasma with cometary gases is important not only in the context of cometary physics but also in the context of astrophysical plasma physics. We may get deep insight into the processes of plasma-neutral gas interaction, excitation and development of hydromagnetic turbulence, and particle acceleration processes. After ionization, the cometary ions begin to rotate around the solar wind magnetic field and form a torus in the velocity space. These ions are then scattered to form a velocity-space shell. This "pickup shell" is clearly identified in the region around the comet Halley [Mukai et al., 1986; Terasawa et al., 1986b; Neugebauer et al., 1986]. It is further found that a substantial number of cometary ions have energies in excess of the expected maximum pickup energy in the spacecraft frame, $E_{max} = 2mV_{sw}^2$ (m and V_{sw} are the mass of ions and the solar wind velocity, respectively) [Gloeckler et al., 1986; Hynds et al., 1986; McKenna-Lawlor et al., 1986; Somogyi et al., 1986; Richardson et al., 1986, 1987]. In the following we shall summarize previous discussion on various possibilities for the acceleration processes around comets [e.g., see Ip and Axford, 1986; Tsurutani et al. 1987].

Second-order Fermi acceleration

A theoretical prediction [Wu and Davidson, 1972; Wu and Hartle, 1974], that pickup ions can excite strong turbulence, has been confirmed by the in situ observations of hydromagnetic turbulence [Tsurutani et al., 1986a,b; Riedler et al., 1986; Saito et al., 1986; Yumoto et al., 1986; Neubauer et al., 1986; Glassmeier et al., 1986; for an updated review, see Lee, 1988]. If the strong turbulence consists of waves propagating in opposite directions, a second-order Fermi process could provide the necessary acceleration [Ip and Axford, 1986, 1987; Gribov et al., 1986; Isenberg, 1987b]. We shall present a quantitative evaluation of the second-order Fermi process in the following sections.

Diffusive shock acceleration

It has been suggested that this process, namely a first order Fermi acceleration process, may work in the foreshock region of the cometary bow shock [Amata and Formisano, 1985]. However, since the cometary bow shocks are weak [Schmidt and Wegmann, 1982; Galeev et al., 1985; Smith et al., 1986; Bame et al., 1986; Johnstone et al., 1986; Neugebauer et al., 1987], this process will not be quite effective in accelerating particles [Ip and Axford, 1986]. In the plasma data from the Halley probes, *Suisei* and *Giotto*, there is some evidence concerning the inefficiency of the diffusive shock acceleration mechanism: If the diffusive shock acceleration is effective, a substantial backstreaming ion fluxes should have been observed coming from the direction of the comet. When *Suisei* was in the immediate upstream region of the Halley's bow shock, on the contrary, the backstreaming ion flux was in the background level [Mukai, private communication, 1986]. Similarly, no clearly identifiable backstreaming protons were observed when *Giotto* was in the magnetically-connected upstream region of comet Halley [Fuselier et al., 1987]. These observations, however, are preliminary. Richardson et al. [1987] and Ip [1987] have suggested that while the second-order Fermi acceleration process is most important in the far upstream region, the diffusive shock acceleration should be working near the bow shock. Further observational and theoretical studies are needed to assess the relative importance of the first- and second-order Fermi acceleration processes.

Acceleration by compressional waves

In the second-order Fermi acceleration process, ions are accelerated by the counterstreaming transverse wave components. If significant compressional (magnetosonic) wave components exist in the turbulence, particles can be accelerated by the stochastic mirror forces exerted by these waves. Tsurutani and Smith [1986b] and Tsurutani et al. [1987] have observed steepened magnetosonic waves in the upstream region of comet Giacobini-Zinner, and have suggested the possiblity of ion acceleration by these waves. However, when the energy density of transverse waves is comparable to that of magnetosonic waves, which seems to be the case, acceleration by transverse waves dominates [Acterberg, 1981; Forman et al., 1986]. The stochastic acceleration effect of magnetosonic waves becomes important, either if the transverse waves are much weaker than magnetosonic waves, of if transverse waves are propagating unidirectionally.

Another possible acceleration process by these magnetosonic waves suggested by Tsurutani et al. [1987] is the shock-drift mechanism working at the steepened edges of these waves. (For the shock-drift acceleration mechanism, see Armstrong et al. [1985].) This mechanism could become quite effective when the wave propagation angle θ_{kB}, namely the angle between the wave vector $\vec{k}$ and the background magnetic field $\vec{B}$, is close to 90°. However, the observed propagation angle, which is less than 30°

[Tsurutani and Smith, 1986b; Tsurutani et al., 1987], does not satisfy this condition.

Acceleration by electrostatic turbulence.

Buti and Lakhina [1987] have discussed heuristically the possibility that lower hybrid wave turbulence excited in the ion pickup region can accelerate cometary ions up to several hundred keV through the nonlinear interaction process. Although this is an interesting possibility, further works are necessary to assess it: Buti and Lakhina have taken the estimation of the maximum energy by Karney and Bers [1977] who solved the motion of ions in a monochromatic wave. It remains to be seen how ions behave in realistic multi-wave turbulence.

Adiabatic heating

As a result of mass loading effect, the solar wind deceleration starts far upstream of the cometary bow shock. Along with the flow deceleration, the picked-up cometary ions are adiabatically compressed and accelerated further. However, this will not be a large effect except very close to the comet [Isenberg, 1987b].

2. Elementary Wave-Particle Interaction Process

We consider hydromagnetic waves propagating parallel or anti-parallel to the average magnetic field $\vec{B}_0$ (in the +X direction). If the wave is monochromatic and circularly polarized, its Y and Z components are written as,

$$B_y = B_\perp \cos(kX-\omega t) \qquad (1)$$
$$B_z = B_\perp \sin(kX-\omega t)$$

where the wavenumber k and the frequency ω satisfy the hydromagnetic dispersion relation,

$$\omega = \pm k V_A \qquad (2)$$

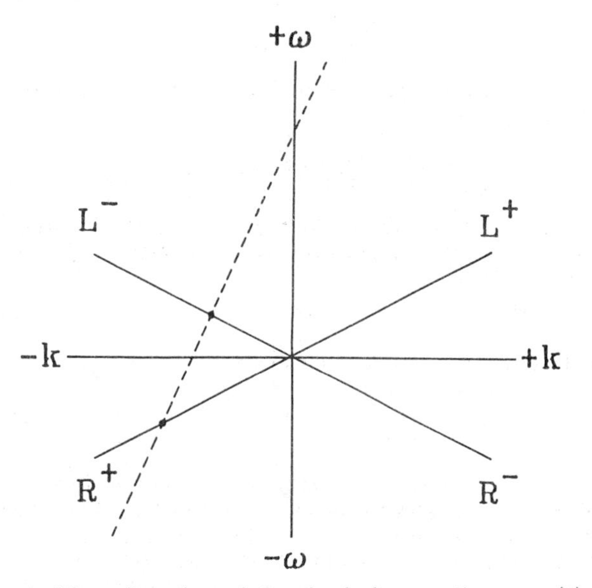

Fig. 1. Wave dispersion relation for hydromagnetic waves (2). Cyclotron resonance condition (3) for ions moving toward the +X direction is shown by a dashed line.

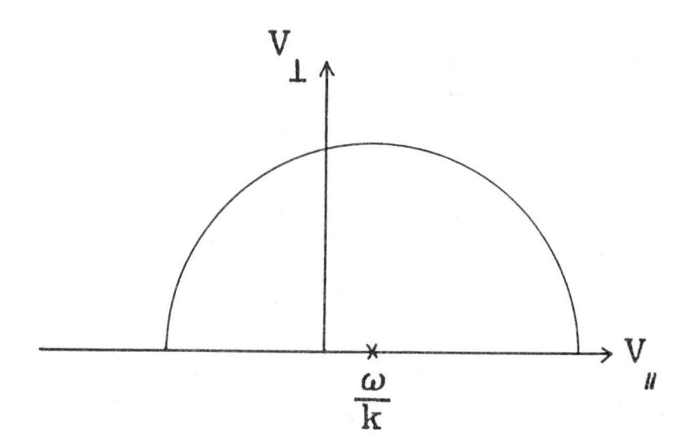

Fig. 2. Energy conservation in the wave rest frame when the waves propagate unidirectionally.

where V_A is the Alfven velocity. In the representation (1) the first quadrant (ω>0 and k>0, Figure 1) of the (ω,k) plane corresponds to left-hand polarized waves propagating in the +X direction (L^+ waves, for abbreviation). The fourth quadrant (ω<0 and k>0), on the other hand, corresponds to right-hand polarized waves propagating in the −X direction (R^- waves). Others, the second and third quadrants, correspond to L^- and R^+ waves, respectively. Waves with an arbitrary spectrum and polarization can be represented by superposition of wave components, each of which is represented by (1).

Strong wave-particle interaction occurs where these waves satisfy the cyclotron resonance condition,

$$\omega = k_{||} V_{||} + \Omega_i \qquad (3)$$

where $V_{||}$ is the velocity component parallel to $\vec{B}_0$ and Ω_i the Larmor frequency of ions. From Fig. 1, it can be seen that the ions moving in the +X direction ($V_{||}$>0) resonantly interact with L^- and R^+ waves. On the other hand, the ions with $V_{||}$<0 resonantly interact with L^+ and R^- waves. (For a recent summary on these interaction modes, see Thorne and Tsurutani [1987].)

For the hydromagnetic waves propagating unidirectionally along $\vec{B}_0$, the wave electric field vanishes in the rest frame of the waves, so that the particle energies are conserved in this frame. In other words, the particle motion is constrained on a circle (Figure 2),

$$\left(V_{||} - \frac{\omega}{k} \right)^2 + V_\perp^2 = const. \qquad (4)$$

where $V_\perp$ is the velocity component perpendicular to $\vec{B}_0$.

An important corollary of the energy conservation (4) is that the energy diffusion of particles does not occur if the waves are propagating unidirectionally (with the constant phase speed V_A). Only pitch-angle diffusion occurs with such waves. For the energy diffusion to occur, existence of counterstreaming waves is essential. Figure 3a and 3b schematically shows how the energy diffusion of particles occurs: If both L^- and R^+ waves exist, ions sometimes move on a circle centered on $(-V_A, 0)$ and sometimes on a circle centered on $(V_A, 0)$. Some "lucky" ions can be accelerated efficiently by the combined effects of L^- and R^+ waves. Of course, there exist "unlucky" ions which are decelerated by the same effect but working in the opposite direction. Since the fate of ions, being lucky or unlucky, is a matter of probability, the behavior of ions in these counterstreaming waves becomes diffusive.

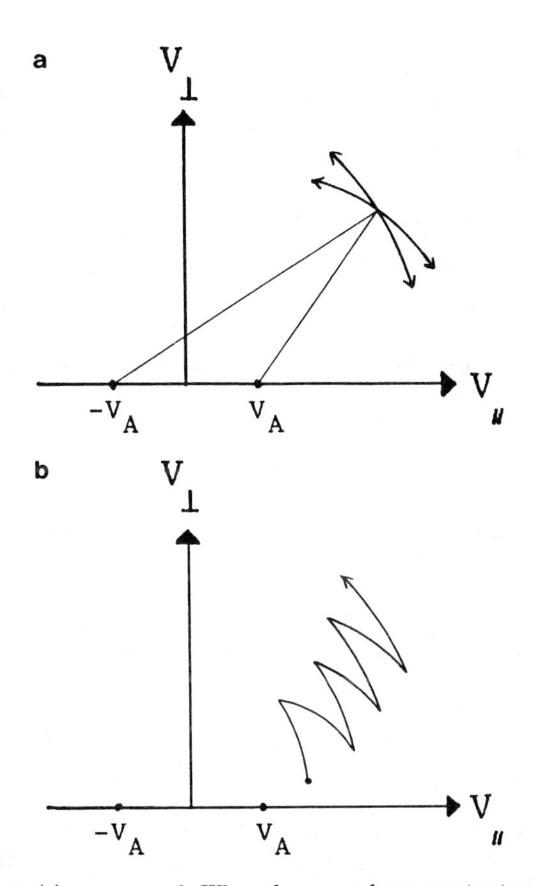

Fig. 3. (a) upper panel: When the waves have counterstreaming components, the particle energy is no longer conserved. (b) lower panel: An example of the particle diffusion in the phase space.

3. Quasi-linear Description of the Second Order Fermi Acceleration Process

When the amplitudes of the waves are not too large, their contribution to the ion motion can be regarded as perturbations to the unperturbed orbits. If we integrate the effect of these perturbations along the unperturbed orbits, we obtain the quasi-linear description of the wave-particle interaction. Let $f \equiv f(V, \mu; t)$ be the distribution function of ions, where V is the magnitude of the velocity $([V_{\parallel}^2 + V_{\perp}^2]^{1/2})$ and μ is the cosine of the pitch angle $(\mu \equiv V_{\parallel}/V)$. The quasi-linear equation for f is [e.g. Isenberg, 1987a],

$$\frac{\partial f}{\partial t} = \frac{\pi q^2}{2m_i c^2} \sum_j \int dk \frac{1}{V_\perp} \hat{G} \left[V_\perp V_A^2 P^j(k) \delta(\omega - kVz - \Omega_i) \hat{G} f \right] \tag{5}$$

where q and m_i are the charge and mass of ions. $P^j(k)$ is the power spectral density of the wave magnetic field with the suffix j running through L^+, L^-, R^+, and R^-. $P^j(k)$ is so normalized that $\int P^j(k)\,dk$ gives the squared amplitude of turbulence $[\delta B^j]^2$. $\hat{G}$ is an operator,

$$\hat{G} = \frac{V_\perp}{V} \left[\frac{\partial}{\partial V} + \left(\frac{k}{\omega} - \frac{\mu}{V}\right) \frac{\partial}{\partial \mu} \right] \tag{6}$$

For ions moving in the +X direction $(\mu>0)$, the resonant interaction occurs with L^- and R^+ waves. In this case, (5) becomes,

$$\frac{\partial f}{\partial t} = \frac{\pi q^2}{2m_i c^2} \frac{V_A^2}{\Omega_i} \frac{1}{V_\perp} \times$$
$$\left\{ \left[\hat{G}^- V_\perp |k| P^{L^-} \hat{G}^- f \right]_{k=k_{L^-}} + \left[\hat{G}^+ V_\perp |k| P^{R^+} \hat{G}^+ f \right]_{k=k_{R^+}} \right\} \tag{7}$$

where

$$\hat{G}^\pm = \frac{V_\perp}{V} \left[\frac{\partial}{\partial V} + \left(\pm\frac{1}{V_A} - \frac{\mu}{V}\right) \frac{\partial}{\partial \mu} \right] \tag{8}$$

and k_{L^-} and k_{R^+} are the resonant wavenumbers defined as,

$$k_{L^-} = -\frac{\Omega_i}{\mu V + V_A} \quad \text{and} \quad k_{R^+} = -\frac{\Omega_i}{\mu V - V_A} \tag{9}$$

Now we introduce the cosine of pitch angles in the wave frames,

$$\mu_- \equiv \frac{\mu V + V_A}{V} \quad \text{and} \quad \mu_+ \equiv \frac{\mu V - V_A}{V} \tag{10}$$

With μ_- and μ_+, we can rewrite (7) as [Skilling, 1975],

$$\frac{\partial f}{\partial t} = \frac{\partial}{\partial \mu_-} \left[\frac{1}{2}(1-\mu_-^2) D_{\mu\mu}^- \frac{\partial f}{\partial \mu_-} \right] + \frac{\partial}{\partial \mu_+} \left[\frac{1}{2}(1-\mu_+^2) D_{\mu\mu}^+ \frac{\partial f}{\partial \mu_+} \right] \tag{11}$$

where the first and second terms represent the effect of pitch angle scattering by the L^- and R^+ waves, respectively. $D_{\mu\mu}^-$ and $D_{\mu\mu}^+$ are the pitch angle diffusion coefficients defined as,

$$D_{\mu\mu}^- \equiv \pi \Omega_i \frac{|k| P^{L^-}(k)}{B_0^2} \bigg|_{k=k_{L^-}}$$
$$D_{\mu\mu}^+ \equiv \pi \Omega_i \frac{|k| P^{R^+}(k)}{B_0^2} \bigg|_{k=k_{R^+}} \tag{12}$$

Equation (7) or (11) has a complex structure and does not have a simple analytic solution. For the case of fast ions, $V \gg V_A$, the energy diffusion is much slower than the pitch-angle diffusion. In this case, we can consider that the ion distribution is almost isotropic, so that the following ordering holds,

$$\left| \frac{1}{V} \frac{\partial f}{\partial \mu} \right| \ll \left| \frac{\partial f}{\partial V} \right| \tag{13}$$

After some manipulation, (7) becomes [Skilling, 1975; Isenberg, 1987a,b],

$$\frac{\partial f}{\partial t} = \frac{1}{V^2} \frac{\partial}{\partial V} \left[V^2 D_{vv} \frac{\partial f}{\partial V} \right] \tag{14}$$

which represents the diffusion in the energy space (The energy E, of course, is $m_i V^2/2$). If the power spectral amplitudes $P^j(k)$ have power law shape,

$$A_j |k|^{-\gamma} \tag{15}$$

the diffusion coefficient D_{vv} is given by,

$$D_{vv} = \xi \left[\frac{q}{m_i c} \right]^{2-\gamma} B_0^{-\gamma} V^{\gamma-1} \frac{2A_+ A_-}{A_+ + A_-} \frac{V_A^2}{\gamma(\gamma+2)} \tag{16}$$

where ξ is a factor of the order of unity. The factor $2A_+A_-/(A_+ + A_-)$ represents the harmonic average of the amplitudes of waves responsible for the scattering. (A_- and A_+ should be replaced by A_{L^-} and A_{R^+} in the phase space region with $\mu>0$, while they should be replaced by A_{R^-} and A_{L^+} in the region with $\mu<0$.) Note that D_{vv} vanishes when the waves are propagating unidirectionally (either A_+ or A_- is zero).

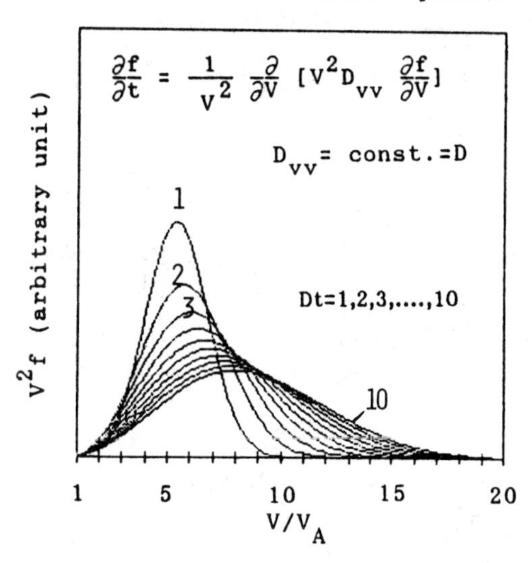

Fig. 4. Solution of the energy diffusion equation (14) (see text).

Figure 4 shows an example: solutions (Green functions) of the diffusion equation (14) at the normalized times, Dt=1, 2, 3, ..., 9, and 10 are shown. Here the diffusion coefficient $D_{vv} \equiv D$ is taken to be independent of velocity. The analytical solution for this choice of D_{vv} has been obtained by Ip and Axford [1986]. At t=0, ions are injected at the velocity V=5 V_A and then subjected to the diffusion process. Owing to the energy diffusion effect, the average velocity and the width of the distribution increase as time elapses. The injection velocity 5 V_A assumed here is close to the value during the ICE encounter with comet Giacobini-Zinner [Tsurutani, private communication, 1988]. The basic physical mechanism, however, does not depend on the choice of the injection velocity.

4. Test Particle Simulation

In the previous section, we calculated the energy diffusion coefficient from the power spectra of the turbulence. For the quasi-linear treatment to hold, the amplitude of the turbulence should be weak enough. For the case of the turbulence around comets, however, the amplitude of the turbulence δB becomes comparable to the averaged field magnitude B_0 [Tsurutani and Smith, 1986a; Glassmeier et al., 1986], and the applicability of quasi-linear theory is not guaranteed a priori. In this section, we shall see how the quasi-linear results are modified by the nonlinear effects. For this purpose, we are conducting a test particle simulation, in which we follow the motions of ions in a given field. Price and Wu [1987] have also taken a similar approach.

We generate a turbulent field by superposing circularly polarized waves (1);

$$B_y^j = \sum_{k=k_{min}}^{k_{max}} B_k^j \cos(kX-\omega_k^j t+\alpha_k^j)$$
$$B_z^j = \sum_{k=k_{min}}^{k_{max}} B_k^j \sin(kX-\omega_k^j t+\alpha_k^j) \tag{17}$$

The suffix j represents polarization and propagation directions, L^+, L^-, R^+, and R^-. The frequency ω_k^j is $\pm kV_A$, where the sign depends on the suffix j (see Fig. 1). α_k^j is an initial phase for each component (in the range from 0 to 2π). To have a random phase relation among components, α_k^j's are determined by a random number generator. The coefficient B_k^j relates to the power spectral density $P^j(k)$ as,

$$\left[B_k^j\right]^2 = P^j(k)\Delta k \tag{18}$$

where Δk relates to the size of the simulation system L_X as,

$$\Delta k \equiv \frac{2\pi}{L_X} \tag{19}$$

The wavenumber k relates to the mode number m in the system as $|k| = m\Delta k$. (The wavelength for waves of the mode number m is L_X/m.) The amplitude of the turbulence δB^j for each j is given by

$$\left[\delta B^j\right]^2 = \sum_{k=k_{min}}^{k_{max}} \left[B_k^j\right]^2 \tag{20}$$

The electric field components corresponding to the magnetic field components are given by,

$$E_y^j = \pm\frac{V_A}{c}B_z^j$$
$$-E_z^j = \pm\frac{V_A}{c}B_y^j \tag{21}$$

where $\pm$ symbol corresponds to the waves propagating in the + or − direction. After calculating the magnetic and electric field components for each mode, we sum them up,

$$B = B^{L^+} + B^{L^-} + B^{R^+} + B^{R^-}$$
$$E = E^{L^+} + E^{L^-} + E^{R^+} + E^{R^-} \tag{22}$$

In these given fields, we follow the motions of ions,

$$m_i\frac{d\vec{V}}{dt} = e\vec{E}(X,t) + \frac{e}{c}\vec{V}\times\vec{B}(X,t)$$
$$\frac{dX}{dt} = V_x \tag{23}$$

Fig. 5a shows the distribution of ions in the velocity space at $\Omega_i t$=200 (time normalized by the ion Larmor frequency). The system size L_X is 512 V_A/Ω_i, which is represented by 1024 grid points. At t=0 we inject 1000 ions, which are distributed uniformly in space and have initial velocities $V_\parallel$=5 V_A and $V_\perp$=0. In this case we set the four wave components, j=L^+, L^-, R^+, and R^-, to have the same amplitude, $[\delta B^j]^2$ = $0.02B_0^2$. We assume here that each component has a power spectrum proportional to $|k|^{-2}$ in the mode number range 5-40. The L^- (R^+) waves resonate with ions of $V_\parallel$=+5 V_A around the mode number 14 (20), which is within the above mode number range. It is evident in Fig. 5a that the energy diffusion as well as the pitch-angle diffusion takes place. Fig. 5b differs from Fig. 5a in that we give finite amplitudes only to wave components, L^- and R^+. Since there are no wave components resonating with ions in the region $V_\parallel$<0, the ions do not diffuse into that region. (Compare Fig. 5a and 5b.) In Fig. 5c (Fig. 5d), two wave components, L^+ and L^- (R^+ and R^-), have finite amplitude. According to the resonance condition ((3), see Fig. 1), one of these two components can efficiently interact with each ion but the other does not. Therefore quasi-linear theory predicts that there occurs no energy diffusion. (Since either A_+ or A_- is zero in (16), the energy diffusion coefficient vanishes.) This explains why the ion distributions in Fig. 5c and 5d have much narrower widths in energy than those in Fig. 5a and 5b. The narrow but finite energy

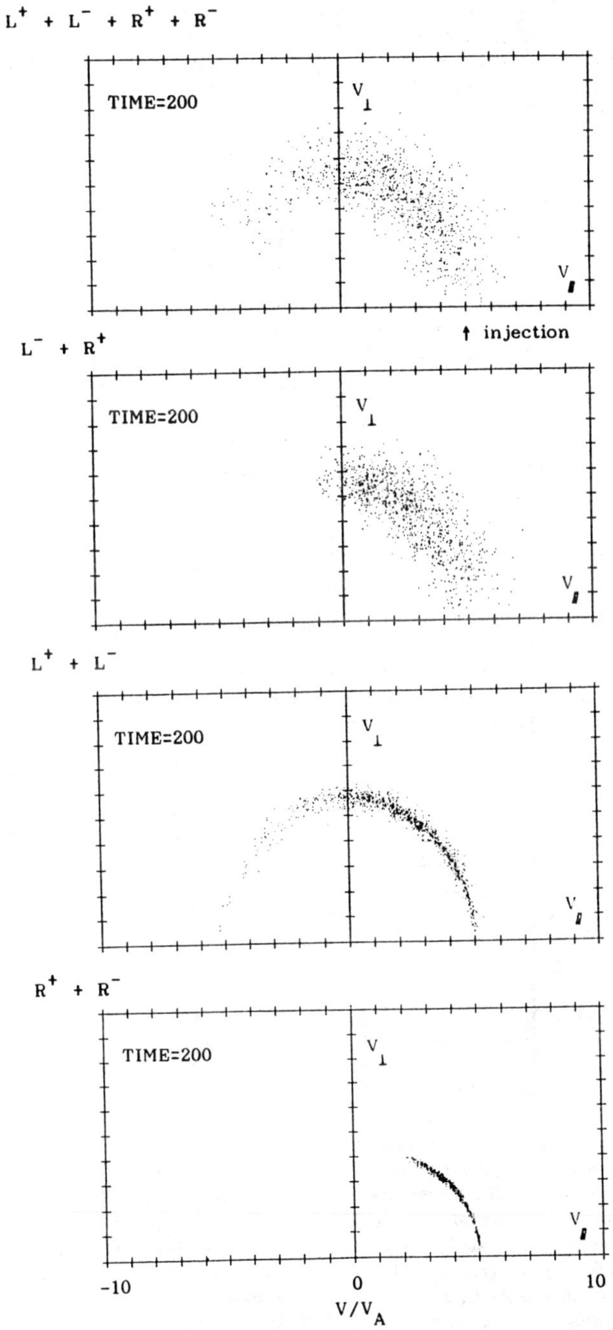

Fig. 5. Results of test particle simulations: Velocity space distributions of ions observed at $\Omega_i t$=200. (a) The top panel: The four wave components, j = L^+, L^-, R^+, and R^-, are set to have the same amplitude, $[\delta B^j]^2$=$0.02B_0^2$. These waves have power spectra proportional to $|k|^{-2}$ in the mode number range 5-40. (b) The second panel: Similar to (a) except that only wave components, L^- and R^+, have finite amplitudes ($[\delta B^j]^2$=$0.02B_0^2$). (c) The third panel: Similar to (a) except that only wave components, L^+ and L^-, have finite amplitudes ($[\delta B^j]^2$=$0.02B_0^2$). (d) The bottom panel: Similar to (a) except that only wave components, R^+ and R^-, have finite amplitudes ($[\delta B^j]^2$=$0.02B_0^2$).

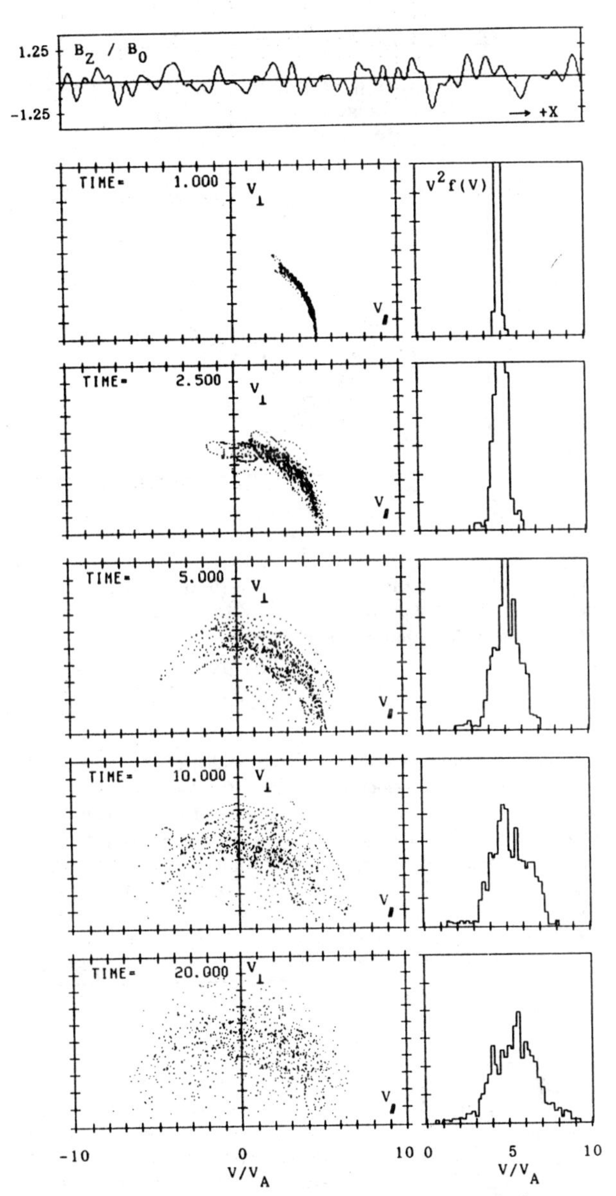

Fig. 6. Results of a test particle simulation. The simulation parameters are similar to those in Fig. 5a, except that the waves have amplitudes of $[\delta B^j]^2$=$0.08B_0^2$ and power spectra proportional to $|k|^{-1}$ in the mode number range 5-40. Top: A snap shot of the waveform. Second to bottom panels: the phase space distribution of ions (left), and the distribution function $f(V)$ multiplied by V^2 (right, in an arbitrary unit). These panels correspond to the observations at times, $\Omega_i t$= 1, 2.5, 5, 10, and 20, respectively.

width appearing in Fig. 5c and 5d is attributable to the nonlinear effect (see later discussion).

We now see the evolution of the distribution of ions in more detail. In Figure 6, the top panel shows a snapshot of the waveform (B_z) taken at $\Omega_i t$=2.5. 1000 ions are injected at t=0 with initial velocities, $V_\parallel$=$5V_A$ and $V_\perp$=0. It is evident in these panels that the pitch angle scattering occurs faster than the energy

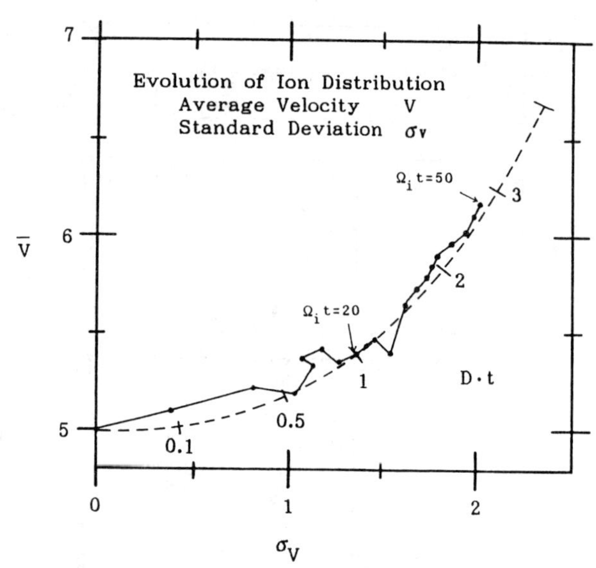

Fig. 7. Evolution of the ion distribution (from the same simulation run as Fig. 6): The average velocity $\bar{V}$ is plotted against the velocity standard deviation σ_v (a solid curve). The quasi-linear result is shown by a dashed curve. $\bar{V}$ and σ_v are measured in the unit of V_A.

diffusion. As time elapses, the energy width of the distribution as well as the average velocity increases. In Figure 7, results of the simulation shown in Fig. 6 is compared with the result of quasi-linear theory. The average velocity $\bar{V}$ is plotted against the velocity standard deviation, σ_v, where $\bar{V}$ and σ_v are observed at $\Omega_i t=$ 2.5, 5, 7.5, 10, ... 47.5, 50 (a solid curve). A dashed curve shows

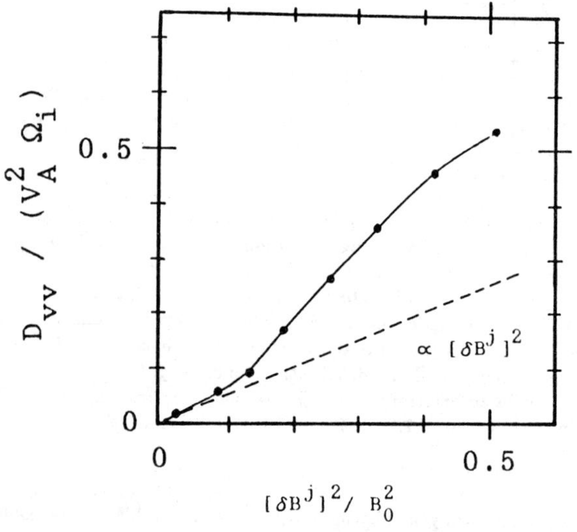

Fig. 8. The observed energy diffusion coefficients are plotted against the wave amplitudes $[\delta B^j]^2$. (The waves have power spectra proportional to $|k|^{-1}$ within the mode number range 5-40.) A dashed line shows the case where the diffusion coefficient is proportional to the wave amplitude squared (the result of quasi-linear theory).

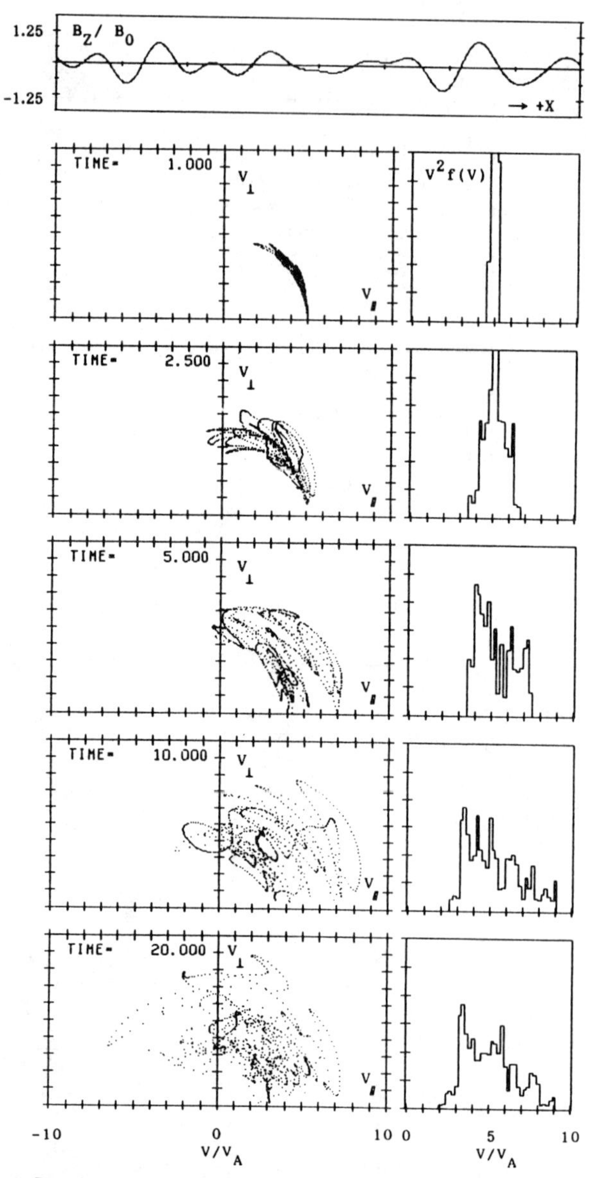

Fig. 9. Results of a test particle simulation. The simulation parameters are similar to those in Fig. 6, except that the wave power spectrum is limited to the mode number range 5-10. Top: A snap shot of the waveform. Second to bottom panels: the phase space distribution of ions (left), and the distribution function $f(V)$ multiplied by V^2 (right). These panels correspond to the observations at times, $\Omega_i t=$ 1, 2.5, 5, 10, and 20, respectively.

the evolutions of the average velocity and the velocity standard deviation calculated from the diffusion equation ((14), see Fig. 4). Time marks on the dashed curve correspond to the normalized times, $Dt=$ 0.1, 0.5, 1, 2, 3, and 4. Since the time characteristics of solid and dashed curves are similar, we conclude that the simulation result fits the result of quasi-linear theory. Note that we can measure the diffusion coefficient in the simulation system from this comparison.

The observed diffusion coefficients in the simulation runs are plotted against the wave amplitudes $[\delta B^j]^2$ in Figure 8. For the weak amplitude waves ($[\delta B^j]^2 < 0.1 B_0^2$), the observed diffusion coefficient is proportional to $[\delta B^j]^2$. This is what is expected if quasi-linear theory holds (see eq.(16)). On the other hand, for the stronger waves, we see enhancement of the energy diffusion coefficient over what quasi-linear theory predicts. We attribute this enhancement of the acceleration to the nonlinear orbit modification by nonresonant waves. To see this, we made another simulation run, where we set the wave amplitudes to be finite only in the nonresonant regime (the mode number 5-10). While the amplitude of each component ($[\delta B^j]^2$) is set equal to that in Fig. 6 ($0.08 B_0^2$), the power amplitude is concentrated in the long wavelength regime and increased there by a factor of ≈ 2.5. The results are shown in Figure 9. For ions to resonate with the given waves, their $V_{\parallel}$ should be larger than $\approx 7 V_A$. Up to the time $\Omega_i t = 5$, there appear no ions which satisfy this resonance condition. However, as seen in the figure, particle acceleration takes place along with the modification of particle positions in the phase space. This is what we call the nonlinear orbit modification.

As stated previously, as long as quasi-linear theory holds, no efficient energy change occurs in the turbulence consisting of L^- and L^+ modes or of R^- and R^+ modes. Nonlinearity also modifies this conclusion. Consider first the interaction of an ion with large amplitude waves consisting of L^- and L^+ modes. If the ion moves toward the +X direction, it shall be resonantly scattered by the L^- waves. If the wave has a large enough amplitude, this ion can be kicked back toward the −X direction after a single encounter. (Although the L^+ wave can modify the motion of this ion nonresonantly, the resonant scattering is much more effective than the scattering by the nonresonant waves even in the nonlinear regime.)

The ion kicked into the −X direction by the L^- wave is now to be resonantly scattered by the L^+ waves. In these processes, "collisions" between the ion and the waves occur in a manner of the "head-on" collision. In these head-on collisions the ion always gets additional energies from the waves. Therefore, we expect a first-order Fermi acceleration process to occur in this situation. Figure 10a shows a trajectory of an ion in the turbulence consisting of L^- and L^+ modes of amplitude $[\delta B^j]^2 = 0.18 B_0^2$. After the injection at $V_{\parallel} = 2 V_A$, this ion is accelerated preferentially. Figure 10b shows the result of a simulation with 1000 ions, whose initial velocity is $V_{\parallel} = 5 V_A$ and $V_{\perp} = 0$. Note that the ion distribution function $V^2 f(V)$ has skewness toward the higher velocity (toward right). The first order Fermi acceleration effect explains this skewness.

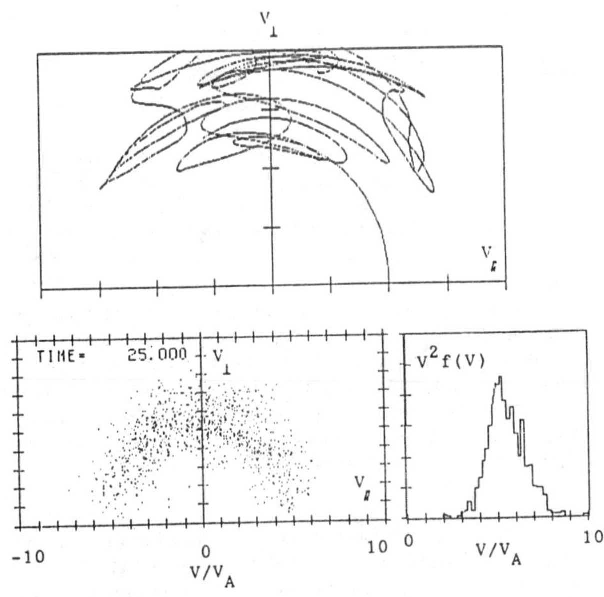

Fig. 10. Results of a test particle simulation for the turbulence consisting of L^+ and L^- modes ($[\delta B^j]^2 = 0.18 B_0^2$). These waves have power spectra proportional to $|k|^{-1}$ within the mode number range 5-40. (a) upper panel: A trajectory of an ion injected at $V_{\parallel} = 2 V_A$ and $V_{\perp} = 0$. (b) lower panel: the phase space distribution of 1000 ions at $\Omega_i t = 25$ (left) and the distribution function $f(V)$ multiplied by V^2 (right).

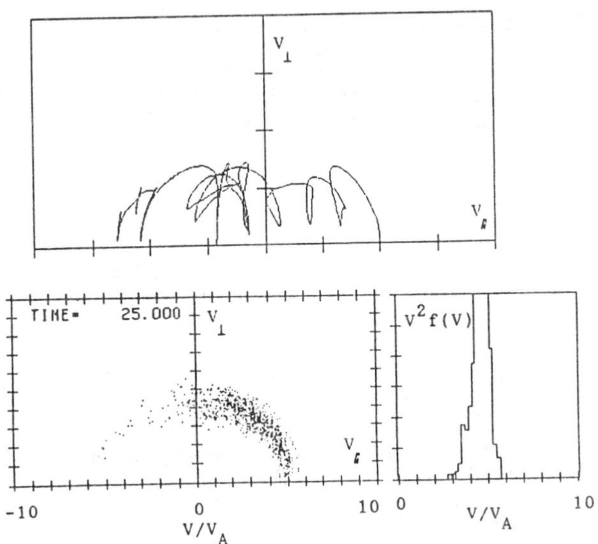

Fig. 11. Similar to Fig. 10 except that it is for a simulation for the turbulence consisting of R^+ and R^- modes ($[\delta B^j]^2 = 0.18 B_0^2$).

If, on the other hand, the turbulence consists of R^+ and R^- modes, the resonant ion-wave interactions occur in a manner of the "tail-on" collision, which results in the energy loss of ions (the first-order Fermi deceleration). Figure 11a shows a trajectory of an ion in the turbulence consisting of R^+ and R^- modes. This ion is preferentially decelerated. Figure 11b shows the results of a simulation with 1000 ions, whose initial velocity is $V_{\parallel} = 5 V_A$ and $V_{\perp} = 0$. As seen in this figure, the ion distribution function has skewness toward the lower velocity, which contrasts the previous case shown in Figure 10b.

5. Summary and Discussion

We have made a quantitative comparison between the results of the quasi-linear treatment and the test particle simulations. We have seen in the simulation results that the pitch angle scattering occurs faster than the energy diffusion. This observation is consistent with what quasi-linear theory predicts and with the recent results by Price and Wu [1987]. It is found further that for the turbulence with a relatively weak amplitude ($[\delta B^j]^2 < 0.1 B_0^2$) the energy diffusion coefficient predicted by quasi-linear theory fit well the simulation results. For the turbulence with a larger amplitude, we have found that the nonlinear effect enhances the energy diffusion coefficient.

We have seen that the quasi-linear energy diffusion does not occur when the turbulence consists of L^+ and L^- waves, or of R^+

and R^- waves. However, if the amplitude of the turbulence becomes sufficiently large, the first-order acceleration (deceleration) occurs nonlinearly in the turbulence consisting of L^+ and L^- waves (R^+ and R^- waves).

Richardson et al. [1987] compared the energy spectra of ions observed in the far upstream region of the comet Giacobini-Zinner with the theoretical model of the second-order Fermi acceleration process by Ip and Axford [1986]. They found that the observed spectra is much softer than the estimation (the acceleration effect is weaker than that assumed in the model). We note that the parameters used by Ip and Axford (f=5 and 1) correspond to the energy diffusion coefficient $D_{vv}=0.15\,V_A^2\Omega_i$ and $0.75\,V_A^2\Omega_i$. The results by Richardson et al. indicate that the actual energy diffusion coefficient D_{vv} is much smaller than $0.15\,V_A^2\Omega_i$. Recently, Ip and Axford [1987] have extended their calculation and concluded that a relatively weak scattering limit characterized by f $\approx$30 (or $0.025\,V_A^2\Omega_i$) may be appropriate in the upstream regions of comets Giacobini-Zinner and Halley. Since the nonlinear effect is unimportant for $D_{vv}<0.1\,V_A^2\Omega_i$ (Fig. 8), we may conclude that quasi-linear theory should be applicable to the process working in the far upstream region of comets Giacobini-Zinner and Halley. Within the cometosheath regions of these comets, on the other hand, stronger turbulence ($\delta B \approx B_0$) were observed. The nonlinear modification of the acceleration efficiency may be important in these regions.

Isenberg [1987b] presented a model of the second-order Fermi acceleration process around comets. He treats the case where particle acceleration occurs between two counterstreaming waves, one of which is the cometary hydromagnetic waves and the other the turbulence intrinsic to the solar wind. The latter turbulence was introduced, since the cometary waves excited by the ion-beam-cyclotron instability propagate predominantly in the upstream direction from the comet. In section 3 we see that the amplitude of the weaker waves determines the overall acceleration efficiency. Because the assumed level of intrinsic turbulence of the solar wind is low, the energy spectra obtained by Isenberg are much softer than what is actually observed. This difficulty may be overcome if we consider the nonlinear dynamics of the cometary waves: When hydromagnetic waves have a sufficiently large amplitude, wave components propagating in the opposite direction from the original waves can grow through nonlinear wave-wave decay processes [see Terasawa et al., 1986a and references therein]. If this is the case, a condition needed for the efficient second-order Fermi acceleration could be realized. However, as we see in the previous sections, appropriate wave modes should exist in the turbulence for the acceleration process to work. A detailed wave mode analysis for the data obtained near the comets is needed to identify the acceleration process responsible for the production of observed high energy ions.

In this paper discussion has been limited to the test particle approach. However, since the energy density of the cometary ions becomes comparable to, or exceeds the solar wind magnetic field energy in the region around and behind the cometary bow shock, self-consistent treatment for a combined system of cometary ions and turbulence should be made. Gary et al. [1987] have presented interesting results on the acceleration process from a self-consistent numerical simulation. They have found that cometary protons are accelerated through the nonresonant interaction with the waves excited by cometary water-group ions. They have interpreted this acceleration in terms of the second-order Fermi acceleration process. However, it remains to be checked whether the condition for the second-order Fermi acceleration process, the existence of counterstreaming waves, is satisfied.

Acknowledgment. The author thanks Drs. T. Abe, A. Nishida, and M. Scholer for valuable discussion. This work was partially supported by Grant-in-Aid 62302011 from the Ministry of Education, Japan.

References

Acterberg, A., On the nature of small amplitude Fermi acceleration, *Astron. Astrophys*, *97*, 259-264, 1981.

Amata, E., and V. Formisano, Energization of positive ions in the cometary foreshock region, *Planet. Space Sci.*, *33*, 1243-1250, 1985.

Armstrong, T. P., M. E. Pesses, R. B. Decker, Shock drift acceleration, in *Collisionless Shocks in the Heliosphere: Reviews of Current Research*, ed. B. T. Tsurutani and R. G. Stone, *Geophysical Monograph 35*, American Geophysical Union, Washington, D. C., pp. 271-285, 1985.

Bame, S. J., and 10 coauthors, Comet Giacobini-Zinner: Plasma description, *Science*, *232*, 356-361, 1986.

Buti, B., and G. S. Lakhina, Stochastic acceleration of cometary ions by lower hybrid waves, *Geophys. Res. Lett.*, *14*, 107-110, 1987.

Forman, M. A., R. Ramaty, and E. G. Zweibel, The acceleration and propagation of solar flare energetic particles, in *Physics of the Sun* ed. P. A. Sturrock, *2*, 249-289, 1986.

Fuselier, S. A., K. A. Anderson, H. Balsiger, K. H. Glassmeier, B. E. Goldstein, M. Neugebauer, H. Rosenbauer, and E. G. Shelley, The foreshock region upstream from the comet Halley bow shock, *Symposium on the Diversity and Similarity of Comets*, ESA *SP-278*, 77-82, 1987.

Galeev, A. A., T. E. Cravens, and T. I. Gombosi, Solar wind stagnation near comets, *Astrophys. J.*, *289*, 807-819, 1985.

Gary, S. P., C. D. Madland, N. Omidi, and D. Winske, Computer simulations of two-ion pickup instabilities in a cometary environment, submitted to *J. Geophys. Res.*, 1987.

Glassmeier, K. H., F. M. Neubauer, M. H. Acuña, and F. Mariani, Strong hydromagnetic fluctuations in the comet P/Halley magnetoshpere observed by the Giotto magnetic field experiment, *Proc. 20th ESLAB Symposium on the Exploration of Halley's Comet*, ESA *SP-250*, *3*, 167-171, 1986.

Gloeckler, G. D., D. Hovestadt, F. M. Ipavich, M. Scholer, B. Klecker, and A. B. Galvin, Cometary pick-up ions observed near Giacobini-Zinner, *Geophys. Res. Lett.*, *13*, 251-254, 1986.

Gribov, B. E., and 16 coauthors, Stochastic Fermi acceleration of ions in the pre-shock region of comet Halley, *Proc. 20th ESLAB symposium on the exploration of Halley's comet*, ESA *SP-250*, *1*, 271-275, 1986.

Hynds, R. J., S. W. H. Cowley, T. R. Sanderson, K.-P. Wenzel, and J. J. Van Rooijen, Observations of energetic ions from comet Giacobini-Zinner, *Science*, *232*, 361-365, 1986.

Ip, W.-H., Cometary ion acceleration processes, *Computer Phys. Commun.*, in press, 1987.

Ip, W.-H., and W. I. Axford, Theories of physical processes in the cometary comae and ion tails, in *Comets*, ed. L. L. Wilkening, The Univ. Arizona Press, Tuscon, pp. 588-634, 1982.

Ip, W.-H., and W. I. Axford, The acceleration of particles in the vicinity of comets, *Planet. Space Sci.*, *34*, 1061-1065, 1986.

Ip, W.-H., and W. I. Axford, A numerical simulation of charged particle acceleration and pitch angle scattering in the turbulent plasma environment of cometary comas, *Proc. 20th Int. Conf. Cosmic Rays*, SH4.2-14, 1987.

Isenberg, P. A., The evolution of interstellar pickup ions in the solar wind, *J. Geophys. Res.*, *92*, 1067-1073, 1987a.

Isenberg, P. A., Energy of pickup ions upstream of comets, *J. Geophys. Res.*, *92*, 8795-8799, 1987b.

Johnstone, A., and 27 coauthors, Ion flow at comet Halley, *Nature*, *321*, 344-347, 1986.

Karney, C. F., and A. Bers, Stochastic ion heating by a perpendicularly propagating electrostatic wave, *Phys. Rev. Lett.*, *39*, 550-554, 1977.

Lee, M. A., ULF waves at comets, this volume, 1988.

McKenna-Lawlor, S., E. Kirsch, D. O'Sullivan, A. Thompson, and K.-P. Wenzel, Energetic ions in the environment of comet Halley, *Nature*, *321*, 347-349, 1986.

Mukai, T., W. Miyake, T. Terasawa, M. Kitayama, and K. Hirao, Ion dynamics and distribution around comet Halley: Suisei observation, *Geophys. Res. Lett.*, *13*, 829-832, 1986.

Neubauer, F. M., and 11 coauthors, First results from the Giotto magnetometer experiment at comet Halley, *Nature*, *321*, 352-355, 1986.

Neugebauer, M., and 10 coauthors, The pick-up of cometary protons by the solar wind, *Proc. 20th ESLAB symposium on the exploration of Halley's comet*, ESA *SP-250*, *1*, 19-23, 1986.

Neugebauer, M., F. M. Neubauer, H. Balsiger, S. A. Fuselier, B. E. Goldstein, R. Goldstein, F. Mariani, H. Rosenbauer, R. Schwenn, and E. G. Shelley, The variation of protons, alpha particles, and the magnetic field across the bow shock of comet Halley, *Geophys. Res. Lett.*, *14*, 995-998, 1987.

Price, C. P., and C. S. Wu, The influence of strong hydromagnetic turbulence on newborn cometary ions, *Geophys. Res. Lett.*, *14*, 856-859, 1987.

Richardson, I. G., S. W. H. Cowley, V. Moore, K. Staines, R. J. Hynds, T. R. Sanderson, K.-P. Wenzel, and P. W. Daly, Spectra and bulk parameters of energetic heavy ions in the vicinity of comet P/Giacobini-Zinner, *Proc. 20th ESLAB Symposium on the exploration of Halley's Comet*, ESA *SP-250*, *3*, 441-445, 1986.

Richardson, I. G., S. W. H. Cowley, R. J. Hynds, C. Tranquille, T. R. Sanderson, K.-P. Wenzel, and P. W. Daly, Observation of energetic water-group ions at comet Giacobini-Zinner: Implications for ion acceleration processes, *Planet. Space Sci.*, *35*, 1323-1345, 1987

Riedler, W., K. Schwingenschuh, Ye. G. Yeroshenko, V. A. Styashin and C. T. Russell, Magnetic field observations in comet Halley's coma, *Nature*, *321*, 288-290, 1986.

Saito, T., K. Yumoto, K. Hirao, T. Nakagawa, and K. Saito, Interaction between comet Halley and the interplanetary magnetic field observed by Sakigake, *Nature*, *321*, 303-307, 1986.

Schmidt, H. U., and Wegmann, R., Plasma flow and magnetic fields in comets, in *Comets*, ed. L. L. Wilkening University of Arizona Press, Tucson, pp. 538-560, 1982.

Skilling, J., Cosmic ray streaming-1, Effect of Alfven waves on particles, *Mon. Not. R. astr. Soc.*, *172*, 557-566, 1975.

Smith, E. J., B. T. Tsurutani, J. A. Slavin, D. E. Jones, G. L. Siscoe, and D. A. Mendis, International Cometary Explorer encounter with Giacobini-Zinner: Magnetic field observations, *Science*, *232*, 382-385, 1986.

Somogyi, A. J., and 33 coauthors, First observation of energetic particles near comet Halley, *Nature*, *321*, 285-288, 1986.

Terasawa, T., M. Hoshino, J.-I. Sakai, and T. Hada, Decay instability of finite-amplitude circularly polarized Alfven Waves: A numerical simulation of stimulated Brillouin Scattering, *J. Geophys. Res.*, *91*, 4171-4187, 1986a.

Terasawa, T., T. Mukai, W. Miyake, M. Kitayama, and K. Hirao, Detection of cometary pickup ions up to 10^7 km from comet Halley: Suisei observation, *Geophys. Res. Lett.*, *13*, 837-840, 1986b.

Thorne, R. M., and B. T. Tsurutani, Resonant interactions between cometary ions and low frequency electromagnetic waves, *Planet. Space Sci.*, in press, 1987.

Tsurutani, B. T., and E. J. Smith, Strong hydromagnetic turbulence associated with comet Giacobini-Zinner, *Geophys. Res. Lett.*, *13*, 259-262, 1986a.

Tsurutani, B. T., and E. J. Smith, Hydromagnetic waves and instabilities associated with cometary ion pickup: ICE observations *Geophys. Res. Lett.*, *13*, 263-266, 1986b.

Tsurutani, B. T., R. M. Thorne, E. J. Smith, J. T. Gosling, and H. Matsumoto, Steepened magnetosonic waves at comet Giacobini-Zinner, *J. Geophys. Res.* , *92*, 11,074-11,082, 1987.

Wu, C. S. and R. C. Davidson, Electromagnetic instabilities produced by neutral particle ionization in the interplanetary space, *J. Geophys. Res.* , *77*, 5399-5406, 1972.

Wu, C. S. and R. E. Hartle, Further remarks on plasma instabilities produced by ions born in the solar wind, *J. Geophys. Res.*, *79*, 283-285, 1974.

Yumoto, K., T. Saito, and T. Nakagawa, Hydromagnetic waves near O^+ (or H_2O^+) ion cyclotron frequency observed by Sakigake at the closest approach to comet Halley, *Geophys. Res. Lett.*, *13*, 825-828, 1986.

PARTICLE SIMULATIONS OF NONLINEAR WHISTLER AND ALFVÉN WAVE INSTABILITIES: AMPLITUDE MODULATION, DECAY, SOLITON AND INVERSE CASCADING

Yoshiharu Omura and Hiroshi Matsumoto

Radio Atmospheric Science Center, Kyoto University, Uji, Kyoto, 611, Japan

Abstract. We present a brief review on previous works and our recent results of particle simulations on nonlinear behavior of large amplitude circularly polarized electromagnetic cyclotron waves propagating parallel to a static magnetic field. Finite amplitude whistler and R-mode Alfvén waves propagating parallel to a static magnetic field are often excited by an electron beam in the magnetosphere and by an ion beam in the cometary environment and in earth's foreshock region, respectively. After saturation of the beam instabilities, whistler mode waves show a smooth inverse cascading process via a modulational instability, while R-mode Alfvén waves show a discrete inverse cascading process via decay instability. It is demonstrated in both cases that a backward travelling wave is excited at the early nonlinear stage followed by an energy exchange between the forward- and backward travelling waves. The energy exchange between the forward and backward travelling waves plays important roles in the smooth inverse cascading process as well as a modulational instability. Depending on parameters, modulational instabilities of finite amplitude whistler mode waves lead to formation of envelope solitons. We also show by simulation that a decay instability controls the nonlinear evolution of R-mode Alfvén waves. Consequences of the wave-wave interactions following the wave-particle interactions are discussed in the context of an energy transfer from resonant beam particles to thermal particles.

1. Introduction

In space plasmas, various types of dynamic nonlinear phenomena are taking place through wave-particle and wave-wave interactions. These interactions play important roles in energy transfer from high energy particles to low energy particles via plasma waves. Owing to increasing capabilities of recent supercomputers, dynamic and complicated processes of the nonlinear phenomena are being clarified via computer simulations. In this paper we review the past theoretical and numerical studies on nonlinear evolution of electromagnetic cyclotron waves, which are commonly observed in space plasmas such as Alfvén waves in the solar wind or VLF whistler mode waves in the magnetosphere. We also show our recent achievements in the simulations of nonlinear evolution of electromagnetic cyclotron waves.

Plasma Wave Instabilities

We may first categorize plasma wave instabilities into two groups, i.e., instabilities through wave-particle interactions and those through wave-wave interactions. In wave-particle interactions, free energy for growing waves is provided from resonant particles which, in most cases, are high energy particles. The instability condition is given by solving a linear dispersion relation. On the other hand, in wave-wave interactions waves are excited receiving energy from a large amplitude pumping wave. In the simplest and most important case, three waves are involved in the interaction, i.e., excited wave (signal), pumping wave (pump), and idling wave (idler). These waves are coupled through nonlinear terms in the dispersion relations. Therefore, nonlinear dispersion relations are essential for the wave-wave interactions. There are a number of good introductions to nonlinear plasma theory [e.g. Sagdeev and Galeev, 1969; Tsytovich, 1970; Davidson, 1972; Karpman; 1975].

Wave instabilities associated with three wave interaction are generally called parametric instabilities. General introductions to parametric instabilities are given elsewhere (e.g., see Kaw et al. [1976] and Cap [1982]). There are a large number of theoretical and numerical studies on purely electrostatic wave-wave interactions [Kruer et al., 1970; Thomson et al., 1973; Kruer and Valeo, 1973; Godfrey et al., 1973; Cohen et al., 1975; Ogino and Takeda, 1975; Ogino et al., 1977; Makino et al., 1979; Freund and Papadopoulos, 1980a; Nicholson et al., 1984]. There are also various types of parametric instabilities which involve electromagnetic waves. We can classify them into three groups based on the wave modes of the pumping waves.

The first group is parametric instabilities driven by a pump electromagnetic wave with a frequency much higher than the electron plasma frequency Π_e. The incident electromagnetic wave (photon) decays into a scattered photon and a low-frequency electrostatic waves such as a Langmuir wave (Raman scattering) or an ion acoustic wave (Brillouin scattering). This three-wave coupling has been of interest for plasma heating via electrostatic waves which can be resonantly excited by two opposed high frequency electromagnetic waves such as lasers [Schmidt, 1973; Cohen, 1974; Cohen et al., 1975].

The second group is parametric instabilities driven by a pump electrostatic wave oscillating near ω_{pe}. This involves an electrostatic coupling among electron plasma waves and ion acoustic waves and electromagnetic waves. A pump electrostatic wave or a Langmuir oscillation can thus be a source of electromagnetic radiation [Freund and Papadopoulos, 1980bc; Pritchett and Dawson, 1983; Akimoto et al., 1987].

The third group is parametric instabilities of electromagnetic cyclotron waves with a frequency below the electron cyclotron frequency. The major topic of this paper concerns this group. Almost all the theoretical and numerical studies for this group have been restricted to parallel propagation along a static magnetic field. Examples of these waves in space plasmas are whistler mode waves in the magnetosphere and Alfvén waves in the solar wind. Decay processes and/or self-trapping of these waves have been studied theoretically by several authors [Sagdeev and Galeev, 1969; Vahala and Montgomery, 1971; Hasegawa, 1972; Lee and Kaw, 1972; Brinca, 1973; Cohen and Dewar, 1974; Lashmore-Davies, 1976; Mio et al., 1976a,b; Ionson and Ong, 1976; Mjølhus, 1976; Goldstein, 1978; Derby, 1978; Sakai and Sonnerup, 1983; Wong and Goldstein, 1986; Longtin and Sonnerup, 1986]. Numerical studies and computer experiments have been performed on the nonlinear instabilities of the electromagnetic cyclotron waves [Hasegawa, 1972; Forslund et al., 1972; Spangler et al., 1985; Terasawa et al., 1986; Spangler, 1986; Machida et al., 1987].

Computer Experiments

We could classify previous simulation studies from a technical point of view. In simulation studies on nonlinear plasma wave interactions, two different models have been used. One is a fluid model where plasma is described by a set of fluid equations such as the MHD equations, or more specific equations such as the nonlinear Schrödinger equation and the Zakharov equations. The other is a particle model where dynamics of individual charged particles is studied by solving the equations of motion as well as the field equations such as the Maxwell equations. A combination of these two different models, a hybrid model, has also been used. A hybrid code simulation, where ions are treated as particles while electrons are treated as a fluid, is effective for plasma waves of low frequency ion modes. The kinetic effects of ion dynamics is retained in the hybrid code simulations. There is another kind of simulations which can retain kinetic effects. It is a Vlasov code simulation, where time evolution of velocity distributions is studies by solving the Vlasov equation. Compared to other simulations, the Vlasov code has been less popular.

In the following we will concentrate on the simulations by a full particle code and a hybrid code. Particle simulations have been performed extensively to study nonlinear wave-particle interactions such as wave excitation and particle heating in space plasmas (see a recent review by Matsumoto[1987]). Particle models are also used in the study of wave-wave interactions [Kruer et al., 1970; Hasegawa, 1972; Forslund et al., 1972; DeGroot and Katz, 1973; Godfrey et al., 1973; Cohen et al., 1975; Matsumoto and Nagai, 1981; Pritchett and Dawson, 1983; Matsumoto and Kimura, 1986; Terasawa et al., 1986; Hada et al., 1987; Machida et al., 1987]. An advantage of a particle model against a fluid model is that it includes kinetic effects such as acceleration and heating of particles associated with corresponding wave-wave interactions.

Electromagnetic Cyclotron Waves

There are two kinds of electromagnetic cyclotron waves with different polarizations. One is a right-hand circularly polarized mode with a frequency from zero to the electron cyclotron frequency. The other is a left-hand circularly polarized mode with a frequency from zero to the ion cyclotron frequency. The former mode is called differently according to its frequency; a whistler mode wave for a frequency much higher than the ion cyclotron frequency and an R(right)-mode Alfvén wave for lower frequencies. This mode is the fast mode and becomes the magnetosonic wave in the case of off-parallel propagation. The latter mode is called an L(left)-mode Alfvén wave. It is the slow mode with a phase velocity less than the Alfvén velocity V_A. In the low frequency limit of $\omega \ll \Omega_i$, where Ω_i is the ion cyclotron frequency, the R- and L-mode Alfvén waves degenerate, and the dispersion relation becomes that of the ideal MHD waves. In the presence of heavy ions, the dispersion curve of the L-mode Alfvén wave is split into several modes with resonance frequencies at cyclotron frequencies of heavy ions.

The whistler mode is one of the most important wave modes in the magnetosphere. They are originated by a lightening, or man-made signals, and they are amplified through the interaction with high energy electrons, which are, as a result of the interaction, precipitated into the ionosphere. The high amplitude low frequency waves (Lion's roar) found in the magnetosheath are also whistler mode waves. On the other hand, Alfvén waves are frequently observed in the solar wind and expected to occur in interstellar and other cosmic plasmas as well. Recently much attention has been paid to Alfvén waves excited by freshly ionized heavy ions from comets in the solar wind. Alfvén waves are generated by high energy ions with temperature anisotropy or by ion beams drifting at a velocity much higher than the thermal velocity. Alfvén mode wave-particle interactions play an important role in the acceleration or heating of ions.

In this paper, we focus our attention on the electromagnetic cyclotron waves propagating parallel to a static magnetic field. What we present is our overall study on the nonlinear behaviour of wave instabilities of these modes involving both wave-particle and wave-wave interactions. We first study the growing phase of whistler mode waves by an electron beam instability, and Alfvén waves by an ion beam instability. We then investigate the subsequent dynamic evolutions of the large amplitude waves through nonlinear wave-wave interactions causing a modulational or a decay instability. Through the nonlinear wave instabilities, the wave number spectra show an inverse cascading. We also briefly describe a formation of envelope solitons which we can find in some of the simulations depending on the parameters.

2. Excitation of Large Amplitude Electromagnetic Cyclotron Waves

In this section we present two different beam instabilities of purely transverse electromagnetic cyclotron waves. One is a whistler mode wave excited by an electron beam, and the other is an R-mode Alfvén wave excited by an ion beam. Both waves are the same electromagnetic cyclotron mode with right-hand polarization, but they are different in the frequency range. For the whistler mode wave, we use an electromagnetic full-particle code with an infinitely massive ion background, neglecting ion dynamics. For the R-mode Alfvén wave, we use an electromagnetic hybrid code with a fluid of massless electrons, neglecting electron dynamics.

Electron Beam Instability

We assume an electron beam formed at the equatorial magnetosphere through injection of high energy particles from the Earth magnetotail region. Such an electron beam with a large pitch angle can excite both whistler mode waves propagating parallel to a static magnetic field through cyclotron resonance and electrostatic waves through Landau resonance. The competing processes of these instabilities were studied by an electromagnetic full particle simulation [Omura and Matsumoto, 1987]. In spite of a strong electrostatic diffusion of the beam at an early stage, electromagnetic pitch angle scattering still proceeds which is followed by a slow growth of whistler mode waves. The saturated amplitude of the whistler waves could be of the order of $10^{-3}\sim10^{-2}B_o$ (B_o : static magnetic field).

We also performed a computer experiment of a pure whistler mode beam instability, where longitudinal electrostatic field is neglected. This simulation model is of significance to check the

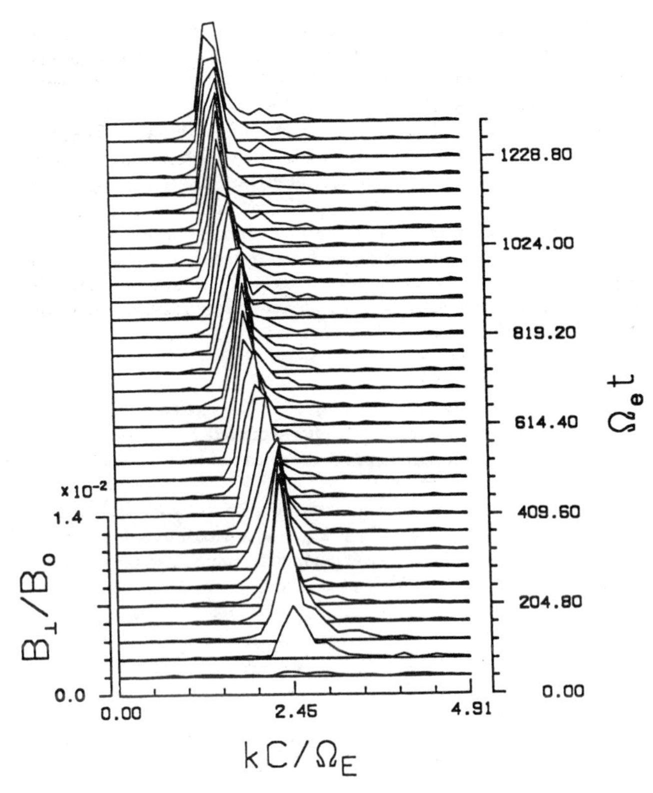

Fig. 1. Time evolution of k-spectra of whistler mode waves excited by an electron beam instability. We see a smooth inverse cascading preserving the sharpness of the spectra. The frequency of the most dominant mode is about $0.5\Omega_e$ initially, and it decreases down to an asymptotic value $0.25\Omega_e$.

validity of several previous studies of whistler mode wave-particle interactions [Nunn, 1974; Helliwell and Crystal, 1973; Matsumoto and Yasuda, 1976; Matsumoto et al., 1980; Vomvoridis and Denavit 1980; Matsumoto and Omura, 1981; Omura and Matsumoto, 1982], because they have also neglected the longitudinal electrostatic field in their theoretical and numerical models.

It is well known that a whistler mode wave interacting with a counter-streaming electron beam grows exponentially and reaches a saturation due to nonlinear phase trapping of beam electrons. The simulations by Matsumoto and Yasuda [1976] and Matsumoto et al. [1980] show the nonlinear trapping and a subsequent quasi-steady state where the excited large amplitude whistler mode wave remains at a constant amplitude. Their simulation model is basically the same as the present simulation except for some minor parameter changes such as a number of superparticles and a form of loss-cone distribution function. Owing to increased capabilities of supercomputers, however, we could trace the time evolution up to $\Omega_e t \approx 1000$ much longer than we did in our previous works, where Ω_e is the electron cyclotron frequency. Parameters assumed in the simulations are the following.

Plasma frequency Π_e	$2\ \Omega_e$
Thermal velocity of background electrons	$0.01\ c$
Parallel beam drift velocity	$-0.168\ c$
Loss cone factor for the subtracted Maxwellian	0.5
Perpendicular beam thermal velocity $V_{\perp T}$	$0.168\ c$
Parallel beam thermal velocity $V_{\parallel T}$	$0.001\ c$
Density ratio of beam and background plasma η	0.01
Grid spacing Δx	$0.04c/\Omega_e$
Number of grid points N_x	1024
Minimum wave number in the system k_{min}	$0.0153\ \Omega_e/c$
Number of superparticles for cold electrons	16384
Number of superparticles for beam electrons	65536
Time step Δt	$0.005\ \Omega_e^{-1}$

where c is the light speed. We have plotted the wave number spectra of the whistler mode waves in Figure 1. Contrary to our expectation, the whistler mode waves excited in the present simulation does not remain at a constant amplitude at a fixed k, but show a very clear inverse cascading, where the wave numbers of most dominant waves decrease in time. As it will be clarified in the following sections, the inverse cascading is a results of strong wave-wave interactions. Therefore, the present simulation is not only interesting for the wave-particle interaction studies but also of great interest as a computer experiment of wave-wave interactions.

Ion Beam Instability

An ion beam, such as reflected ions back-streaming from Earth's bow shock or cometary pick-up heavy ions in the solar wind, can excite several different wave modes. Several simulations have been performed on the ion beam instabilities [Winske and Leroy, 1984; Hoshino and Terasawa, 1985; Rogers et al., 1985; Winske and Gary, 1986; Omidi and Winske, 1986; Matsumoto et al., 1987; Kojima et al., 1988; Kaya et al., 1988]. As for a cometary ion beam in the solar wind, there exist several unstable wave modes, such as longitudinal beam modes, anomalous cyclotron resonant R-modes, cyclotron resonant L-modes and non-resonant modes. In Figure 2 we present an $\omega-k$ dispersion diagram of these unstable modes. We have assumed three species of ions, i.e., solar wind protons, α particles, and cometary heavy ions with the following parameters.

Plasma frequency of H^+	$6023\ \Omega_{H^+}$
Plasma frequency of He^{++}	$753\ \Omega_{H^+}$
Plasma frequency of H_2O^+	$145\ \Omega_{H^+}$
Light speed c	$6742\ V_A$
Thermal velocity of H^+, He^{++}	$0.578\ V_A$
Drift velocity of H_2O^+	$6.36\ V_A$
Perpendicular thermal velocity of H_2O^+	$6.36\ V_A$
Parallel thermal velocity of H_2O^+	$0.004\ V_A$
Electron β_e	1.0

where V_A and Ω_{H^+} are an Alfvén velocity and a proton cyclotron frequency, respectively.

Using the hybrid code, we performed several simulation runs for different wave number ranges and studied the competing processes of these co-existing instabilities [Matsumoto et al., 1987]. The R-mode Alfvén waves at relatively high frequencies are the most dominant electromagnetic waves in the early stage of the interaction, and attain a large amplitude of the order of $0.1B_o$. In Figure 3 we show wave number spectra of the R-mode Alfvén waves and their time evolution. We see a clear inverse cascading similar to that of the whistler mode waves seen in Figure 1. However, the detailed structure of the spectra is different from that of the whistler mode waves. The whistler mode waves show a smooth transition of wave spectra in k-space, while the R-mode Alfvén waves show a discontinuous transition with a discrete jump. The difference depends on the mechanisms of the wave-wave instabilities, i.e., on whether they are modulational or decay instabilities. We will discuss the difference further in Section 4.

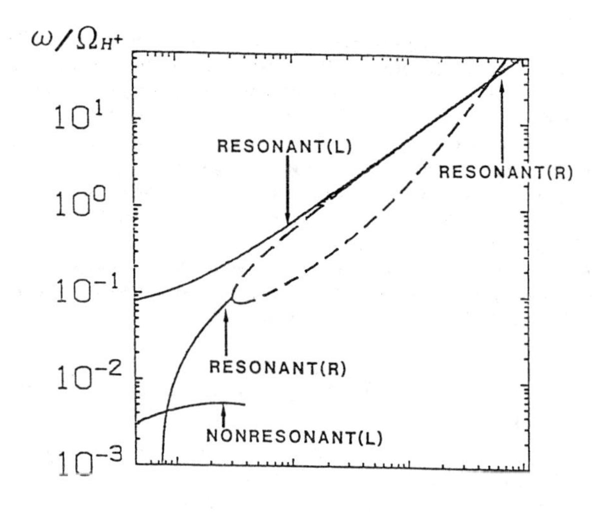

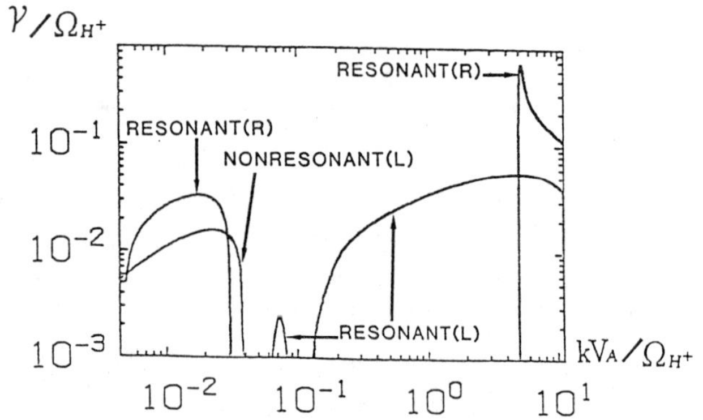

Fig. 2. Dispersion relation of unstable plasma waves in the presence of cometary heavy ions for a solar wind plasma around a comet.

3. Parametric Instabilities of Electromagnetic Cyclotron Waves

In the previous section, we have seen excitation of large amplitude whistler mode waves and Alfvén waves through interactions with electron and ion beams, respectively. As the amplitudes become large, the nonlinear wave-wave interaction becomes dominant, and the wave energy is transferred to other wave modes. Such instabilities of a finite amplitude wave are generally called parametric instabilities, whose controlling parameter is the amplitude of the wave.

We can find such instabilities in various type of plasma waves under a wide range of plasma parameters. For example, there is a nonlinear interaction between high frequency Langmuir and low frequency ion-acoustic waves. Stability of these electrostatic waves can be studied within the context of the Zakharov equations [e.g., Nicholson, 1983]. The most general instability in this case involves the single finite-amplitude Langmuir wave, two other Langmuir waves, and one low frequency wave. The stability analysis proceeds by assuming that the amplitudes of the two other Langmuir waves and the low frequency wave are infinitesimal.

Very similarly, we can study stability of finite amplitude electromagnetic cyclotron waves by assuming one finite-amplitude Alfvén wave with wave number k_o, one longitudinal acoustic wave with wave number k and two other transverse electromagnetic waves with k_o-k and k_o+k. These waves with k_o-k and k_o+k are called Stokes and anti-Stokes waves in an analogy to a phenomenon of nonlinear optics called "stimulated Brillouin scattering" [Terasawa et al., 1986].

Among various types of wave-wave instabilities, there are two major instabilities which have been studied intensively for the last two decades. One is a modulational instability, the other is a decay instability. In the simulation of the modulational instability, we also found another type of nonlinear instability where the energy of the forward travelling wave at k_o is transferred to the backward travelling wave at $-k_o$, and vice versa. We will discuss these three parametric instabilities of electromagnetic cyclotron waves in the following subsections.

Modulational Instability

A modulational instability is characterized by a long wavelength of modulation compared with the wavelength of a "pump" or parent electromagnetic wave, i.e., $k \ll k_o$. The modulation of the amplitude is equivalent to generation of the two side bands with wave numbers $k_o - k$ and $k_o + k$. It is noted that k is regarded as a wave number of an unstable longitudinal sound wave which is excited by a parallel force $F_\parallel$ of the form,

$$F_\parallel = -\frac{1}{n}\frac{\partial}{\partial x}\left[\frac{B_w^2}{2\mu_o}\right] \tag{1}$$

where the static magnetic field is taken along the x-axis, and n, μ_o and B_w are the number density of electrons and ions, permeability in a vacuum and amplitude of the transverse wave magnetic field, respectively. Since the amplitude modulation excites longitudinal waves by the $F_\parallel$, it is very important to determine whether the modulation is stable or not.

The stability of the amplitude modulation of the electromagnetic cyclotron waves is generally studied within the context of the nonlinear Schrödinger equation. The equation was derived by Karpman and Krushkal [1969], and for a more general form by Brinca [1973] with an assumption of a nonlinear dispersion relation as

$$\omega = \Omega(k_o) + q(k_o)\, a^\beta \quad (\beta > 0) \tag{2}$$

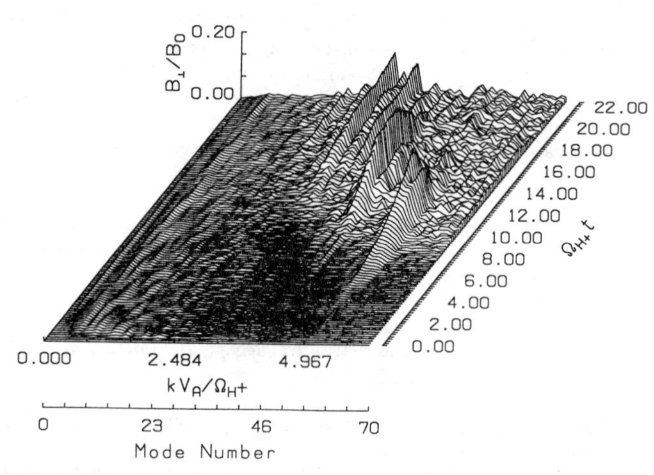

Fig. 3. Time evolution of k-spectra of R-mode Alfvén waves and their time evolution.

where $\omega = \Omega(k_o)$ is the linear dispersion relation and

$$q(k_o) = \left[\frac{\partial \omega}{\partial (a^\beta)}\right]_{a=0} \quad (3)$$

Karpman and Krushkal assumed $\beta = 2$ in their derivation. We assume a wave form modulated around an equilibrium state with a wave of amplitude a_o propagating along the x axis with a frequency ω_o and a wave number k_o,

$$\psi(x,t) = \phi(x,t)\exp[i(\omega_o t - k_o x)] \quad (4)$$

and

$$\phi(x,t) = a(x,t)\exp[i\theta(x,t)] \quad (5)$$

We perform the Taylor expansion of (5) around the equilibrium state. Differentiating ψ with respect to t and x, and introducing new variables, $\xi = x - v_g t$ and $\tau = t$, we obtain the nonlinear Schrödinger equation

$$-i\frac{\partial\phi}{\partial\tau} + \frac{1}{2}v_g{}'\frac{\partial^2\phi}{\partial\xi^2} - q_o(a^\beta - a_o^\beta)\phi = 0 \quad (6)$$

where $q_o = q(k_o)$ and

$$v_g = \frac{\partial\Omega(k_o)}{\partial k_o}, \quad v_g{}' = \frac{\partial^2\Omega(k_o)}{\partial k_o^2} \quad (7)$$

$v_g{}'$ and q_o represent the dispersion and nonlinear effects, respectively. We assume a modulational perturbation about the equilibrium $a = a_o$ and $\theta = 0$ of the form

$$a^\beta - a_o^\beta = a_1\exp[i(\omega\tau - k\xi)], \quad \theta = \theta_1\exp[i(\omega\tau - k\xi)] \quad (8)$$

where $a_1 \ll a_o^\beta$, and $\theta_1 \ll 1$. Separating the real and imaginary parts of (6), and substituting (8) into (6), we obtain a dispersion relation for the modulational perturbation (ω,k)

$$\omega^2 = \left(\frac{kv_g{}'}{2}\right)^2\left(k^2 + 2\beta a_o^\beta\frac{q_o}{v_g{}'}\right) \quad (9)$$

When $q_o/v_g{}' < 0$, the modulational perturbation becomes unstable in the range

$$k^2 < -2\beta a^\beta\frac{q_o}{v_g{}'} \quad (10)$$

The maximum temporal growth rate γ_m is

$$\gamma_m = \beta|q_o|a_o^\beta \quad (11)$$

and occurs at

$$k_m = \left[-\frac{\beta a_o^\beta q_o}{v_g{}'}\right]^{1/2} \quad (12)$$

When $q_o/v_g{}' > 0$, the right-hand side of (9) is always positive, and the modulation is stable.

Based on the above analysis, let us examine the modulational instability for the whistler mode. It is well known that $v_g{}'$ of a whistler mode wave is positive for the frequency range $0 < \omega < \Omega_e/4$ and negative for $\Omega_e/4 < \omega < \Omega_e$, where Ω_e is the electron cyclotron frequency. Therefore the stability changes at $\Omega_e/4$ depending on the sign of nonlinear frequency shift characterized by q_o.

Tam [1969] examined the nonlinear frequency shift for whistler waves propagating parallel to the static magnetic field in a cold plasma with immobile ions. His result shows the nonlinear frequency shift q_o is zero. Therefore he predicted no modulational instability takes place in this case.

Hasegawa [1972] calculated the nonlinear frequency shift for electromagnetic cyclotron waves propagating parallel to the static magnetic field in a cold plasma with mobile ions as

$$q_{oi} = \frac{k_o V_A{}^2}{4v_g} \quad (13)$$

Therefore he predicted a modulational instability in the frequency range $\Omega_e/4 < \omega < \Omega_e$.

Brinca [1973] calculated the nonlinear frequency shift due to relativistic effects in addition to the contribution from mobile ions. The nonlinear shift by the relativistic effects is expressed by

$$q_{or} = -\frac{\omega_o^3}{2\Pi_e^2}\left(1 - \frac{\omega_o}{\Omega_e}\right) \quad (14)$$

Therefore he concludes that the whistler mode waves are modulationally unstable for the frequency range

$$\left(\frac{\Omega_i\Pi_e^2}{4}\right)^{1/3} > \omega > \frac{\Omega_e}{4} \quad (15)$$

for a cold dense plasma which satisfy $\Pi_e > 1/4\,(\Omega_e^3/\Omega_i)^{1/2}$.

For low frequency Alfvén waves, equation (13) is valid in a low β plasma, and we have $q_{oi} > 0$. R-mode Alfvén waves have positive $v_g{}'$, and they are modulationally stable. On the other hand, L-mode Alfvén waves have negative $v_g{}'$, and they are modulationally unstable. In a finite β plasma, a number of theoretical studies reached the same results [Sakai and Sonnerup, 1983; Wong and Goldstein, 1986; Longtin and Sonnerup, 1986]. Modulational instability occurs if $V_p > C_s$ for L-mode Alfvén waves, and if $V_p < C_s$ for R-mode Alfvén waves, where V_p and C_s are the phase velocity of the unperturbed wave and the unperturbed sound speed, respectively.

We have performed another one-dimensional particle simulation starting with a finite amplitude whistler mode wave propagating parallel to the static magnetic field. Plasma parameters are basically the same as in the simulation presented in Figure 1, except that no electron beam is assumed. In this particle simulation, the relativistic effect is neglected. Though we did not trace the ion motion, we a priori assumed that the ions move along the static magnetic field quickly enough to cancel the electrostatic charge perturbation produced by the nonlinear effect. Technically this corresponds to neglect of the electrostatic field parallel to the static magnetic field. A justification of this treatment is as follows. The charge density fluctuation associated with the electrostatic waves caused by the nonlinear effect is neutralized in a time scale of Π_i^{-1}, where Π_i is the ion plasma frequency. This condition is met for a high density plasma where $\Pi_i \approx \omega$ is satisfied.

We have initially assumed a finite amplitude whistler mode wave with *mode* = +32 (*mode*: number of spatial wave oscillations in the system) as seen in the top panel of Figure 4. The frequency ω of the wave is $0.5\Omega_e$. As time goes on, an amplitude modulation develops as seen in Figure 4. This is an example of the modula-

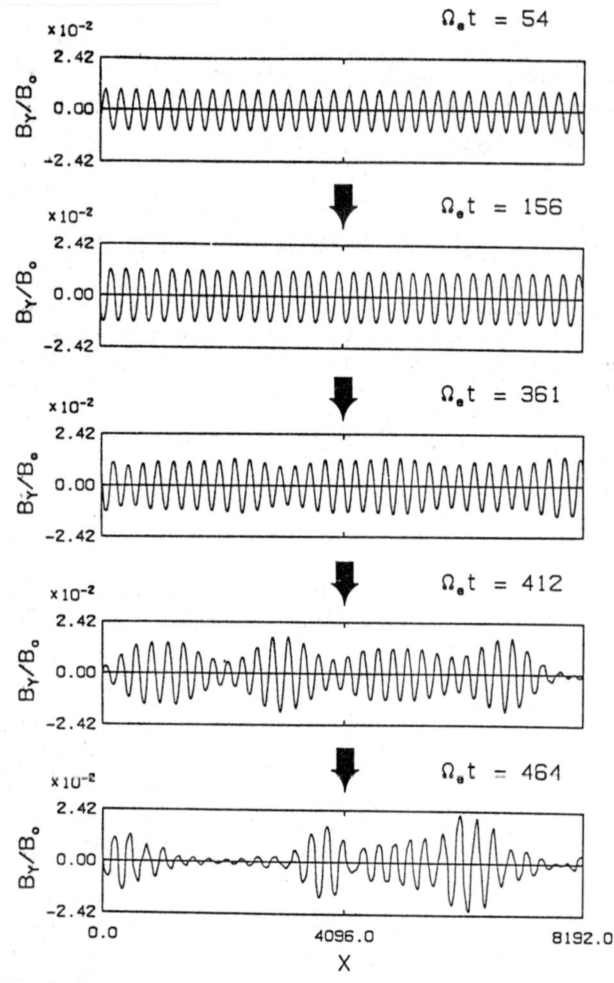

Fig. 4. Spatial profile of B_y field at different times showing development of a modulational instability.

tional instability discussed above. In Figure 5, we present a time evolution of wave number spectra of the whistler mode wave shown in Figure 4. We have decomposed the wave by the spatial helicities, i.e. right and left polarized spirals, which correspond to backward and forward travelling waves, respectively. Although we have assumed a single forward travelling wave at $\Omega_e t = 0.0$, a backward wave soon appears at the same wave number as the initial forward travelling wave, namely at *mode* = −32. This is another type of strong wave-wave interaction other than modulational and decay instabilities, and will be discussed in the following subsection.

After several energy oscillations between the forward and backward travelling waves with the same absolute wave number, the modulational instability develops. The combination of the energy oscillations between the forward and backward travelling waves and the modulational instability can cause the wave energy transferred to the lower wave numbers as we will discuss in Section 4. Based on many simulation runs with different parameters, We found that the wave energy transfer stops at $\omega = \Omega_e/4$. This implies that the modulational instability occurs for a whistler mode wave in the frequency range of $\Omega_e/4 < \omega < \Omega_e$, where $v_g' < 0$. The condition for the modulational instability, i.e., q_o/v_g' requires that there must exist a nonlinear dispersion relation which gives $q_o>0$.

The positive nonlinear frequency shift $q_o > 0$ agrees with Hasegawa's results [1972]. He assumed quasi-neutrality in the longitudinal motion of plasma, which is equivalent to our treatment of neglecting electrostatic fields. However, we should note that the nonlinear frequency shift given by Hasegawa is based on the cold plasma dispersion. We may have to take into account the nonlinear effects induced by the thermal kinetic effect, because an effective thermalization occurs in the presence of the wave energy exchange between the backward and forward travelling waves, as discussed in the following subsection. Therefore we may have to consider a nonlinear frequency shift q_o as a function of a parallel thermal velocity $V_{\parallel T}$ of electrons and a wave number k_o as

$$q_{oT} = q(k_o, V_{\parallel T}). \tag{16}$$

We performed several simulation runs with different parallel thermal velocities of background electrons. We found a tendency that the modulational instability develops more quickly with a larger thermal velocity.

Interaction of Backward and Forward Travelling Waves

An interaction of backward and forward travelling electromagnetic cyclotron waves is important, because it causes an effective heating of thermal particles in the parallel direction [Omura et al., 1988]. In Figure 5, we found an energy transfer from the forward travelling wave at k_o to the backward travelling wave at $-k_o$. Figure 6 shows an early time development of the wave-wave interaction between forward and backward travelling waves. The backward travelling wave develops as the forward travelling wave becomes small, and once the backward travelling wave attain a certain level of amplitude, the wave energy begins to be transferred between forward and backward travelling waves repeatedly at relatively short period as seen in the spectra at $\Omega_e t = 130 \sim 180$.

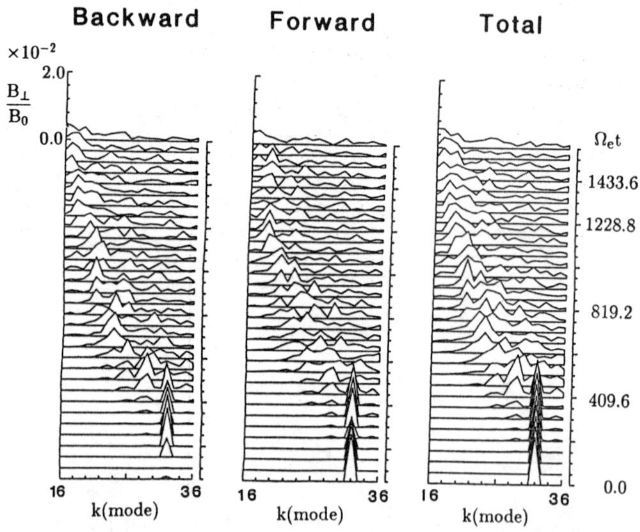

Fig. 5. Time evolution of k-spectra of a finite amplitude whistler mode wave initially travelling in the forward direction. Panels from left to right are backward, forward and total components of transverse magnetic field.

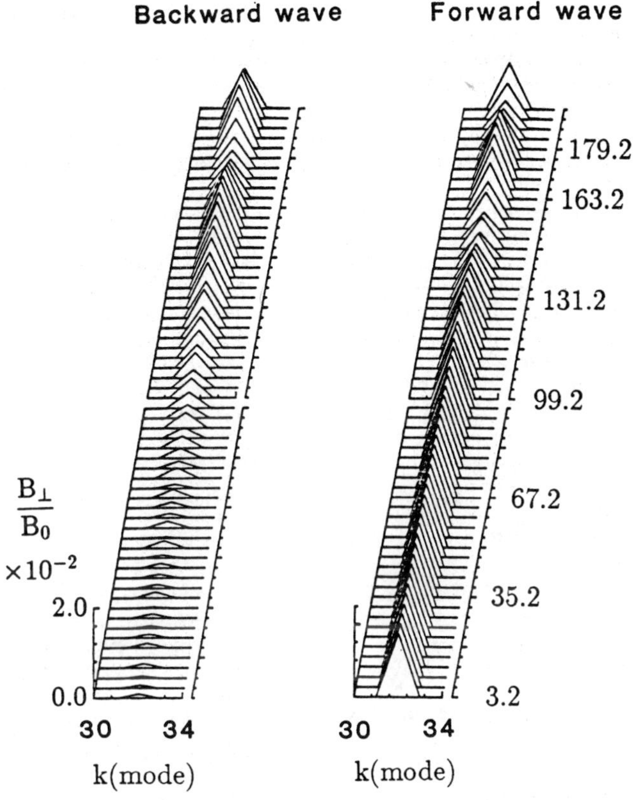

Fig. 6. Early phase of wave-wave interaction shown in Figure 5. Back-and-forth energy exchange is seen.

The back-and-forth energy exchange can be explained by an energy oscillation between forward and backward travelling waves through a coupling of $\vec{v}_{\perp}\times\vec{B}_{\perp}$ forces exerting each other. They attempt to share the energy evenly, but the initial imbalance of the energy causes the back-and-forth oscillation of energy transfer. The mechanism of the back-and-forth energy exchange is understood by solving the Maxwell equation and the momentum equation of cold electrons. We assume two transverse electromagnetic waves propagating forward and backward along the static magnetic field with amplitude B_f (forward) and B_b(backward), respectively. Retaining the nonlinear terms associated with $\vec{v}\times\vec{B}$ force and the current formed by the first order density perturbation n_1 and the first order velocity perturbation, we can obtain the following relation

$$\frac{dB_b}{dt} = -C_1\, n_1 B_f \sin\psi \tag{17}$$

$$\frac{dB_f}{dt} = C_1\, n_1 B_b \sin\psi \tag{18}$$

where C_1 is a constant. The phase ψ is given by

$$\psi = \psi_n + \psi_b - \psi_f \tag{19}$$

where ψ_n, ψ_b and ψ_f are the phases of n_1, B_b and B_f, respectively. From the equation of continuity and the momentum equation, we also obtain

$$\frac{d^2 n_1}{dt^2} = -C_2\, B_f B_b \cos\psi \tag{20}$$

where C_2 is a constant. The above equations (17) (18) and (20) have oscillating solutions which satisfy the following relation,

$$|B_f|^2 + |B_b|^2 = const. \tag{21}$$

Therefore we have the energy oscillation between forward and backward travelling waves.

In Figure 7 we plotted time histories of parallel and perpendicular thermal energies of electrons. The perpendicular thermal energy is the kinetic energy of background electrons which perform a sloshing motion to support the wave propagation. The decrease of the perpendicular thermal energy means a decrease of the kinetic energy of the whistler mode waves. During the initial backward and forward energy exchange process, background electrons are strongly thermalized in the parallel direction. This is because of the longitudinal force (1) which appears in the presence of the backward and forward travelling electromagnetic waves [Omura et al., 1988]. The backward and forward travelling waves at wave

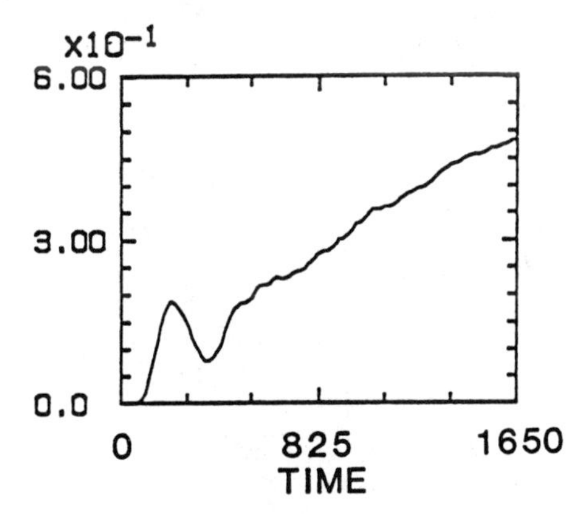

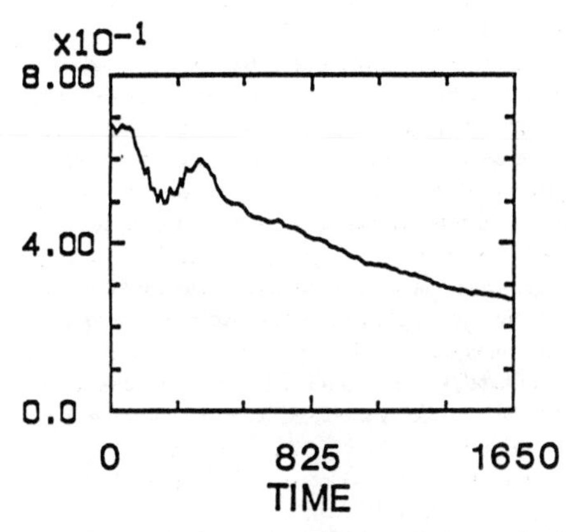

Fig. 7. Time history plots of parallel and perpendicular thermal energies of the background electrons, corresponding to the simulation results shown in Figures 4, 5 and 6.

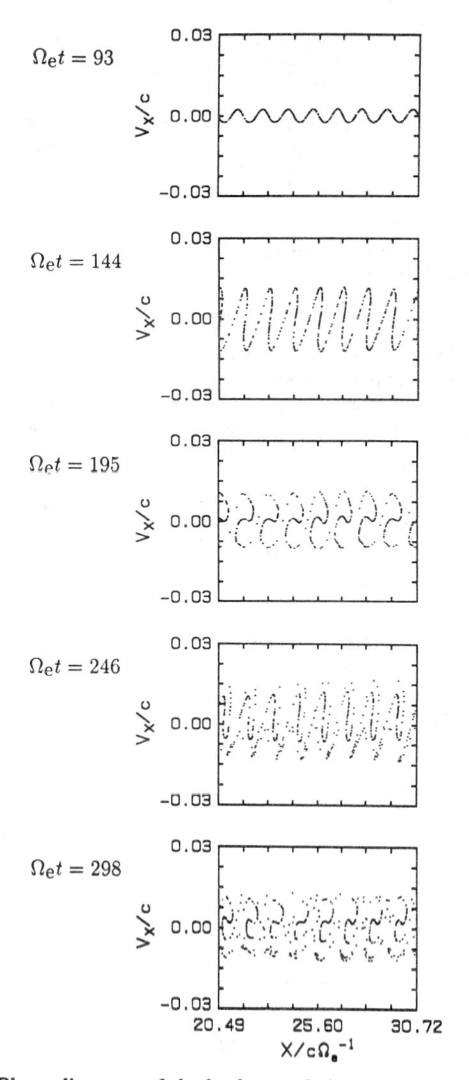

Fig. 8. Phase diagrams of the background electrons in the $x - v_z$ phase space.

numbers k_o and $-k_o$ form a standing wave, which gives rise to a static longitudinal force with wave number $2k_o$. This acts as a static potential and traps background electrons. The electrons are accelerated in the parallel directions and rotates in the $x - v_z$ phase space as shown in Figure 8. The longitudinal force becomes most effective when the amplitudes of the forward and backward travelling waves are of the same order, and it disappears when either forward or backward wave is dominant. Since we have the back-and-forth energy oscillation, some of the accelerated electrons are detrapped from the static potential and move freely in the phase space, which leads to the thermalization. After the electrons are thermalized in the parallel direction, the modulational instability slowly develops.

Decay Instability

In contrast to a modulational instability characterized by $k \ll k_o$, a decay instability is characterized by $k > k_o$, where k is the wave number of an unstable sound wave. This means that an excited wave with wave number $k_o - k$ (< 0) travels in the backward direction. The excited wave should be a normal mode satisfying the the three wave resonance condition , i.e., ($k_o - k$, $\omega_o - \omega$) must be a normal mode. Therefore the decay instability is more selective than a modulational instability. A recent theoretical study by Wong and Goldstein [1986] gives characteristics of the decay and modulational instabilities as follows. For R-mode Alfvén waves the decay instability is dominant if β ($\equiv (C_s/V_A)^2$) < 1, while the modulational instability dominates for sufficiently high beta. The growth rate of the decay instability of left-hand waves is greater than the modulational instability at all values of beta.

Terasawa et al. [1986] confirmed, by a hybrid code simulation, a decay instability of finite-amplitude circularly polarized Alfvén waves. They observed an energy transfer from the parent R-mode Alfvén wave, which was given a priori at t=0, to two daughter Alfvén-like waves and a sound-like wave. They also observed the daughter Alfvén waves are also unstable for further decay, and that the wave energy is continuously transferred to the longer wavelength regime, i.e., an inverse cascading process.

In the previous section, we found that a similar decay process occurs in the R-mode Alfvén waves excited by an ion beam. Wave energies are mainly transferred from forward travelling waves to backward travelling waves with smaller wave numbers. Comparing the spectra in Figures 1 and 3, we see a clear difference in the evolution of the spectra. In the presence of the modulational instabilities as in Figure 1, the transformation of wave energy to lower k is very smooth, while energy transfer in the decay instability takes place in a discrete manner. We have decomposed the spectra into forward- and backward-travelling waves, and plotted their evolution in Figure 9, to confirm energy transfer from the forward to backward-travelling waves.

4. Inverse Cascading of Electromagnetic Cyclotron Waves

In fluid dynamics, spatially large scale perturbations are often converted to small scale perturbations. This process where wave energy at long wavelength is transferred to wave energy at short wavelength is called a cascading process. However, a two-dimensional hydromagnetic turbulence has a property that energy is transferred to a longer wavelength. This unique property is called an inverse-cascading process. In Section 2, we have seen nonlinear wave-wave interactions through which wave energies at short wavelength are transferred to those at longer wavelength, i.e., from large wave number to smaller wave number waves. Therefore, this process of energy transfer via wave-wave interaction is called an inverse cascading process in an analogy with the hydromagnetic turbulence. However, it should be noted that the inverse cascading process is a rather common phenomenon as in a weak turbulence theory developed in plasma physics [Hasegawa et al., 1979].

In the following we study the inverse cascading process of the electromagnetic cyclotron waves propagating parallel to the static magnetic field. The simulation results suggest that there are several different mechanisms in the inverse cascading.

Inverse Cascading via Modulational Instability

In the examples of large amplitude whistler simulations, shown in Figures 1, 4 and 5, finite amplitude whistler mode waves were unstable against the modulational instability. However, in addition to the modulational instability, we observed a very clear inverse cascading preserving an initial sharpness of the spectra. The inverse cascading of the whistler mode waves stops at the frequency of $\Omega_e/4$, where the stability of amplitude modulation changes as

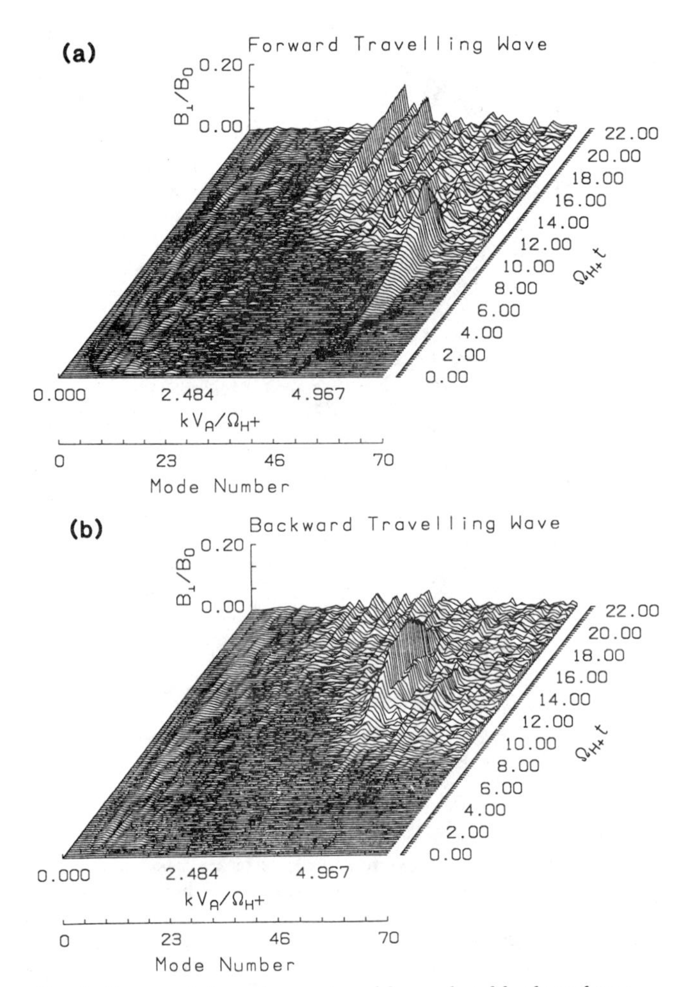

Fig. 9. Time evolution of k-spectra of forward and backward travelling R-mode Alfvén waves, decomposed from the k-spectra in Figure 3.

discussed in Section 3. This suggests that the modulational instability plays an important role in the inverse cascading. However, the theory of the modulational instability does not predict the inverse cascading process. Another important process, which was not considered in the previous theories, should be necessary. It is the energy transfer between the forward and backward travelling waves, which we observed at the initial stage in the simulation starting with a finite amplitude whistler mode wave in a cold electron plasma.

We now examine a three-wave resonance condition when the back-and-forth energy exchange and modulational instability proceed simultaneously. After the energy oscillation between the forward and backward travelling waves, we have two waves (ω_o, k_o) and ($\omega_o, -k_o$). Each wave suffers the modulational instability producing additional two sidebands ($\omega_o \pm \delta\omega, k_o \pm \delta k$). Thus we have six coexisting waves in the system. Among them the combination of (ω_o, k_o) and ($\omega_o - \delta\omega, -k_o + \delta k$) or that of ($\omega_o, -k_o$) and ($\omega_o - \delta\omega, k_o - \delta k$) gives the maximum growth rate of the parametric three-wave coupling instabilities. For this combination, the longitudinal density perturbation produced by the nonlinear force (1) has the minimum phase velocity $|\delta\omega \,/\, (2k_o - \delta k)|$. The lower the phase velocity of the density perturbation gives higher coupling efficiency, and thus higher growth rate of the instability. Through this process, the wave energy is transferred from the forward travelling wave (k_o, ω_o) to the backward travelling daughter wave ($-k_o + \delta k$, $\omega_o - \delta\omega$) through a longitudinal virtual wave ($2k_o - \delta k$, $\delta\omega$). Similar energy transfer from the backward travelling wave to the forward travelling daughter wave takes place. Thus the smooth inverse cascading process is the results of both the modulational instability and the back-and-forth energy oscillation.

Under certain parameters where modulational instability is prohibited the smooth inverse cascading was not observed. Instead, a decay process with a discrete jump in frequency was observed for a whistler mode wave with an amplitude of the same order of the static magnetic field.

Inverse Cascading via Decay Instability

We have studied the decay instability of the electromagnetic cyclotron waves in Section 3. Through a decay instability, the wave energy is transferred to the waves of lower wave numbers travelling in the opposite direction. If the decay process makes the daughter wave sufficiently intense, the daughter wave is also unstable for a further decay. Thus the wave energy is transferred to the wave at much lower wave numbers. Such an inverse cascading process via repeated decay processes is found in a particle simulation by Terasawa et al. [1986]. We also found the repeated decay processes in the simulation of R-mode Alfvén waves excited by an ion beam (see Figures 3 and 9). The inverse cascading via decay processes shows an energy transfer with a discrete jump to the lower wave numbers.

5. Envelope Soliton of Electromagnetic Cyclotron Waves

In this section, we show that the nonlinear Schrödinger equation derived in Section 3 has a soliton solution, which is well known as a solution to the Korteweg-de Vries (K-dV) equation. We also show an example of particle simulation, where the nonlinear evolution of the electromagnetic cyclotron waves leads to a formation of envelope solitons.

Nonlinear Schrödinger Equation

In section 3, we have shown the nonlinear Schrödinger equation (6) for the electromagnetic cyclotron waves. There we assumed a wave packet with a constant amplitude a_o as an equilibrium state. The wave packet (a_o) is one of the solutions of the equation, and the stability of which against a small modulation ($|a - a_o| \ll a$) was examined. In this section, we discuss another solution of the nonlinear Schrödinger equation, i.e., solitary waves or solitons. We rewrite the nonlinear Schrödinger equation (6) assuming $a_o \ll a$, and $\beta = 2$.

$$i\phi_\tau + p\phi_{\xi\xi} + q|\phi|^2\phi = 0 \qquad (22)$$

where $p = -v_{go}'/2$ and $q = q_o$, and the subscript $\xi\xi$ means the second order derivative with respect to ξ. Assuming a solution of the form

$$\phi = a\exp(i\omega\tau) \qquad (23)$$

we have

$$a_{\xi\xi} = \frac{\omega}{p}\,a\,(1 - \frac{q}{\omega}a^2) \qquad (24)$$

First of all, we assume $\omega > 0$, and we obtain the following solutions for different signs of p and q [Watanabe, 1985].

For $p > 0$ and $q > 0$, we have

$$a = (\frac{2\omega}{q})^{1/2} \operatorname{sech} \left\{ (\frac{\omega}{p})^{1/2} \xi \right\} \tag{25}$$

where the amplitude a has a form of a solitary wave. For $p<0$ and $q>0$, we have

$$a = -(\frac{\omega}{q})^{1/2} \tanh \left\{ (-\frac{\omega}{2p})^{1/2} \xi \right\} \tag{26}$$

where the amplitude has a form of a shock wave. For $\omega < 0$, it is obvious that (25) is valid for $p < 0$ and $q < 0$, and (26) for $p > 0$ and $q < 0$. Therefore, we have a soliton solution of the following form for $pq > 0$

$$\phi = a_o \operatorname{sech} \left\{ (\frac{q}{2p})^{1/2} \xi \right\} \exp \left\{ i \frac{q a_o^2}{2} \tau \right\} \tag{27}$$

where $a_o = (2\omega/q)^{1/2}$. Noting that the phase variation $\exp i(k_o x - \omega_o t)$ should be multiplied to (27), (27) represents an envelope soliton with a nonlinear frequency shift of $\omega = qa_o^2/2$. Since $\xi = x - v_{go} t$, the envelope propagates at the group velocity v_{go}. It should be noted that the condition for the soliton solution $pq > 0$ is identical to the condition for the modulational instability discussed in Section 3. Therefore, the soliton solution can be a result of the modulational instability, if the nonlinear characteristics are not changed during the transition process. However, the transition process itself is very difficult to be analyzed by a theoretical formulation and thus numerical simulations play an important role in this respect.

Previous Study on the Envelope Solitons of Alfvén Waves

There have been a number of studies on the envelope solitons of electromagnetic cyclotron waves. The modulational instability and envelope-solitons of Alfvén waves propagating parallel to the static magnetic field are studied by Mio et al. [1976]. They derived the modified nonlinear Schrödinger equation, and solved the equation, obtaining the conditions for modulational instabilities and formation of envelope-solitons. Properties of Alfvén solitons in a finite-beta plasma are examined by Spangler and Sheerin [1982]. They derived a derivative nonlinear Schrödinger equation, which is solved by a Pseudo-potential method. They found an envelope soliton which possesses a width-amplitude relationship. Spangler et al. [1985] solved the derivative nonlinear Schrödinger equation numerically and studied properties of nonlinear Alfvén waves and solitons. They found that development of amplitude modulation of Alfvén waves strongly depends on the polarization of the wave and plasma β value. Ovenden et al.[1983] studied the interaction of circularly-polarized Alfvén waves with the surrounding plasma in high speed solar wind streams. They compared the characteristics of Alfvén solitons with observational results obtained from Helios I and II.

Formation of Envelope Solitons

The time evolution of the modulational instability shown in Section 3 depends on the initial wave amplitude as well as the thermal velocity of the background electrons. We have performed a number

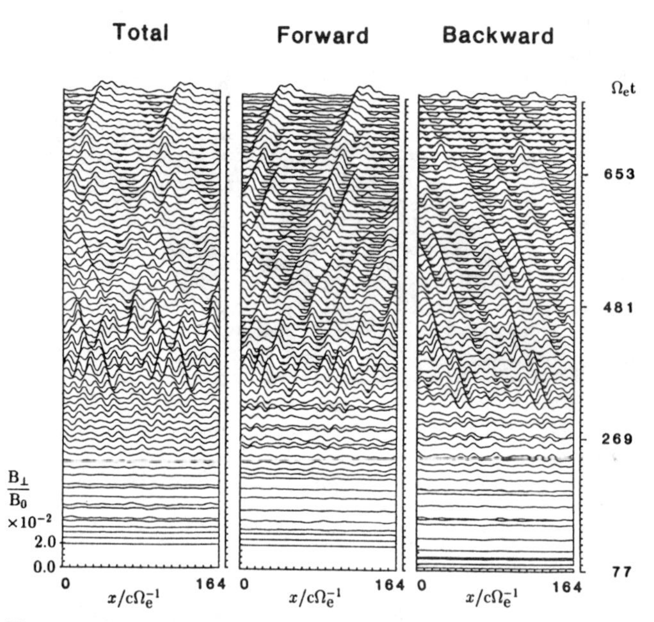

Fig. 10. Formation of envelope solitons: spatial profile of the amplitude of the whistler mode wave magnetic field, and their time evolution. Backward and forward travelling components and their summation are plotted from right to left, respectively. Two cycles of the periodic simulation system are plotted in each panel, i.e., the real simulation system is $81.92x/c\Omega_e^{-1}$.

of simulation runs with different parameters, and found that there is a tendency that the wave number spectra are broadened in the k space for the smaller initial thermal velocity. When the wave spectra becomes broad in the process of the inverse cascading process, we find a formation of solitary wave packets, or envelope solitons in the x real space. It is noted that the condition of the modulational instability $q_o/v_{go}' < 0$ allows a soliton solution of the nonlinear Schrödinger equation. In the simulations involving the modulational instability such as shown in Figures 4 and 5, we found formation of envelope solitons at later phase of the interaction. We show one example of such simulations with the following parameters: $\Pi_e = 2\Omega_e$, $V_e = 0.001c$, $B_w = 0.03B_o$, $mode = 32$, $\omega = 0.587$ $\delta x = 0.08c/\Omega_e$ and $N_x = 1024$. Figure 10 shows the spatial amplitude variation of forward and backward travelling components of B_y and their summation. We can find a formation of solitary wave packets travelling forward and backward directions after a development of the amplitude modulation.

6. Conclusions

In this paper, we presented a brief review on previous works and our recent achievement of particle simulations on nonlinear fate of circularly polarized electromagnetic cyclotron waves propagating parallel to the static magnetic field. Two different wave modes are studied using two different simulation codes. One is an electron cyclotron wave, or a whistler mode wave, studied by an electromagnetic full-particle code. The other is an ion cyclotron wave, or an Alfvén wave, studied by an electromagnetic hybrid code. We first studied excitation of these modes by an electron beam and an ion beam, respectively. Both modes saturate at large amplitudes and further lead to a series of nonlinear wave-wave interactions, i.e., energy transfer to the backward travelling waves followed by the back-and-forth energy oscillation and modulational or decay instabilities, inverse cascading and in some cases formation of envelope solitons. It was demonstrated that the combination of the back-and-forth energy oscillation and the modulational instability plays an important role to allow the smooth inverse cascading process for whistler mode waves.

We interpreted the inverse cascading processes of the finite-amplitude whistler and Alfvén waves in terms of physical elementary processes. As we have seen in Sections 3 and 4, the wave-particle instabilities are followed by wave-wave instabilities in a single run of the simulations. We can view series of these nonlinear wave instabilities as energy transfer from high energy particles to thermal particles. The electromagnetic cyclotron waves are excited by the high energy electrons or ions receiving free energy from high energy resonant particles. The waves grow up to a sufficient amplitude to cause wave-wave instabilities such as back-and-forth energy oscillation and modulational or decay instabilities. These transverse wave-wave interactions give rise to longitudinal perturbations, which thermalize the cold particles. Through the inverse cascading, the wave energy is transferred to lower wave number modes as well as thermal particles. Therefore, the free energy of high energy particles are transferred to the thermal particles via nonlinear processes of the electromagnetic cyclotron waves.

We have demonstrated that computer experiments using particle codes help us to understand complicated nonlinear processes in space plasmas. However, the present models are one-dimensional and restricted to the waves at parallel propagation. The parallel propagation is a reasonable assumption because the growth rate of the beam instabilities maximizes in the parallel direction. The one-dimensional system, however, comes from a technical limitation. In order to incorporate the wave-wave interactions properly, we need a very large simulation system with sufficiently large number of grid points.

A special attention should be given to the discreteness of the simulation system for the study of wave-wave interactions. For all kinds of simulations, the wave equations are solved by a difference scheme. Therefore the wave number spectra are not continuous, even if the simulation model is virtually infinite because of the assumed periodicity of the system. The wave numbers allowed in the simulation system are multiple of the minimum wave number $k_{min}=2\pi/L$, where L is the system length of the one-dimensional model. Therefore wave-wave interactions are not adequately studied unless the k-mode resolution, which is given by k_{min}, is sufficiently small. Actually, the wave-wave interactions observed in the present simulations could not be reproduced in simulations with the system of smaller size.

Because of the limitation of computer capability, most of the simulation studies as well as the present simulations have been performed assuming the one-dimensional system. This prohibits existence of some of the plasma wave instabilities such as self-focussing instabilities or filamental instabilities. As a future study, two-dimensional system should be taken for the studies. Needless to say, for the two-dimensional system , we should take enough number of grid points for each dimension.

Acknowledgments. Some of the simulations in the present work were performed in collaboration with H. Tanaka, Y. Iwane and H. Kojima. Computations were done at the Data Processing Center, Kyoto University. Suggestions by two referees C. F. Kennel and T. Tamao were of great help for revision of the manuscript. This research was supported by Grant-in-Aid of MONBUSHO #61460049, #62611519 and #62740243.

References

Fluid Model Simulations of Wave-Wave Interactions

Kruer, W.L, and E.J. Valeo, Nonlinear evolution of the decay instability in a plasma with comparable electron and ion temperatures, *Phys. Fluids, 16*, 675, 1973.

Makino, M., T. Ogino, and S. Takeda, Computer simulation of modulational instability for the electron plasma wave, *Japanese J. Appl. Phys., 18*, 145, 1979.

Nicholson, D.R., G.L. Payne, R.M. Downie, and J.P. Sheerin, Solitons vs. parametric instabilities during ionospheric heating, *Phys. Rev. Lett., 52*, 2152, 1984.

Ogino, T., and S. Takeda, Computer simulation for the parametric instabilities in Plasmas, *J. Phys. Soc. Japan, 38*, 1133, 1975.

Spangler, S.R., J.P. Sheerin, and G.L. Payne, A numerical study of nonlinear Alfvén waves and solitons, *Phys. Fluids, 28*, 104, 1985.

Spangler, S.R., The evolution of nonlinear Alfvén waves subject to growth and damping, *Phys. Fluids, 29*, 2535, 1986.

Thomson, J.J, R.J. Faehl, and W.L. Kruer, Mode-coupling saturation of the parametric instability and electron heating, *Phys. Rev. Lett., 31*, 918, 1973.

Particle Model Simulations of Wave-Wave Interactions

Akimoto, K., H.L. Rowland and K. Papadopoulos, Electromagnetic radiation from strong langmuir turbulence, preprint, LANL, Los Alamos, 1987

Cohen, B.I., M.A. Mostrom, D.R. Nicholson, A.N. Kaufman, and C.E. Max, Simulation of laser beat heating of a plasma, *Phys. Fluids, 18*, 470, 1975.

DeGroot, J.S., and J.I. Katz, Anomalous plasma heating induced by a very strong high-frequency electric field, *Phys. Fluids, 16*, 401, 1973.

Forslund, D.W., J.M. Kindel, and E.L. Lindman, Parametric excitation of electromagnetic waves, *Phys. Rev. Lett., 31*, 249, 1972.

Godfrey, B.B., K.A. Taggart, and C.E. Rhoades Jr., Computer simulation of the saturation of the parametric instability in the weak turbulence regime, *Phys. Fluids, 16*, 2279, 1973.

Hada, T., C.F. Kennel, and T. Terasawa, Excitation of compressional waves and the formation of shocklets in the earth's foreshock, *J. Geophys. Res., 92*, 4423, 1987.

Hasegawa, A., Theory and computer experiment on self-trapping instability of plasma cyclotron waves, *Phys. Fluids, 15*, 870, 1972.

Kruer, W.L., P.K. Kaw, J.M. Dawson, and C. Oberman, Anomalous high-frequency resistivity and heating of a plasma, *Phys. Rev. Lett., 18*, 987, 1970.

Machida, S., S.R. Spangler, and C.K. Goertz, Simulation of amplitude-modulated circularly polarized Alfvén waves for beta less than one, *J. Geophys. Res., 92*, 7413, 1987.

Matsumoto, H., and K. Nagai, Steepening, soliton, and Landau damping of large amplitude magnetosonic waves : Particle code computer simulation, *J. Geophys. Res., 86*, 10068, 1981.

Matsumoto H., and T. Kimura, Nonlinear excitation of electron cyclotron waves by a monochromatic strong microwave: Computer simulation analysis of the MINIX Results, *Space Power, 6*, 187, 1986.

Omura, Y., and H. Matsumoto, Competing processes of whistler and electromagnetic instabilities in the magnetosphere, *J. Geophys. Res., 92*, 8649, 1987.

Omura, Y., H. Usui, and H. Matsumoto, Parallel heating associated with interaction of forward and backward electromagnetic cyclotron waves, *J. Geomag. Geoelectr., 40*, 949, 1988.

Pritchett, P.L., and J.M. Dawson, Electromagnetic radiation from beam-plasma instabilities, *Phys. Fluids, 26*, 1114, 1983.

Terasawa, T., M. Hoshino, J. Sakai, and T. Hada, Decay instability of finite-amplitude circularly polarized Alfvén waves: A numerical simulation of stimulated Brillouin scattering, *J. Geophys. Res., 91*, 4171, 1986.

Studies of Ion Beam Instabilities

Hoshino, M., and T. Terasawa, Numerical study of the upstream wave excitation mechanism, 1, Nonlinear phase bunching of beam ions, *J. Geophys. Res., 90*, 57, 1985.

Kaya, N., H. Matsumoto, and B. T. Tsurutani, Test particle simulation study of whistler wave packets observed near Comet Giacobini-Zinner, *submitted to Geophys. Res. Lett.*, 1988.

Kojima, H., H. Matsumoto, Y. Omura, and B. T. Tsurutani, Nonlinear evolution of high-frequency R-mode waves excited by water group ions near comets: Computer experiments, *submitted to Geophys. Res. Lett.*, 1988.

Matsumoto H., Y. Omura, H. Kojima and B. T. Tsurutani, Linear analysis and computer simulation of wave instabilities driven by cometary ions, *Proceeding of Chapman Conference on Plasma Waves and Instabilities in Magnetosphere and at Comets*, ed. by H. Oya and B. T. Tsurutani, 26, 1987.

Omidi, N., and D. Winske, Simulation of the solar wind interaction with the outer regions of the coma, *J. Geophys. Res., 91*, 397, 1986.

Rogers, B., S. P. Gary, and D. Winske, Electromagnetic hot ion beam instabilities: Quasi-linear theory and simulation, *J. Geophys. Res., 90*, 9494, 1985.

Winske, D., and M.M. Leroy, Diffuse ions produced by electromagnetic ion beam instabilities, *J. Geophys. Res., 89*, 2673, 1984.

Winske, D., and S. P. Gary, Electromagnetic instabilities driven by cool heavy ion beams, *J. Geophys. Res., 91*, 6825, 1986.

Theory of Wave-Wave Interactions

Brinca, A.L., Whistler modulational instability, *J. Geophys. Res., 78*, 181, 1973.

Cohen, Space-time interaction of opposed transverse waves in a plasma, *Phys. Fluids, 17*, 496, 1974.

Cohen, R.H., and R.L. Dewar, On the backscatter instability of solar wind Alfvén waves, *J. Geophys. Res., 79*, 4174, 1974.

Derby, N.F.Jr., Modulational instability of finite-amplitude circularly polarized Alfvén Waves, *Astrophys. J., 224*, 1013, 1978.

Freund, H.P., and K. Papadopoulos, Oscillating two-stream and parametric decay instabilities in a weakly magnetized plasma, *Phys. Fluids, 23*, 139, 1980a.

Freund, H.P., and K. Papadopoulos, Spontaneous emission of radiation from localized Langmuir perturbation, *Phys. Fluids, 23*, 732, 1980b.

Freund, H.P., and K. Papadopoulos, Radiation from a localized Langmuir oscillation in a uniformly magnetized plasma, *Phys. Fluids, 23*, 1546, 1980c.

Goldstein, M.L., An instability of finite amplitude circularly polarized Alfvén waves, *Astrophys. J., 219*, 700, 1978.

Hasegawa, A, C.G. Maclennan, and Y. Kodama, Nonlinear behavior and turbulence spectra of drift waves and Rossby waves, *Phys. Fluids, 22*, 2122, 1979.

Ionson, J.A., and R.S.B Ong, The long time behavior of a finite amplitude shear Alfvén wave in a warm plasma, *Plasma Phys., 18*, 809, 1976.

Karpman, V.I., and E.M. Krushkal, Modulated waves in nonlinear dispersive media, *Soviet Physics JETP, 28*, 277, 1969.

Lashmore-Davies, C.N., Modulational instability of a finite amplitude Alfvén wave, *Phys. Fluids, 19*, 587, 1976.

Lee, Y.C., and P.K. Kaw, Parametric instabilities of ion cyclotron waves in a Plasma, *Phys. Fluids, 15*, 911, 1972.

Longtin, M., and B.U.O. Sonnerup, Modulation instability of circularly polarized Alfvén waves, *J. Geophys. Res., 91*, 6816, 1986.

Mio, K., T. Ogino, K. Minami, and S. Takeda, Modified nonlinear Schrodinger equation for Alfvén waves propagating along the magnetic field in cold plasmas, *J. Phys. Soc. Japan, 41*, 265, 1976a.

Mio, K., T. Ogino, K. Minami, and S. Takeda, Modulational instability and envelope-solitons for nonlinear Alfvén waves propagating along the magnetic field in plasmas, *J. Phys. Soc. Japan, 41*, 667, 1976b.

Mjølhus, E., On the modulational instability of hydromagnetic waves parallel to the magnetic field, *J. Plasma Phys., 16*, 321, 1976.

Ogino, T., M. Makino, and S. Takeda, Modulational instability of electron plasma and ion plasma waves, *J. Phys. Soci. Japan, 43*, 295, 1977.

Ovenden, C.R., H.A. Shah, and S. J. Schwarts, Alfvén solitons in the solar wind, *J. Geophys. Res., 88*, 6095, 1983.

Schmidt, Resonant excitation of electrostatic modes with electromagnetic waves, *Phys. Fluids, 16*, 1676, 1973.

Sakai, J., and B.U.O. Sonnerup, Modulational instability of finite amplitude dispersive Alfvén waves, *J. Geophys. Res., 88*, 9069, 1983.

Spangler, S.R., and J.P. Sheerin, Properties of Alfvén solitons in a finite-beta plasma, *J. Plasma Phys., 27*, 193, 1982.

Tam, C.K.W., Amplitude dispersion and nonlinear instability of whistlers, *Phys. Fluids, 12*, 1028, 1969..

Vahala, G., and D. Montgomery, Parametric amplification of Alfvén waves, *Phys. Fluids, 14*, 1137, 1971.

Wong, H.K., and M.L. Goldstein, Parametric instabilities of circularly polarized Alfvén waves including dispersion, *J. Geophys. Res., 91*, 5617, 1986.

Studies of Whistler Mode Wave-Particle Interaction

Helliwell, R.A., and T.L. Crystal, A feedback model of cyclotron interaction between whistler-mode wave and energetic electrons in the magnetosphere, *J. Geophys. Res., 78*, 7357, 1973.

Matsumoto, H., and Y. Yasuda, Computer simulation of nonlinear interaction between a monochromatic whistler wave and an electron beam, *Phys. Fluids, 19*, 1513, 1976.

Matsumoto, H., K. Hashimoto and I. Kimura, Dependence of coherent whistler interaction on wave amplitude, *J. Geophys. Res., 85*, 644, 1980.

Matsumoto, H., and Y. Omura, Cluster- and channel-effect phase bunchings by whistler waves in the nonuniform geomagnetic field, *J. Geophys. Res., 86*, 779, 1981.

Nunn, D, A self-consistent theory of triggered VLF emissions, *Planet. Space Sci., 22*, 349, 1974.

Omura, Y., and H. Matsumoto, Computer simulations of basic processes of coherent whistler wave-particle interactions in the magnetosphere, *J. Geophys. Res., 87*, 4435, 1982.

Vomvoridis, J. L., and J. Denavit, Nonlinear evolution of a monochromatic whistler wave in a nonuniform magnetic field, *Phys. Fluids, 23*, 174, 1980.

General Introduction to Nonlinear Plasma Wave Instabilities

Cap, F. F., *Handbook on Plasma Instabilities, Volume 3*, Academic Press, 1982.

Davidson, R.C., *Methods in Nonlinear Plasma Theory,* Academic Press, 1972.

Karpman, *Non-Linear Waves in Dispersive Media,* Pergamon Press, 1975.

Kaw, P.K., W.L. Kruer, C.S. Liu and K. Nishikawa, *Advances in Plasma Physics, vol. 6,* John Wiley & Sons, 1976.

Matsumoto, H., Numerical simulations of plasma waves in Geospace, *Physica Scripta, T18*, 188, 1987..

Nicholson, D. R., *Introduction to Plasma Theory,* John Wiley & Sons, 1983.

Sagdeev, R. Z., and A. A. Galeev, *Nonlinear plasma theory,* W. A. Benjamin, 1969.

Tsytovich, V.N., *Nonlinear Effects in Plasma,* Plenum Press, 1970.

Watanabe, S., *Introduction to Soliton Physics,* in Japanese, Baifukan, 1985.

REVIEW OF IONOSPHERIC TURBULENCE

M. Temerin

Space Sciences Laboratory, University of California, Berkeley, California 94720

P. M. Kintner

School of Electrical Engineering, Cornell University, Ithaca, New York 14850

Abstract. The Navier-Stokes equation and its plasma analog indicate that the exchange of energy is produced by the quadratic term of the convective derivative. If one assumes that the energy exchange is local in wave number space, then it is possible to demonstrate that turbulent spectra take on the form of power laws. Since many observations of plasma wave spectra in the ionosphere and magnetosphere yield power laws, the turbulence hypothesis is an attractive interpretation. The least complex environment to study space plasma turbulence is the equatorial ionosphere. Here the Rayleigh-Taylor process injects energy as large coherent structures across a large region of k space, from 50 m to 100 km, and drift waves develop on the steep density gradients. In the most intense cases the drift waves cascade to short wavelengths. Hence the explanation of equatorial observations requires a hierarchy of processes. The high latitude ionosphere is not unstable to the Raleigh-Taylor process but nonetheless there is ample evidence for turbulent processes. The most compelling evidence is the existence of large scale density and electric field spatial irregularities throughout the auroral zone. Satellite evidence for the irregularities exists throughout the auroral zone and over the altitude range of 450 km to 13,000 km. The measurement techniques cover the wavelength range from 10 m to 800 m. In some cases the spectral index is the order of $-5/3$, implying a Kolmogorov process associated with fluid velocity shears (Kelvin-Helmholtz instability). In yet other cases the electric field fluctuations are associated with magnetic field fluctuations implying the existence of kinetic Alfvén waves. In other cases the index is much smaller, the order of -0.5, which does not agree with any of the local theories. In these cases better agreement may be obtained by considering nonlocal effects such as ionospheric coupling.

Introduction

A strong motivation for studying ionospheric turbulence comes from the satellite observations that the strongest electric field and density fluctuations occur over a broad range of low frequencies. In addition, ground-based observers see spread F and strong radio star and satellite radio beacon scintillations. Pictures of the aurora during active times show patterns strongly resembling fluid turbulence (Figure 1). The wave power spectra associated with such phenomena are often characterized by a power law such that the power as a function of frequency is given by $P(\nu) \sim C\,\nu^m$ where m is some negative number the order of -1 to -3 or so. This is very reminiscent of Navier-Stokes turbulence where it is known that in the inertial range the power as a function of wave number obeys a power law. Thus it is tempting to interpret the satellite observations of low-frequency fluctuations as the Doppler shift of spatial irregularities which obey fluid equations similar to those that govern Navier-Stokes turbulence in ordinary fluids. Because of the low β of the plasma, the equations governing the plasma flow in the ionosphere are, in fact, similar to those of two-dimensional Navier-Stokes fluids with, however, the added complications due to neutral collisions, currents, magnetospheric-ionospheric coupling, parallel electric field inhomogeneities and transient effects due to propagating Alfvén waves. Because of these added complications no clear inertial range may exist in some cases and the complete understanding of the low-frequency spectrum must then be based on a detailed examination of the sources of free energy and of dissipation.

Low-frequency turbulence is a common feature of the equatorial, auroral and polar ionosphere. The equatorial ionosphere is a much simpler system and the basic instability mechanisms are fairly well understood. This is not the case for the auroral ionosphere. In the auroral ionosphere some of the important problems involve magnetospheric-ionospheric coupling: to what extent does turbulence generated in the ionosphere extend into the magnetosphere, or vice versa, is a question of current interest. The interpretation of satellite data is complicated by some simple uncertainties, concerning which some progress has recently been made, such as the extent to which the measured spectra reflect propagating waves or Doppler shifted spatial irregularities.

Navier-Stokes Turbulence

The word "turbulence" has different meanings to different people. In this review we wish to be more specific both theoretically and experimentally. Turbulence will be defined as a process where the linear Fourier modes of a system become coupled. That is, the Fourier modes exchange energy at a rate comparable or in excess of the rate that energy is externally channeled into a

Fig. 1. View of the aurora during active times (from Kelley and Kintner, 1978).

single Fourier mode or that energy is removed viscously from a single Fourier mode. An obvious example of this process can be observed in the common coffee cup. Suppose that the cup is vibrated. Depending on the vibrational axis and on the vibrational frequency standing waves may be established on the surface of the coffee. The various possible standing waves are the normal modes of the system. To a good approximation they are independent and exchange no energy. On the other hand, if the coffee is stirred and cream is introduced as a passive tracer, we see a rich variety of structures not described by single normal modes. Vortices form, stretch and then fold upon themselves. This is turbulence.

One convenient method for examining turbulence in fluids is to examine the Navier-Stokes equation which can be expressed as

$$\frac{\partial \mathbf{v}}{\partial t} + \mathbf{v} \cdot \nabla \mathbf{v} = -\nabla P + \mu \nabla^2 \mathbf{v} \tag{1}$$

and

$$\nabla \cdot \mathbf{v} = 0$$

The Fourier series form of this equation is

$$\left[\frac{\partial}{\partial t} + \mu k^2\right] v_i(\mathbf{k}, t) = -\frac{i}{2} P_{ijm}(\mathbf{k}) \sum_{\mathbf{p}+\mathbf{q}=\mathbf{k}} v_j(\mathbf{p}, t) v_m(\mathbf{q}, t) \tag{2}$$

where

$$P_{ijm}(\mathbf{k}) = k_j(\delta_{im} - k_i k_m / k^2) + k_m(\delta_{ij} - k_i k_j / k^2)$$

and the indices i,j,m refer to the cartesian components of v and k. The important point to notice here is that the time rate of change of $\mathbf{v}$ associated with mode $\mathbf{k}$ depends on both the viscous damping term μk^2 and on the nonlinear coupling from modes $\mathbf{p}$ and $\mathbf{q}$ which is on the right hand side of eq. (2) and has a the magnitude of kv^2. The ratio of nonlinear coupling (the inertial terms) to the viscous damping is called the Reynolds number and has the value $R = v/\mu k$. For μ independent of k this implies that for small k the inertial terms dominate and for larger k the viscous terms dominate.

The usual path for the analysis of turbulence is to examine the dependence of the modal energy $E(k) \sim v(k)^2$ on k for some assumptions. A common assumption is that the inertial terms only interact locally. In the absence of viscosity, the kinetic energy defined as

$$U = \sum_k |(v(\mathbf{k})|^2$$

is invariant. If the fluid is stirred on a large scale, energy cascades to smaller scales. At the smallest scales the viscous terms are dominant and kinetic energy is dissipated. In the inertial range, where only the right side of (2) is important, simple dimensional arguments give the form of the power spectrum. This leads to one dimensional power spectra of the form $E(k) \sim k^{-5/3}$ for three-dimensional systems [Kolmogorov, 1941]. Power spectra of this form have in fact been observed in large scale fluids.

The neutral fluid result is similar to the results acquired by assuming an unstable plasma with some form of energy input. An end result of the calculation is generally a power law. Different ranges of k space may have different power laws and the power laws can frequently be compared with experiment. The overall responsibility of a scientist is to develop a convincing argument that an observed power law corresponds with a specific plasma process.

Because of the low β of the ionosphere, it is often convenient to treat the ionosphere as a two-dimensional fluid [Seyler et al., 1975; Ott, 1978] with flows only perpendicular to the magnetic field. In the absence of electrical currents (1) describes such a flow. Taking the curl of (1) then gives

$$\frac{\partial \omega}{\partial t} + \mathbf{v} \cdot \nabla \omega = \mu \nabla^2 \omega \tag{3}$$

where $\boldsymbol{\omega} = \nabla \times \mathbf{v}$ is the vorticity. In the absence of dissipation the flow is given by

$$v = c(\mathbf{E} \times \mathbf{B})/B^2$$

and since $E = -\nabla\Phi$

$$\omega = -\frac{c}{B} \nabla^2 \Phi = \frac{4\pi c \rho}{B}$$

Thus ω can also be thought of as a line charge density. The two-dimensional flow can be thought of as a set of line charges interacting by moving according to $c(\mathbf{E} \times \mathbf{B})/B^2$, or, in other words, the vorticity is carried by the flow.

In Fourier space (3) can be written as

$$\left[\frac{\partial}{\partial t} + \mu k^2\right] \omega(k, t) = \sum_{\mathbf{p}+\mathbf{q}=\mathbf{k}} M(\mathbf{p}, \mathbf{q}) \omega(\mathbf{p}, t) \omega(\mathbf{q}, t) \tag{4}$$

where

$$M(\mathbf{p}, \mathbf{q}) = \frac{\mathbf{b}}{2} \cdot (\mathbf{p} \times \mathbf{q}) \left[\frac{1}{\mathbf{p}^2} - \frac{1}{\mathbf{q}^2}\right] \tag{5}$$

and $\mathbf{b}$ is a unit vector perpendicular to the plane of fluid flow.

Eqs. (4) and (5) have two inviscid invariants: the energy and the enstrophy, which are defined as

$$U = \sum_k |v(\mathbf{k})|^2$$

and

$$\Omega = \sum_k |\omega(k, t)|$$

respectively.

Kraichnan [1967] showed that if such a fluid is stirred at some intermediate scale, energy has a reverse cascade to large scales while enstrophy cascades to smaller scales with the result that the spectrum goes as $E(k) \sim k^{-5/3}$ for large scales and as $E(k) \sim k^{-3}$ for small scales.

The above provides two examples of how the nature of the turbulence is reflected in the form of the power spectrum. Experimentally the spectrum of density or electric field fluctuations is the single most important piece of information as to the nature

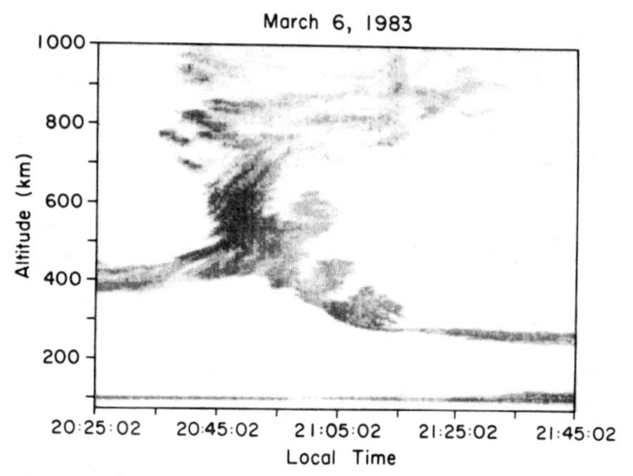

Fig. 2. An Jicamarca backscatter power map associated with equatorial spread F phenomena (from Kelley et al., 1986).

of the wave instabilities and turbulence in the ionosphere. However, it is not the only piece of information that is important nor does a broad, smoothly varying power spectrum necessarily indicate that turbulent processes (i.e., the mixing of scales through the inertial terms in (1)) are dominant. The injection of energy by linear instabilities at various scales or the steepening of linear structure (steepened structures are phase coherent and phase information is lost in the power spectrum) may also produce smoothly varying power spectra similar to those produced by turbulent processes. It is often not clear which processes are dominant in producing the observed power spectra and in some cases several processes may be important for understanding the power spectra. We will now discuss briefly some examples of turbulence from the equatorial ionosphere and, in somewhat more detail, examples from the auroral and polar ionosphere and ionosphere and magnetosphere.

Equatorial F Region

Due to the relative success of neutral fluid turbulence theory it is important to investigate the extent to which such concepts are applicable to space plasmas. One of the least complicated nontrivial turbulent plasmas occurs in the equational F region. Here an interchange instability (Raleigh-Taylor) creates fluctuations in plasma density and potential which span 6 orders of magnitude in spatial scale and twelve orders of magnitude in power spectral density. The outer scales are likely determined by seeding of the instability by atmospheric motions (gravity waves).

Some of the range of scales can be seen in Figure 2, which is a VHF backscatter map of non-thermal signals from the Jicamarca radar. The echoing structure spans hundreds of km, while the scatters themselves are only 3 m in size (less than the ion gyroradius).

From about 50 km to 100 m the primary structuring source is the Rayleigh-Taylor process. To first order the growth rate is independent of wave number in this range. This means that, unlike fluid turbulence, which has a neutral subrange within which energy only cascades from large to small scales, a plasma has the possibility to inject energy across a wide range of k space. This may compete with energy cascade and greatly modify the resulting spectrum.

Experimentally there seems to be four regimes in the equatorial F region which are organized by turbulence strength and altitude. The altitude axis may in fact be better labeled ion collision frequency since the change of this parameter with altitude most likely controls the observed spectral changes. The four regimes are summarized in Table I.

TABLE I. Altitudinal Dependence of Fluctuation Models

	High Altitude (ν_{in} small)	Low Altitude (ν_{in} large)
Weak Fluctuations	k^{-2+} Spectrum Anomalous Diffusion	k^{-2+} Spectrum Classical Diffusion
Strong Fluctuations	$k^{-5/3}$ Spectrum Anomalous Diffusion	Coherent Steepened Structures

The bottom left panel is perhaps the most interesting. As shown in Figure 3 [LaBelle and Kelley, 1986] the observed spectrum (a) is actually in good quantitative agreement with the spectrum of density irregularities predicted by Sudan [1983] and Sudan and Keskinen [1984]. Sudan and Keskinen argued that if the eddy growth rate Γ exceeds the growth rate γ in some range of k space, a Kolmogorov-like spectrum results. One problem with the fit shown in Figure 3 is that it requires an enhanced diffusion coefficient 400 times the classical diffusion. LaBelle and Kelley [1986] and LaBelle et al. [1986] argue that this diffusion could be due to drift waves at smaller scales. Such a

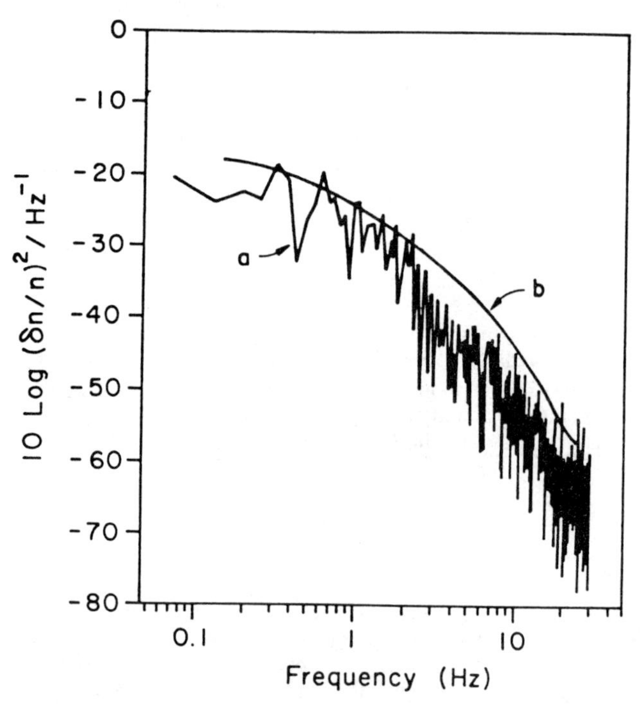

Fig. 3. A comparison of theory and experiment. Curve a is a spectrum of density fluctuation from a rocket flight into a region of equatorial spread F. Curve b is the spectrum of density irregularities predicted by Sudan and Keskinen [1984] (from LaBelle and Kelley, 1986).

high diffusion coefficient results in a sharp break in the spectrum where the damping exceeds growth and the classic "viscous" subrange exists. The density and potential in the viscous range below 100m obey a Boltzman relationship ($\delta n/n = e\phi/k_b T_e$), highly indicative of drift waves [LaBelle et al., 1986].

Though the agreement between the theory and the experiment shown in Figure 3 is encouraging, other spectra often show a $k^{-(2+)}$ relation. Costa and Kelley [1978] and Fejer and Kelley [1980] have pointed out that such spectra result from observed phase coherent steepened density structures. Such steepened and also anisotropic structures are also seen in computer simulations [Keskinen et al., 1980; Zalesak et al., 1982; Zargham and Seyler, 1987]. Kelley et al. [1987] have shown that the steepened and anisotropic structures produced by computer simulations agree well with data. There is evidence then that turbulence is "approached" via development of coherent structures which bifurcate into smaller structures. Only if the process is driven hard is the coherence lost with the result of a Kolmogorov-like spectrum.

High Latitude Ionosphere

The high latitude ionosphere, including the auroral zone and polar cap, is a complex and active environment. Here we will focus on one aspect of the high latitude ionosphere – spatial irregularities and waves above 450 km altitude. Although these waves were discovered almost two decades ago [Maynard and Heppner, 1969; Barrington et al., 1971; Laaspere et al., 1971; Kelley and Mozer, 1972], our understanding of their physics is less mature compared to processes within the equatorial ionosphere, partly because of the increased complexity of the high latitude ionosphere and partly because of the more limited radar coverage at high altitudes.

The physical characteristics of high latitude turbulence have been inferred from measurements by the spacecraft OV1-17 [Kelley and Mozer, 1972; Temerin, 1979b], Hawkeye [Kintner, 1976; Gurnett and Frank, 1977], S3-3 [Temerin, 1978; Temerin et al., 1981], DE-1 [Gurnett et al., 1984], and Viking [Kintner et al., 1987]. Together they have established that low frequency fluctuations commonly exist outside the plasmasphere with peak amplitudes within the auroral zone. The low frequency fluctuations are found over an altitude range of at least as low as 450 km and at least as high as 23,000 km. From wavelengths of 37 m to 10 m the electric field fluctuations have very slow phase velocities, much less than the ion acoustic speed. Density fluctuations coexist with the electric field fluctuations. The density fluctuations are also known to have phase velocities much less than the ion acoustic speed over the wavelength wave range of 800 m to 20 m. In the regions of largest amplitude there are also magnetic field fluctuations with Poynting vectors suggestive of Alfvén waves.

Interpreting satellite data is sometimes not easy. One well known problem is that it is difficult to distinguish between spatial and temporal variations. This means that it is sometimes not clear whether the measured satellite power spectra are due to finite frequency propagating waves or spatial turbulence. Another problem is the origin of the turbulence. The region of auroral field lines can be thought to consist of three parts: a large, fairly homogeneous equatorial magnetosphere, a rather dense ionosphere, and a region along auroral field lines of low plasma density connecting the other two regions through field-aligned currents. All these regions can produce turbulence. Both the ionospheric and magnetospheric regions have relatively large inertia and can sometimes be treated separately. However, a more correct treatment has to take into account the coupling between these regions. In addition, regions of field-aligned currents are themselves unstable and may generate turbulence through current-driven instabilities.

We will concentrate on satellite measurements of electric fields. Such measurements are equivalent to measurements of the plasma flow perpendicular to the magnetic field since $\mathbf{v} = c(\mathbf{E}\times\mathbf{B})/B^2$. Some of the first systematic measurements of high latitude electric field turbulence were made by the OV1-17 satellite [Kelley and Mozer, 1972]. These measurements established that electric field fluctuations in the range of the 10 Hz to 500 Hz (satellite frame, of course) existed throughout the auroral zone and polar cap. The spectral index was -1.46 ± 0.2. Later, Temerin [1979b] established that the electric field of the fluctuations was polarized perpendicular to the magnetic field.

Using data from the Hawkeye satellite at altitudes of a few thousand kilometers, Kintner [1976] found a spectral index of -1.89 ± 0.26 for small amplitude fluctuations. These results were in fairly good agreement with the OV1-17 results. However, Kintner also found that in regions of stronger fluctuations associated with shears in the auroral zone convection, the spectral index decreased to -2.80 ± 0.34 and that there were also magnetic field fluctuations with a spectral index of -4.02 ± 0.59. The ratio of the electric to magnetic fluctuation is an important clue. Figure 4 shows this ratio as a function of frequency during four different occasions [Kintner and Seyler, 1985]. The results rule out two explanations as the cause of these fluctuations: field-aligned currents without associated parallel electric fields and electromagnetic waves [Kintner, 1976]. For both these mechan-

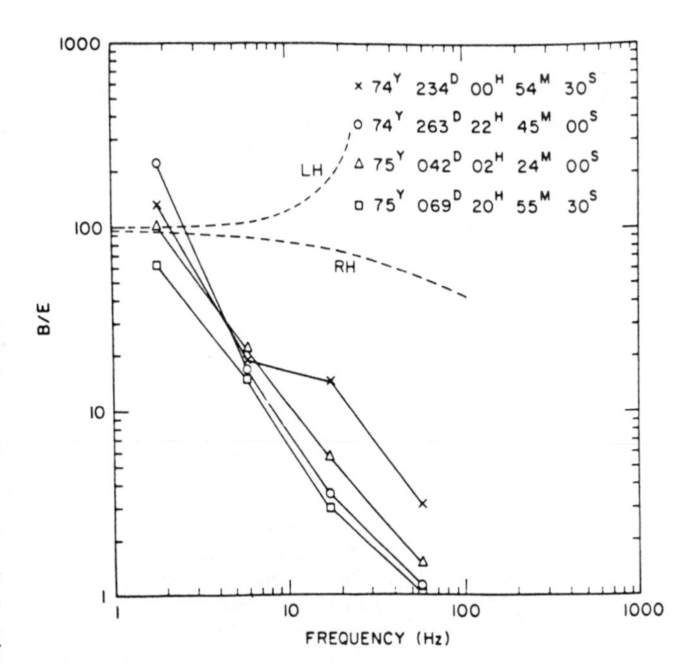

Fig. 4. The calculated index of refraction ($n = cB/E$) during four different times taken from the HAWKEYE satellite. If the electric and magnetic fluctuations were produced by parallel propagating EM waves their index of refraction should behave as the dashed lines for the left-hand and right-hand modes (from Kintner and Seyler, 1985).

isms one would expect a constant ratio of B/E. Gurnett et al. [1984], who acquired similar data from the DE-1 satellite, suggested two additional explanations. They suggested that the fluctuations could be due either to field-aligned currents with associated parallel electric fields or to obliquely propagating kinetic Alfvén waves. Such waves have a fairly large electrostatic component. We would like to point out that such waves are consistent with the data in Figure 4. In the low β ($\beta < m_e/m_i$) plasma of the high altitude auroral zone, the Alfvén dispersion relation is given by

$$\omega = k_{\parallel} V_A \left[1 + \frac{c^2 k_{\perp}^2}{\omega^2} \right] \quad \text{for} \quad \omega \ll \Omega_i$$

and the magnetic to electric field ratio by [Temerin et al., 1986]

$$B/E = \frac{\omega_{pi}}{\omega_{ci}} \frac{\omega_{pe}}{ck} \left[1 + \frac{\omega_p^2}{c^2 k^2} \right]^{-1/2}$$

$$\approx \frac{\omega_{pi}\,\omega_{pe}}{\omega_{ci}\,c\,k_{\perp}} \quad \text{for} \quad ck_{\perp} > \omega_p$$

Thus $B/E \sim k^{-1}$, which is a fairly good fit to the case within the auroral zone convection. The condition $ck_{\perp} > \omega_p$ can be met in the auroral zone at altitudes of a few thousand kilometers for wavelengths of a few kilometer or less. Such wavelengths produce Doppler shifts in the range above 1 Hz.

As has been mentioned, an important concern in the analysis of space plasma data is distinguishing between spatial and temporal fluctuations. One way to resolve this problem is to take advantage of the finite length of the electric field antenna. Temerin [1978] showed that much of the low-frequency electric field turbulence on auroral field lines is due to spatial structures with electric field perpendicular to the magnetic field which have been Doppler shifted by the relative motion of the satellite with respect to the plasma. That this was the case was made evident by the characteristic signatures that such Doppler shifted structures produce on frequency-time spectrograms. Fuselier and Gurnett [1984] have shown that similar features in the ISEE frequency-time spectrograms can also be interpreted by invoking Doppler shift effects.

Data from the S3-3 satellite show examples of unexpectedly flat power spectra showing evidence of spatial turbulence. We will show here some examples of these data in perhaps more detail than can be justified in a review paper because we would like to take this opportunity to present some new data. An example of an electric field power spectrum taken from the S3-3 satellite is shown in Figure 5. The S3-3 was a polar orbiting satellite with three orthogonally oriented pairs of spherical double probe electric field antennas. The satellite spun slowly (~3 rotations per minute) in a cartwheel fashion in polar orbit which had an apogee of a little over 8000 km and a perigee of 240 km. The pairs of spheres separated from each other by wires formed two orthogonally-oriented 37-m-long radial antennas.

Figure 5 shows the power spectrum of data from one of the radial antennae which had the smaller noise level and the larger gain. The data from 0-18.6 kHz were telemetered broadband in a dedicated FM-FM telemetry channel and digitized at 44,000 samples/second. To produce Figure 5 the samples were averaged four at a time and then .37 s of data containing 4096 samples were fast Fourier transformed to produce power spectra of the

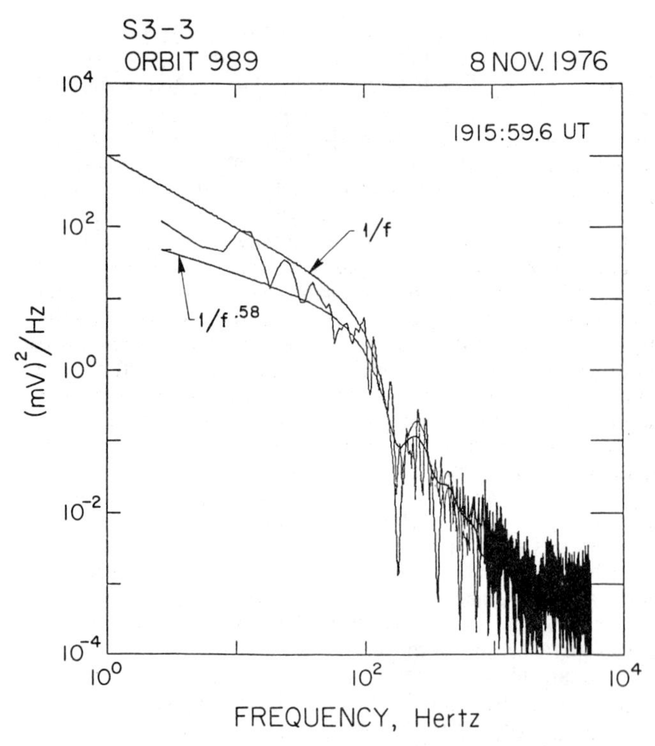

Fig. 5. Example of an electric field power spectrum from the S3-3 satellite. The curve labeled $f^{-.58}$ is the best fit to the power spectrum and takes into account finite antenna effects and bulk plasma drifts. The f^{-1} curve is a guide to the eye and uses the satellite velocity as the relative satellite-plasma velocity.

data between 0-5500 Hz. The detector had a high pass filter at 8.6 Hz. Thus the data below about 10 Hz are not a reliable though the power spectra were adjusted for the nominal gain change at low frequencies. Also it should be noted that at higher frequencies there is a capacitive coupling between the plasma and the antenna which together with the internal capacitance of the detector produces a voltage divider network which attenuates the higher frequency signals. The attenuation at a given frequency depends on the ratio of the resistive to capacitive coupling of the probe with the plasma at that frequency which in turn depends on the plasma density and temperature. These values are not completely known.

The power spectrum in Figure 5 follows an approximately $f^{-.6}$ power law between 10 and 100 Hz and an approximately $f^{-2.6}$ power law between 100 Hz and 2000 Hz at which point the detector reaches its intrinsic noise level. However, it is important to notice that this change in the power law is an artifact of the finite size of the double probe antenna. As pointed out by Temerin [1978, 1979a], the double probe has a complicated but very characteristic response for wavelengths of the order of the double probe separation distance (37 m in this case) or smaller. When this response of the detector is taken into account all the features of the power spectrum can be fitted by a single power law if one assumes that the spectrum is due to the Doppler shift of near zero frequency turbulence. The apparent break in the spectrum corresponds to the Doppler shift of a wave whose length is equal to the effective double probe antenna length. In

Figure 5 the part of the power spectrum above 150 Hz corresponds to the Doppler shifts of structures with scales less than the antenna length. Because of the finite antenna effect this higher frequency portion of the spectrum has an apparent power law index 2 less than the low frequency portion. In addition, biteout features in the spectrum occur when the effective antenna length corresponds to an integer multiple of the Doppler shifted wavelength [Temerin, 1978, 1979a].

The measured power spectrum depends not only on the length of the antenna but also on its orientation both with respect to the magnetic field (this determines the effective antenna length) and with respect to the projection of the velocity vector of the satellite on the plane normal to the magnetic field. Though the orientation of the antenna with respect to the magnetic field is well known, the orientation of the antenna with respect to the projected velocity vector is not well known since the velocity that is relevant is the velocity with respect to the plasma and not necessarily the nominal satellite velocity. Thus under the assumption that the measurement is of near-zero frequency Doppler-shifted two-dimensional turbulence with a single power law, there are still four free parameters that can be used to fit the observed power spectrum: (1) the overall power level, (2) the power law index, and (3,4) the two components of velocity of the plasma in the plane normal to the magnetic field. In Figure 5 the best fit using these four parameters is labelled $f^{-.58}$, that is, by the power law index that best fits the power spectrum had the power spectrum not been affected by finite antenna effects at higher frequencies. In addition there is a one parameter fit labelled f^{-1}. This is intended as a guide to the eye. Only the overall power was fitted: the power index was assumed to be one and the relative velocity was assumed to be the satellite velocity. The largest effect of fitting the velocity is to change the depth of the biteout features in the power spectrum. This is because the direction of the double probe antenna must coincide fairly precisely with the direction of the satellite velocity for the biteouts to be prominent. Usually the fitted velocity is close to the satellite velocity.

The remarkable aspect of this spectrum is the goodness of the fit over the frequency range of 10 Hz to 2000 Hz (corresponding to Doppler shifts of spatial features with scales of 600 m to 3 m) using a single power law. The other remarkable feature of the spectrum is the small value (.58) of the power law index. Such a spectrum, were it to be integrated to infinite frequency (infinitesimal scales), would diverge, giving infinite power in the electric field. Thus there should be a break in the spectrum at smaller spatial scales which in these cases cannot be detected because of the noise level of the instrument.

Such small values of the power law index are typical for turbulence characterized by the "fingerprint" pattern of biteouts on frequency-time spectrograms previously reported [Temerin, 1978]. More examples of electric field data together with their fast Fourier transforms are shown in Figures 6 and 7. Figure 6 shows power spectra from four consecutive segments of data for the same orbit as in Figure 5. Note that the frequency of the biteout features and the break in the measured power spectrum change as they should as the satellite rotates thus changing its effective antenna length perpendicular to the magnetic field.

The low-frequency turbulence does not always show a fingerprint pattern in frequency-time spectrograms or such small values of the power law index. The examples in Figure 8 and 9 show data with larger power law indices. It's clear that if the power law index is equal to 2.83 as in example C, the fingerprint pattern will not be seen because the power will reach the noise level before the first biteout in the power spectrum.

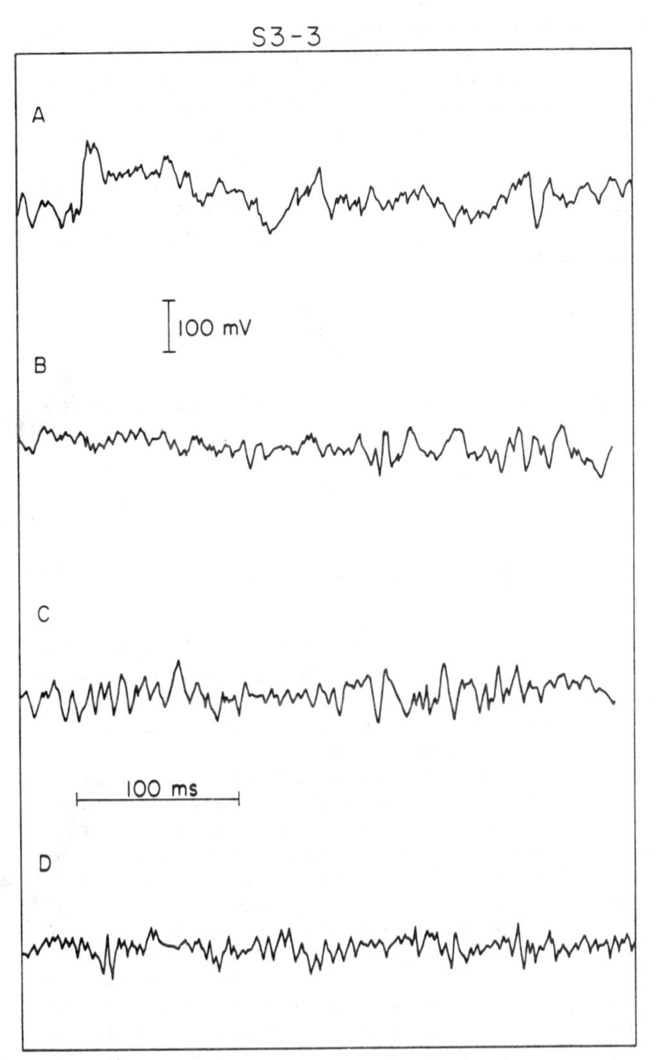

Fig. 6. Examples of electric field data from the S3-3 satellite.

A more direct way to distinguish spatial and temporal variations is to use two separate detectors. This has been done on the Viking satellite [Kintner et al., 1987] using an Langmuir probe interferometer technique [Kintner et al., 1984]. The plasma wave interferometer on the Viking satellite consists of two Langmuir density probes separated by 80 meters on an axis perpendicular to the spacecraft spin axis. The telemetry rate gives a bandwidth of 428 Hz for each probe.

Figure 10 shows an example of data from the interferometer when the satellite was located at an altitude of 7100 km, a magnetic local time of 20:50 and an invariant latitude of 83°. Because of the satellite rotation the antenna axis was oriented nearly along the magnetic field twice each spin. At these times the signal from both probes is virtually the same (panels 1 and 3 of Figure 10). When, however, the probes are oriented perpendicular to the magnetic field (panels 2 and 4 of Figure 10), the signal of the forward probe leads the signal of the trailing probe by a time close to the time it takes the satellite to move the probe separation distance of 80 meters (1.4×10^{-2} s at 5.7 km/s).

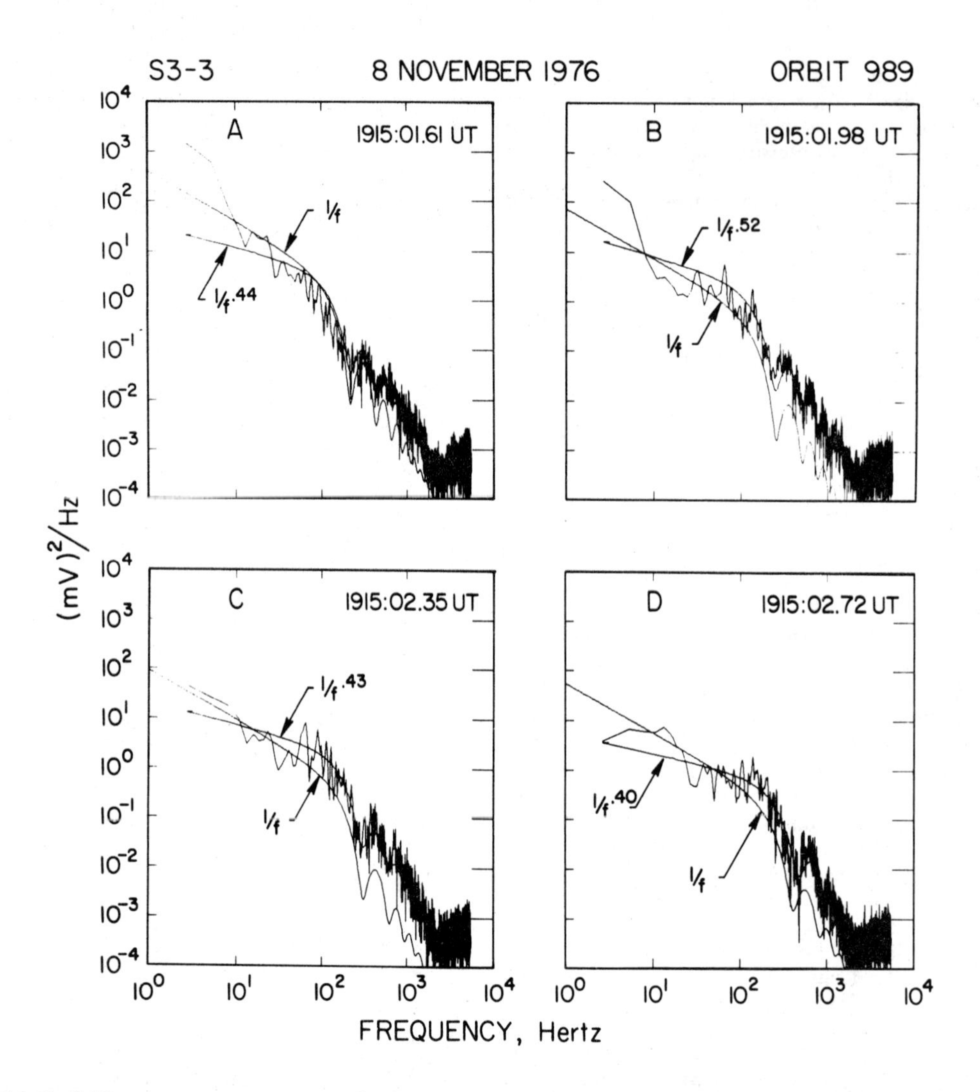

Fig. 7. Electric field power spectra corresponding to the data in Fig. 6. Note the small absolute value of the power law index.

This implies that the density fluctuations are field-aligned and have a phase velocity much smaller than the spacecraft velocity. These results are similar to those previously found by the S3-3 satellite for the electric field.

By using a more quantitative cross-spectral analysis technique, the coherency and the relative phase angle as a function of frequency can be found for the data in Figure 10. These results can be interpreted to show that the power spectrum between 5 Hz and 200 Hz is dominated by the Doppler shift of spatial density structure with spatial scales between 1 km and 25 m.

The density interferometer and electric field double probe results that show clear examples of very slow phase velocity field-aligned structures in the auroral and polar ionosphere are typical of regions of low-intensity turbulence with density fluctuations of the order of 1% and of electric field fluctuations of a few mV/m. The example shown in Figure 10, for instance, is taken from the polar cap poleward of the evening side auroral oval. In the more active regions of the auroral oval fluctuations two orders of magnitude larger can occur in both the density and the electric field. Figure 11 shows an example of data for the Viking satellite taken in an active region associated with ion conic heating. Fluctuations in the electric field of over 100 mV/m occur on time scales of a few milliseconds.

In such active regions the electric field double probe data and the density interferometer data no longer show the characteristic features indicative of spatial irregularities. One possible reason for this is the existence of fast temporal fluctuations due to propagating waves. The existence of large amplitude propagating waves is indicated, for instance, by flickering aurora [Kunitake and Oguti, 1984; Temerin et al., 1986]. However, in other cases a more likely explanation is the existence of large convective ($c(\mathbf{E} \times \mathbf{B})/B^2$) velocities associated with the large turbulent electric fields. In the case of low amplitude turbulence the satellite velocity is typically the largest velocity and the frequency spread of the power spectrum is due to the Doppler shift of spatial turbulence due to the satellite motion. However, 100 mV/m fluctuating electric fields that are typical at the 13,500 km altitude of the Viking satellite in the auroral zone correspond to

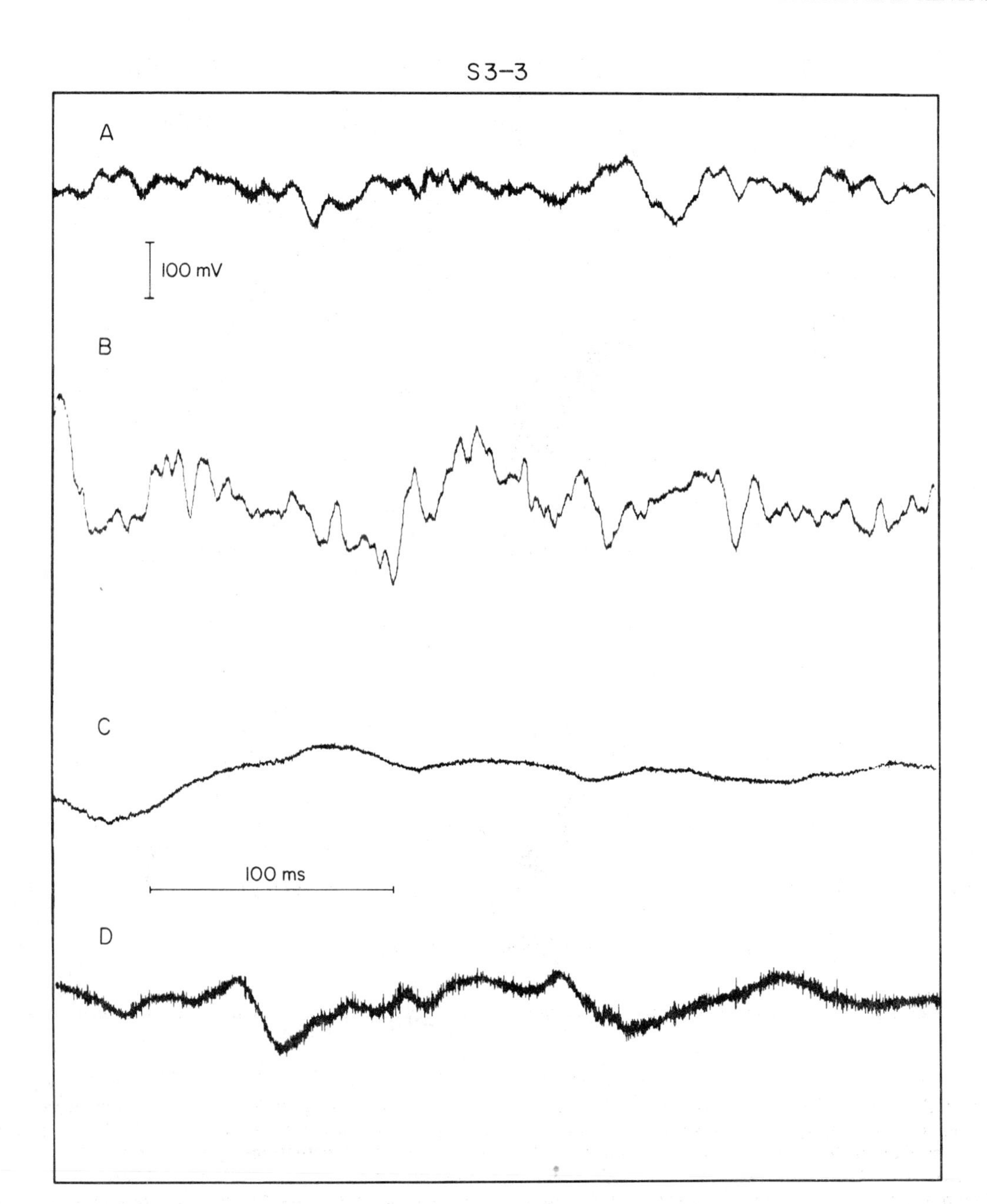

Fig. 8. More examples of electric field data from the S3-3 satellite.

convective velocities of about 100 km/s. Such velocities are much larger than the satellite velocity, so it is unlikely that the Doppler shift from the satellite velocity is the dominant cause of the frequency spread of the power spectrum. Instead, the frequency spread may be due to the Doppler shift of small scale structures embedded in large scale convective flows. This implies that the nonlinear convective terms are dominant and frequencies that one would infer from linear dispersion relations are no longer applicable in describing the power spectrum of the low frequency turbulence.

Theories of High Latitude Turbulence

Theories of high latitude turbulence can be divided into three parts: those which invoke instabilities in the ionosphere, those which invoke instabilities in the outer magnetosphere and those which invoke instabilities along magnetic field lines joining the ionosphere with the outer magnetosphere. In all cases it is necessary to consider magnetosphere-ionosphere coupling since the different inhomogeneous regions along auroral field lines are unlikely to undergo similar instabilities at the same time.

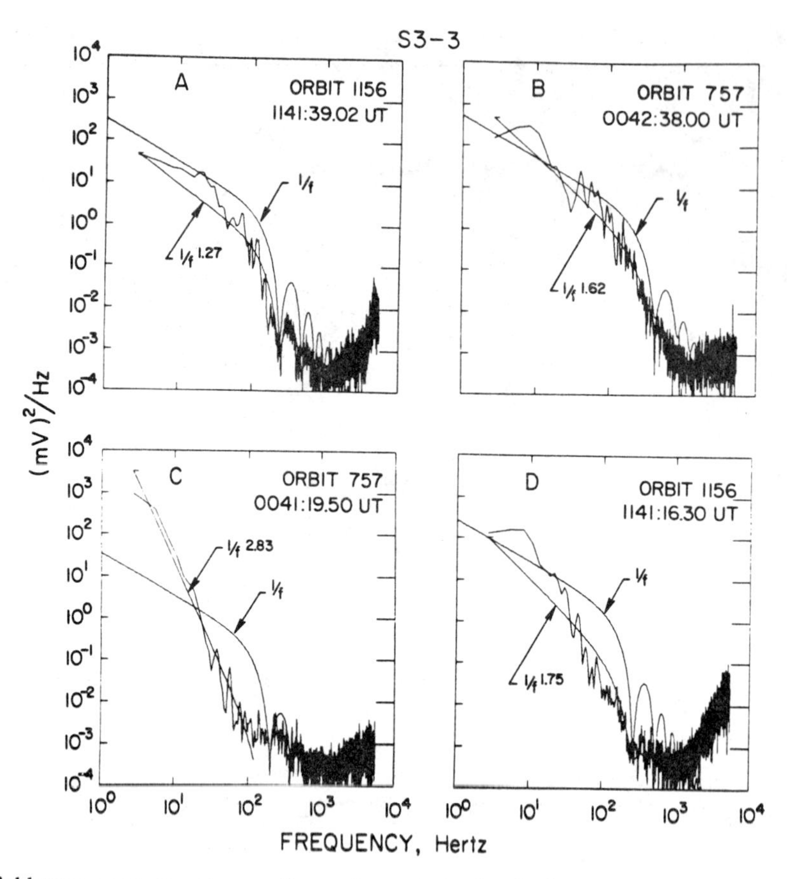

Fig. 9. Electric field power spectra corresponding to the data in Fig. 8. Note the variety of power law indices in these cases.

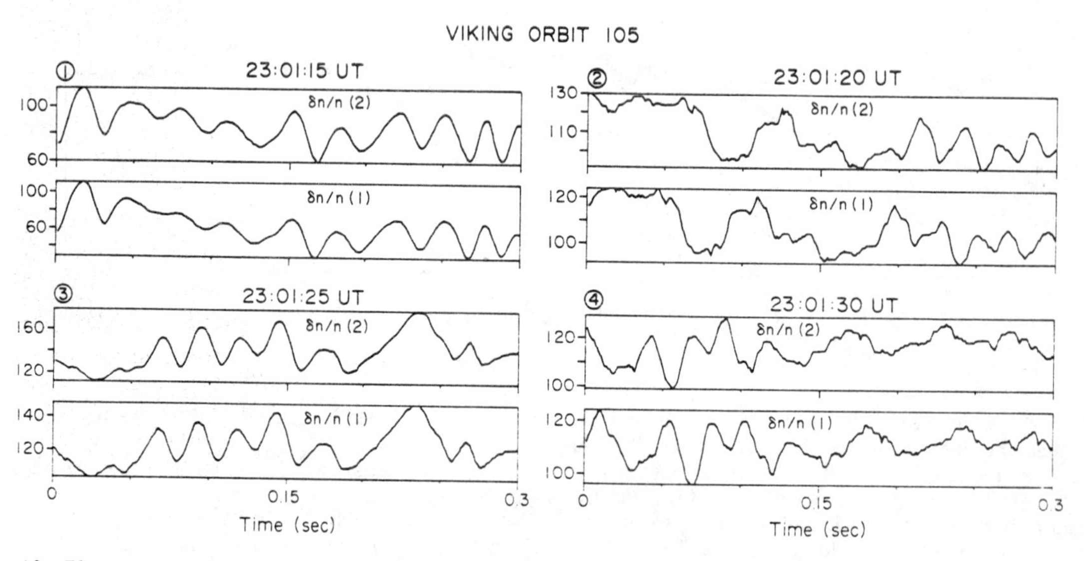

Fig. 10. Plasma density wave interferometer results from the Viking satellite. In panels 1 and 3 the two density probes are aligned with the magnetic field. In panels 2 and 4 one of the density probes leads the other. These results show that the density perturbations are field-aligned and have phase velocities much smaller than the satellite velocity (from Kintner et al., 1987).

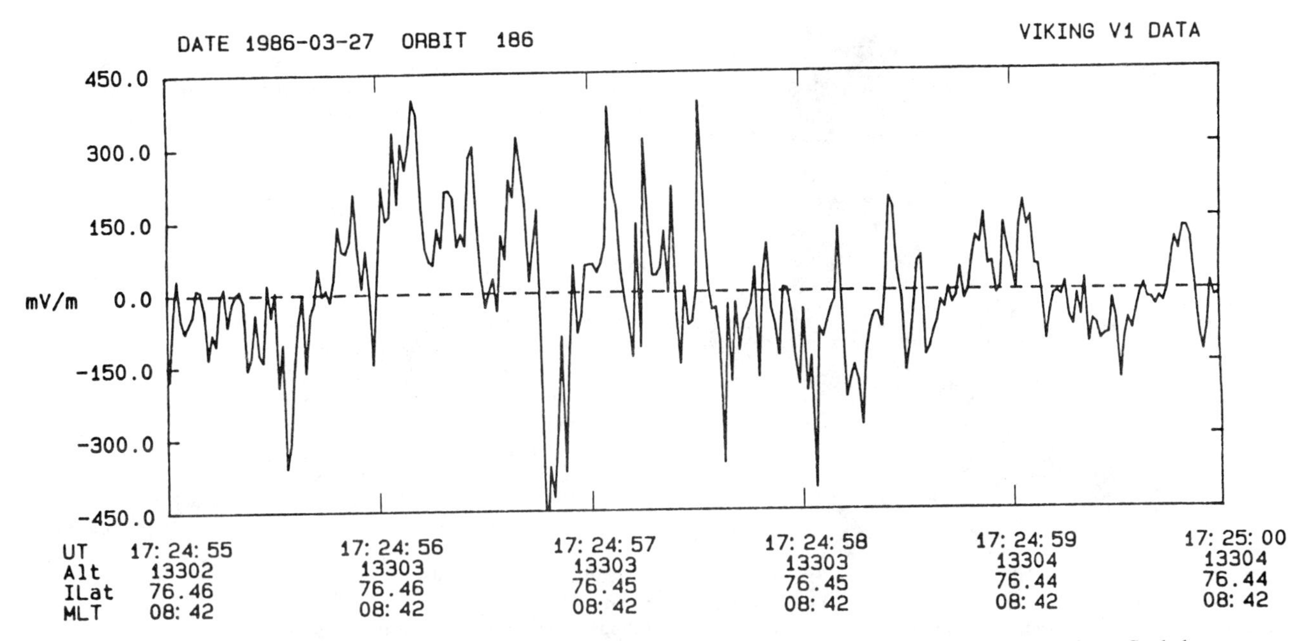

Fig. 11. An example of Viking low-frequency electric field data from a region of ion conic generation. Such large electric fields are common in the Viking data (from data analyzed for the event reported by Hultqvist et al., 1988).

The ionospheric instabilities depend on collisions between the neutral atmosphere and the plasma. One such instability is the **E×B** instability. It depends on the relative motion between a plasma density enhancement and the neutral background as could perhaps happen if a neutral wind were blowing across an auroral arc. In the rest frame of the neutral background such a neutral wind implies an electric field perpendicular to the plasma density gradient. The basic instability mechanism is simple. Because of current continuity, regions of enhanced plasma density (higher conductivity) have smaller electric fields and these convect slower at the $c(\mathbf{E}\times\mathbf{B})/B^2$ velocity. Thus on the front side of the plasma density gradient, plasma density enhancements move slower and thus fall back into regions of higher density which leads to stability while on the back side of density enhancements the same behavior leads to instability as regions of higher density fall back into regions of lower density.

Mitchell et al. [1985] have simulated the behavior of the **E×B** instability. Figure 12 shows the evolution of this instability in the absence of magnetosphere-ionosphere coupling. When magnetospheric coupling in the form of increased inertia is included the instability develops slower and in a more isotropic manner, as shown in Figure 13. Though some possibly important effects of magnetosphere-ionosphere coupling such as parallel electric fields and finite Alfvén wave propagation times are not included in the simulation, the importance of magnetospheric coupling is well illustrated. The nonlinear development of the instability produces an inverse cascade in k_y, the direction of the background electric field. The resultant turbulence is not isotropic because of the continuous presence of the neutral wind which imposes a unique direction on the system.

In the above discussion of the **E×B** instability, turbulence was driven by the relative convection of the plasma with respect to the neutral background. This brings to mind the old question: can the ionosphere regulate magnetospheric convection? [Coroniti and Kennel, 1973]. The usual assumption is that the large scale flows in the high latitude ionosphere are driven by the magnetosphere which, in turn, is driven by the interaction with the solar wind. This suggests that the smaller scale turbulent flows in

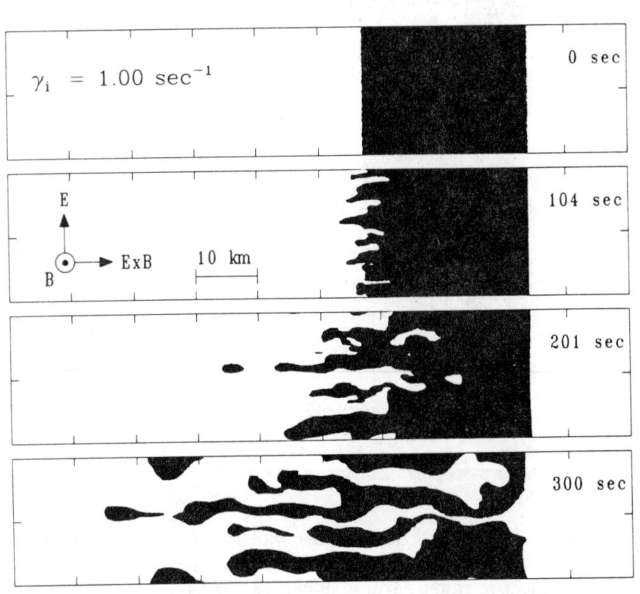

Fig. 12. Results from the simulated evolution of the **E × B** instability without initial effects due to magnetospheric coupling. The shaded region represents those areas whose density is greater than 2.5 times the background density. The simulation grid is periodic in the direction of **E** and is moving with the **E × B** velocity (from Mitchell et al., 1985).

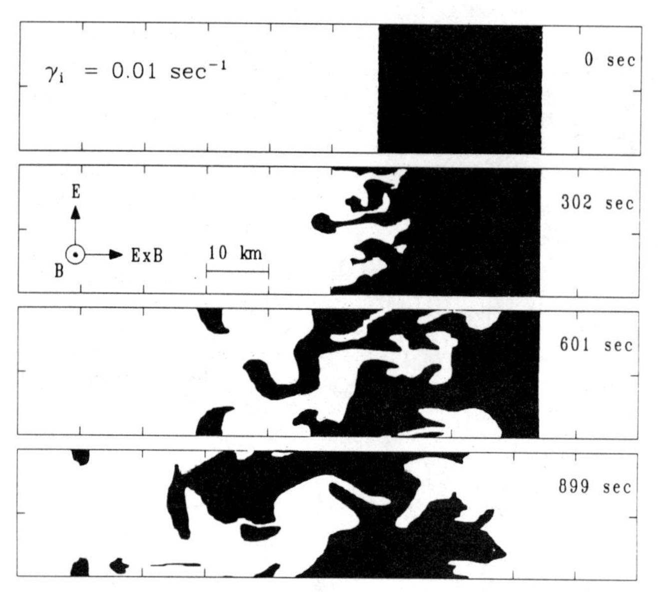

Fig. 13. A simulation similar to that in Figure 12 but with inertial effects due to magnetospheric coupling included. Note that the instability develops more slowly and is qualitatively different (from Mitchell et al., 1985).

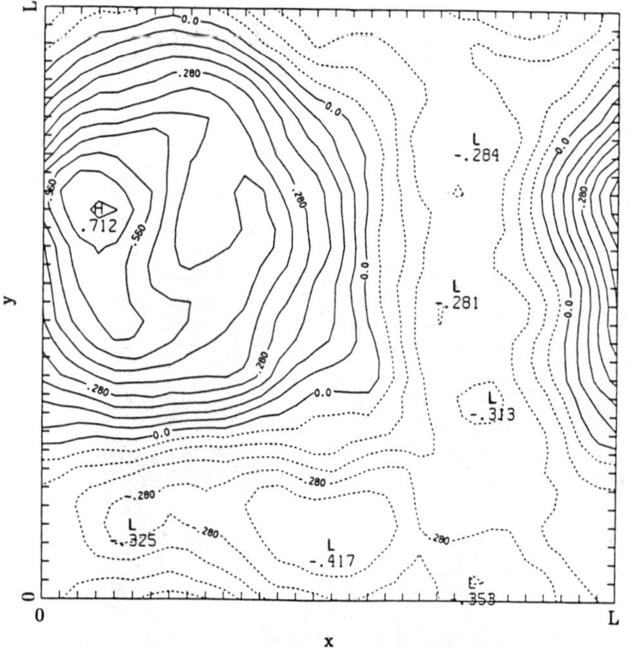

Fig. 14. The randomly selected inertial state (t = 0) for the two-dimensional numerical calculations. Streamlines or contours of constant electric potential are shown in the equatorial plane of the magnetosphere. The circulation direction is clockwise along solid lines and counter-clockwise along dotted lines. The high and low values and some selected contour values are indicated. The x axis maps to a line in the ionosphere that is locally tangent to a contour of constant magnetic latitude; the mapped y axis is tangent to an ionospheric meridian of constant magnetic local time of longitude (from Lotko and Schultz, 1988).

the ionosphere may also result from the mapping of turbulent flows in the outer magnetosphere. Lotko et al. [1987] and Lotko and Schultz [1988] have considered the effect of the ionosphere on turbulent flows in the outer magnetosphere. In the outer magnetosphere, turbulence is presumably generated by Kelvin-Helmholtz instabilities due to sheared flows. However, in their study Lotko and Schultz [1988] do not drive the instability but instead let a given initial turbulent state decay. The plasma is treated as a two-dimensional incompressible Navier-Stokes fluid with a viscosity caused, presumably, by microscale processes. In the simulation the viscosity causes small-scale structures to damp leading to an isotropic k^{-3} power law energy spectrum at large k similar to that expected for small scales in the inertial range of two-dimensional turbulence.

The difference between the simulation and ordinary two-dimensional Navier-Stokes flows is that large scales are also damped. This is because the perpendicular electric field associated with the large scale flows maps to the ionosphere where energy is dissipated by ohmic heating caused by cross-field currents. Small-scale electric fields, however, do not map effectively because they cause large parallel currents which lead instead to parallel electric fields as described by Chiu and Cornwall [1980] and Lyons [1980] and verified by Weimer et al. [1985]. Also the mapping of the large scale electric fields to the ionosphere can introduce an anisotropy in the large-scale turbulence because the mapping of magnetic field lines from the equatorial magnetosphere to the ionosphere doesn't preserve the latitudinal-longitudinal aspect ratio. Figures 14, 15 and 16 show an example of a simulation that models the decay of magnetospheric turbulence. Figure 14 shows the initial streamlines or contours of constant electric potential, while Figure 15 shows the

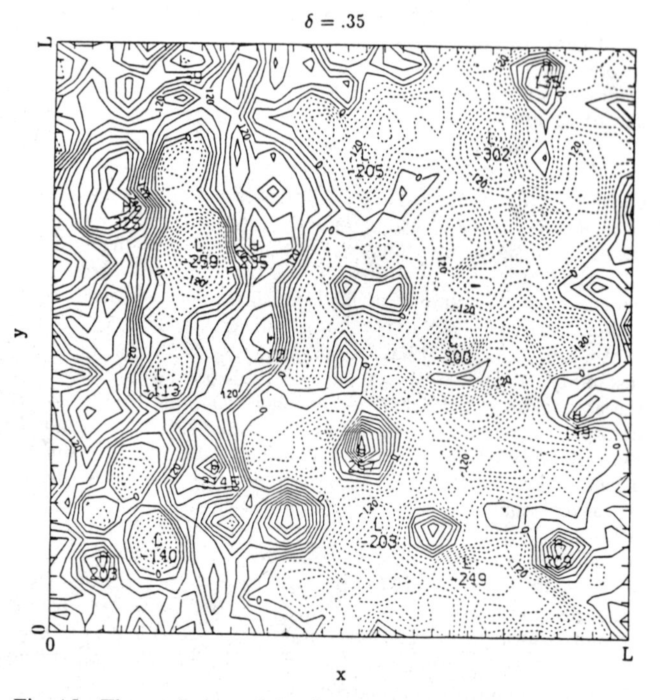

Fig. 15. The evolution of the flow shown in Figure 14. Streamline distribution after time γt = 10 where γ is the resistive damping rate associated with ionospheric drag on the convecting magnetic field lines. High and low values should be multiplied by 10^{-4} (from Lotko and Schultz, 1988).

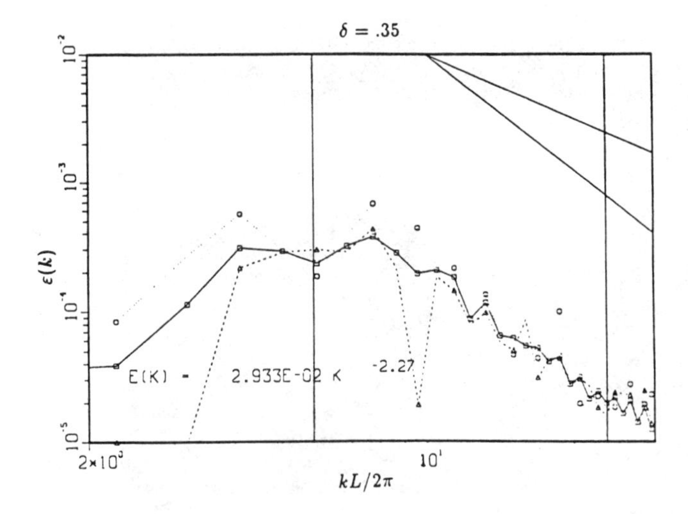

Fig. 16. Energy spectrum for the flow represented in Fig. 15. Dotted line – averaged over values within 30° of $\theta = 0°$ of $k = k_x$; dashed line – averaged over values within 30° of $\theta = \pi/2$ or $k = k_y$; solid line – averaged over all angles. The straight slanted lines in the upper right corner refer to $k^{-5/3}$ and k^{-3} power laws (from Lotko and Schultz, 1988).

streamlines after time $\gamma t = 10$. The energy spectrum is shown in Figure 16. For small values of k the power is substantially reduced by ionospheric damping and is distributed anisotropically, as can be seen since the dotted line in Figure 16 represents the spectrum in the latitudinal direction while the dashed line represents the longitudinal spectrum. Because of this damping the energy spectrum does not have the $k^{-5/3}$ relation one would expect from the inverse cascade at small values of k. A similar simulation, but taking into account magnetic perturbations, has been performed by Song and Lysak [1988] who find that the inclusion of magnetic perturbations increases the rate of cascade of energy to small scales.

One method of generating turbulence in both the outer magnetosphere and at intermediate altitudes is through shear flows. Keskinen et al. [1988] have analyzed the nonlinear evolution of the Kelvin-Helmholtz instability in the high latitude ionosphere. The instability develops differently depending on whether ionosphere-magnetosphere coupling is included. Figure 17 shows the development of the Kelvin-Helmholtz instability without ionospheric coupling, while Figure 18 shows the development with ionospheric coupling. It is apparent that the development is qualitatively different, and subjectively one would say more realistic, with the ionospheric effects included.

While the mechanism mentioned above and other similar mechanisms all contribute to turbulence seen in the ionosphere and along auroral field lines, none is adequate to explain the

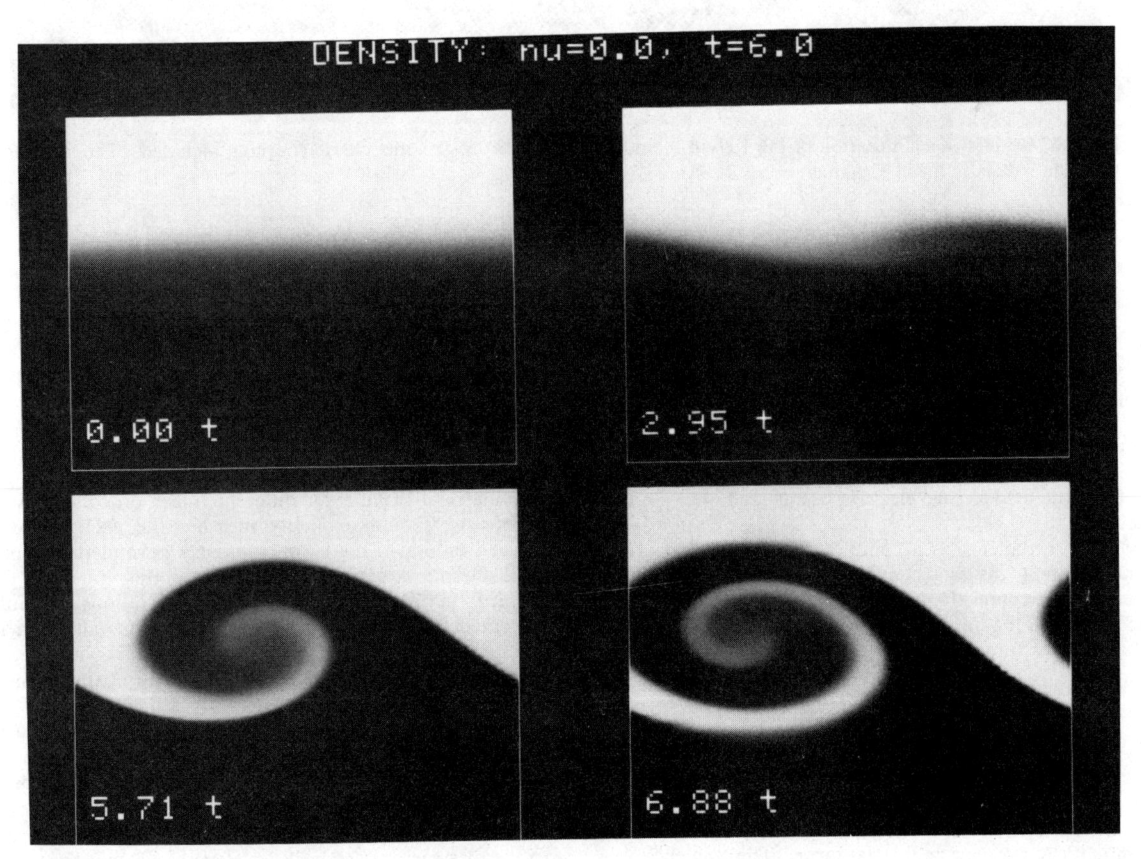

Fig. 17. Results of a simulation of the Kelvin-Helmholtz instability without ionospheric coupling. The shading represents density (from Keskinen et al., 1988).

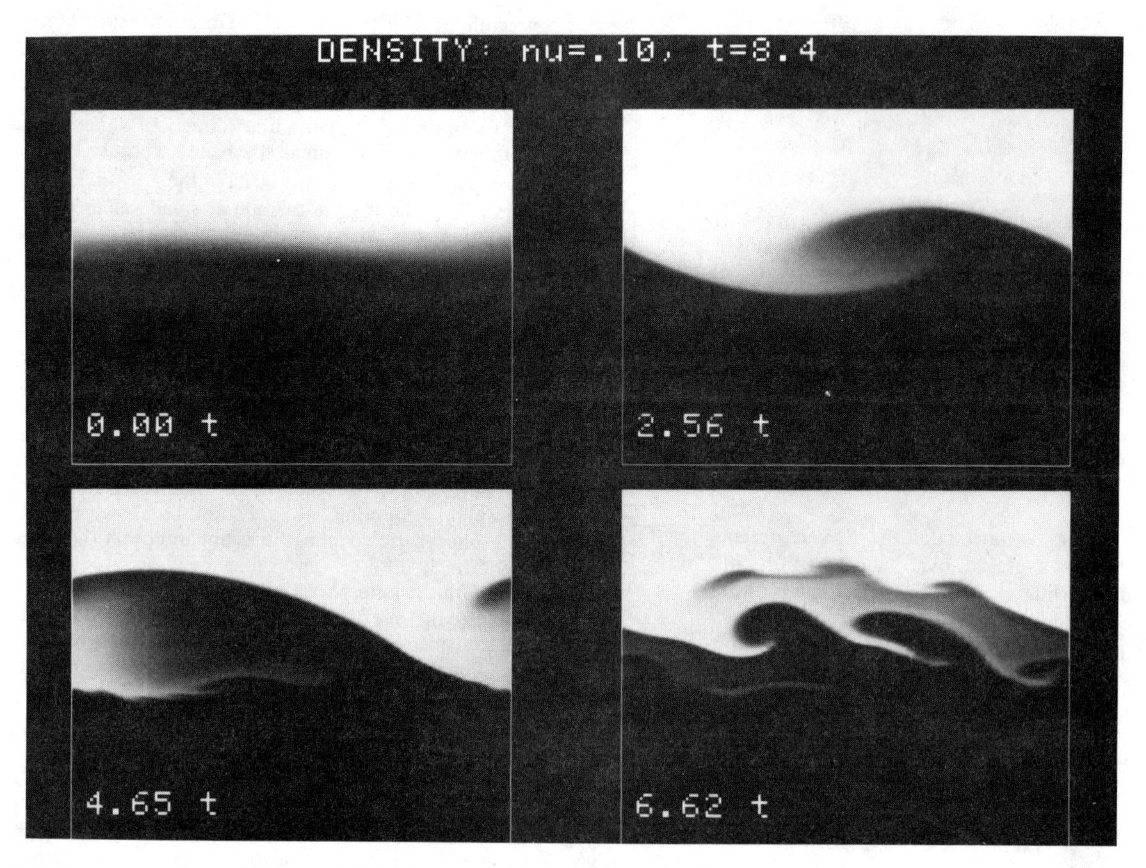

Fig. 18. Results of a simulation of the Kelvin-Helmholtz instability with ionospheric coupling included. The shading represents density (from Keskinen et al., 1988).

large small-scale electric field seen at intermediate altitudes on auroral field lines. As already mentioned, the Viking satellite typically saw electric fields with amplitudes greater than 100 mV/m in the auroral zone. Because such electric fields are larger than one would expect from the mapping of either ionospheric or magnetospheric electric fields to intermediate altitudes, the electric fields seen at intermediate altitudes are probably generated locally through instabilities in the field-aligned currents. At these altitudes low frequency turbulence is associated with the generation of ion conics, field-aligned electron beams, and other auroral phenomena.

High latitude turbulence is a complex subject because of the variety of different instability and damping mechanism and because of the interconnection of diverse ionospheric and magnetospheric regions along auroral field lines. These problems provide a challenge for future research.

Summary

In this paper we have suggested that turbulence should be defined by considering the nonlinear coupling of the normal modes of a system. If energy is exchanged between the normal modes at a rate which is fast compared to damping rates, then the system is turbulent over an inertial range. The Navier-Stokes equation and its plasma analog indicate that the exchange of energy is produced by the quadratic term of the convective derivative. If one assumes that the energy exchange is local in wave number space, then it is possible to demonstrate that turbulent spectra take on the form of power laws. Since many observations of plasma wave spectra in the ionosphere and magnetosphere yield power laws, the turbulence hypothesis is an attractive interpretation.

On the other hand one should be careful of simply assuming that a power law spectrum implies a turbulent process. It may instead be the saturated state of a linear process with no mode coupling or a mixture of different linear processes. A rigorous experiment should explicitly measure the exchange of energy between modes. This is unfortunately beyond the scope of any space plasma experiment. To resolve this dilemma one may appeal to simulations. Fluid plasma simulations do confirm the concept of mode coupling in many situations although in other situations they indicate the development of coherent nonlinear structures. Obviously the evaluation of any experimental result should carefully consider all alternatives.

The least complex environment to study space plasma turbulence is the equatorial ionosphere. Here the Rayleigh-Taylor process injects energy as large coherent structures across a large region of k space, from 50 m to 100 km, and drift waves develop on the steep density gradients. In the most intense cases the drift waves cascade to short wavelengths. Hence the explanation of equatorial observations requires a hierarchy of processes.

The high latitude ionosphere is not unstable to the Raleigh-

Taylor process but nonetheless there is ample evidence for turbulent processes. The most compelling evidence is the existence of large scale density and electric field spatial irregularities throughout the auroral zone. Satellite evidence for the irregularities exists throughout the auroral zone and over the altitude range of 450 km to 13,000 km. The measurement techniques cover the wavelength range from 10 m to 800 m. In some cases the spectral index is the order of $-5/3$, implying a Kolmogorov process associated with fluid velocity shears (Kelvin-Helmholtz instability). In yet other cases the electric field fluctuations are associated with magnetic field fluctuations implying the existence of kinetic Alfvén waves. In other cases the index is much smaller, the order of -0.5, which does not agree with any of the local theories. In these cases better agreement may be obtained by considering nonlocal effects such as ionospheric coupling.

Since drift waves are important within the context of equatorial turbulence they may also be important in the high latitude, high altitude ionosphere/magnetosphere. For drift waves to grow steep density gradients are necessary. Large density cavities with associated gradients are known to be produced within auroral acceleration regions. After the auroral event is over the density gradients remain and are convected throughout the polar cap and auroral zone. In this quiescent environment the drift wave instability has hours and perhaps days to grow and develop into a turbulent state. This hypothesis has yet to be examined in detail because it is necessary first to compare large scale density measurements with measurements of spatial irregularities and then to determine whether the density fluctuations are acting as a passive scalar or are participating in the momentum equations. Hopefully this will be an active area for future research.

The appeal of turbulence as an explanation for many observations of plasma waves stems from the widespread occurrence of turbulence in neutral fluids. To the extent that the ionosphere and magnetosphere are described by a fluid approximation we expect turbulence to be important. Proving this hypothesis with rigor is at best a difficult task and it is fair to state that even in the compelling examples there is room for doubt. The experimental evidence is suggestive but not conclusive.

Acknowledgments. We thank C. Seyler for helpful discussions. This research was supported at Berkeley by Office of Naval Research Research contract N00014-81-C-0006 and National Science Foundation grant ATM-8517737. The work at Cornell was funded by ONR grant N00014-81-K-0018.

References

Barrington, R. E., T. R. Hartz, and R. W. Harvey, Diurnal distribution of ELF, VLF, and LF noise at high latitudes as observed by Alouette 2, *J. Geophys. Res.*, *76*, 5278, 1971.

Chiu, Y. T., and J. M. Cornwall, Electrostatic model of a quiet auroral arc, *J. Geophys. Res.*, *85*, 543, 1980.

Coroniti, F. V., and C. F. Kennel, Can the ionosphere regulate magnetospheric convection? *J. Geophys. Res.*, *78*, 2837, 1973.

Costa, E., and M. C. Kelley, On the role of steepened structures and drift waves in equatorial spread F, *J. Geophys. Res.*, *83*, 4359, 1978.

Fejer, B. G., and M. C. Kelley, Ionospheric irregularities, *Rev. Geophys.*, *18*, 401, 1980.

Fuselier, S. A., and D. A. Gurnett, Short wavelength ion waves upstream of the earth's bow shock, *J. Geophys. Res.*, *89*, 91, 1984.

Gurnett, D. A., and L. A. Frank, A region of intense plasma wave turbulence on auroral field lines, *J. Geophys. Res.*, *82*, 1031, 1977.

Gurnett, D. A., R. L. Huff, J. D. Menietti, J. L. Burch, J. D. Winningham, and S. D. Shawhan, Correlated low frequency electric and magnetic noise along the auroral field lines, *J. Geophys. Res.*, *89*, 8971, 1984.

Hultqvist, B., R. Lundin, K. Stasiewicz, L. Block, P.-A. Lindqvist, G. Gustafsson, H. Keskinen, A. Bohnsen, T. A. Potemra, and L. J. Zanitti, Simultaneous observations of upward moving field-aligned energetic electrons and ions on auroral zone field lines, *J. Geophys. Res.*,, in press, 1988.

Kelley, M. C., and P. M. Kintner, Evidence for two-dimensional inertial turbulence in a cosmic-scale low-β plasma, *Astrophys. J.*, *220*, 339, 1978.

Kelley, M. C., and F. S. Mozer, A satellite survey of vector electric fields in the ionosphere at frequencies of 10 to 500 Hz, *J. Geophys. Res.*, *77*, 4158, 1972.

Kelley, M. C., J. LaBelle, E. Kudeki, B. G. Fejer, Sa. Basu, Su. Basu, K. D. Baker, C. Hanuise, P. Argo, R. F. Woodman, W. E. Swartz, D. T. Farley, and J. W. Meriwether, Jr., The Condor Equatorial Spread F campaign: overview and results of the large-scale measurements, *J. Geophys. Res.*, *91*, 5487, 1986.

Kelley, M. C., C. E. Seyler, and S. Zargham, Collisional interchange instability 2. A comparison of the numerical simulations with the in situ experimental data, *J. Geophys. Res.*, *92*, 10,089, 1987.

Keskinen, M. J., S. L. O. Ossakow, and P. K. Chaturvedi, Preliminary report of numerical simulations of intermediate wavelength collisional Rayleigh-Taylor instability in equatorial spread F, *J. Geophys. Res.*, *85*, 1775, 1980.

Keskinen, M. J., H. G. Mitchell, J. A. Fedder, P. Satyanarayana, S. T. Zalesak, and J. D. Huba, Nonlinear evolution of the Kelvin-Helmholtz instability in the high latitude ionosphere, *J. Geophys. Res.*, *93*, 137, 1988.

Kintner, P. M., Observations of velocity shear turbulence, *J. Geophys. Res.*, *81*, 5114, 1976.

Kintner, P. M., J. LaBelle, M. C. Kelley, L. J. Cahill, Jr., T. Moore and R. Arnoldy, Interferometric phase velocity measurements, *Geophys. Res. Lett.*, *11*, 19, 1984.

Kintner, P. M., and C. E. Seyler, The status of observations and theory of high latitude ionospheric and magnetospheric plasma turbulence, *Space Sci. Rev.*, *41*, 91, 1985.

Kintner, P. M., M. C. Kelley, G. Holmgren, H. Keskinen, G. Gustafsson, and J. LaBelle, Detection of spatial density irregularities with the Viking plasma wave interferometer, *Geophys. Res. Lett.*, *14*, 467, 1987.

Kolmogorov, A. N., The local structure of turbulence in incompressible viscous fluids for very high Reynolds numbers, *Compti. Ren. Acad. Sci. U.S.S.R.*, *30*, 301, 1941.

Kraichnan, R. H., Inertial ranges in two-dimensional turbulence, *Phys. Fluids*, *10*, 1417, 1967.

Kunitake, M., and T. Oguti, Spatial-temporal characteristics of flickering spots in flickering auroras, *J. Geomagn. Geoelectr.*, *36*, 121, 1984.

Laaspere, T., W. C. Johnson, and L. C. Semprebon, Observations of auroral hiss, LHR noise and other phenomena in the frequency range 20 Hz-540 kHz on Ogo 6, *J. Geophys. Res.*, *76*, 4477, 1971.

LaBelle, J., and M. C. Kelley, The generation of kilometer scale irregularities in equatorial spread F, *J. Geophys. Res.*, *91*, 5504, 1986.

LaBelle, J., M. C. Kelley, and C. E. Seyler, An analysis of the role of drift waves in equatorial spread F, *J. Geophys. Res.*, *91*, 5513, 1986.

Lotko, W., B. U. O. Sonnerup, and R. L. Lysak, Nonsteady boundary layer flow including ionospheric drag and parallel electric fields, *J. Geophys. Res., 92,* 8635, 1987.

Lotko, W., and C. G. Schultz, Internal shear layers in auroral dynamics, in *Magnetosphere-Ionosphere Plasma Models, AGU Monograph,* in press, 1988.

Lyons, L. R., Generation of large-scale regions of auroral currents, electric potentials, and precipitation by the divergence of the convection electric field, *J. Geophys.Res.,85,*17,1980.

Maynard, N. C., and J. P. Heppner, Variations in electric fields from polar orbiting satellites, *NASA/GSFC X Doc 612-69-374,* 1969.

Mitchell, H. G., J. A. Fedder, M. J. Keskinen, and S. T. Zalesak, A simulation of high latitude F-layer instabilities in the presence of magnetosphere-ionosphere coupling, *Geophys. Res. Lett., 12,* 783, 1985.

Ott, E., The theory of Rayleigh-Taylor bubbles in the equatorial ionosphere, *J. Geophys. Res., 83,* 2066, 1978.

Seyler, C. E., Y. Salu, D. Montgomery, and G. Knorr, Two-dimensional turbulence in inviscid fluids or guiding center plasmas, *Phys. Fluids, 18,* 803, 1975.

Song, Y., and R. L. Lysak, Turbulent generation of auroral currents and fields – a spectral simulation of 2-D MHD turbulence, in *Magnetosphere-Ionosphere Plasma Models, AGU Monograph,* in press, 1988.

Sudan, R. N., Unified theory of Type I and Type II irregularities in the equatorial electrojet, *J. Geophys. Res., 88,* 4853, 1983.

Sudan, R. N., and M. J. Keskinen, Unified theory of the power spectrum of intermediate wavelength ionospheric electron density fluctuations, *J. Geophys. Res., 89,* 9840, 1984.

Temerin, M., The polarization, frequency, and wavelength of high-latitude turbulence, *J. Geophys. Res., 83,* 2609, 1978.

Temerin, M., Doppler shift effects on double-probe measured electric field power spectra, *J. Geophys. Res., 84,* 5929, 1979a.

Temerin, M., Polarization of high latitude turbulence as determined by analysis of data from the OV1-17 satellite, *J. Geophys. Res., 84,* 5935, 1979b.

Temerin, M., C. Cattell, R. Lysak, M. Hudson, R. B. Torbert, F. S. Mozer, R. D. Sharp, and P. M. Kintner, The small-scale structure of electrostatic shocks, *J. Geophys. Res., 86,* 11,278, 1981.

Temerin, M., J. McFadden, M. Boehm, C. W. Carlson, and W. Lotko, Production of flickering aurora and field-aligned electron flux by electromagnetic ion cyclotron waves, *J. Geophys. Res., 91,* 5769, 1986.

Weimer, D. R., C. K. Geortz, D. A. Gurnett., N. C. Maynard, and J. L. Burch, Auroral zone electric fields from DE1 and 2 at magnetic conjunctions, *J. Geophys. Res., 90,* 7479, 1985.

Zalesak, S. T., S. L. Ossakow, and P. K. Chaturvedi, Nonlinear equatorial spread F: The effect of neutral winds and background pedersen conductivity, *J. Geophys. Res., 87,* 151, 1982.

Zargham, S., and C. E. Seyler, Collisional interchange instability 1. Numerical simulations of intermediate scale irregularities, *J. Geophys. Res., 92,* 10,073, 1987.

DISCRETE ELECTROMAGNETIC EMISSIONS IN PLANETARY MAGNETOSPHERES

Roger R. Anderson and William S. Kurth

Department of Physics and Astronomy, The University of Iowa, Iowa City, Iowa 52242

Abstract. A common feature of electromagnetic plasma instabilities in planetary magnetospheres explored to date is the occurrence of discrete emissions. The first concentrated studies of whistlers and discrete emissions were conducted using ground based very-low-frequency receivers in the early part of this century. Many of these emissions were given names such as "chorus", "risers", and "hooks", which attempted to describe the sound of the received signals when they were routed through an amplifier and speaker system. We now know that the majority of these emissions are generated by whistler-mode plasma instabilities in the Earth's magnetosphere. Despite the advanced state of our understanding of plasma instabilities, the reasons that the waves organize themselves into intense narrowband wave packets and the detailed explanation of the frequency variations are still the subject of study (experimentally, theoretically and by computer simulation) and debate. Discrete whistler-mode "chorus" emissions have been observed in the magnetospheres of Earth, Jupiter, Saturn, and Uranus. Although differing in some details, the basic character of the emissions, consisting of a series of narrowband tones each rising or falling in frequency on a time scale ranging from tenths of a second to several seconds, is essentially the same. Another type of discrete emission called "lion roars" also occurs in the Earth's magnetosheath with characteristics very similar to chorus except it tends to occur at a lower frequency. Typical energies for electrons resonating with waves from these various discrete emissions range from a few hundred eV to as high as several hundred keV. In the Earth's magnetosphere discrete whistler-mode emissions have been shown to play an important role in the pitch-angle scattering and loss of energetic electrons from the radiation belt. Electromagnetic ion cyclotron waves in the Earth's magnetosphere have intense narrowband features similar in some respects to chorus. At frequencies above both the electron cyclotron frequency and the electron plasma frequency, terrestrial auroral kilometric radiation, which propagates predominantly in the free space (R-X) mode, has a very complex frequency-time structure consisting of many discrete emissions. A similar type of radio emission at Jupiter called S-burst has been studied for many years by radio astronomers and is characterized by discrete tones drifting downward in frequency. Whether these free space mode emissions are fundamentally similar to the discrete ion cyclotron and whistler-mode emissions is an open question. Discrete electromagnetic emissions have also been observed to be triggered in the magnetosphere by signals from ground transmitters as well as from lightning generated whistlers. Numerous active experiments, including ground transmitters, ionospheric heaters, and artificially injected particle beams and plasma, have produced discrete electromagnetic emissions in the Earth's ionosphere and magnetosphere. Both the observations and interpretations of discrete emissions will be reviewed.

Introduction

Naturally occurring electromagnetic radiation in the ELF-VLF frequency range observed in the Earth's magnetosphere can typically be classified as whistlers, hiss, or discrete emissions. A common feature of electromagnetic plasma instabilities in planetary magnetospheres explored to date is the occurrence of discrete emissions. The purpose of this paper is to review the present state of knowledge and observations of discrete electromagnetic emissions detected in the magnetospheres of Earth and the outer planets of the solar system. We will review the early history of discrete emission observations in Section 2. Section 3 will concentrate on ground observations made since the 1940's. In Section 4 we will cover the satellite observations of discrete electromagnetic emissions and related phenomena at the Earth. In Section 5 we will review observations at other planets. Active experiments studying wave propagation, wave-wave interactions, wave-particle interactions, and the generation and triggering of discrete emissions are covered in Section 6. Presentations of various theories proposed to explain the phenomena associated with discrete emissions are contained in Section 7. Computer simulations of wave-particle interactions and the generation and triggering of discrete emissions will be reviewed in Section 8. Section 9 discusses related phenomena observed at frequencies above both the electron cyclotron frequency and the plasma frequency. A summary of the paper and thoughts on future directions of research related to discrete electromagnetic emissions are included in the final section.

History

The first known report on observations of discrete electromagnetic emissions was made by Preece [1894]. Operators listening to telephone receivers connected to telegraph wires during a display of aurora borealis in March 1894 heard tweeks and possibly whistlers and dawn chorus. Observations of whistlers in Austria dating back to 1886 were reported by Fuchs [1938]. The whistlers were heard on a 22-km telephone line without amplification. Barkhausen [1919] discussed new phenomena discovered using amplifiers connected to metallic probes inserted into the ground at points several hundred meters apart. At certain times strange whistling sounds could be heard and Barkhausen suggested that they were correlated with meteorological influences. Eckersley [1925] described disturbances of a musical nature that were heard when a telephone or any other audio-recorder system were connected to a large aerial. The disturbances' time duration varied from a small fraction of a second up to a fifth of a second.

Burton [1930] and Burton and Boardman [1933a and b] reported on audio frequency atmospherics observed as a result of studying submarine cable interference. They identified various types of musical and non-musical atmospherics including what we now call "chorus", "high-latitude chorus", and "risers". They also reported a positive correlation between swishes and magnetic disturbances. For further details on the history of discrete emission research prior to the 1940's, the reader is referred to Chapter 2 of the Helliwell [1965] book, Whistlers and Related Ionospheric Phenomena.

Ground Observations

Storey [1953] conducted an investigation into the nature and origin of whistlers using a home-made spectrum analyzer. This paper resulted in the first interpretation of the known properties of whistlers including the fact that they are produced by waves originating in lightning flashes. Storey deduced many of the propagation characteristics of whistler-mode waves later corroborated by observations. Although Storey's paper dealt primarily with whistlers, other types of audio-frequency atmospherics were noted including dawn chorus, steady hiss, and isolated rising whistlers. All three types were found to be associated with magnetic activity and possibly to each other.

From the mid-1950's on, discrete emissions received much attention from scientists. Spectrograms of triggered single and multiple risers were included in the collection of spectra of unusual whistlers and other events prepared by Dinger [1956]. Allcock and Martin [1956] reported the simultaneous (less than 0.1 second difference) occurrence of "dawn chorus" at places 600 km apart. Storey [1953] had described "dawn chorus" as consisting of a multitude of rising whistles and sounding like a rookery heard from a distance. Morgan [1957] reviewed whistlers, atmospherics (spherics), tweeks, and dawn chorus. Morgan [1957] described dawn chorus as being strongest at local dawn and consisting of a chorus of overlapping, rising tones (chirps) generally confined to the mid-audio range. The typical tones rose slowly, then rapidly, and then again slowly.

Allcock [1957] reported on the characteristics of dawn chorus observed at Wellington, New Zealand, (45° South geomagnetic latitude) and found no general correlation with whistlers, a strong correlation with simultaneous local magnetic variations, a pronounced diurnal variation in activity with the peak occurring near 0400 local time, and possibly a weak seasonal variation in activity. Pope [1957] found that the diurnal variation in "dawn chorus" observed at College, Alaska, (65° North geomagnetic latitude) had a maximum at 1400 hours local time. When the observations at several different geomagnetic latitudes were compared, Pope [1957] found that the local time of maximum in the diurnal variation of the occurrence of "dawn chorus" increased with increasing geomagnetic latitude. He therefore suggested that it should more properly be called just "chorus". Watts [1957a and b] reported the observations of slowly rising chorus elements out of the top of a very intense hiss band detected at Boulder, Colorado, during a strong geomagnetic storm period.

The first comprehensive and readily available survey of VLF emissions was the report by Gallet [1959a]. Gallet divided discrete VLF emissions into eight major classes. A copy of Gallet's classification table is shown in Figure 1.

Allcock [1957], Pope [1960], and Crouchley and Brice [1960] using data from stations between 40° and 65° geomagnetic latitude found that the local time of diurnal maximum in chorus activity was a linear function of geomagnetic latitude.

Another extensive set of VLF emissions spectra are contained in Chapter 7 of Helliwell [1965]. For his atlas of VLF emissions, Helliwell developed a classification scheme that was a modification and an extension of that developed by Gallet

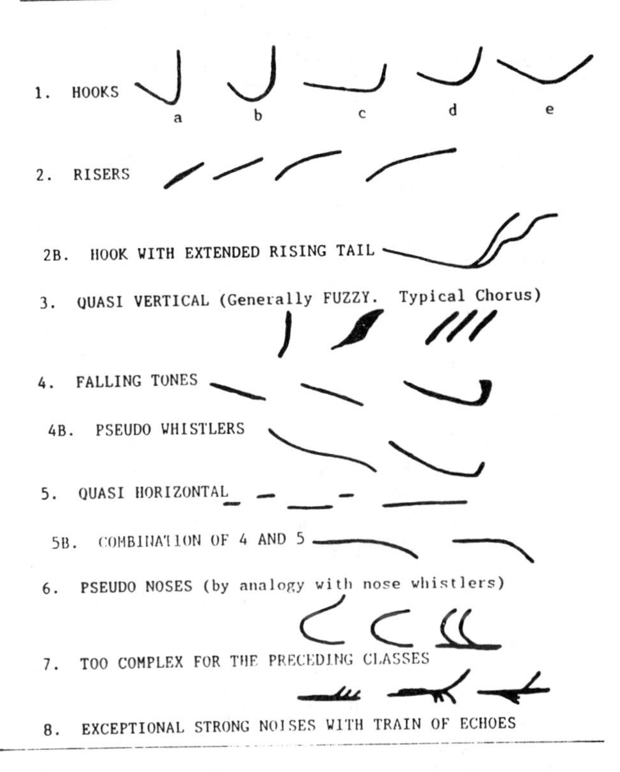

Fig. 1. Classes of Discrete VLF Emissions from Table II of Gallet [1959a].

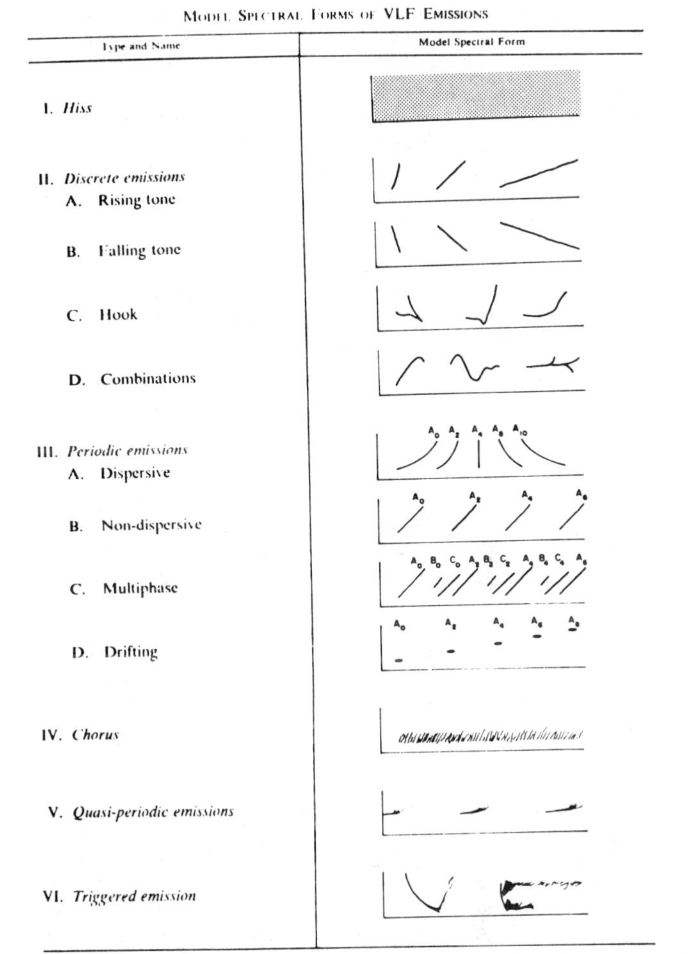

Fig. 2. Classification by type, name, and model spectral form for VLF emissions from Chapter 7 of Helliwell [1965].

[1959a]. Figure 2 depicts the classification scheme used by Helliwell. A detailed description of each type of VLF emission can be found on page 207 of Helliwell [1965]. Helliwell [1965] also noted that for chorus often a background continuum of hiss was present.

Pope [1963] reported on a high-latitude study of chorus carried out using data from several Alaskan stations. The maximum occurrence of chorus was found to be in the 60° to 70° geomagnetic latitude range. Seasonal variations were found to depend intricately on latitude. Below the auroral zone, the diurnal maximum depended on "eccentric geomagnetic latitude". These results suggested that the occurrence of chorus depended on the far field of the earth rather than on that in the ionosphere close to the earth's surface. This led Pope to speculate that chorus was generated far from the earth in the vicinity of the equatorial plane. The local time of the maximum increased with increasing latitude. The rate of change of frequency for chorus varied greatly and the over-all average was about 3 kHz/second. Chorus was found to correlate with geomagnetic activity but the correlation could be either positive or negative depending on both the season and local time. In early work at mid-latitudes, Storey [1953] and Allcock [1957] had found that the occurrence of VLF emissions showed a positive correlation with magnetic activity. At higher latitudes this is not always the case. Pope [1963] found that the correlation between chorus and magnetic activity was positive during the solstice periods but negative during equinox periods. Pope suggested that that the negative correlation might result from ionospheric absorption which is known to be correlated with magnetic activity. Crouchley and Brice [1960] reported that at Macquarie Island (61° South geomagnetic latitude) there was a negative correlation between chorus and magnetic activity. They speculated that the region of maximum chorus activity moves toward the equator during disturbed periods. Yoshida and Hatanaka [1962a and b] also investigated the relation between VLF emissions and magnetic activity at different latitudes. They found that the curve of emission activity with respect to K_p reaches a peak at a certain value of K_p that decreases with latitude. At middle-latitude stations (like Boulder) peak activity occurs at K_p = 6. At high-latitude stations the peak was found to be at K_p values of 1 to 3.

Tokuda [1962] found that sudden increases in chorus strength observed at College, Alaska, were usually preceded by negative bays in the H-component of the magnetic field. Helliwell [1965] showed that during moderate geomagnetic disturbances absorption events measured by a riometer were closely related to chorus intensity increases.

Observations, primarily concentrated in the mid-latitude region, of both naturally occurring and triggered VLF emissions made on the ground, aboard rockets, and aboard satellites were reviewed by Rycroft [1972]. Hayakawa et al. [1986b] presented the statistical characteristics of mid-latitude VLF emissions (both unstructured hiss and structured emissions) observed at Moshiri Observatory (35° North geomagnetic latitude, L=1.6) in Japan from 1974 to 1984. Both types of emissions were found to be correlated with geomagnetic disturbances. From the time delays of the emission events behind the associated geomagnetic disturbances, they estimated the resonant electron energy for VLF hiss to be ~5 keV at L=3-4 and for the structured emissions (primarily chorus) it was considerably larger, such as ~20 keV at L~4.

Ungstrup and Jackerott [1963] used 4 1/2 years' worth of data from Godhavn, Greenland, (79.9° North geomagnetic latitude) to study "polar chorus". "Polar chorus" was observed usually below 1500 Hz, and the diurnal variation of its occurrence usually showed a peak at 0900 local time independent of the time of the year. The seasonal variation of occurrence showed a maximum in the summer months. A negative correlation was observed between the daily measurement of local geomagnetic activity and the number of hours per day containing chorus.

Other types of discrete electromagnetic emissions observed by ground stations included periodic emissions [Gallet, 1959a; Pope and Campbell, 1960; Lokken et al., 1961; Brice, 1962; and Helliwell, 1963], quasi-periodic emissions (also known as "long period VLF pulsations" [Watts et al., 1963; Carson et al., 1965], 'fast hisslers' [Sieren, 1975], and triggered emissions (see discussion in following paragraphs). Very complete reviews of the various VLF emissions observed on the ground up to the early 1960's are contained in Brice [1964a] and Helliwell [1965]. Kimura [1967] reviewed the observations and theories of VLF emissions detected on the ground and by satellite in the period 1957 to 1967. Particular attention was paid to chorus and hiss and their relationships to other geomagnetic phenomena. Characteristics of artificially stimulated emissions and quasiperiodic emissions were also studied in detail.

The interaction between ELF-VLF emissions and magnetic pulsations has been studied by many investigators using ground-based observations. Quasi-periodic ELF-VLF emissions were studied by Kitamura et al. [1969], Sato et al. [1974], Sato and Kokubun [1980], Sato and Fukunishi [1981], and Sato et al. [1981]. Sato et al. [1981] used simultaneous data from Syowa, Antarctica, (L=6) and from the ISIS 1 and 2 satellites and found a one-to-one correlation in the intensity modulation of the Quasi-periodic emissions. Sato and Kokubun [1981] studied regular period ELF-VLF pulsations and Sato [1984] investigated short period pulsations. Sato [1984] used simultaneous data from the Syowa Station and the ISIS 2 satellite and found one-to-one correspondence in the periodic emissions over a wide latitude range from L~3.5 to L~14.0. Kokubun et al. [1981] reported on correlated observations of VLF chorus bursts and impulsive magnetic variations at L=4.5. Fraser-Smith and Helliwell [1980] investigated the stimulation of geomagnetic pulsations by naturally occurring repetitive VLF activity.

Triggered emissions have received much attention observationally and theoretically (see Section 7) and in simulations (see Section 8). Helliwell [1963] reported cases of VLF emissions triggered by whistlers. Helliwell et al. [1964] reported the discovery of VLF emissions triggered by whistler-mode signals from U. S. Navy VLF stations. They reported that while dashes frequently triggered emissions, dots seldom did so. Similar observations of "artificially stimulated emissions" (ASE) were reported in Helliwell [1965], Kimura [1967, 1968], Carpenter [1968], Carpenter et al. [1969], and Lasch [1969].

The frequency-time behavior of ASE's was studied in detail by Stiles and Helliwell [1975]. Stiles and Helliwell [1977] studied the amplitude behavior of several hundred VLF whistler mode pulse signals and their associated ASE's and found that the observations were in qualitative agreement with a model that attributes signal growth and ASE's to an interaction between coherent waves and counterstreaming gyroresonant electrons.

Electron precipitation has been found to be associated with discrete VLF emissions. During a period of enhanced activity at Siple Station a one-to-one correlation was found between short bursts of Bremsstrahlung X rays (which arise from the precipitation of energetic electrons) and bursts of VLF emissions by Rosenberg et al. [1971]. Foster and Rosenberg [1976] showed that the bursts were triggered by the low-frequency tail of whistlers propagating from the northern hemisphere. The observed ~0.2-s time delay between the reception of a VLF burst and the associated X-ray burst was calculated to be related to the difference in wave and particle propagation times from a burst generation region near the equator to the detectors at Siple Station if the waves and electrons left the interaction region in opposite directions. Detailed analysis of the data supported the Rosenberg et al. [1971] and Foster et al. [1976] hypotheses that the observations were due to an equatorial cyclotron resonance interaction occurring at the outer edge of the plasmapause on the L=4.2 field line. Simultaneous observations of Bremsstrahlung X-rays and VLF radio wave emissions measured at Roberval, Quebec, Canada, and Siple Station, Antarctica, revealed a detailed correlation between electron microbursts precipitated in one hemisphere and chorus elements of rising frequency recorded at the conjugate point [Rosenberg et al., 1981]. Their results suggested that microburst generation regions are located within 20° of the equator on subauroral field lines.

One-to-one correlations were observed at L~4 between bursts of ducted VLF noise (primarily chorus) in the ~2 to 4 kHz range and optical emissions at λ4278 by Helliwell et al. [1980b]. The optical observations were made at Siple Station and the wave observations were made both at Siple Station and at its conjugate station, Roberval. The observations were made in the austral winter of 1977 and most of the events occurred near dawn during or shortly following substorm events. The data were found to be consistent with scattering of electrons into the loss cone over Siple by emissions that were triggered by waves propagating away from the equator after reflection in the ionosphere over Siple. Doolittle and Carpenter [1983] presented the first observations of ionospheric optical emissions correlated with VLF waves at the conjugate location, Roberval. Since most whistlers recorded at Siple or Roberval originate in the northern hemisphere, Roberval provided observations of the direct precipitation induced during the first pass of a wave as it propagates southward.

'Fast hisslers' are very brief bursts of auroral hiss dispersed in the whistler mode that occasionally occur during substorms. They were first observed by Sieren [1975] in broadband VLF data recorded at Byrd Station, Antarctica, during substorm 'breakup' (expansion) phases.

The characteristics of dawn-side mid-latitude VLF emissions associated with substorms were deduced from two-station direction finding measurements by Hayakawa et al. [1986a]. Ground stations at Brorfelde, Denmark, (L~3) and Chambon-la-Foret, France, (L~2) were used in this study. The emissions exhibited a regular frequency drift, usually with the frequency increasing with local time, which they interpreted as being related to the energy dispersion of injected electron drifting from the midnight sector to the generation region in the dawn sector.

A number of investigators have used the fine structure of the ground-observed VLF chorus and other discrete emissions to deduce the type and location of the wave-particle interaction processes in the magnetosphere generating the emissions [Dowden, 1971a, b; Smirnova, 1984; Sazhin et al., 1985].

In this section we have concentrated on ground observations of discrete electromagnetic emissions and related phenomena associated with their interactions with the magnetospheric plasma and energetic particles. Since many of the emissions are in the whistler-mode or are related to whistlers and other whistler-mode emissions, the propagation characteristics of whistler-mode signals eventually received on the ground are important for understanding many of the observations. The theory of whistler propagation is well reviewed by Walker [1976] and Hayakawa and Tanaka [1978] extensively review ground-based studies of low-latitude whistler propagation. In situ observations from spacecraft as well as results from active experiments will be discussed in the following two sections. Theoretical and simulation studies of the generation and propagation of the discrete emissions and the wave-particle interactions associated with them will be covered in Sections 7 and 8.

Satellite Observations at Earth

The first observations of VLF emissions from a satellite were made using a magnetometer on the Vanguard III satellite [Cain et al., 1961; Cain et al., 1962]. Few emissions were observed because of the low inclination of the orbit. Frequent emission activity in the VLF range was observed at higher latitudes on Alouette 1 [Barrington and Belrose, 1963; Barrington et al., 1963a] and Injun III [Gurnett, 1963; Gurnett and O'Brien, 1964]. One of the more surprising results was the observations of chorus below 1 kHz at all latitudes including the equatorial region. On the ground such emissions were confined to the high latitude region. Gurnett and O'Brien repeatedly observed the simultaneous occurrences of VLF electromagnetic emissions, auroral optical emissions, and particle precipitation into the atmosphere. Another observation unique to spacecraft was the observation of triggered emissions near the lower hybrid resonance frequency [Barrington et. al., 1963b; Brice et al., 1964; Brice, 1964a (Chapter VII)].

Oliven and Gurnett [1968] found using Injun III data that microbursts of E > 40 keV precipitating electrons were always accompanied by a group of VLF chorus emissions, but chorus was not always accompanied by microbursts. One-to-one (chorus burst to electron microburst) correspondence between individual bursts was rare. Similar results were later reported by Holzer et al. [1974] using OGO 6 data who suggested that the lack of exact coincidence was probably due to the propagation characteristics of chorus.

Holzer et al. [1974] also suggested that the low-latitude chorus boundary at the orbit of OGO 6 was close to the field line marking the low-altitude boundary of the plasmasphere. Carpenter et al. [1969b] and Anderson and Gurnett [1973] have also shown that chorus in the outer magnetosphere tends to end very near the plasmapause. VLF observations from the Ariel 3 satellite combined with VLF observations from the Moshiri ground station in Japan during two magnetic storms also showed that the chorus-type morning emissions were observed outside the plasmapause while the narrow-banded hiss-type emissions occurred within the plasmasphere [Hayakawa et al., 1977].

Taylor and Gurnett [1968] presented a morphological study of VLF emissions observed with the Injun 3 satellite. The most intense emissions (ELF hiss and then chorus) occurred from 55° to 75° invariant latitude during local day (peaking at 65° and 0800 to 1000 magnetic local time). The region of most intense VLF emissions was found to move to lower latitudes during geomagnetically disturbed times. Dunckel and Helliwell [1969], using OGO 1 data acquired during a geomagnetically quiet nine-month period, observed whistler-mode emissions (primarily chorus and mid-latitude hiss) within L=10 for all local times except for L>5 in the midnight to dawn sector. From the fact that the upper-frequency limit of most of the emissions was proportional to the minimum electron cyclotron frequency along the magnetic field line passing through the satellite, they concluded that the source of the emissions was near the equatorial plane. Estimates of the average intensity of the emissions agreed with the intensity required to explain the precipitation of electrons through pitch angle scattering by whistler-mode waves calculated from Kennel and Petschek [1966]. The emissions observed beyond L=8 on the dayside were burst-like and their occurrence tended to decrease with increasing radial distance. The location of the maximum intensity in that region agreed with that found for 1-keV electron fluxes which they interpreted as giving strong support to theories of generation based on electron cyclotron resonance. The spatial extent and frequency of occurrence of ELF chorus and hiss has also been studied by Russell et al. [1969] using OGO 3 data.

Poynting flux measurements of hiss, chorus, whistlers, saucers, and other electromagnetic emissions were obtained on the low-altitude (~680-~2600 km) polar-orbiting Injun V satellite using measurements of the electric and magnetic fields from sensors orthogonal to each other and to the ambient magnetic field [Gurnett et al., 1969; Mosier and Gurnett, 1969; Gurnett et al., 1971]. ELF hiss and chorus were generally observed to be downgoing but occasionally they were observed to be upgoing at invariant latitudes less than 60°.

The close correspondence between hiss and chorus noted above by Dunckel and Helliwell [1969] has been frequently evident both in ground-based and satellite observations of VLF emissions. An example is shown in Figure 3 which contains a high-time resolution spectrogram obtained from the wideband data on Explorer 45 (S^3-A) [Anderson and Gurnett, 1973]. Several hiss bands are evident including one from about 2 kHz to 4 kHz and another from about 5.4 kHz to 6 kHz. Many discrete electromagnetic chorus emissions are observed which initially abruptly rise in frequency, level off for a short period of time, abruptly rise again, and then abruptly fall in frequency. It appears that the discrete chorus elements begin at a frequency near the most intense part of the lower hiss band. The maximum frequency for most of the discrete chorus elements lies in the frequency range of the upper hiss band. Koons [1981] reported that in data from the SCATHA satellite, chorus emissions are often observed to start at frequencies that are within a hiss band.

Tsurutani and Smith [1974] carried out an extensive investigation of chorus using the OGO 5 search coil magnetometer data. Chorus was detected in conjunction with magnetospheric substorms throughout the region from L=5 to L=9 but only during the postmidnight hours and only within ± 15° of the geomagnetic equator where it was the most intense. The chorus occurred in narrow frequency bands (up to 200 Hz bandwidth) over the frequency range from less than 1/4 to 3/4 of the electron cyclotron frequency (Ω_e) at the equator except that it was strongly attenuated in a narrow band near $\Omega_e/2$. Banded chorus was reported first by Burtis and Helliwell [1969] in a study of the VLF emissions observed by OGO 1 and OGO 3. Studies of VLF banded emissions, including chorus, detected by the short electric antenna on OGO 5 were presented in Coroniti et al. [1971]. Burtis and Helliwell [1976] studied more than 400 hours of broadband data from OGO 3 in order to determine the occurrence patterns and normalized frequency of magnetospheric chorus in the region $4 \lesssim L \lesssim 10$. They found that magnetospheric chorus occurred mainly from 0300 to 1500 local time, at higher L at noon than at dawn, and moved to lower L during geomagnetic disturbances, all in accord with ground observations of VLF chorus. Burtis and Helliwell found that chorus was bandlimited and that the center frequency varied as L^{-3}. Maeda and Smith [1981] found that off-equatorial observations of two-banded chorus showed that the emissions above $\Omega_e/2$ were very weak in contrast to those observed near the equator. They concluded that the dominant sources of these two-banded chorus emissions were within 5° of the equator.

Narrowband chorus without structure and falling tones were most frequently observed by Tsurutani and Smith [1974], but rising tones, hooks, and 5-15 second quasi-periodic groups of chorus elements were also observed. A high-time resolution spectrogram of whistler-mode chorus obtained from the wideband data on Explorer 45 is shown in Figure 4. In this example taken at L=4.6 and about 8° above the geomagnetic equator, predominantly rising tones are observed. The conspicuous gap just below 4 kHz is at about one-half of the local electron cyclotron frequency. Figures 3 and 4 of Anderson and Gurnett [1973] show examples of banded chorus that contain predominantly nearly vertical emissions (some followed by rising diffuse noise) with bandwidths on the order of a kHz.

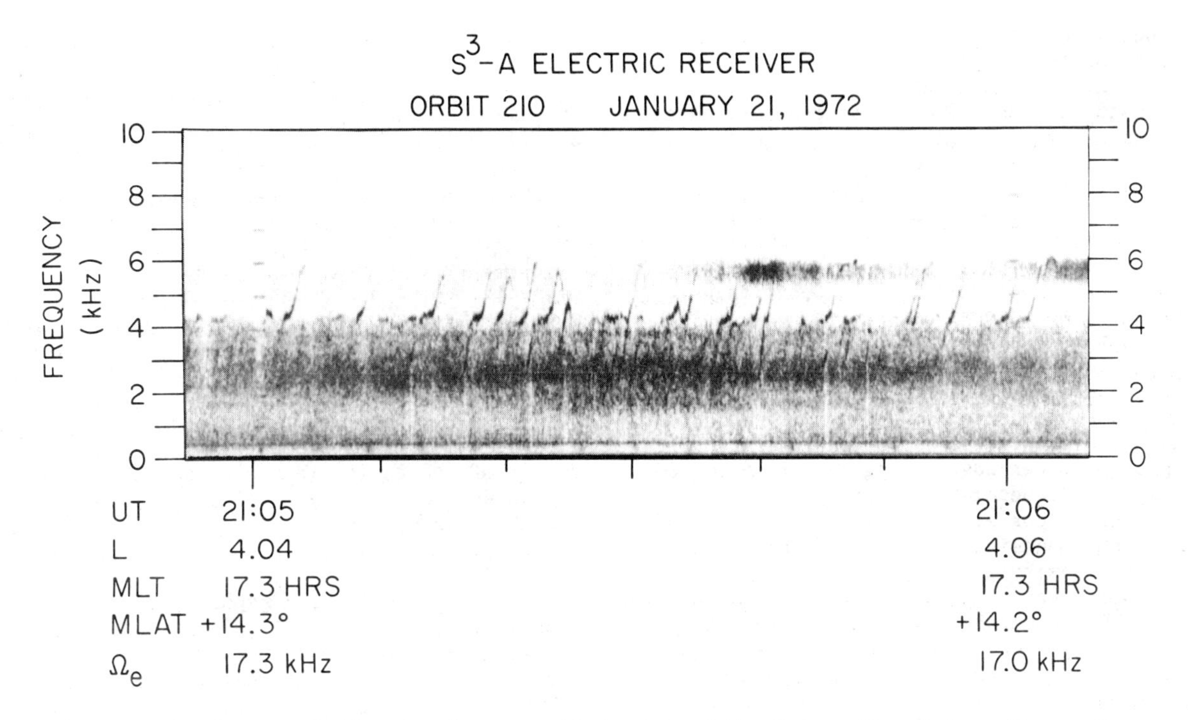

Fig. 3. A high-time resolution spectrogram obtained from the University of Iowa Plasma Wave Experiment wideband data on Explorer 45 (S^3-A). The frequency range shown is from 0 kHz to 10 kHz and the time period displayed is slightly longer than one minute. Several hiss bands are evident including one from about 2 khz to 4 khz and another from about 5.4 kHz to 6 kHz. Many discrete electromagnetic chorus emissions are observed which initially abruptly rise in frequency, level off for a short period of time, abruptly rise again, and then abruptly fall in frequency. It appears that the discrete chorus elements begin at a frequency near the most intense part of the lower hiss band. The maximum frequency for most of the discrete chorus elements lies in the frequency range of the upper hiss band.

Tsurutani and Smith [1974] noted that the distribution of chorus as a function of local time and L was strikingly similar to the distribution of enhanced, trapped, and precipitated substorm electrons with energies > ~(10-40) keV. They attributed the postmidnight occurrence of chorus to the eastward curvature and gradient drift of the injected electrons. Cyclotron resonant interactions are expected to be strongest at the equator as they observed. The confinement to within 15° of the equator was attributed to Landau damping by low-energy (1-10 keV) auroral electrons. They suggested that the attenuation band at one-half the electron cyclotron frequency might result

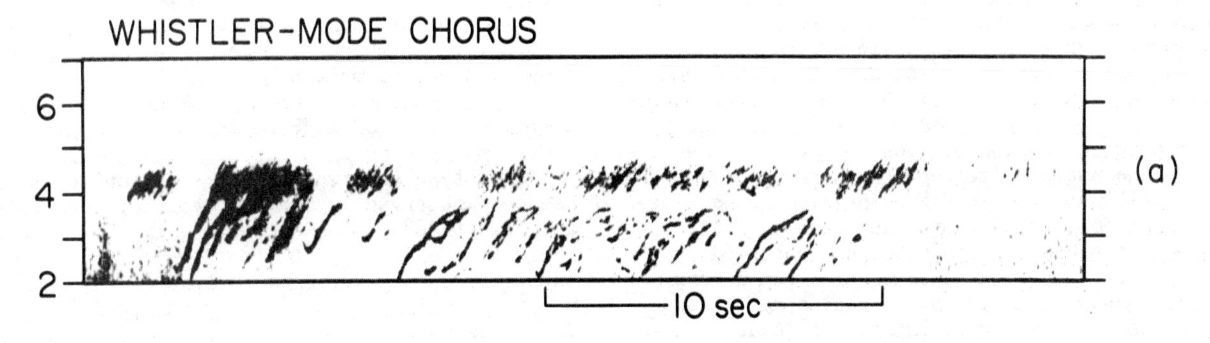

Fig. 4. A high-time resolution spectrogram of whistler-mode chorus obtained from the University of Iowa Plasma Wave Experiment wideband data on Explorer 45. The frequency range shown is from 2 kHz to 7 kHz and a 30-second time period is displayed. The conspicuous gap just below 4 kHz is at one-half of the local electron cyclotron frequency.

from Landau damping by electrons that have energy corresponding to cyclotron resonance but are traveling in the same direction as the waves.

A subsequent study of OGO 5 search coil data by Tsurutani and Smith [1977] found that chorus occurs in principally two magnetic latitude regions. Equatorial chorus is detected near the equator (its occurrence peaks at |magnetic latitude|<5°) and high-latitude chorus is found at magnetic latitudes greater than 15°. Equatorial chorus has an abrupt onset in the postmidnight sector and a second enhancement from dawn to noon, a pattern that they noted was similar to energetic electron precipitation. While equatorial chorus occurs primarily during substorms, high-latitude chorus is not strongly dependent on geomagnetic disturbances and often occurs during prolonged quiet periods. Using data from the OGO 6 search coil magnetometer, Thorne et al. [1977] showed a strong correlation between the onset of the substorm expansive phase and the enhancements of ELF chorus emissions in the postmidnight and early morning sectors. They suggested that the ELF chorus was generated through cyclotron resonance with anisotropic electrons injected into the nightside outer radiation zone at the substorm onset.

Burton and Holzer [1974] investigated the origin and propagation of chorus in the outer magnetosphere by determining the wave normals using the OGO 5 triaxial search coil magnetometer data. Their conclusions were that chorus was generated within 25° of the equatorial plane on the dayside and within 2° on the nightside, that chorus was generated by a Doppler-shifted cyclotron resonance with 5 to 150 keV electrons but only when the distribution was peaked perpendicular to $\hat{B}$ and the anisotropy exceeded a critical value, and that in the source the wave normals were contained within an unstable generation cone of 20° half angle about $\hat{B}$. Goldstein and Tsurutani [1984] also used the OGO 5 triaxial search coil magnetometer data to determine the wave normal directions of chorus near the equatorial source region. They concluded that wave growth was maximum for waves propagating parallel to $\hat{B}$ and the emission occurred in a narrow beam.

Comparisons of the spectra of chorus and simultaneous electron pitch angle distributions obtained on Explorer 45 (S^3-A) directly confirmed the Kennel and Petschek [1966] electron cyclotron resonance instability predictions for the association between chorus emission frequency and electron pitch angle anisotropy [Anderson, 1976]. This is illustrated in Figure 5 which shows wideband data containing chorus and simultaneous electron pitch angle distributions observed on November 26, 1971. A rapid change from a very anisotropic pitch angle distribution to a less anisotropic pitch angle distribution exactly coincides with the chorus spectra changing from having frequency components almost only above $\Omega_e/2$ to having lower frequency components both above and below $\Omega_e/2$. The Kennel and Petschek theory requires more anisotropic distributions for the generation of higher frequency emissions. Further studies of Explorer 45 observations of VLF emissions associated with enhanced magnetospheric electrons were reported in Anderson and Maeda [1977]. They found that the features observed indicated that the VLF emissions are produced by low-energy (1 to 10 keV) electrons which penetrate into the dusk-midnight sector of the magnetosphere from the geomagnetic tail during magnetic storms and substorms and drift eastward outside the plasmasphere.

Isenberg et al. [1982] presented simultaneous observations of energetic electrons and dawnside chorus observed in geosynchronous orbit by the SCATHA satellite. For the five days studied in June of 1979, every encounter with injected clouds of energetic electrons was accompanied by chorus activity. They concluded that dawnside chorus was generated by substorm-injected, anisotropic clouds of electrons with energies between 10 and 100 keV.

An experimental study reported by Cornilleau-Wehrlin et al. [1985] of the relationship between energetic electrons and ELF waves observed on board GEOS offered additional support for the Kennel and Petschek [1966] quasi-linear theory. In this study they calculated the particle anisotropy and temporal growth rate expected from the particle measurements and compared them directly to the critical anisotropy (dependent only on the wave frequency and the electron cyclotron frequency) and the measured wave spectrum. Excellent agreement was found between the theoretical predictions and the observations.

Estimations of the wave distribution function for electromagnetic waves including chorus and hiss observed on board GEOS-1 were presented in Lefeuvre et al. [1981]. The wave energy for both these naturally occurring emissions was generally concentrated within two wave packets whose wave normals were approximately in the same off-meridian plane and oriented in the same way relative to the direction of the earth's magnetic field. Hayakawa et al. [1984] analyzed the wave normal directions of magnetospheric chorus emissions observed on the geostationary GEOS-2 spacecraft which was located in the equatorial region at L=6.6. The chorus was generated in conjunction with substorms and at local times from around midnight to afternoon. They found that different types of chorus are associated with different wave normals.

A new type of discrete VLF emission at $L < 4$ has been observed in the magnetosphere by Poulsen and Inan [1988]. The emission elements are confined to a bandwidth of 1 to 5 kHz and the local cutoff frequency varies with L and equals ~0.2 to 0.5 the equatorial electron cyclotron frequency. While the discrete and burst-like nature of the emissions are very similar to chorus emissions observed at higher L, the dispersion of individual elements is often different from typical chorus and the emissions are observed inside as

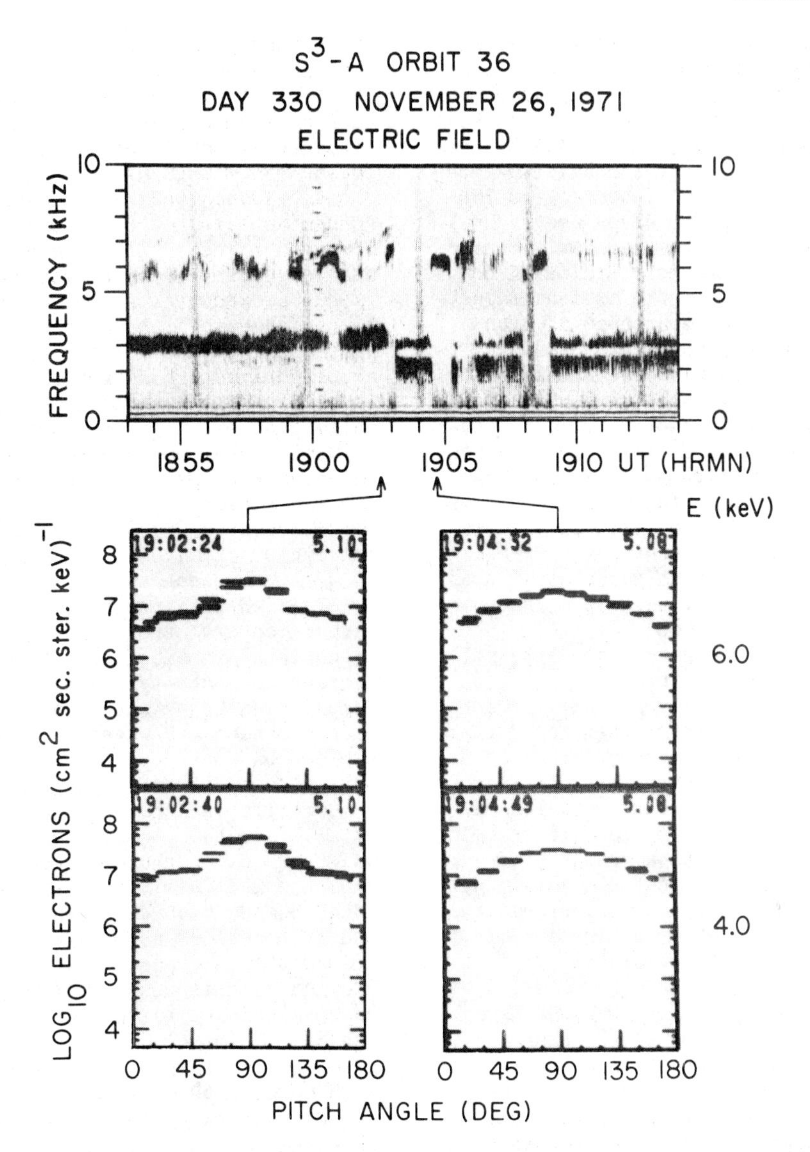

Fig. 5. Comparison between chorus spectra and electron pitch angle distributions observed on Explorer 45 (S^3-A). The top panel contains an electric field frequency-time spectrogram from 1853 UT to 1914 UT on November 26, 1971. The frequency range is from 0 to 10 kHz. The lower portion of the figure contains the electron pitch angle distributions acquired at ~1902:30 UT (left side) and ~1904:40 UT (right side) for 4.0 keV (bottom) and 6.0 keV (top). Note that the rapid change from a very anisotropic pitch angle distribution to a less anisotropic pitch angle distribution exactly coincides with the chorus spectra changing from having frequency components almost only above $\Omega_e/2$ to having lower frequency components both above and below $\Omega_e/2$.

well as outside the plasmapause. A quasi-electrostatic (short wavelength whistler-mode) broadband burst of VLF emissions was observed by Maeda and Anderson [1982] to be associated with the enhancement of anisotropic ring-current electrons from the inner edge of the plasma sheet.

While chorus has been probably the most studied of the discrete electromagnetic emissions observed by satellite experiments, other discrete electromagnetic emissions, both naturally occurring and artificially stimulated, have also been observed. Intense, sporadic bursts of narrowband low-frequency electromagnetic noise called "lion roars" are frequently observed in the earth's magnetosheath. They were first observed by Smith et al. [1969] in the search coil magnetometer data from OGO's 1, 3, and 5. Smith and Tsurutani [1976] reported that "lion roars" are the strongest whistler-mode signals found in the magnetosheath. The occurrence of lion roars was found to be related to the level of geomagnetic activity as measured by K_p. In a survey of 411 1-hour intervals when OGO 5 was in the magnetosheath, Smith and Tsurutani [1976] found that the probability of occurrence (an occurrence being three or more intense lion roars in the hour) ranged from 10% in magnetically quiet intervals to 75% during disturbed periods. The waves are right-hand circularly polarized with propagation essentially along the ambient magnetic field. A correlation between decreases in the ambient magnetic field magnitude and the occurrence of lion roars was observed. Anderson et al. [1982] showed that lion roars in the magnetosheath near the magnetopause were well correlated with the ultra-low-frequency (ULF) magnetic field fluctuations. They were observed coincident with the low $|B|$ phase of the magnetic waves. Plasma wave magnetic noise bursts resembling lion roars have also been detected in the high-latitude magnetosheath near the polar cusp [Gurnett and Frank, 1972a, 1978] and in the distant magnetotail near the neutral sheet [Gurnett et al., 1976; Anderson, 1984; Cattell et al., 1986].

Tsurutani et al. [1982a] used a complete set of ISEE plasma wave, plasma, and field data to identify the plasma instability responsible for ELF electromagnetic lion roars. They found that lion roars detected close to the magnetopause are generated by the cyclotron instability of anisotropic thermal electrons when the local plasma critical energy falls to values close to or below the electron thermal energy as a result of decreases in the ambient magnetic field magnitude. They speculated that the high beta, low critical energy regions conducive to the generation of lion roars were due to a nonoscillatory drift mirror wave.

Long duration lion roars ($\gtrsim$5 min) are often detected immediately behind or in the nearby downstream region of subsolar quasi-perpendicular bow shocks [Rodriguez, 1985]. The long duration lion roars are associated with the relatively ordered magnetosheath conditions as indicated by lower-than-normal magnetic field variances and higher-than-normal magnetic field intensities.

ELF electromagnetic noise emissions with discrete frequency structure were observed on OGO 3 in the outer plasmasphere very close to the magnetic equator by Russell et al. [1970]. The waves existed between twice the proton cyclotron frequency and half the lower hybrid resonance frequency, were observed only within 2° of the magnetic equator, and were found to be in bounce resonance with energetic electrons. Gurnett [1976] used wideband data from IMP 6 and Hawkeye 1 to investigate this intense electromagnetic noise and showed that it consisted of a complex superposition of many harmonically spaced lines. He concluded that ion cyclotron harmonic resonances provided a much more satisfactory explanation of the harmonic structure of the equatorial noise. These waves, now referred to as electromagnetic ion Bernstein mode emissions [Fredricks, 1968], have also been observed on DE-1 in close association with a highly anisotropic low-energy (5-50 eV) ion population [Olsen et al., 1987; Gurnett and Inan, 1988].

Observations of discrete electromagnetic ULF emissions by GEOS have been reported by Perraut et al. [1978; 1982] and Young et al. [1981]. Perraut et al. [1978] concluded that they had observed magnetosonic waves, ion Bernstein waves, and ion cyclotron harmonic waves. Young et al. [1981] and Roux et al. [1982] studied wave particle interactions associated with ion cyclotron waves near the helium gyrofrequency. Perraut et al. [1982] concluded that the waves near the equator having harmonically related structured emissions were magnetosonic waves.

Temerin and Lysak [1984] identified narrow-banded ELF waves with frequencies between the local hydrogen and singly-charged helium gyrofrequencies observed on S3-3 as electromagnetic ion cyclotron waves. The waves were generated in or just below the auroral acceleration region by accelerated electron beams and propagated in the Alfvén-ion cyclotron branch of the cold electromagnetic dispersion relation as modified by the presence of both hydrogen and helium ions. Temerin and Lysak [1984] also concluded that the narrow-banded ELF waves associated with inverted V electrons observed by Gurnett and Frank [1972b] were also electromagnetic ion cyclotron waves.

Cornilleau-Wehrlin et al. [1978] and Neubert et al. [1983b] reported observations on GEOS-1 of whistler-mode turbulence generated by signals from the NLK (Jim Creek, Washington) transmitter at 18.6 and 18.65 kHz. Wave-normal directions and interactions associated with the Omega Norway transmitter signals received on GEOS have been studied by numerous investigations including Ungstrup et al. [1978], Lefeuvre et al. [1982], and Neubert et al. [1983a]. The first satellite based observations of emission triggering by high-power ground-based communications (as opposed to navigation) transmitters were reported by Inan and Helliwell [1982] using the Linear Wave Receiver on DE-1.

Space limitations prevent us from a complete survey of all research on discrete electromagnetic emissions observed in the Earth's magnetosphere. A number of reviews should guide the interested reader to more detailed information than we have been able to present. Kimura [1967] reviewed observations and theories of VLF emissions researched during the first decade of space exploration. Gendrin [1975] extensively reviewed waves and wave-particle interactions in the magnetosphere. Matsumoto [1979] reviewed both the experimental and theoretical aspects of coherent nonlinear effects in whistler-mode interactions in the magnetosphere. Wave Instabilities in Space Plasmas [Palmadesso and Papadopoulos, 1979] contains many pertinent articles and reviews on various aspects of research (including observational, theoretical, and laboratory and computer simulations) related to discrete electromagnetic

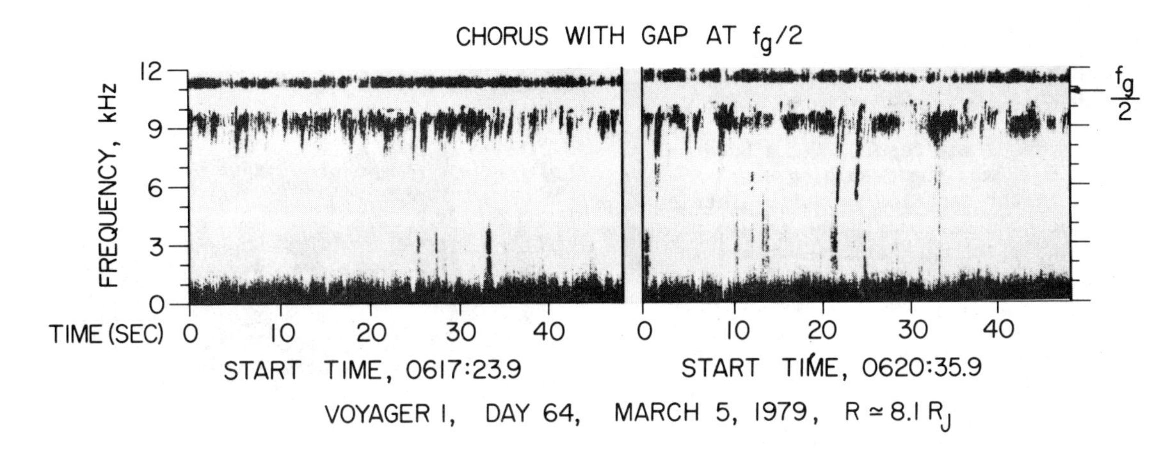

Fig. 6. Two wideband spectrograms from the Io torus at Jupiter showing many of the features of chorus in the Jovian Magnetosphere. the obvious components are a narrow band of emission just above $\Omega_e/2$, a gap at $\Omega_e/2$, structured, rising tones below the gap, some of which begin at very low frequencies, near the band of hiss at the bottom of the spectrum, and two narrowband emissions defining the primary range of frequencies the rising tones cover.

emissions. Magnetospheric plasma wave research up to 1979 has been reviewed by Shawhan [1979a, b]. Anderson [1982] reviewed progress made through plasma wave research activities resulting from the International Magnetospheric Study. Research on plasma waves in planetary magnetospheres (including Earth, Jupiter, and Saturn) for the 1979-1982 quadrennium was reviewed by Anderson [1983]. Research activities on waves and instabilities during the 1983-1986 quadrennium have been reviewed by Inan [1987]. A review of plasma wave observations obtained on DE-1 during its first six years of operation are contained in Gurnett and Inan [1988].

Observations at Other Planets

One of the fundamental contributions of the Voyager mission to the outer planets was the confirmation that the plasma wave spectra of Jupiter, Saturn, and Uranus are not unlike that of the Earth [Scarf et al., 1979; Gurnett et al., 1981a; Gurnett et al., 1986a; Scarf et al., 1987]. This similarity includes the presence of whistler mode chorus at each of the outer planets in forms which are often virtually indistinguishable from terrestrial chorus. Other discrete emissions observed at the outer planets include lightning whistlers at Jupiter [Gurnett et al., 1979], and narrowband electromagnetic emissions at Jupiter [Gurnett et al., 1983], Saturn [Gurnett et al., 1981b], and Uranus [Kurth et al., 1986]. In this section we shall describe, briefly, the Voyager observations of these emissions and will attempt to put them in context with the terrestrial observations.

Chorus was reported near L=8 in the Jovian magnetosphere by Coroniti et al. [1980] and was subsequently observed at Saturn [Gurnett et al., 1981a] and Uranus [Gurnett et al., 1986a; Scarf et al., 1987]. In each case, the chorus was easily identified by its characteristic discrete, rising tones in the frequency range of about 2 to 12 kHz, generally at about one-half the electron cyclotron frequency. The narrow-band whistler-mode chorus emissions observed on the outbound pass from Uranus produced the most intense wave-particle interactions so far detected by Voyager [Scarf et al., 1987; Coroniti et al., 1987]. The most thorough description of non-terrestrial chorus was given by Coroniti et al. [1984], based on a large quantity of high-resolution wideband data available at Jupiter.

As seen in Figure 6, the Jovian chorus appears to be the most complex of the examples at the outer planets, perhaps because only a few examples were available at Saturn and Uranus due to telemetry constraints. The Jovian chorus can include a quasi-continuous, narrowband emission just above $\Omega_e/2$ separated by a deep spectral gap from the lower frequency band of risers. Within the band of risers a hiss-like emission was often observed which consists of two very narrow tones which roughly defined the upper and lower frequency limits of the band of rising chorus elements [Coroniti et al., 1984]. While such spectra are probably not the norm at Earth, Coroniti et al. were able to find evidence in the literature for emissions similar to each of these components at Earth. In fact, Figure 3 shows chorus emissions observed at Earth which begin in a low frequency hiss band and terminate in a higher frequency hiss band. On the other hand, a totally satisfactory explanation of the detailed frequency-time structure of these components is not yet available for the terrestrial counterparts.

Coroniti et al. [1984] suggested a number of possible explanations for the features seen in Figure 6. The twin tones may be electrostatic or resonance cone whistlers. Coroniti et al. speculated that the Landau absorption of obliquely propagating chorus may result in the spectral gap just below $\Omega_e/2$ and the resulting quasi-linear plateau in the parallel velocity distribution of the electrons may permit the growth of electrostatic and electromagnetic whistlers in the narrow band just above Ω_e.

Inan et al. [1983] compared the Jovian chorus to terrestrial chorus from the point of view of understanding the differences in amplification in the two environments. They found the interaction length for whistler mode waves interacting with resonant electrons to be 2 - 5 times larger than at Earth. Even though the wave intensities are similar at the two magnetospheres, the wave intensity required to reach the threshold of nonlinearity at Jupiter is 5 - 100 times lower than at Earth. The explanation for these apparently contradictory findings is thought to be the fact that the fluxes of resonant keV electrons is about a factor of 100 greater at Jupiter, hence, temporal growth rates are greater at Jupiter; the latter were measured to exceed 750 dB/sec.

Of course, the significance of the detection of chorus is that there must be sufficient fluxes of resonant electrons present to drive the waves and the waves will pitch-angle scatter the electrons, causing them to enter the loss cone and be lost to the atmosphere. At Jupiter Coroniti et al. [1980] reported that the chorus would resonate with few keV electrons and that the whistler mode chorus along with electron cyclotron harmonic emissions could make up a sizable fraction of the energy in precipitating fluxes of electrons needed to account for the observed auroral luminosity at Jupiter.

An interesting effect attributable to chorus has been reported by Reinleitner et al. [1984] in which bursts of electrostatic noise at high frequencies have been reported at Jupiter and Saturn (as well as at the Earth) which show evidence of an interaction with a chorus wave at lower frequencies. Figure 7 [Figure 8 of Reinleitner et al., 1984] shows waveform analyses which indicate a modulation of the high frequency electrostatic bursts by a lower frequency wave which is at a frequency consistent with chorus. This effect is seen generally in the outer magnetosphere at both Jupiter and Saturn. The mechanism involves trapping of electrons in Landau resonance with the chorus wave and subsequent acceleration of the trapped electrons to form a beam.

Another discrete whistler mode emission observed at Jupiter was of importance in the confirmation of atmospheric lightning at the planet [Gurnett et al., 1979]. In this case, we are referring to lightning whistlers which have a well-recognized frequency-time shape as a result of the dispersion due to propagating through the magnetized plasma of the Io torus. Figure 8 [Figure 1 of Kurth et al., 1985] shows examples of whistlers observed by Voyager 1 as it passed through the Io torus. The whistler observations at Jupiter are summarized by Kurth et al. [1985].

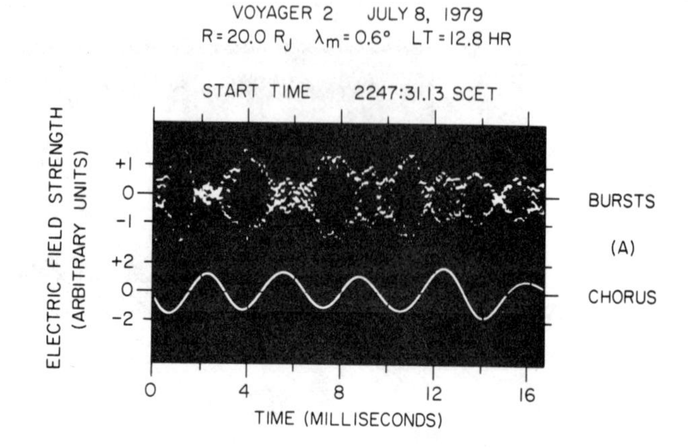

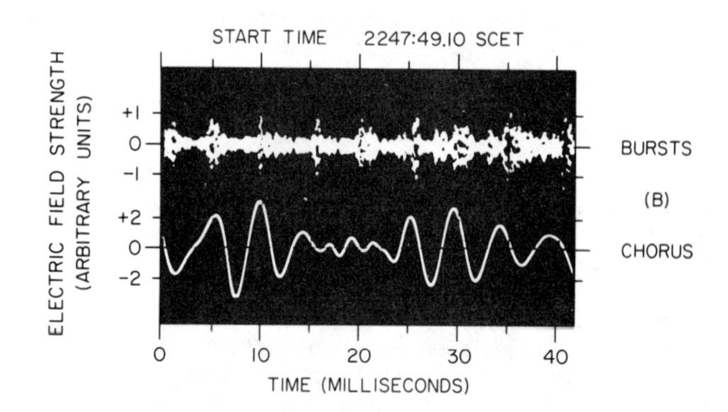

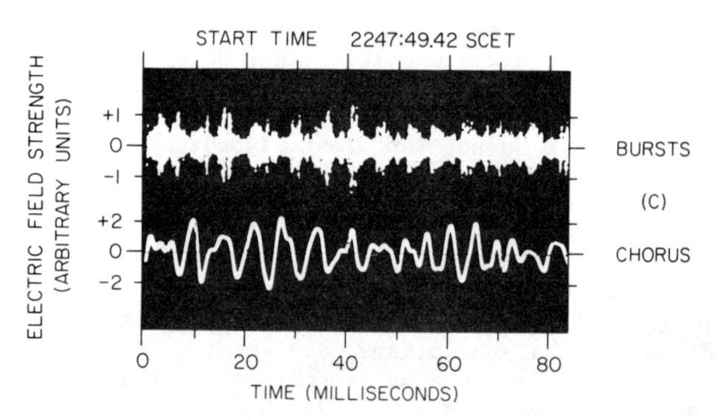

Fig. 7. Analyses of high frequency electrostatic bursts and low frequency chorus waves in the Jovian magnetosphere [Reinleitner et al., 1984] showing the modulation of the high frequency bursts by the chorus waves. The mechanism for this phenomenon involves the trapping and acceleration of electrons by the chorus wave which creates a distribution function unstable to the electrostatic bursts.

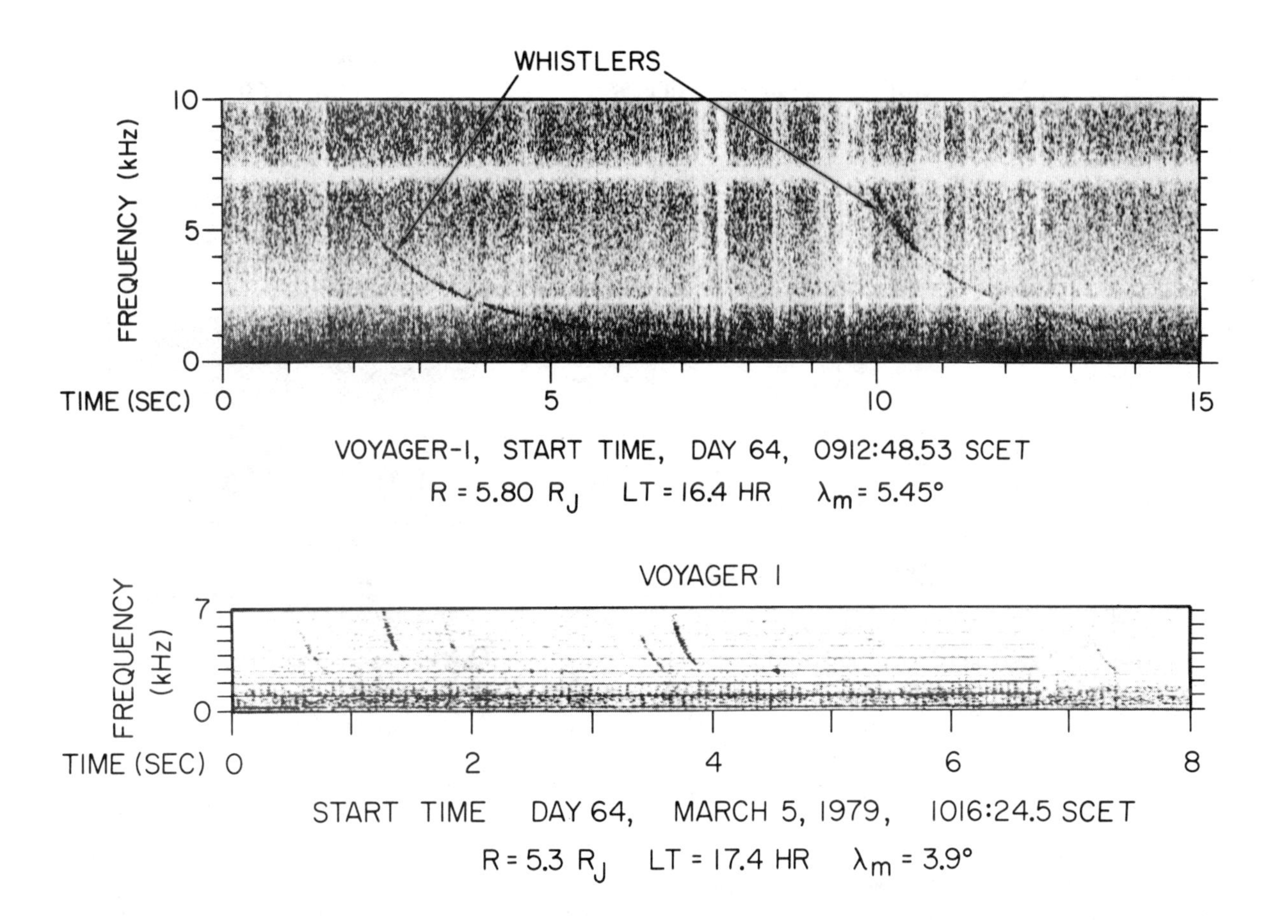

Fig. 8. These two panels show examples of lightning whistlers observed in the Io torus at Jupiter [Kurth et al., 1985]. The differences in dispersion are due to the propagation of the whistlers through varying portions of the high density torus. Those with the greatest dispersion have passed from one hemisphere, completely through the torus to the opposite hemisphere. Those with much smaller dispersions have been observed by Voyager 1 within the torus, but in the same hemisphere as the source of the whistlers.

Scarf et al. [1981] and Boruki et al. [1982] utilized the whistler observations to derive estimated lightning stroke rates in the Jovian atmosphere, a quantity which is important in understanding the chemistry of the atmosphere. Other uses of the lightning whistlers, particularly their dispersion, is to infer information of the electron density along a magnetic flux tube, the path of the whistlers. Tokar et al. [1982a, b] used this information to model the concentration of light ions in the vicinity of the Io torus, an important contributor to our understanding of the chemistry and physics of the Io-Jovian interaction, especially because of the non-local nature of the information provided by the whistler dispersions.

During the same era in which it was becoming clear that the continuum radiation spectrum at the Earth was much more complicated than originally thought [Kurth et al., 1981], consisting in many cases of numerous narrowband electromagnetic tones, Voyager was discovering that the outer planets also exhibited the narrowband emissions [Gurnett et al., 1981a; Gurnett et al., 1983]. Later, it was also found that narrowband emissions could be found at Uranus [Gurnett et al., 1986a; Kurth et al., 1986]. Figure 9 [Figure 5 from Gurnett et al., 1983] illustrates the complex structure of these narrowband emissions at Jupiter in a spectacular event which consisted of families of bands which moved to lower frequencies as a function of time. The generally accepted generation mechanism for these emissions is mode conversion from intense upper hybrid resonance emissions; recent evidence [Jones et al., 1987] has been uncovered at the Earth supporting a linear conversion process [see, for example, Oya, 1971; Jones, 1976; Budden, 1979]. The source upper hybrid bands are found near the plasmasphere and the magnetopause at Earth and

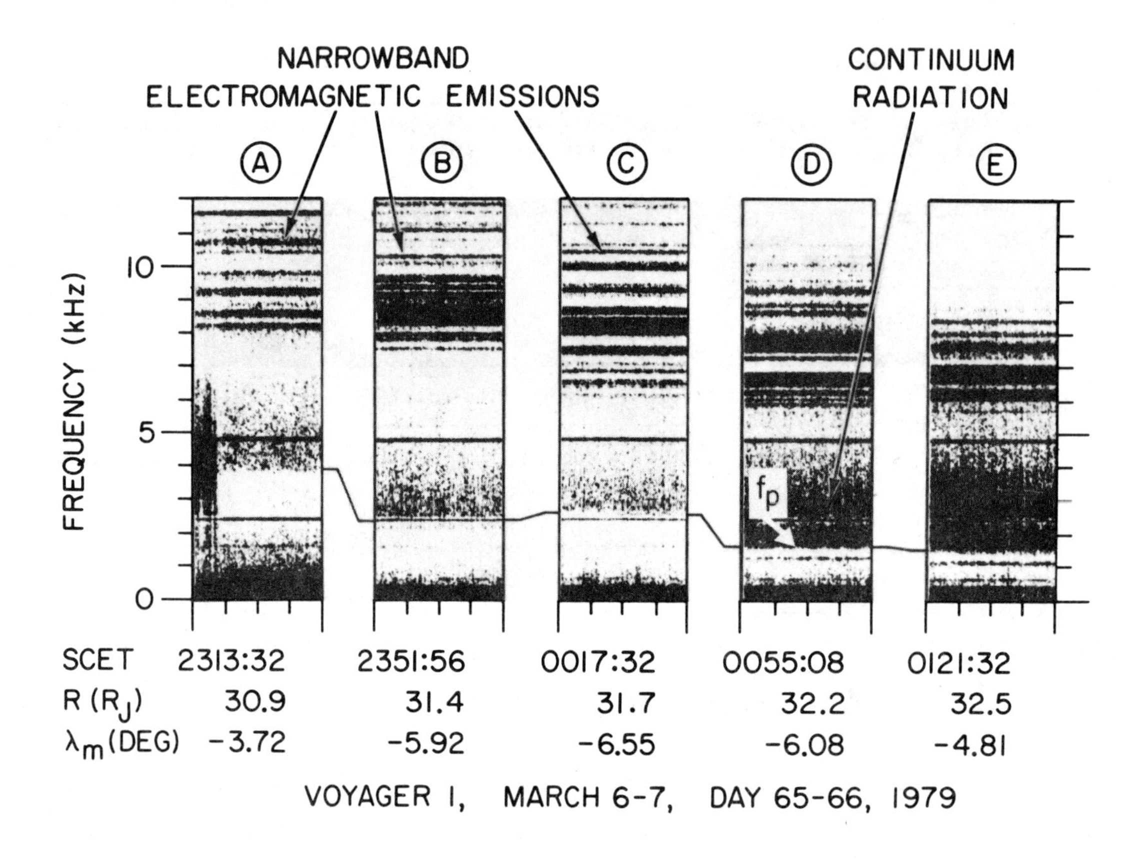

Fig. 9. Examples of the numerous narrowband electromagnetic emissions observed in the Jovian magnetosphere [Gurnett et al., 1983]. In this series of spectrograms, the bands can be seen to move to generally lower frequencies as a function of time, indicating, perhaps, a drift of the source region to larger distances, hence, lower critical frequencies in the plasma.

near the edge of the Io torus and magnetopause at Jupiter.

A series of observations at Saturn found evidence for similar narrowband emissions at Saturn, however, a line near 5 kHz seems to dominate the spectrum there [Gurnett et al., 1981b]. Gurnett et al. speculate that the source could be located on the Dione L-shell and Jones [1983] has performed ray tracing indicating the source is near a steep plasma density gradient near L=10.4 at about 4° latitude. The Uranian emissions are centered near 5 kHz and are somewhat different in appearance from the previously mentioned examples in that they are extremely bursty, having durations in some cases of order 1 second (see Figure 10 [Figure 2 of Kurth et al., 1986]). While it is possible the source, itself, could be responsible for the bursty nature of the Uranian emissions, Kurth et al. [1986] speculated that the rapidly rotating magnetosphere with its large dipole tilt could account for the burstiness simply by rotating a narrow beam of emission past the observer.

Active Experiments

The triggering of VLF emissions in the ionosphere by whistler-mode signals transmitted from ground transmitters was first reported by Helliwell et al., [1964]. In order to study wave-particle interactions in the magnetosphere under controlled conditions, a high-power VLF transmitter connected to a 21 km dipole was installed by Stanford University at Siple Station, Antarctica, [Helliwell and Katsufrakis, 1974]. The length of the antenna was increased to 42 km in 1983 [I. Kimura, private communication, 1988]. Active experiments using this transmitter (and its successors) have yielded many significant advances in wave propagation, wave-wave interaction, and wave-particle interaction research [Helliwell, 1974 and 1977]. Helliwell [1974] stated that applications of wave injection experiments included (1) study of emission mechanisms, (2) control of energetic particle precipitation, (3) diagnostics of cold and hot plasma, and (4) VLF communications. Ground observations of the

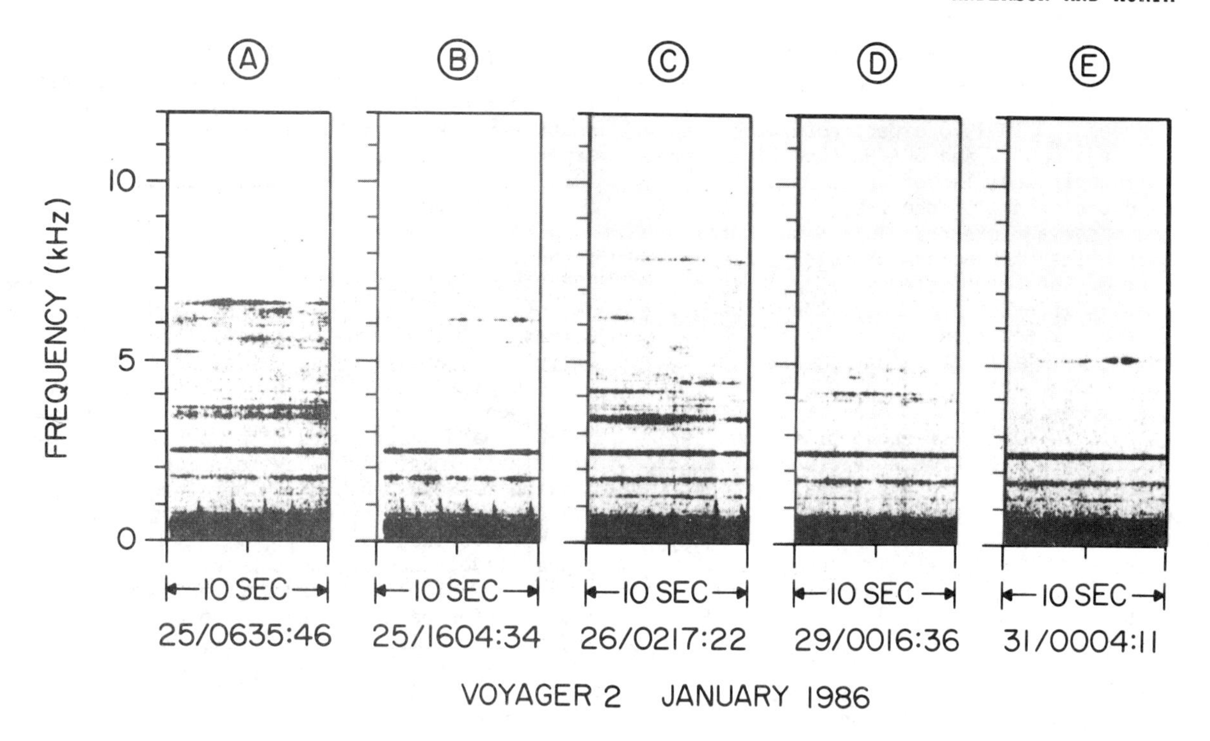

Fig. 10. Examples of sporadic narrowband radio emissions observed on the nightside of Uranus by Voyager 2 [Kurth et al., 1986]. It is likely that these emissions are related to the narrowband emissions shown in Figure 9 from Jupiter, although the very bursty nature of the Uranian emissions is somewhat puzzling. It is speculated that the rapid rotation of the highly tilted dipole field may be responsible for the burstiness.

output of the injected waves and of the associated triggered emissions made at magnetically conjugate sites have revealed many interesting facts about wave growth and triggering in the magnetosphere [Helliwell, 1979]. Among the first interesting observations were that pulses at least 150 ms long (using a 100 kW transmitter at 5.5 kHz, the predicted efficiency being 4%) were required to trigger emissions and that the interaction produced growth in time rather than in space [Helliwell and Katsufrakis, 1974]. Their observations showed that the input wave triggers an oscillation that grows exponentially with time, with growth rates of the order of 100 dB/s, until the input signal terminates or the gain reaches about 30 dB. They found these results contrary to theories of cyclotron-resonance interaction (e.g., Liemohn [1967]) that did not include feedback and predicted spatial growth of the traveling waves. In an experiment conducted using Siple transmissions, Helliwell et al. [1980a] found that when the input power to the transmitting antenna was reduced below a 'threshold' value, growth and triggering cease. Another key parameter in the ability of Siple transmissions to trigger emissions was found to be the spectral purity of the transmitted signal [Helliwell, 1983]. Helliwell found that if the width of the input signal spectrum exceeded ~10 Hz, there was a significant loss of gain observed in the received emissions.

VLF transmitter experiments conducted at Siple Station, Antarctica, have shown that long ($\gtrsim$1 s) keydown signals injected into the magnetosphere often generated sidebands as a result of nonlinear interactions with energetic particles [Park, 1981]. Raghuram et al. [1977] earlier had shown that when an injected monochromatic signal suppresses hiss in a narrow band, frequently a narrow band of enhanced noise below the quiet band appeared.

Helliwell et al. [1986] reported on an experiment which used the Siple transmitter to generate band-limited VLF noise in order to simulate the interaction of natural magnetospheric hiss with energetic radiation belt particles. While the observed spectrum of the generated noise at times closely resembled that of naturally occurring hiss, at other times discrete emissions were generated from the incoherent noise signal. This implied that magnetospheric chorus can be triggered by hiss signals as has been implied by numerous satellite observations.

Garnier et al. [1982] studied stimulated wave-particle interactions during high-latitude ELF wave injection experiments into the outer magnetosphere. Transmitters from the University of Paris and from the Aerospace Corporation used a 10.6 km power line in Norway tuned to the transmitter frequency. In 1978 experiments were conducted when the GEOS-2 and SCATHA satellites were near the magnetic meridian of the transmitter. Emissions correlated with the transmissions, however, were observed less than 10% of the total transmission time. Phenomena observed included enhancement of the transmitted wave, generation of waves at a different frequency, the frequency shifting or other modification of natural emissions, and artificially stimulated emissions.

The occurrence, origin, and propagation of discrete electromagnetic emissions have also been investigated by a number of sub-orbital sounding rocket experiments. A VLF electric and magnetic field experiment was flown on a sounding rocket in February 1970 from ESRANGE (invariant latitude of 64.8°) near Kiruna, Sweden, in order to study the generation of VLF signals in connection with aurora and the propagation of these signals in the ionosphere [Ungstrup, 1975]. The payload penetrated a proton glow aurora and flew over an auroral arc. Strong narrowband VLF signals in the frequency range 3.0-3.4 kHz with structure similar to that of chorus were observed around 180 km altitude near apogee when the payload crossed the magnetic field lines connected to the auroral arc and in a height-limited range (130-162 km on the upleg and 153-120 km on the downleg) inside the region with glow aurora.

Gendrin et al. [1970a, b] reported on several experiments using rockets that were launched from the Kerguelen Islands along the magnetic field lines, up to an altitude of ~400 km, to study the wave-particle interactions that take place during natural VLF emissions. The emissions observed had large oblique incidences. Above 120 km the VLF emissions were circularly polarized but below 120 km they tended to be elliptically polarized. During dawn chorus events, they observed flux increases of newly injected electrons, both trapped and precipitated. A fluctuation of the trapped electron intensities was detected in coincidence with a VLF periodic emission and a hydromagnetic wave with the same periodicity.

The University of Minnesota developed a series of Electron Echo sounding rocket experiments that injected electron beams to study the dynamics of particles in the earth's magnetic field and to study wave-particle interactions [Winckler, 1974]. Whistler-mode waves associated with the injected beams were observed and could be classified by type dependent on pitch angle and gun energy of the injected electrons [Cartwright and Kellogg, 1974; Monson et al., 1976]. Both the frequency and amplitude of the emissions were strongly dependent on the time from the start of the pulse and pitch angle. Structured emissions including descending tones were observed for certain conditions.

During the Space Shuttle STS-3 in March of 1982, electromagnetic VLF wave emissions were detected by the Plasma Wave Experiment receivers on the Plasma Diagnostics Package (PDP) that were generated by electron beams injected from the shuttle by the Fast Pulse Electron Gun (FPEG) experiment [Shawhan et al., 1984; Reeves et al., 1988] A variety of emissions were observed which depended on whether the FPEG generated a DC, modulated, or pulsed beam. Numerous electromagnetic VLF emissions were also observed by the PDP Plasma Wave Experiment during FPEG operations on the Spacelab 2 (SL-2) flight in July and August of 1985 [Gurnett et al., 1986; Bush et al., 1987; Farrell et al., 1988].

Neubert et al. [1986] discussed the waves generated by injected electron beams from the SEPAC (Space Experiments with Particle Accelerators) experiment flown on the Spacelab 1 shuttle mission in November 1983. They found that the VLF emissions in the 0.75- to 10-kHz range could be characterized by a ω^{-n} power law and that the signal level depended on the beam angle to the magnetic field. The strongest emissions were observed for parallel beams. Neubert et al. [1986] identified the drift wave instability as the likely generation mechanism for the VLF waves.

A controlled in situ study of VLF wave-particle interactions was carried out using the Stanford University Siple Station, Antarctica, transmitter and the University of Iowa Plasma Wave Experiments on Explorer 45 (S^3-A) and IMP 6 [Inan et al., 1977]. Emissions stimulated by the transmitter pulses were observed on three of twenty-five passes where transmissions were attempted when the spacecraft were near the equator and within ± 30° longitude of the magnetic field line passing through Siple.

New observations of non-ducted coherent VLF waves from ground-based transmitters (Siple Station, Antarctica, and the Omega navigation network) using the Stanford University VLF Wave Injection Experiment on the ISEE-1 satellite were reported by Bell and Inan [1981] and Bell et al. [1981]. The propagation paths from the source to the satellite were deduced on the basis of the group time delay and Doppler shift. Although many different paths were observed, it was found that emissions are triggered by the later arriving pulses that have traversed the geomagnetic equator. Examples of discrete electromagnetic emissions triggered by ground transmitter signals are shown in Figure 11. The spectrogram of wideband data from the VLF Wave Injection Experiment on the ISEE-1 satellite shows discrete electromagnetic emissions being triggered in the magnetosphere by signals from the Omega navigation transmitters. The amplitudes of the triggered emissions are about 20 dB greater than the amplitudes of the transmitter signals [T. F. Bell, private communication, 1988]. Several excellent examples of triggered emissions observed on DE-1 are contained in Gurnett and Inan [1988].

Studies of wave-particle interactions using the EXOS-B (JIKIKEN) satellite and the Siple Sta-

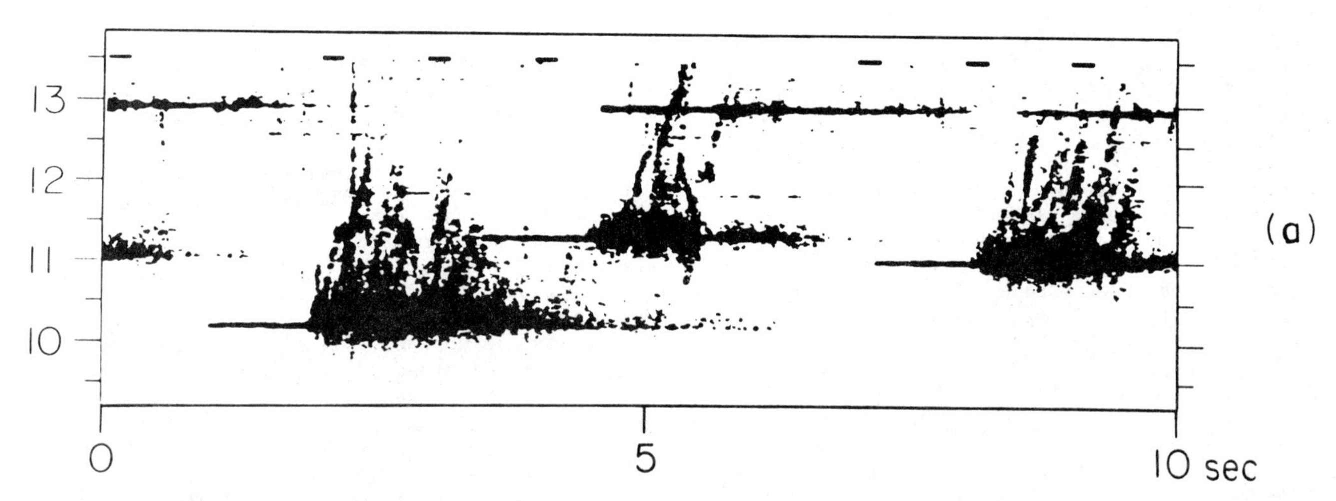

Fig. 11. A 10-second spectrogram of wideband data from the Stanford University VLF Wave Injection Experiment on the ISEE-1 satellite which shows discrete electromagnetic emissions being triggered in the magnetosphere by signals from the Omega navigation transmitters. The spectrogram covers the frequency range from about 9 Khz to 14 kHz. The amplitudes of the triggered emissions are about 20 dB greater than the amplitudes of the transmitter signals [T. F. Bell, Private Communication, 1988].

tion, Antarctica, VLF transmitter have yielded some interesting observations. Triggered emissions associated with the Siple transmissions were observed on EXOS-B only on five passes which were confined to a six day period following a large magnetic storm [Kimura et. al, 1983; Bell et. al., 1983]. Although the pitch angle anisotropy for electrons from 85 eV to 6.9 keV measured on EXOS-B was low when the Siple triggered emissions were observed, the electron flux at resonant energies in the equatorial region was four or five times larger than that on other non-triggering days.

ELF and VLF signals radiated by the 'polar electrojet antenna' were observed in an ionospheric heating experiment reported by Barr et al. [1986]. The heating facility near Tromso, Norway, radiated 270 MW using a frequency of 2.759 MHz that was modulated with frequencies in the range from 223 Hz to 5.44 kHz. ELF and VLF signals over this frequency range were observed ay Lycksele, Sweden, 554 km from the heating transmitter.

The Active Magnetospheric Particle Tracer Explorers (AMPTE) IRM spacecraft performed a series of eight chemical releases in the solar wind, magnetosheath, and near-Earth magnetotail. A variety of plasma waves were observed in conjunction with each of the releases [Gurnett et al., 1985; Häusler et al., 1986; Gurnett et al., 1986b; Koons and Anderson, 1988]. Whistler-mode emissions were most notable in the late phase of the lithium magnetotail releases.

The use of coherent variable frequency VLF signals injected into the magnetosphere for plasma diagnostics was investigated by Carlson et al. [1985]. When the injected signals from Siple Station, Antarctica, were ramped from 1 to 8 kHz, upper and lower cutoffs were observed in the signals received at the conjugate site, Roberval, Quebec, Canada. Carlson et al. [1985] interpreted the upper cutoff characteristics in terms of off-equatorial gyroresonant interaction regions and ducted propagation limited to frequencies below half the local electron cyclotron frequency. The observed lower cutoff frequencies varied systematically with transmitted ramp slope which suggested to them a threshold in the resonant electron number density above which rapid temporal growth and saturation could occur. They used this concept to develop a hot plasma diagnostic technique which, for an assumed $g(\alpha)E^{-n}$ electron distribution, provided an estimate of the energy dependence n. Tests of this concept as well as emission triggering by ramped signals were also discussed.

Theory

Most theories for the generation of electromagnetic discrete emissions in the magnetosphere assume the energy is derived primarily from the kinetic energy of streams of charged particles trapped on the lines of force of the earth's magnetic field [Helliwell, 1965]. Conversion mechanisms fall into two major categories depending on whether the longitudinal motion or the transverse motion of the charged particles is the controlling factor. Mechanisms depending on the longitudinal motion include Čerenkov radiation and an amplification process analogous to a traveling-wave tube. Mechanisms of emission generation associated with the transverse motion of the charged particles include cyclotron radiation and the transverse resonance instability.

Bekefi and Brown [1961] contains an excellent review of observations of radio-frequency emissions from both extra-terrestrial plasmas and laboratory plasmas and of the numerous mechanisms that could produce them. Direct generation by the Čerenkov process was first considered by Ellis [1957, 1959] and Eidman [1958].

Cyclotron radiation from electrons was first studied at very low frequencies by Eidman [1958] and later extensively by Dowden [1962a, b, c; 1963]. Dowden [1962a] postulated that "hooks" might be produced by Doppler-shifted cyclotron radiation from bunches of electrons traveling away from the observer's hemisphere in a helix about a line of force. He further suggested that the particle bunch might mirror which would give rise to a series of hooks of similar shape separated in time by the period of oscillation of the bunch. Brice [1962] discussed these postulations as they applied to "hooks" and found that the mechanism was not valid for his data.

A basic difficulty with the Čerenkov and cyclotron generation mechanisms for discrete emissions is that the calculated intensities derived from them for the audio frequency range are too low to account directly for the observed intensities [Maeda and Kimura, 1962, 1963; Taylor and Shawhan, 1974; and Sieren, 1975]. These difficulties led to the search for more efficient conversion mechanisms. Gallet [1959a, b], Gallet and Helliwell [1959], Bell and Helliwell [1960], and Rydbeck and Askne [1963] proposed a traveling-wave-tube (TWT) mechanism in which a stream of electrons spiraling in the earth's magnetic field performs the function of the beam in the tube, and the whistler-mode wave replaces the slow wave traveling on the helix of the tube. An alternative amplification mechanism was proposed by Maeda and Kimura [1962; 1963] in which amplification takes place on a charged particle beam carrying a slow cyclotron wave. The necessary condition for amplification in all of these theories is that the phase velocity of the wave be approximately equal to the longitudinal particle velocity in the streams. These amplification mechanisms are frequency selective since the phase velocity is frequency dependent.

Brice [1960] discussed the application of the TWT mechanism to whistlers and whistler trains. Magnetic bunching (bunching in phase space) was suggested as a mechanism for producing coherent radiation from particles organized in bunches and which could explain very-low-frequency whistler-mode triggered emissions [Brice, 1963] and hooks, risers, and falling tones [Hansen, 1963]. This mechanism, now referred to as the transverse resonance plasma instability, was examined in detail by Bell and Buneman [1964] who derived the conditions for the instability. The relationships between energy and pitch angle changes for the transverse resonance interaction with whistler-mode waves were derived by Brice [1964b]. Brice concluded that the nonconvective transverse resonance plasma instability was the most plausible explanation for the generation of long enduring quasi-constant frequency emissions [Brice, 1964c] and for the diurnal variation of chorus [Brice, 1964d]. The proton and electron energies required for both longitudinal and transverse resonance as a function of geomagnetic latitude were calculated by Brice [1964a, c] and are displayed in Figure 1 of Brice [1964c].

Bell [1964] proposed an extension of the TWT mechanism in which the energy for amplification of the wave is derived from the transverse motions of the electrons instead of from the longitudinal motion. As with the original TWT mechanism, oblique propagation and matching of the longitudinal component of the electron velocity with the wave phase velocity were required. Bell found that the gyrating electrons would encounter significant variations of the electric and magnetic fields of the wave when the gyro-radius was of the order of a wavelength in the medium and growth rates comparable to those calculated for the transverse resonance instability could be realized.

Helliwell [1963] proposed that the observed triggering of VLF emissions by whistlers was controlled by packets of electromagnetic waves echoing in the whistler-mode. The wave packets were hypothesized to organize temporarily the particles in existing streams of charge so that their radiation was coherent. The resultant temporary increase in radiation would be seen on the ground as a short burst of noise. In an attempt to theoretically determine the number of particles that a given whistler-mode wave can trap, Bell [1965] studied stationary solutions to the nonlinear Vlasof-Boltzmann equations represented by one-dimensional electromagnetic waves in a hot magnetoplasma.

Kimura [1967] reviewed the observations and theories of VLF emissions detected on the ground and by satellite in the period 1957 to 1967. He then proposed a theoretical interpretation for the offset phenomenon of the artificially stimulated emissions based on cyclotron instability by an electron beam. Kimura also calculated the energies of the beam particles needed for gyroresonance and the frequencies of the emissions that would be generated as a function of radial distance using appropriate assumptions about the magnetospheric parameters. From this he evaluated the energies of particles needed for the generation of chorus and hiss.

Helliwell [1967] found that some spectral forms of discrete VLF emissions (such as inverted hooks) were not satisfactorily explained by theories of generation based solely on gyroresonance between energetic streaming electrons and whistler-mode waves traveling in the opposite direction. He proposed an extension of this idea in which spatial variations of the electron cyclotron frequency and the Doppler-shifted wave frequency were matched. This would maximize the coupling time between a resonant electron and the wave and therefore also the output wave intensity. Dowden [1971a] used this criterion to derive a method to determine the electron energy spectrum and structure from ground-observed VLF emission spectrograms. Dowden [1971a] applied this method to the analysis of 75 discrete emissions from a single 3-hour event and found that over the range of resonant longitudinal energies scanned by the emissions, 6-60 keV, the average energy spectrum was of the form $E^{-2.2}$, in general agreement with satellite measurements in the same region (L=3.3).

A self-consistent solution of the quasi-linear theory of gyroresonant interactions was obtained by Roux and Solomon [1971] in order to predict the spectral shape and intensity of VLF waves in the magnetosphere. They found that the relaxation of the warm electron anisotropy due to quasi-linear effects changed the growth rate of the waves. Their results agreed with the gross characteristics of VLF waves observed on satellites.

Helliwell and Crystal [1973] proposed a feedback model of cyclotron interaction between whistler-mode waves and energetic electrons in the magnetosphere for the generation of narrowband VLF emissions. In this model, streaming energetic resonant electrons are temporarily phase-bunched by whistler-mode waves, causing transverse currents which act like circularly polarized antennas, producing stimulated Doppler-shifted radiation. The interaction takes place in an "emission cell" located at or near the equatorial plane. The inclusion of feedback between the stimulated radiation and the incoming particles provides a self-consistent description of fields and currents in time and space. Helliwell and Crystal found that the predictions of the model were in good agreement with experimental observations. This model was discussed further by Helliwell [1974], Nunn [1975], and Helliwell and Crystal [1975]. Helliwell and Inan [1982] proposed a new feedback model that included an interaction region centered on the equator which was treated as an unstable feedback amplifier with a delay line. Improvements over the previous model included taking into account the inhomogeneity of the medium, the effects of the energetic particle distribution, and the phase advance or retardation and frequency change during growth as well as not requiring a predetermined interaction length. This new feedback model was found to be able to account for the observed signal growth and emission triggering in controlled experiments using the Siple Station, Antarctica, VLF transmitter.

Kennel and Petschek [1966] extensively studied whistler-mode, ion cyclotron-mode and magnetosonic-mode instabilities in a very successful attempt to explain the observed limit on stably trapped particle fluxes in the earth's magnetosphere. Whistler-mode waves are right-hand circularly polarized waves above the ion cyclotron frequency and below the electron cyclotron frequency. Ion cyclotron waves are anisotropic Alfvén waves with left-hand circular polarization and frequencies near the ion cyclotron frequency. Magnetosonic waves are right-hand circularly polarized waves at frequencies below the ion cyclotron frequency. Kennel and Petschek [1966] found that Doppler-shifted cyclotron resonant interactions with whistler-mode waves (for electrons) and ion-cyclotron waves (for ions) were the dominant mechanisms for the pitch angle diffusion of particles in the magnetosphere. They found that whether or not a whistler-mode emission at frequency ω was unstable depended only on whether or not the pitch angle anisotropy exceeded $1/(\Omega_e/\omega-1)$. However, its rate of growth or damping depended both on the pitch angle anisotropy and on the fraction of electrons that were resonant. Similar stability and growth conditions were found for the ion cyclotron waves with regard to the ion distributions. For ion distributions to be unstable, the pitch angle anisotropy had to exceed $1/(\Omega_i/\omega-1)$ where Ω_i is the ion cyclotron frequency. In contrast, they found that the magnetosonic mode was unstable when the ion pitch angle anisotropy was sufficiently negative.

Using the concepts of Kennel and Petscek [1966], Das [1968] studied the triggering of emissions, particularly by Morse pulses, by calculating the growth rate as a function of frequency for the disturbed plasma after the pulse had passed. The nonlinear effect of the pulse on the electron distribution was examined. The results were encouraging especially for explaining why dashes were so much more effective at triggering than dots. Qualitatively Das was also able to predict enhancements of the growth rate in narrow frequency bands at the edges of the spectrum of the triggering pulse and how hooks and risers could result from the nonuniform geomagnetic field. Rycroft [1972] reviewed the observations of both naturally occurring and triggered VLF emissions and discussed the theoretical interpretation of the majority of the emissions arising from the transverse resonance cyclotron instability in the equatorial plane in the vicinity of the plasmapause. He estimated the energy of the gyroresonant electrons that would be precipitated from the radiation belts as a consequence of the wave-particle interactions arising from this instability.

Dungey [1972] offered an excellent review and discussion of instabilities in the magnetosphere from a theoretical point of view and essentially beginning with the basics. In a complementary paper, Gendrin [1972] also presented an excellent review of gyroresonant wave-particle interaction theories and their application to experimental data.

Matsumoto and Kimura [1971] conducted a theoretical study of the whistler-mode cyclotron instability both in linear and in nonlinear regimes in conjunction with the generation of VLF emissions in the magnetosphere. They found that the whistler triggered emissions excited around the upper cutoff frequencies of whistlers may be explained by the whistler-mode cyclotron instability and a model distribution function inferred from satellite data. Whistler triggered emissions resulting from the nonlinear interaction between a whistler-mode wave and the particles resonating with it were studied by Dysthe [1971]. He found that the interaction caused a change in the amplification or absorption by the plasma and a phase-bunching of the resonant and near-resonant particles which gave rise to a current. This current has roughly the same frequency and wave number structure as the wave causing it and thus gives rise to a new whistler-mode emission by acting like an antenna. Brinca [1972a] analyzed the growth rates of whistlers with arbitrary frequency and direction of propagation in a cold plasma permeated by a dilute energetic electron population and discussed the results' application to artificially stimulated emissions. He found that the stimulated signals triggered by nonducted whistlers will in general be oblique. Brinca [1972b] studied analytically whistler side-band growth due to nonlinear wave-particle interactions. One result in particular, the identification of the onset of artificially stimulated emissions with the creation of whistler sidebands, gave good agreement with observations. The whistler sideband instability is the unstable growth of other whistler waves with frequencies near the main wave frequency resulting from resonant electrons in a large-amplitude whistler wave acquiring a correlation in phase of their perpendicular velocities with respect to the wave magnetic field. Denavit and Sudan [1975a, b] analyzed both theoretically and by computer simulation the whistler sideband instability and the evolution of large-amplitude whistler wave packets in a hot collisionless plasma. They found that the phase-correlated trapping of electrons caused new emissions and elongated the wave packet in the direction of emergence of the resonant electrons.

Matsumoto et al. [1974] examined longitudinal and transverse phase bunching in whistler-mode wave-particle interactions. Longitudinal phase bunching is caused by a longitudinal acceleration while transverse phase bunching is caused by a transverse acceleration. For a plasma composed of relatively high pitch angle electrons, longitudinal phase bunching prevails and the electrons are not accelerated or decelerated continuously. Consequently, only a small periodic perturbation of the amplitude of the original whistler-mode wave is expected. For a plasma in which low pitch electrons dominate, the acceleration of $v_\perp$ and the tight transverse phase bunching forms a rapidly growing resonant current yielding a considerable damping of the original whistler-mode wave. They pointed out that possible cases in the magnetosphere where electrons with small pitch angles might exist include artificial beam injection from a rocket- or satellite-borne electron gun and the acceleration of thermal electrons by a transient field-aligned electric field. Phase-bunching and other nonlinear processes occurring in gyroresonant wave-particle interactions were also extensively studied by Gendrin [1974].

The effect of a succession of narrowband Morse pulses propagating in the whistler-mode on a steady state distribution function of electrons trapped in the Earth's magnetic field was studied by Ashour-Abdalla [1972] using a one-dimensional Fokker-Planck equation. Corresponding growth rates were computed as a function of both time and frequency. The behavior of the growth rates led Ashour-Abdalla to postulate that a minimum delay time (inversely related to the power in the signal) is required before triggering can occur. Many of the observational characteristics of artificially stimulated emissions could be explained by the findings in this study.

Nunn [1971] extensively investigated the triggering of VLF emissions by magnetospheric whistler Morse pulses. Nunn studied the behavior of resonant particles in a whistler train in an inhomogeneous medium and found that second order resonant particles became stably trapped in the wave. After one or two trapping periods these particles dominated the resonant particle distribution function and produced large currents. The in-phase component of the current produced growth rates n times larger than the linear value when the particles had been trapped for n trapping periods. The reactive component of the current caused a steady change in the wave frequency. When a realistic zero-order distribution function was used in conjunction with a magnetospheric whistler pulse in the equatorial region, Nunn's results replicated a number of the detailed features experimentally observed.

Sudan and Ott [1971] also presented a theory of triggered VLF emissions based on resonant electrons being phase correlated with the wave magnetic field by a finite length whistler train moving in the opposite direction. The time for the phase correlation was of the order of the period of oscillation of a particle trapped in the effective "potential well" of the wave. The phase-correlated electrons were subject to an instability in the form of an emitted whistler with a growth rate given by

$$\gamma/\omega \cong (n_c/n_o)^{2/5} (\langle V_c \rangle/c)^{2/5} (\Pi_e/\Omega_e)^{2/5}$$

where n_c is the number density of phase-correlated particles, n_o is the cold plasma number density, $\langle V_c \rangle$ is the mean transverse velocity, c is the speed of light, Π_e is the electron plasma frequency, and Ω_e is the electron cyclotron frequency. The emitted frequency varied according to $\omega = k(\omega)v_\parallel - \Omega_e$ where $v_\parallel$ is the zero order longitudinal velocity of the resonant electrons, and the wave number k is a function of frequency ω as determined through the whistler dispersion relation. Sudan and Ott [1971] found good agreement between their theory and observations and showed that both triggered risers and hooks could be explained.

Kimura [1974] reviewed the observations and theoretical works in regard to the interrelation between VLF and ULF emissions. Carson et al. [1965] were the first to hypothesize that the source of oscillations for observed quasi-periodic emissions was hydromagnetic waves. Kimura [1974] offered tentative interpretations of VLF and ULF emissions which are closely associated through a modulation of the electron distributions. Sazhin [1978] developed a self-consistent quasilinear model of the interaction between VLF emissions and geomagnetic pulsations and obtained an explicit expression for the modulation frequency dependence.

The nonlinear aspects, both experimental and theoretical, of wave-particle interactions in the magnetosphere were reviewed by Gendrin [1975]. He also discussed the kinds of experiments that could be made at high altitudes with or in conjunction with spacecraft in order to arrive at a better understanding of magnetospheric processes involving waves and particles. Wave-particle interactions in the outer magnetosphere were reviewed by Fredricks [1975].

Yamamoto [1976] investigated the nonlinear development of the plasma-beam instability in a magnetized plasma. This study showed that the amplitude of an unstable whistler wave in the presence of a small electron beam first increases at the linear growth rate until the electrons are trapped. Then, after overshooting, it approaches a steady state.

Brinca [1977] studied the influence of a CW (continuous wave) transmission on magnetospheric noise. His analysis of the modifications introduced in a turbulent whistler noise spectrum with the injection of a coherent whistler led to a nonlinear dispersion equation for the stochastic modes. These modes are submitted to real frequency shifts and corrections to their growth rates were found to be in qualitative agreement with observations made in Siple Station VLF wave injection experiments which showed the creation of noise-free bands when CW whistler modes are transmitted [Helliwell, 1979].

Roux and Pellat [1976; 1978] studied trapped and untrapped particle trajectories in the adiabatic approximation for a finite amplitude, quasi-monochromatic whistler-mode wave propagating along geomagnetic field lines. They found that particles which are detrapped at the triggering wave termination have been phase organized and may then act coherently for a while, giving rise to an emission with either rising or falling tones, depending on the sign of the inhomogeneity variation. These emissions can in turn trap electrons and the emission process can be self-sustained. Dowden [1982] found that an additional wave at a slightly different frequency from that which originally trapped the electrons can detrap these electrons even if the additional wave is 20 dB weaker than the trapping wave if the frequency difference is appropriate.

A theory for two-banded chorus generation by energetic electrons during substorms was developed by Curtis [1978]. He suggested that the emission band just below $\Omega_e/2$ was whistler-mode noise and that the emission band just above $\Omega_e/2$ was the lowest harmonic of the ordinary mode. Both bands were assumed to be generated locally at the observation point and to be excited by power law-distributed energetic electrons with a variable spectral index and a weak loss cone feature. The calculated growth rates yielded a frequency spectrum similar to that of the observed emissions. Gendrin and Roux [1980] studied theoretically the energization and diffusion of helium ions under the influence of hydromagnetic ion cyclotron waves. They considered a plasma consisting of three different ion populations: a thermal isotropic population containing both hydrogen and helium ions and an energetic hydrogen population with a positive temperature anisotropy ($T_\perp > T_\parallel$). They showed that for small concentrations of helium ions, of the order of 1 to 10%, these ions could be energized by their interaction with the ion cyclotron wave up to and above suprathermal energies ($E \gtrsim 20$ eV) and on some occasions even up to the order of the Alfven energy of the cold plasma population ≈ 5 keV). Gendrin and Roux [1980] discussed their theoretical results in light of observations from GEOS experimenters that showed a close association between the occurrence of ion cyclotron ULF waves and the presence of thermal or supra-thermal helium ions in the equatorial region of the magnetosphere. Mauk et al. [1981] interpreted their ATS-6 observations of a gap in the electromagnetic ion cyclotron wave spectrum at the helium cyclotron frequency and of strongly cyclotron phase bunched helium ions as resulting from the absorption of the waves through cyclotron resonance by cool ambient populations of helium ions.

In order to explain results from a computer simulation experiment studying the dependence of coherent nonlinear whistler interaction on wave amplitude, Matsumoto et al. [1980] developed a theory based on the trapping of the resonant hot electron beam by a monochromatic whistler wave. O'Neil et al. [1971] earlier had shown that a small cold electron beam immersed in a background plasma will cause a single wave to grow exponentially at the linear growth rate until the beam electrons are trapped. Then the wave amplitude stops growing and begins to oscillate about a mean value. During the trapping process the beam electrons are bunched in space and a power spectrum of the higher harmonics of the electric field is produced. The theoretical results of Matsumoto et al. [1980] showed that the saturation level of the whistler wave is independent of the wave initial amplitude but is determined by the beam and background plasma conditions. Their simulation and theoretical results explained observations from several experiments using the Siple VLF transmitter and suggested that similar future experiments could be used to remotely diagnose beam properties.

Koons [1981] developed a model for the generation of chorus based on numerous published observations that narrowband hiss emissions were simultaneously present with ELF chorus emissions outside the plasmasphere and data from the SCATHA satellite which showed that chorus emissions often start at frequencies within a hiss band. Electrons in a narrow range of energies and pitch angles would be organized in phase by the Doppler-shifted cyclotron resonance with the larger-amplitude spectral components in the hiss band. The bandwidth of the cyclotron resonance was found to be narrow enough so that the electrons would not be dephased by waves in adjacent portions of the highly structured spectrum. The amplitude of the hiss was sufficient to significantly phase bunch the electrons in a calculated interaction time. The chorus emission would then be generated as the phase-bunched electrons moved adiabatically along the geomagnetic field line.

Vomvoridis et al. [1982] examined analytically and by computer simulation a nonlinear mechanism for the generation of magnetospheric VLF emissions that could account for the triggering of monochromatic emissions by signals of sufficient strength and duration (while the background noise and weak short signals were not amplified) and the occurrence of frequency changes after the emissions reached a sufficiently large amplitude. Their mechanism depended on the simultaneous propagation and amplification of wave packets along geomagnetic lines that would maintain the nonuniformity ratio $R \propto \Delta B_0/B_w$ in the regime $|R| \approx 0.5$ which corresponded to maximum amplification. (B_0 is the geomagnetic field strength and B_w is the wave magnetic field strength.) For a constant frequency, this condition yielded triggering thresholds which are related to the properties of the magnetosphere. For a varying frequency $\omega(t)$, their results yielded the condition $\partial\omega/\partial t \propto \omega_t^2$ for the large-amplitude portion of the risers, where $\omega_t \propto B_w^{1/2}$ denotes the trapping frequency of the wave.

The nonlinear gyroresonance interaction in the magnetosphere between energetic electrons and coherent VLF waves propagating at an arbitrary angle ψ with respect to the earth's magnetic field B_0 was investigated by Bell [1984]. Bell [1984] used an extension of the phase trapping model developed in earlier work by Bell [1965], Dysthe [1971], and Nunn [1974a]. He found that near the geomagnetic equatorial plane, the threshold value of wave amplitude necessary to produce phase trapping was proportional to the gradient of ψ along B_0. For interactions far from the magnetic equator, the variations in ψ were less important than gradients in B_0 for determining phase trapping efficiencies. His predictions of higher threshold values for phase trapping for nonducted waves were in general agreement with experimental data concerning VLF emission triggering by nonducted waves. Using this theory, Bell [1986] calculated the wave magnetic field amplitude thresholds for nonlinear trapping of energetic gyroresonant and Landau resonant electrons by nonducted waves in the magnetosphere.

A quasistatic theory of triggered VLF emissions was presented by Nunn [1984b] who simplified the self-consistent nonlinear interaction of cyclotron resonant electrons with a narrowband wave field by removing the time variable. He considered only a saturated 'quasi-static' emission and modeled self-consistently a nonlinear emission generating region. Nunn [1984b] found that very high pitch angles were important even with low anisotropy distribution functions. For example, risers are driven by higher pitch angle particles (~77°) than are fallers (~60°). He also found that the generation region solutions required a high degree of power loss (~80-95% of input power). The nonlinear current generated in the duct radiates most of the power into unducted modes. This nonlinear unducting provides a saturation mechanism that serves to stabilize the generation region structure.

Thorne and Tsurutani [1981] and Tsurutani et al. [1982a] showed that lion roars were closely related to magnetic field decreases which occur within large scale magnetosheath structures set up by a mirror instability. They concluded that an electron cyclotron resonance instability was responsible for the generation of lion roars. Moreira [1983] used a linear hot plasma dispersion relation for parallel and oblique propagation to show that the electron cyclotron instability generating lion roars was absolute as opposed to convective. He also showed that the observed lion roars corresponded to a nonlinear stage of a saturated whistler wave instability.

Dawnside mid-latitude VLF emissions were investigated in terms of the quasi-linear electron cyclotron instability by Hayakawa et al. [1985; 1986a, b]. They constructed a model to explain the morphology of VLF emissions based on their determinations of the resonant energies, theoretical electron drift orbits, and observed local time dependences. They surmised that electrons in a wide energy range are injected during geomagnetic disturbances around the midnight sector and then drift eastward. Lower energy (~5 keV) electrons would tend to drift closer to the Earth, resulting in the dawnside enhancement of VLF hiss within the plasmasphere. Furthermore, these electrons would continue around and enter the duskside asymmetric plasmaspheric bulge and generate VLF hiss there. The higher energy (~20 keV) electrons would tend to drift at L shells farther away from the Earth and would be responsible for the generation of structured VLF emissions around dawn due to an increase in plasma density from the sunlit ionosphere. However, the orbits of such high energy electrons are forbidden from entering the duskside of the magnetosphere and no duskside peak in the occurrence of the structured emissions would be expected, which is in agreement with experimental observations.

Brinca and Tsurutani [1988] investigated low-frequency electromagnetic waves stimulated by two coexisting newborn ion species. Although their study emphasized the cometary environment, they pointed out that the results were useful for interpretation of other observations of low-frequency wave activity elsewhere in space. Gary and Madland [1988] also investigated electromagnetic ion instabilities in a cometary environment.

Much additional material on theoretical investigations of discrete electromagnetic emissions may be found in Plasma Waves in Space and Laboratory [Thomas and Landmark, 1970], a review article on plasma instabilities in the magnetosphere by Hasegawa [1971], Nonlinear and Turbulent Processes in Physics [Sagdeev, 1983], and in a review article on waves and instabilities by Inan [1987]. A very valuable source of information on most plasma wave instabilities is Stix [1962] treatise, The Theory of Plasma Waves.

Simulations

Matsumoto and Kimura [1971] carried out computer simulations in order to study the nonlinear evolution of the whistler-mode cyclotron instability. They showed that the change of frequency with time of whistler triggered emissions as well as characteristics of ASE are well explained by resonant nonlinear behavior of whistler-mode cyclotron instabilities. A computer simulation performed by Sudan and Denavit [1973] showed how very-low-frequency emissions could be triggered in the magnetosphere and then propagate as whistlers. The nonlinear interaction between obliquely propagating whistlers and energetic electrons in fundamental gyroresonance was studied through numerical simulation by Brinca [1975]. Phase bunching occurred in the perpendicular velocities of the resonant electrons analogous to the parallel whistler case. Brinca found that the computed organized current brought about by the bunching suggested that obliquely propagating whistlers had the capability to trigger emissions through the gyroresonance interaction.

Computer simulations of the whistler sideband instability which results from electrons being correlated in phase by a large amplitude whistler wave propagating parallel to the ambient magnetic field were performed by Denavit and Sudan [1975a] using a one dimensional particle code. Unstable growth of sidebands at various frequencies around the main frequency were obtained dependent on the plasma parameters chosen. Further computer simulations were performed by Denavit and Sudan [1975b] to study the effects of phase-correlated electrons on whistler wave packet propagation. These simulations showed the elongation of large-amplitude wave packets and provided the wave number and frequency characteristics of the new emissions generated.

Computer simulations using a one-dimensional electromagnetic nonrelativistic particle code were used by Ossakow et al. [1972b] to study whistler turbulence in plasmas driven by initially anisotropic bi-Maxwellian and loss-cone electron distributions. They found that the linear growth saturates when the growth rate is of the order of the particle trapping frequency. Ossakow et al. [1972c] carried out both theoretical and computer simulation studies of whistler instabilities in three types of collisionless plasmas. Initial bi-Maxwellian, Maxwellian with a loss cone, and hot Maxwellian superimposed on a more dense cold isotropic background electron distributions were used. The computer simulation experiments were performed using a nonrelativistic electromagnetic particle code with three velocity dimensions and one spatial dimension. For all three initial conditions it was found that initially the total wave magnetic energy grows in agreement with linear theory and the electron distribution isotropizes with an accompanying falling off of high k number, initially unstable waves. Then the total wave magnetic energy saturates and there is a residual kinetic energy anisotropy together with a further switching off of the higher mode numbers for the remainder of the simulation. Similar results were obtained by Ossakow et al. [1973a, b] for a magnetospheric-type plasma--a hot Maxwellian component with 60° loss cone was superimposed on a cold isotropic electron background.

Nunn [1974a] conducted a computational study of nonlinear resonant particle trajectories in wave fields consisting of an array of narrowband waves of closely spaced frequencies and wave numbers in a theoretical study of banded chorus. He looked at two analogous systems, cyclotron resonance with a whistler wave field and Landau resonance with an electrostatic wave field. He found that the wave array was able to trap particles in much the same way as a single mode. Inhomogeneity played an important role in his analysis because it caused the energy of trapped particles to change. The nonlinear resonant particle current preserved the model structure of the wave field and did not change the frequency of individual modes. Nunn found that most of the energy went into the mode at one end of the array, depending on the direction of the inhomogeneity. Nonlinear resonant particle excitation of a broadband signal was found to cause spectral structuring to develop automatically. The calculated spectral structure agreed well with that of a banded chorus element analyzed by Coroniti et al. [1971]. However, Stiles [1975] offered an alternate interpretation for the narrow lines in the spectra. He suggested that the apparent single modes could be more easily explained by the assumption of closely spaced (~30 ms) continuously rising (or falling) tones such as that presented by Shaw and Gurnett [1971] to explain harmonically banded whistlers.

A self-consistent computer program was constructed by Nunn [1974b] which followed in detail the time development of a Morse pulse as it crossed the equator in order to study triggered VLF emissions. Nunn's results showed that the nonlinear trapping of electrons in an inhomogeneous medium is fully able to account for the triggering instability. Nunn's model was able to reproduce a number of triggered emissions with characteristics similar to what had been observed. Das and Kulkarni [1975] investigated nonlinear wave-particle interactions in the whistler mode in a non-uniform magnetic field. The equations describing the time development of the amplitude and phase of the wave packet were solved numerically by computing the resonant particle current in a self-consistent manner. The effect of the second order resonant particles arising due to non-uniformity of the ambient magnetic field was found to be dominant near the equatorial plane of the Earth. The growth of the waves because of trapped particles was found to be substantial for triggering an emission and the changes in phase led to the frequency-time structure. The results of Das and Kulkarni [1975] were also capable of reproducing the frequency-time structure of VLF emissions in the case of a Morse pulse.

A computer experiment on the self-consistent coherent nonlinear interaction between a large amplitude whistler-mode wave propagating along an external magnetic field was carried out by Matsumoto and Yasuda [1976]. The results of this simulation apply to triggered very low frequency emissions and are useful for analyzing controlled wave-particle interaction experiments using artificial beam injection in space.

Vomvoridis and Denavit [1979] presented computer simulations of the correlation in phase of test electrons in a monochromatic whistler-mode wave propagating along a nonuniform external magnetic field. The dynamics of the resonant electrons depended on the inhomogeneity (nonuniformity) ratio

$$R = \left(3 v_r - \frac{v_\perp^2 - \kappa v_r^2}{\Omega_e / k}\right) \frac{d\Omega_e/dz}{2\omega_t^2} \quad .$$

Here, the resonant velocity $V_r = (\omega - \Omega_e)/k$. An electron density proportional to Ω^κ was assumed; z is the coordinate along the external ambient magnetic field; $V_\perp$ is the electron velocity perpendicular to the external magnetic field B_0; and $\omega_t = (kV_\perp \Omega B_w/B_0)^{1/2}$ is the small amplitude trapping frequency of resonant electrons in the wave field B_w. For $|R| < 1$, where electron trapping is possible, they found a hole (or an island) of test particles in phase plots of the distribution function. This caused the appearance of a transverse current coherent with the wave which was most intense for $|R| \lesssim 0.6$.

Self-consistent computer simulations and theoretical considerations of the growth (or damping) and frequency shift of a monochromatic whistler-mode wave propagating along a nonuniform external magnetic field with a constant gradient ($d\Omega/dz$ = constant) were presented by Vomvoridis and Denavit [1980]. They found that the nonlinear evolution of the wave could be described by three growth rates which represented (1) the effects of the resonant electrons similar to the uniform field case, (2) the effect of the nonuniformity on the untrapped resonant electrons, and (3) the effect of the nonuniformity on the trapped electrons.

The dependence of coherent nonlinear whistler interaction on wave amplitude was investigated by self-consistent computer simulations by Matsumoto et al. [1980] in order to understand experiments using the Siple VLF transmitter. The simulations revealed that for sufficiently small initial wave amplitudes, the monochromatic whistler wave propagating parallel to the external magnetic field grows exponentially at the linear growth rate until the hot beam electrons are phase trapped in the whistler field. Then the wave amplitude begins to oscillate about a mean value and eventually approaches an ergodic saturation level which is almost the same regardless of the initial wave amplitude. If the initial wave amplitude is larger than this saturation level, the whistler wave shows neither growth nor damping but merely shows an amplitude oscillation about its initial level.

Basic processes of coherent whistler-mode wave-particle interactions in the magnetosphere were also studied by Omura and Matsumoto [1982] using self-consistent computer simulations. Nonlinear processes of wave growth in a uniform magnetic field were examined in detail. When the inhomogeneity of the magnetic field was taken into account, they found that untrapped resonant electrons as well as trapped electrons played a significant role in the wave evolution. Matsumoto and Omura [1983] carried out another simulation study of VLF triggered emissions to investigate quantitative changes of the velocity distribution function of resonant electrons caused by the combined action of nonlinear phase trapping and geomagnetic inhomogeneity following the theory of Roux and Pellat [1978]. They found that the contributions of electrons trapped in the middle of the whistler wave train were as important as that by electrons trapped at the wave front. Matsumoto and Omura [1983] also found that deformation of the velocity distribution function was caused both by trapped electrons and by untrapped electrons perturbed by the whistler triggering wave.

Nonlinear pitch angle scattering and trapping of energetic particles during Landau resonance interactions with whistler-mode waves in the magnetosphere were studied in a test particle simulation study conducted by Tkalcevic et al. [1984]. They used the time averaged equations of motion derived by Inan and Tkalcevic [1982]. Their results showed that for typical parameters in the magnetosphere, the precipitation fluxes for Landau resonance interactions were much smaller than those induced in gyroresonance interactions.

The role of the loss cone in the formation of impulsive bursts of precipitation was investigated by Davidson [1986]. He showed that the spontaneous growth of VLF plasma waves, followed by pitch angle diffusion of the electrons that interact with the waves, can lead to bursts of precipitation in the auroral regions. His results were in good agreement with observations for precipitation bursts at energies above 25 keV.

A test particle model of the cyclotron resonance interaction of waves and trapped radiation belt particles was used by Inan [1986] to estimate the energetic electron fluxes precipitated by Jovian VLF chorus waves observed by Voyager 1 and 2 near the Io torus. He also estimated the effects in the Jovian ionosphere of the chorus-induced precipitation by using existing ionospheric models. Inan [1986] then proposed and discussed a possible radio beacon occultation experiment for detecting the effects.

The competing processes of whistler-mode and electrostatic-mode instabilities induced by an electron beam were studied by Omura and Matsumoto [1987] who used both a linear growth rate analysis and an electromagnetic particle simulation. In the linear growth rate analysis, they found that the growth rate for the electrostatic instability was always larger than the whistler-mode growth rate. A short simulation run also showed that a monoenergetic beam could give very little energy to whistler-mode waves because of the competition with faster growing electrostatic waves. However, a long simulation run starting with a warm electron beam showed that whistler-mode waves were excited in spite of small growth rates and the coexisting quasi-linear electrostatic diffusion process. Omura and Matsumoto [1987] emphasized that a proper understanding of the physics involved in the simulations of two coexisting instabilities with different time scales required many grid points as well as an enormous number of time steps.

Related Phenomena

Discrete spectral features also occur in other electromagnetic modes at frequencies well above the audio frequency range where the majority of research has concentrated. At frequencies above both the electron cyclotron frequency and the electron plasma frequency, terrestrial auroral kilometric radiation (AKR), which propagates predominantly in the free space (R-X) mode, has a very complex frequency-time structure consisting of many discrete emissions. This unexpected feature was first observed in the Plasma Wave Experiment wideband data on ISEE-1 and -2 [Gurnett et al., 1979]. Spectra of wideband data from ISEE-2 containing auroral kilometric radiation in the frequency range from 125 kHz to 135 kHz are shown in Figure 12. The quite evident fine structure consisting of many discrete narrowband rising and falling elements is very similar to that observed at much lower frequency in high time resolution chorus spectra such as that shown in Figure 4. Figure 13 displays spectra of simultaneous wideband data from ISEE-1 and -2 containing auroral kilometric radiation in the frequency range from 125 kHz to 135 kHz [Figure 8 in Gurnett and Anderson, 1981]. These two high resolution spectrograms show the apparent triggering of rising emissions by a narrowband drifting feature. These triggering effects are remarkably similar to the triggering of whistler-mode emissions by ground VLF transmitters as shown in Figure 11, by hiss bands as shown in Figure 3, and by whistlers. Additional spectrograms showing the discrete fine structure in auroral kilometric radiation at frequencies ranging from 62 kHz to 540 kHz are shown in Gurnett et al. [1979] and Gurnett and Anderson [1981]. Dynamic spectra from EXOS-B indicating fine structure in auroral kilometric radiation are contained in Morioka et al. [1981] and Oya and Morioka [1989]. Although unexpected in the auroral kilometric radiation, the occurrences of bursts with rapidly drifting center frequencies is a common feature in other astrophysical radio sources such as solar radio bursts [Kundu, 1965] and Jovian radio emissions [Warwick, 1967]. For example, a type of radio emission from Jupiter called S-burst has been studied for many years by radio astronomers and is characterized by discrete tones drifting downward in frequency [Gallet, 1961]. Bursty radio emissions from Uranus in the frequency range from 70 kHz to 1.2 MHz were observed by Evans et al. [1987] who noted that they were very similar to terrestrial AKR. This similarity suggested to them that much of the same physics was occurring at the two planets. Whether or not these free space mode emissions observed at Earth and Uranus are fundamentally similar to the discrete ion cyclotron and whistler-mode emissions is an open question.

Stimulated electromagnetic emission (SEE) in ionospheric heating experiments [Thide et al., 1982; 1983] is a high frequency phenomena with characteristics similar to the emissions and sidebands generated in some of the active experiments in the VLF range discussed in Section 6. Strong and systematically occurring spectral features both up-shifted and down-shifted from the pump frequency have been observed. A rather complicated but viable process has been discussed by Leyser and Thide [1988] for the generation of SEE that could possibly be applied to the audio frequency phenomena under certain circumstances. The ponderomotive force associated with the powerful electromagnetic pump wave leads to large-scale density depletions at the standing wave maxima. Langmuir waves are excited through the parametric decay instability in these density

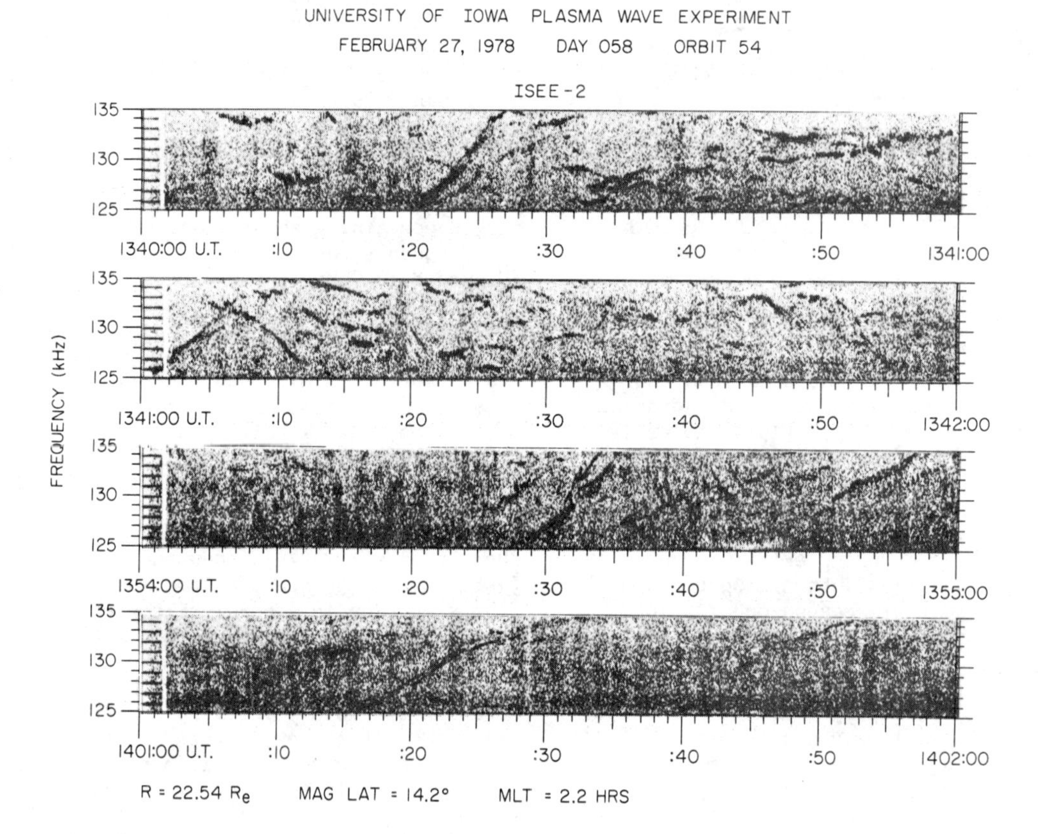

Fig. 12. Four 1-minute spectra of wideband data from the University of Iowa Plasma Wave Experiment on ISEE-2 containing auroral kilometric radiation in the frequency range of 125 kHz to 135 kHz. The fine structure consisting of many discrete narrowband rising and falling elements is very similar to that observed at much lower frequency in high time resolution chorus spectra such as that shown in Figure 4.

cavities. The Langmuir waves must then undergo mode conversion to electromagnetic waves in order to be detected as SEE. The nonlinear mixing and excitation of sidebands at high frequencies in the ionosphere have also recently been studied experimentally by Ganguly and Gordon [1986] and theoretically by Lalita and Tripathi [1988]. Some of the characteristics are quite similar to those observed at much lower frequencies and discussed in Section 6. For example, the power in the sidebands was found to be a strong nonlinear function of the incident pump strength, showing the presence of a threshold.

Summary

This review has shown that much progress has been made, especially in the past three decades, in measuring the characteristics of discrete electromagnetic emissions and in determining and understanding the wave-particle interactions which generate them. An important element leading to this progress has been the in situ satellite observations which have provided measurements of the waves, particles, plasma, and fields in the generation and interaction regions of the ionospheres and magnetospheres of the Earth, Jupiter, Saturn, and Uranus. An amazing observational result is the high degree of similarity of discrete emissions (especially chorus) at all planetary magnetospheres despite drastic differences in plasma sources, energy sources, magnetospheric configurations, etc.

Improved ground-observing facilities with new technological instrumentation such as direction-finders have been important for increasing our knowledge of the characteristics of the terrestrial emissions including their direction of propagation and the location of their exit points form the ionosphere into the atmosphere. Active experiments using both research transmitters and existing navigation and communications transmitters in the VLF range have aided greatly in investigations of the wave-particle interaction processes. Interesting and useful results have also been obtained from numerous experiments

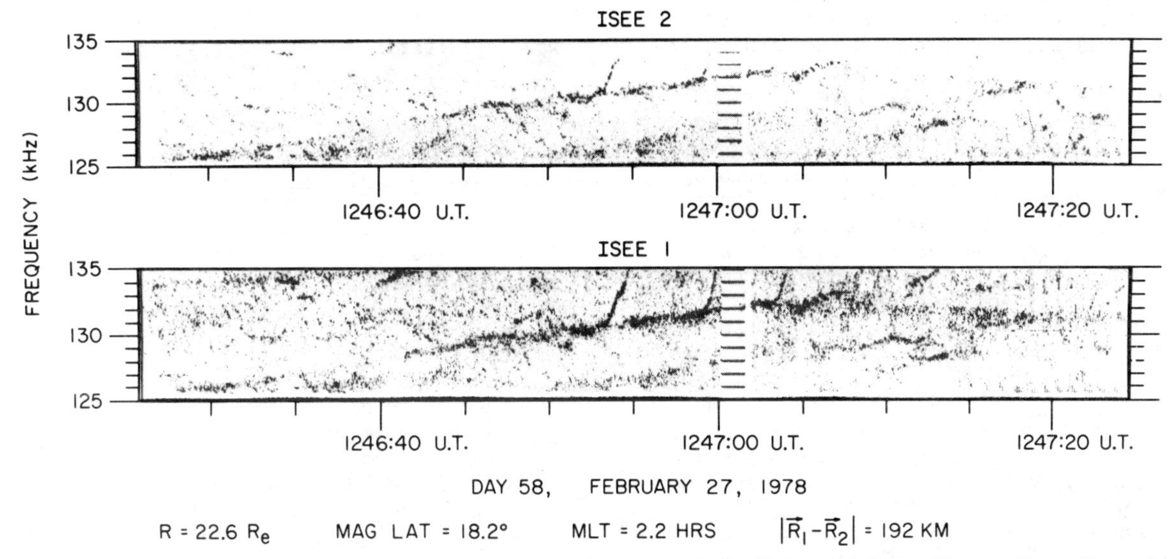

Fig. 13. Spectra of simultaneous wideband data from the University of Iowa Plasma Wave Experiments on ISEE-1 and -2 containing auroral kilometric radiation in the frequency range of 125 kHz to 135 kHz (Figure 8 in Gurnett and Anderson, 1981). Each panel contains one minute's worth of data. These high resolution spectrograms show the apparent triggering of rising emissions by a narrowband drifting feature. These triggering effects are remarkably similar to the triggering of whistler-mode emissions by ground VLF transmitters as shown in Figure 11 and by hiss bands as shown in Figure 3.

where particle beams or plasma have been injected into the existing magnetospheric plasma.

The vast amount of interesting observations has stimulated scientists to analyze numerous instabilities and wave-particle interactions theoretically and by computer simulation. In many cases, very good agreement has been found between the results of these investigations and the observations. This includes in many instances being able to duplicate the intricate frequency-time structure of many of the emissions observed. Excellent predictions on the flux and spectra of precipitated particles associated with the discrete emissions have also been obtained for a variety of magnetospheric parameters.

Despite the progress described in this review, many unanswered or only partially answered questions remain that require further investigations, both observationally and theoretically. In some cases, a single phenomena has been explained by two or more competing theories with apparently successful results. Additional observational details are thus needed to discern which theories are really appropriate. Some of the theoretical and computer simulation studies have had to make assumptions on details of the wave or magnetospheric parameters that were not known or available. Correlated observations in the magnetosphere with high time resolution particle experiments providing full three dimensional distribution functions and with high time resolution plasma wave instruments capable of measuring the full wave distribution function (three dimensional electric and magnetic including phase and wavelength measurements) are needed to fully understand the wave-particle interactions generating the discrete emissions (both the naturally occurring and the artificially stimulated ones).

The increased use of large high speed super computers (along with the continuing advancements and improvements in such computers) which will allow long time duration full three dimensional simulations with realistic plasma parameters should allow even better investigations and understanding of the various discrete electromagnetic emission phenomena.

One theme inherent in many of the observational, theoretical, and simulation papers is the potential value of controlled whistler-mode injection via ground transmitter experiments for the remote diagnosis of various plasma parameters. With assistance in the planning from interested theoreticians and computer simulators, new experiments should be conducted to test such ideas and to continue the related studies of wave-wave and wave-particle interactions. Another important set of experiments to be carried out include in situ injection and observation of waves and particle beams from spacecraft launched into the ionosphere and magnetosphere.

Continued and increasing support from the various national science and space physics funding agencies for the continued analysis of the very valuable already existing data bases, operation of existing satellites and ground observing facilities, and the development of new and improved satellite experiments (both active and observational) and ground transmitter and observational facilities along with adequate support for theoretical and simulation investigations are all necessary and should be strongly encouraged if we are to continue the significant progress made in the past three decades as indicated by the papers covered in this review. Continued support and encouragement for meetings and conferences that bring the researchers from many nations together are also very important for efficient progress and increased understanding of the discrete emission phenomena and the techniques for investigating them.

Acknowledgments. The authors wish to express their sincere thanks to the many people who have assisted in the preparation of this manuscript including D. A. Gurnett for his support and who has been primarily responsible for many of the plasma wave investigations from which we have used data, T. F. Bell and U. S. Inan for providing illustrations and comments, M. Brown and R. Huff for their help in producing illustrations, C. Frank and T. Thompson for their assistance with the literature search and references, B. T. Tsurutani and I. Kimura for their encouragement and useful comments, and especially to Jean Hospodarsky for her ever-efficient typing and production of the manuscript. We are most grateful for the many extra hours of effort put forward by Jean and the excellent work that she did in helping us complete this manuscript. This research was supported by NASA/GSFC contract NAG5-28701 and contract 957723 with the Jet Propulsion Laboratory and by NASA/GSFC grants NAG5-310 and NAG5-1093 with additional support from NASA grants NGL-16-001-002 and NGL-16-001-043 and the Office of Naval Research.

References

Allcock, G. Mck., A study of the audio-frequency radio phenomenon known as "dawn chorus", Australian J. Phys., 10(2), 286-298, 1957.

Allcock, G. Mck., and L. H. Martin, Simultaneous occurrence of "dawn chorus" at places 600 km apart, Nature, 178, 937-938, 1956.

Anderson, R. R., Wave-particle interactions in the evening magnetosphere during geomagnetically disturbed periods, Ph.D. Thesis, Dept. of Physics and Astronomy, The University of Iowa, Iowa City, Iowa, pp. 107, 1976.

Anderson, R. R., Current status of IMS plasma wave research, Rev. Geophys. and Space Phys., 20, 631-640, 1982.

Anderson, R. R., Plasma waves in planetary magnetospheres, Rev. Geophys. and Space Phys., 21, 474-494, 1983.

Anderson, R. R., Plasma waves at and near the neutral sheet, Proceedings of the Conference on Achievements of the IMS, Eur. Space Agency Spec. Publ., ESA SP-217, 199-204, 1984.

Anderson, R. R., and Donald A. Gurnett, Plasma wave observations near the plasmapause with the S^3-A satellite, J. Geophys. Res., 78, 4756, 1973.

Anderson, R. R., and K. Maeda, VLF emissions associated with enhanced magnetospheric electrons, J. Geophys. Res., 82, 135-146, 1977.

Anderson, R. R., C. C. Harvey, M. M. Hoppe, B. T. Tsurutani, T. E. Eastman, and J. Etcheto, Plasma waves near the magnetopause, J. Geophys. Res., 87, 2087, 1982.

Ashour-Abdalla, M., Amplification of whistler waves in the magnetosphere, Planet. Space Sci., 20, 639-662, 1972.

Barkhausen, H., Zwei mit Hilfe der neuen Verstaerker entdeckte Erscheinungen, Physik. Z., 20(1919), 401, 1919.

Barr, R., P. Stubbe, M. T. Rietveld, and H. Kopka, ELF and VLF signals radiated by the "polar electrojet antenna": Experimental results, J. Geophys. Res., 91, 4451-4459, 1986.

Barrington, R. E., and J. S. Belrose, Preliminary results from the very-low-frequency receiver aboard Canada's Alouette satellite, Nature, 198, 651-656, 1963.

Barrington, R. E., J. S. Belrose, and D. A. Kelley, VLF noise bands observed by the Alouette 1 satellite, J. Geophys. Res., 68, 6539-6541, 1963a.

Barrington, R. E., J. S. Belrose, and G. L. Nelms, Ion composition and temperatures at 1000 km as deduced from simultaneous observations of a VLF plasma resonance and topside sounding data from the Alouette 1 satellite, J. Geophys. Res., 68, 1647-1664, 1963b.

Bekefi, G., and S. C. Brown, Emissions of radio-frequency waves from plasmas, Am. J. Phys., 29, 404-428, 1961.

Bell, T. F., Wave-particle gyroresonance interactions in the Earth's outer ionosphere, Ph.D. Thesis, Stanford Electronics Laboratories, Stanford University, Stanford, Calif., 1964.

Bell, T. F., Nonlinear Alfvén waves in a Vlasov plasma, Phys. Fluids, 8, 1829-1939, 1965.

Bell, T. F., The nonlinear gyroresonance interaction between energetic electrons and coherent VLF waves propagating at an arbitrary angle with respect to the earth's magnetic field, J. Geophys. Res., 89, 905-918, 1984.

Bell, T. F., The wave magnetic field amplitude threshold for nonlinear trapping of energetic gyroresonant and Landau resonant electrons by nonducted VLF waves in the magnetosphere, J. Geophys. Res., 91, 4365-4379, 1986.

Bell, T. F., and O. Buneman, Plasma instability in the whistler mode caused by a gyrating electron stream, Phys. Rev., 133, A1300-A1302, 1964.

Bell, T. F., and R. A. Helliwell, Traveling-wave amplification in the ionosphere, (Proc. Sym. Phys. Proc. in Sun-Earth Environment, July 20-21, 1959), DRTE No. 1025, Defense Res. Telecommunications Estab., Ottawa, Canada, 215-222, 1960.

Bell, T. F., and U. S. Inan, ISEE-1 observations in the magnetosphere of VLF emissions triggered by nonducted coherent VLF waves during VLF wave-injection experiments, Adv. Space Res., 1, 203, 1981.

Bell, T. F., U. S. Inan, and R. A. Helliwell, Nonducted coherent VLF waves and associated triggered emissions observed on the ISEE-1 satellite, J. Geophys. Res., 86, 4649-4670, 1981.

Bell, T. F., U. S. Inan, I. Kimura, H. Matsumoto, T. Mukai, and K. Hashimoto, EXOS-B/Siple Station VLF wave-particle interaction experiments: 2. Transmitter signals and associated emissions, J. Geophys. Res., 88, 295-309, 1983.

Boruki, W. J., A. Bar-Nun, F. L. Scarf, A. F. Cook II, and G. E. Hunt, Icarus, 52, 492, 1982.

Brice, N. M., Traveling wave amplification of whistlers, J. Geophys. Res., 65, 3840-3842, 1960.

Brice, N. M., Discussion of paper by R. L. Dowden, "Doppler-shifted cyclotron radiation from electrons: A theory of very low frequency emissions from the exosphere", J. Geophys. Res., 67, 4897-4899, 1962.

Brice, N. M., An explanation of triggered VLF emissions, J. Geophys. Res., 68, 4626-4628, 1963.

Brice, N., Discrete VLF emissions from the upper atmosphere, Tech. Rep. 3412-6, Radiosci. Lab., Stanford Univ., Stanford, Calif., 1964a.

Brice, N. M., Fundamentals of very low frequency emission generation mechanisms, J. Geophys. Res., 69, 4515-4522, 1964b.

Brice, N., Maximum duration of discrete very low frequency emissions, J. Geophys. Res., 69, 4698-4700, 1964c.

Brice, N., A qualitative explanation of the diurnal variation of chorus, J. Geophys. Res., 69, 4701-4703, 1964d.

Brice, N. M., R. L. Smith, J. S. Belrose, and R. E. Barrington, Recordings from satellite Alouette 1, Triggered very low frequency emissions, Nature, 203, 926-927, 1964.

Brinca, A., L., On the stability of obliquely propagating whistlers, J. Geophys. Res., 77, 3495-3507, 1972a.

Brinca, A. L., Whistler side-band growth due to nonlinear wave-particle interaction, J. Geophys. Res., 77, 3508-3523, 1972b.

Brinca, A. L., On the triggering capability of obliquely propagating whistlers, J. Geophys. Res., 80, 203-206, 1975.

Brinca, A. L., On the influence of a CW whistler on magnetospheric noise, Planet. Space Sci., 25, 879-885, 1977.

Brinca, A. L., and B. T. Tsurutani, Survey of low-frequency electromagnetic waves stiumulated by two coexisting newborn ion species, J. Geophys. Res., 93, 48-58, 1988.

Budden, K. G., The theory of radio windows in the ionosphere and magnetosphere, J. Atmos. Terr. Phys., 42, 287, 1979.

Burtis, W. J., and R. A. Helliwell, Banded chorus--A new type of VLF radiation observed in the magnetosphere by OGO 1 and OGO 3, J. Geophys. Res., 74, 3002-3010, 1969.

Burtis, W. J., and R. A. Helliwell, Magnetospheric chorus: Occurrence patterns and normalized frequency, Planet. Space Sci., 24, 1007-1024, 1976.

Burton, E. T., Submarine cable interference, Nature, 126(3167), 55, 1930.

Burton, E. T., and E. M. Boardman, Effects of solar eclipse on audio frequency atmospherics, Nature, 131, 81-82, 1933a.

Burton, E. T., and E. M. Boardman, Audio frequency atmospherics, Proc. IRE, 21(10), 1476-1494, 1933b. Also published in Bell System Tech. J., 12, 498-516, 1933b.

Burton, R. K., and R. E. Holzer, The origin and propagation of chorus in the outer magnetosphere, J. Geophys. Res., 79, 1014, 1974.

Bush, R. I., G. D. Reeves, P. M. Banks, T. Neubert, P. R. Williamson, W. J. Raitt, and D. A. Gurnett, Electromagnetic fields from pulsed electron beam experiments in space: Spacelab-2 results, Geophys. Res. Lett., 14, 1015-1018, 1987.

Cain, J. C., I. R. Shapiro, J. D. Stolarik, and J. P. Heppner, A note on whistlers observed above the ionosphere, J. Geophys. Res., 66, 2677-2680, 1961.

Cain, J. C., I. R. Shapiro, and J. D. Stolarik, Whistler signals observed with the Vanguard III satellite, J. Phys. Soc. Japan, 17 (Suppl. A-II, Intl. Conf. on Cosmic Rays and the Earth Storm, Part II, 84-87, 1962.

Carlson, C. R., R. A. Helliwell, and D. L. Carpenter, Variable frequency VLF signals in the magnetosphere: Associated phenomena and plasma diagnostics, J. Geophys. Res., 90, 1507-1521, 1985.

Carpenter, D. L., Ducted whistler-mode propagation in the magnetosphere; A half-gyrofrequency upper intensity cutoff and some associated wave growth phenomena, J. Geophys. Res., 73, 2919, 1968.

Carpenter, D. L., C. G. Park, H. A. Taylor, Jr., and H. C. Brinton, Multi-experiment detection of the plasmapause from EOGO satellites and Antarctic ground stations, J. Geophys. Res., 74, 1837-1847, 1969a.

Carpenter, D. L., K. Stone, and S. Lasch, A case of artificial triggering of VLF magnetospheric noise during the drift of a whistler duct across magnetic shells, J. Geophys. Res., 74, 1848-1855, 1969b.

Carson, W. B., J. A. Koch, J. H. Pope, and R. M. Gallet, Long-period very low frequency emission pulsations, J. Geophys. Res., 70, 4293-4303, 1965.

Cartwright, D. G. and P. J. Kellogg, Observations of radiation from an electron beam artificially injected into the ionosphere, J. Geophys. Res., 79, 1439-1457, 1974.

Cattell, C. A., F. S. Mozer, R. R. Anderson, E. W. Hones, Jr., and R. D. Sharp, ISEE observations of the plasma sheet boundary, plasma sheet and neutral sheet: 2. Waves, J. Geophys. Res., 91, 5681-5688, 1986.

Cornilleau-Wehrlin, N., R. Gendrin, and R. Perez, Reception of the NLK (Jim Creek) transmitter onboard GEOS-1, Space Sci. Rev., 22, 443-451, 1978.

Cornilleau-Wehrlin, N., J. Solomon, A. Korth, and G. Kremser, Experimental study of the relationship between energetic electrons and ELF waves observed on board GEOS: A support to quasi-linear theory, J. Geophys. Res., 90, 4141-4154, 1985.

Coroniti, F. V., R. W. Fredricks, C. F. Kennel, and F. L. Scarf, Fast time resolved spectral analysis of VLF banded emissions, J. Geophys. Res., 76, 2366-2381, 1971.

Coroniti, F. V., W. S. Kurth, F. L. Scarf, S. M. Krimigis, C. F. Kennel, and D. A. Gurnett, Whistler mode emissions in the Uranian radiation belts, J. Geophys. Res., 92, 15,234-15,248, 1987.

Coroniti, F. V., F. L. Scarf, C. F. Kennel, W. S. Kurth, and D. A. Gurnett, Detection of Jovian whistler mode chorus; Implications for the Io torus aurora, Geophys. Res. Lett., 7, 45, 1980.

Coroniti, F. V., F. L. Scarf, C. F. Kennel, and W. S. Kurth, Analysis of chorus emissions at Jupiter, J. Geophys. Res., 89, 3801, 1984.

Crouchley, J., and N. M. Brice, A study of "chorus" observed at Australian stations, Planet. Space Sci., 2, 238-245, 1960.

Curtis, S. A., A theory for chorus generation by energetic electrons during substorms, J. Geophys. Res., 83, 3841-3848, 1978.

Das, A.C., A mechanism for VLF emissions, J. Geophys. Res., 73, 7457-7471, 1968.

Das, A. C., and V. H. Kulkarni, Frequency-time structure of VLF emissions, Planet. Space Sci., 23, 41-52, 1975.

Davidson, G. T., Pitch angle diffusion in morningside aurorae. 1. The role of the loss cone in the formation of impulsive bursts of precipitation, J. Geophys. Res., 91, 4413-4427, 1986.

Denavit, J., and R. N. Sudan, Whistler sideband instability, Phys. Fluids, 18, 575-584, 1975a.

Denavit, J., and R. N. Sudan, Effect of phase-correlated electrons on whistler wave packet propagation, Phys. Fluids, 18, 1533-1541, 1975b.

Dinger, H. E., Whistling atmospherics, NRL Rept. 4825, Naval Res. Lab., Wash., D.C., 1956.

Doolittle, J. H., and D. L. Carpenter, Photometric evidence of electron precipitation induced by first hop whistlers, Geophys. Res. Lett., 10, 611-614, 1983.

Dowden, R. L., Doppler-shifted cyclotron radiation from electrons: A theory of very low frequency emissions from the exosphere, J. Geophys. Res., 67(5), 1745-1750, 1962a.

Dowden, R. L., Cyclotron theory of very-low-frequency discrete emissions, Nature, 195 (4846), 1085-1086, 1962b.

Dowden, R. L., Method of measurement of electron energies and other data from spectrograms of VLF emissions, Australian J. Phys., 15(4), 490-503, 1962c.

Dowden, R. L., Very low frequency emissions from the exosphere, Ionospheric Prediction Service Research Rept., University of Tasmania, Hobart, Tasmania, 1963.

Dowden, R. L., Electron energy spectrum and structure deduced from analysis of VLF discrete emissions by using the Helliwell criteria, J. Geophys. Res., 76, 3034-3045, 1971a.

Dowden, R. L., VLF Discrete emissions deduced from Helliwell's theory, J. Geophys. Res., 76, 3046-3054, 1971b.

Dowden, R. L., Detrapping by an additional wave of wave-trapped electrons, J. Geophys. Res., 87, 6237-6242, 1982.

Dunckel, N., and R. A. Helliwell, Whistler-mode emissions on the OGO 1 satellite, J. Geophys. Res., 74, 6371-6385, 1969.

Dungey, J. W., Instabilities in the magnetosphere, in Solar-Terrestrial Physics/1970, ed. E. R. Dyer, Proc. International Symposium on Solar-Terrestrial Physics, Leningrad, May 1970, pp. 219-235, D. Reidel Publ. Co., Dordrecht-Holland, 1972.

Dysthe, K. B., Some studies of triggered whistler emissions, J. Geophys. Res., 76, 6915-6931, 1971.

Eckersley, T. L., A note on musical atmospheric disturbances, Phil. Mag., 49(5), 1250-1260, 1925.

Eidman, V. Ya., The radiation from an electron moving in a magnetoactive plasma, J. Exptl. Theoret. Phys. (USSR), 34, 131-138, 1958.

Ellis, G. R. A., Low-frequency radio emission from aurorae, J. Atmos. Terrest. Phys., 10, 302-306, 1957.

Ellis, G. R. A., Low frequency electromagnetic radiation associated with magnetic disturbances, Planet. Space Sci., 1, 253-258, 1959.

Evans, D. R., J. H. Romig, and J. W. Warwick, Bursty radio emissions from Uranus, J. Geophys. Res., 92, 15,206-15,210, 1987.

Farrell, W. M., D. A. Gurnett, P. M. Banks, R. I. Bush, and W. J. Raitt, An analysis of whistler-mode radiation from the Spacelab 2 electron beam, J. Geophys. Res., 93, 153-161, 1988.

Foster, J. C., and T. J. Rosenberg, Electron precipitation and VLF emissions associated with cyclotron resonance interactions near the plasmapause, J. Geophys. Res., 81, 2183-2192, 1976.

Foster, J. C., T. J. Rosenberg, and L. J. Lanzerotti, Magnetospheric conditions at the time of enhanced wave-particle interactions near the plasmapause, J. Geophys. Res., 81, 2175-2182, 1976.

Fraser-Smith, A. C., and R. A. Helliwell, Stimulation of Pc1 geomagnetic pulsations by naturally occurring repetitive VLF activity, Geophys. Res. Lett., 7, 851, 1980.

Fredricks, R. W., Structure of generalized ion Bernstein modes from the full electromagnetic dispersion relation, J. Plasma Phys., 2, 365, 1968.

Fredricks, R. W., Wave-particle interactions in the outer magnetosphere: A review, Space Sci. Rev., 17, 741, 1975.

Fuchs, J., Discussion, A report to the National Academy of Sciences--National Research Council, Wash., D.C., Natl. Acad. of Sci.--Natl. Res. Coun., Pub. 581, 105, 1938.

Gallet, R. M., The very low-frequency emissions generated in the earth's exosphere, Proc. IRE, 47(2), 211-231, 1959a.

Gallet, R. M., Propagation and production of electromagnetic waves in a plasma, Nuovo Cimento, Supplement 13(1), 234-256, 1959b.

Gallet, R. M., Radio observations of Jupiter, in Planets and Satellites, ed. G. P. Kuiper, p. 500, University of Chicago Press, Chicago, 1961.

Gallet, R. M., and R. A. Helliwell, Origin of "very-low-frequency emissions", J. Res. NBS, 63D, 21-27, 1959.

Ganguly, S., and W. E. Gordon, Nonlinear mixing in the ionosphere, Geophys. Res. Lett., 13, 503-505, 1986.

Garnier, M., G. Girolami, H. C. Koons, and M. H. Dazey, Stimulated wave-particle interactions during high-latitude ELF wave injection experiments, J. Geophys. Res., 87, 2347, 1982.

Gary, S. P. and C. D. Madland, Electromagnetic ion instabilities in a cometary environment, J. Geophys. Res., 93, 235-241, 1988.

Gendrin, R., Gyroresonant wave-particle interactions, in Solar-Terrestrial Physics/1970, ed. E. R. Dyer, Proc. International Symposium on Solar-Terrestrial Physics, Leningrad, May 1970, pp. 236-269, D. Reidel Publ. Co., Dordrecht-Holland, 1972.

Gendrin, R., Phase-bunching and other non-linear processes occuring in gyroresonant wave-particle interactions, Astrophys. Space Sci., 28, 245-266, 1974.

Gendrin, R., Waves and wave-particle interactions in the magnetosphere: A review, Space Sci. Rev., 18, 145-200, 1975.

Gendrin, R., and A. Roux, Energization of helium ions by proton-induced hydromagnetic waves, J. Geophys. Res., 85, 4577-4586, 1980.

Gendrin, R., C. Berthomier, H. Cory, A. Meyer, B. Sukhera, and J. Vigneron, Very-low-frequency and particle rocket experiment at Kerguelen Islands: 1. Very-low-frequency measurements, J. Geophys. Res., 75, 6153-6168, 1970a.

Gendrin, R., J. Etcheto, and B. de la Porte des Vaux, Very-low-frequency and particle rocket experiment at Kerguelen Islands: 2. Particle measurements, J. Geophys. Res., 75, 6169-6181, 1970b.

Goldstein, B. E., and B. T. Tsurutani, Wave normal directions of chorus near the equatorial source regions, J. Geophys. Res., 89, 2789-2810, 1984.

Gurnett, D. A., Very low frequency electromagnetic emissions observed with the ONR/SUI satellite Injun III, Dept. of Phys. and Astron., State Univ. of Iowa, Iowa City, Iowa, 1963.

Gurnett, D. A., Plasma wave interactions with energetic ions near the magnetic equator, J. Geophys. Res., 81, 2765-2770, 1976.

Gurnett, D. A. and R. R. Anderson, The kilometric radio emission spectrum: Relationship to auroral acceleration processes, AGU Geophysical Monograph 25, Physics of Auroral Arc Formation, S.-I. Akasofu and J. R. Kan, editors, American Geophysical Union, Washington, D.C., 341-350, 1981.

Gurnett, D. A., and L. A. Frank, VLF hiss and related plasma observations in the polar magnetosphere, J. Geophys. Res., 77, 172-190, 1972a.

Gurnett, D. A., and L. A. Frank, ELF noise bands associated with auroral electron precipitation, J. Geophys. Res., 77, 3411-3417, 1972b.

Gurnett, D. A., and L. A. Frank, Plasma waves in the polar cusp: Observations from Hawkeye 1, J. Geophys. Res., 83, 1447, 1978.

Gurnett, D. A., and U. S. Inan, Plasma wave observations with the Dynamics Explorer 1 spacecraft, Rev. Geophys., 26, 285-316, 1988.

Gurnett, D. A. and B. J. O'Brien, High-latitude geophysical studies with satellite Injun 3: Part 5, Very-low-frequency electromagnetic radiation, J. Geophys. Res., 69, 65-89, 1964.

Gurnett, D. A., R. R. Anderson, F. L. Scarf, R. W. Fredricks, and E. J. Smith, Initial results from the ISEE-1 and -2 plasma wave investigation, Space Sci. Rev., 23, 103, 1979.

Gurnett, D. A., R. R. Anderson, B. Häusler, G. Haerendel, O. H. Bauer, R. A. Treumann, H. C. Koons, R. H. Holzworth, and H. Lühr, Plasma waves associated with the AMPTE artificial comet, Geophys. Res. Lett., 12, 851-854, 1985.

Gurnett, D. A., R. R. Anderson, P. A. Bernhardt, H. Lühr, G. Haerendel, O. H. Bauer, H. C. Koons, and R. H. Holzworth, Plasma waves associated with the first AMPTE magnetotail barium release, Geophys. Res. Lett., 13, 644-647, 1986.

Gurnett, D. A., L. A. Frank, and R. P. Lepping, Plasma waves in the distant magnetotail, J. Geophys. Res., 81, 6059-6071, 1976.

Gurnett, D. A., W. S. Kurth, and F. L. Scarf, Plasma waves near Saturn: Initial results from Voyager 1, Science, 212, 1981a.

Gurnett, D. A., W. S. Kurth, and F. L. Scarf, Narrowband electromagnetic radiation from Saturn's magnetosphere, Nature, 292, 733, 1981b.

Gurnett, D. A., W. S. Kurth, and F. L. Scarf, Narrowband electromagnetic emissions from Jupiter's magnetosphere, Nature, 302, 385, 1983.

Gurnett, D. A., W. S. Kurth, F. L. Scarf, and R. L. Poynter, The first plasma wave observations at Uranus, Science, 233, 106, 1986a.

Gurnett, D. A., W. S. Kurth, J. T. Steinberg, P. M. Banks, R. I. Bush, and W. J. Raitt, Whistler-mode radiation from the Spacelab 2 electron beam, Geophys. Res. Lett., 13, 225-228, 1986b.

Gurnett, D. A., S. R. Mosier, and R. R. Anderson, Color spectrograms of very-low-frequency Poynting flux data, J. Geophys. Res., 76, 3022-3033, 1971.

Gurnett, D. A., G. W. Pfeiffer, R. R. Anderson, S. R. Mosier, and D. A. Cauffman, Initial observations of VLF electric and magnetic fields with the Injun 5 satellite, J. Geophys. Res., 74, 4631-4648, 1969.

Gurnett, D. A., R. R. Shaw, R. R. Anderson, W. S. Kurth, and F. L. Scarf, Whistlers observed by Voyager 1: Detection of lightning on Jupiter, Geophys. Res. Lett., 6, 511, 1979.

Hansen, S. F., A mechanism for the production of certain types of very-low-frequency emissions, J. Geophys. Res., 68, 5925-5936, 1963.

Hasegawa, A., Plasma instabilities in the magnetosphere, Rev. Geophys., 9, 703, 1971.

Häusler, B., L. J. Woolliscroft, R. R. Anderson, D. A. Gurnett, R. H. Holzworth, H. C. Koons, O. H. Bauer, G. Haerendel, R. A. Treumann, P. J. Christiansen, A. G. Darbyshire, M. P. Gough, S. R. Jones, A. J. Norris, H. Lühr, and N. Klöcker, Plasma waves observed by the IRM and UKS spacecraft during the AMPTE solar wind lithium releases: Overview, J. Geophys. Res., 91, 1283-1299, 1986.

Hayakawa, M., and Y. Tanaka, On the propagation of low-latitude whistlers, Rev. Geophys. Space Phys., 16, 111-123, 1978.

Hayakawa, M., K. Bullough, and T. R. Kaiser, Properties of storm-time magnetospheric VLF emissions as deduced from the Ariel 3 satellite and ground-based observations, Planet. Space Sci., 25, 353-368, 1977.

Hayakawa, M., Y. Tanaka, S. S. Sazhin, and T. Okada, An interpretation of dawnside mid-latitude VLF emissions in terms of quasi-linear electron cyclotron instability, in Nonlinear and Environmental Electromagnetics, ed. by H. Kikuchi, Elsevier Science Pub., Amsterdam-New York, 33-42, 1985.

Hayakawa, M., Y. Tanaka, S. S. Sazhin, T. Okado, and K. Kurita, Characteristics of dawnside mid-latitude VLF emissions associated with sub-storms as deduced from the two-stationed direction finding measurement, Planet. Space Sci., 34, 225-243, 1986a.

Hayakawa, M., Y. Tanaka, S. Shimakura, and A. Iizuka, Statistical characteristics of medium-latitude VLF emissions (unstructured and structured): Local time dependence and the association with geomagnetic disturbances, Planet. Space Sci., 34, 1361-1372, 1986b.

Hayakawa, M., Y. Yamanaka, M. Parrot, and F. Lefeuvre, The wave normals of magnetospheric chorus emissions observed on board GEOS 2, J. Geophys. Res., 89, 2811-2821, 1984.

Helliwell, R. A., Whistler-triggered periodic VLF emissions, J. Geophys. Res., 68, 5387-5395, 1963.

Helliwell, R. A., Whistlers and Related Ionospheric Phenomena, Stanford University Press, Stanford, California, 1965.

Helliwell, R. A., A theory of discrete VLF emissions from the magnetosphere, J. Geophys. Res., 72, 4773-4790, 1967.

Helliwell, R. A., Controlled VLF wave injection experiments in the magnetosphere, Space Sci. Rev., 15, 781-802, 1974.

Helliwell, R. A., Active very low frequency experiments on the magnetosphere from Siple Station, Antarctica, Phil. Trans. Roy. Sci. London B, 279, 213, 1977.

Helliwell, R. A., Siple Station experiments on wave-particle interactions in the magnetosphere, in Wave Instabilities in Space Plasmas, edited by P. J. Palmadesso and K. Papadopoulous, pp. 191-203, D. Reidel, Hingham, Mass., 1979.

Helliwell, R. A., Controlled stimulation of VLF emissions from Siple Station, Antarctica, Radio Sci., 18, 801-814, 1983.

Helliwell, R. A., and T. L. Crystal, A feedback model of cyclotron interaction between whistler-mode waves and energetic electrons in the magnetosphere, J. Geophys. Res., 78, 7357-7371, 1973.

Helliwell, R. A., and T. L. Crystal, Reply, J. Geophys. Res., 80, 4399-4400, 1975.

Helliwell, R. A., and U. S. Inan, VLF wave growth and discrete emission triggering in the magnetosphere: A feedback model, J. Geophys. Res., 87(A5), 3537-3550, 1982.

Helliwell, R. A., and J. P. Katsufrakis, VLF wave injection into the magnetosphere from Siple Station, Antarctica, J. Geophys. Res., 79, 2511-2518, 1974.

Helliwell, R. A., D. L. Carpenter, and T. R. Miller, Power threshold for growth of coherent VLF signals in the magnetosphere, J. Geophys. Res., 85, 3360-3366, 1980a.

Helliwell, R. A., D. L. Carpenter, U. S. Inan, and J. P. Katsufrakis, Generation of band-limited VLF noise using the Siple transmitter: A model for magnetospheric hiss, J. Geophys. Res., 91, 4381-4392, 1986.

Helliwell, R. A., J. Katsufrakis, M. Trimpi, and N. Brice, Artificially stimulated very-low-frequency radiation from the ionosphere, J. Geophys. Res., 69, 2391-2394, 1964.

Helliwell, R. A., S. B. Mende, J. H. Doolittle, W. C. Armstrong, and D. L. Carpenter, Correlations between λ4278 optical emissions and VLF wave events observed at L ~4 in the Antarctic, J. Geophys. Res., 85, 3376-3386, 1980b.

Holzer, R. E., T. A. Farley, R. K. Burton, and M. C. Chapman, A correlated study of ELF waves and electron precipitation on OGO 6, J. Geophys. Res., 79, 1014, 1974.

Inan, U. S., Jovian VLF chorus and Io torus aurora, J. Geophys. Res., 91, 4543-4550, 1986.

Inan, U. S., Waves and instabilities, Rev. of Geophys., 25, 588, 1987.

Inan, U. S. and R. A. Helliwell, DE-1 observations of VLF transmitter signals and wave-particle interactions in the magnetosphere, Geophys. Res. Lett., 9, 917-920, 1982.

Inan, U. S., and S. Tkalcevic, Nonlinear equations of motion for Landau resonance interactions with a whistler mode wave, J. Geophys. Res., 87, 2363-2367, 1982.

Inan, U. S., T. F. Bell, D. L. Carpenter, and R. R. Anderson, Explorer 45 and IMP 6 observations in the magnetosphere of injected waves from the Siple Station VLF transmitter, J. Geophys. Res., 82, 1177-1187, 1977.

Inan, U. S., R. A. Helliwell, and W. S. Kurth, Terrestrial vs Jovian VLF chorus; A comparative study, J. Geophys. Res., 88, 6171-6180, 1983.

Isenberg, P. A., H. C. Koons, and J. F. Fennell, Simultaneous observations of energetic electrons and dawnside chorus in geosynchronous orbit, J. Geophys. Res., 87, 1495-1503, 1982.

Jones, D., Source of terrestrial non-thermal continuum radiation, Nature, 260, 686, 1976.

Jones, D., Source of Saturnian myriametric radiation, Nature, 306, 453, 1983.

Jones, D., W. Calvert, D. A. Gurnett, and R. L. Huff, Observed beaming of terrestrial myriametric radiation, Nature, 328, 391, 1987.

Kennel, C. F., and H. E. Petschek, Limit on stably trapped particle fluxes, J. Geophys. Res., 71, 1-28, 1966.

Kimura, I., On observations and theories of the VLF emissions, Planet. Space Sci., 15, 1427-1462, 1967.

Kimura, I., Triggering of VLF magnetospheric noise by a low-power (~100 Watts) transmitter, J. Geophys. Res., 73, 445-447, 1968.

Kimura, I., Interrelation between VLF and ULF emissions, Space Sci. Rev., 16, 389-411, 1974.

Kimura, I., H. Matsumoto, T. Mukai, K. Hashimoto, T. F. Bell, U. S. Inan, R. A. Helliwell, and J. P. Katsufrakis, EXOS-B/Siple Station VLF wave-particle interaction experiments: 1. General description and wave-particle correlations, J. Geophys. Res., 88, 282-294, 1983.

Kitamura, T., J. A. Jacobs, T. Watanabe, and R. B. Fling, Jr., An investigation of quasi-periodic VLF emissions, J. Geophys. Res., 74, 5652-5664, 1969.

Kokubun, S., K. Hayashi, T. Oguti, K. Tsuruda, S. Machida, T. Kitamura, O. Saka, and T. Watanabe, Correlations between very low frequency chorus bursts and impulsive magnetic variations at L 4.5, Can. J. Phys., 59, 1034, 1981.

Koons, H. C., The role of hiss in magnetospheric chorus emissions, J. Geophys. Res., 86, 6745-6754, 1981.

Koons, H. C., and R. R. Anderson, A comparison of the plasma wave spectra for the eight AMPTE chemical releases, J. Geophys. Res., 93, 10,016-10,024, 1988.

Kundu, M. R., Solar Radio Astronomy, Interscience Publishers, New York, 1965.

Kurth, W. S., D. A. Gurnett, and R. R. Anderson, Escaping nonthermal continuum radiation, J. Geophys. Res., 86, 5519, 1981.

Kurth, W. S., D. A. Gurnett, and F. L. Scarf, Sporadic narrowband radio emissions from Uranus, J. Geophys. Res., 91, 11,958, 1986.

Kurth, W. S., B. D. Strayer, D. A. Gurnett, and F. L. Scarf, A summary of whistlers observed by Voyager 1 at Jupiter, Icarus, 61, 497, 1985.

Lalita, and V. K. Tripathi, Nonlinear mixing and excitation of sidebands in the ionosphere, J. Geophys. Res., 93(8), 8689-8695, 1988.

Lasch, S., Unique features of VLF noise triggered in the magnetosphere by Morse-code dots from NAA, J. Geophys. Res., 74, 1856-1858, 1969.

Lefeuvre, F., T. Neubert, and M. Parrot, Wave normal directions and wave distribution functions for ground-based transmitter signals observed on GEOS 1, J. Geophys. Res., 87, 6203-6217, 1982.

Lefeuvre, F., M. Parrot, and C. Delannoy, Wave distribution functions estimation of VLF electromagnetic waves observed onboard GEOS 1, J. Geophys. Res., 86, 2359-2375, 1981.

Leyser, T. B., and B. Thide, Effect of pump-induced density depletions on the spectrum of stimulated electromagnetic emissions, J. Geophys. Res., 93(8), 8681-8688, 1988.

Liemohn, H. B., Cyclotron-resonance amplification of VLF and ULF whistlers, J. Geophys. Res., 72, 39-55, 1967.

Lokken, J. E., J. A. Shand, S. C. Wright, L. H. Martin, N. M. Brice, and R. A. Helliwell, Stanford-Pacific Naval Laboratory conjugate point experiment, Nature, 192, 319-320, 1961.

Maeda, K., and R. R. Anderson, A broadband VLF burst associated with ring current electrons, J. Geophys. Res., 87, 9120-9128, 1982.

Maeda, K., and I. Kimura, Amplification of the VLF electromagnetic wave by a proton beam through the exosphere, J. Phys. Soc. Japan, 17 (Suppl. A-II), 92-95, 1962.

Maeda, K., and I. Kimura, Origin and mechanism of VLF emissions, in Space Science Research III, Wiley, New York, 1963.

Maeda, K., and P. H. Smith, VLF-emissions associated with ring current electrons: Off-equatorial observations, Planet. Space Sci., 29, 825-835, 1981.

Matsumoto, H., Nonlinear whistler-mode interaction and triggered emissions in the magnetosphere: A review, in Wave Instabilities in Space Plasmas, edited by P. J. Palmadesso and K. Papadopoulos, pp. 163-190, D. Reidel, Hingham, Mass., 1979.

Matsumoto, H., and I. Kimura, Linear and nonlinear cyclotron instability and VLF emissions in the magnetosphere, Planet. Space Sci., 19, 567-608, 1971.

Matsumoto, H., and Y. Omura, Computer simulation studies of VLF triggered emissions deformation of distribution function by trapping and

detrapping, Geophys. Res. Lett., 10, 607-610, 1983.
Matsumoto, H., and Y. Yasuda, Computer simulation of nonlinear interaction between a monochromatic whistler wave and an electron beam, Phys. Fluids, 19, 1513-1522, 1976.
Matsumoto, H., K. Hashimoto, and I. Kimura, Two types of phase bunching in the whistler mode wave-particle interaction, J. Geomag. Geoelectr., 26, 365-383, 1974.
Matsumoto, H., K. Hashimoto, and I. Kimura, Dependence of coherent nonlinear whistler interaction on wave amplitude, J. Geophys. Res., 85, 644-652, 1980.
Mauk, B. H., C. E. McIlwain, R. L. McPherron, Helium cyclotron resonance within the earth's magnetosphere, Geophys. Res. Lett., 8, 103-106, 1981.
Monson, S. J., P. J. Kellogg, and D. G. Cartwright, Whistler mode plasma waves observed on Electron Echo 2, J. Geophys. Res., 81, 2193-2199, 1976.
Moreira, A., Stability analysis of magnetosheath lion roars, Planet. Space Sci., 31, 1165-1170, 1983.
Morgan, M. G., Whistlers and dawn chorus, Ann. IGY, 1957-1958, 3, 315-328, 1957.
Morioka, A., H. Oya, and S. Miyatake, Terrestrial kilometric radiation observed by satellite Jikiken (EXOS-B), J. Geomagn. Geoelec., 33, 37, 1981.
Mosier, S. R., and D. A. Gurnett, VLF measurements of the Poynting flux along the geomagnetic field with the Injun 5 satellite, J. Geophys. Res., 74, 5675-5687, 1969.
Neubert, T., F. Lefeuvre, M. Parrot, and N. Cornilleau-Wehrlin, Observations on GEOS-1 of whistler mode turbulence generated by a ground-based VLF transmitter, Geophys. Res. Lett., 10, 623-626, 1983a.
Neubert, T., W. W. L. Taylor, L. R. O. Storey, N. Kawashima, W. T. Roberts, D. L. Reasoner, P. M. Banks, D. A. Gurnett, R. L. Williams, and J. L. Burch, Waves generated during electron beam emissions from the space shuttle, J. Geophys. Res., 91, 11,321-11,329, 1986.
Neubert, T., E. Ungstrup, and A. Bahnsen, Observations on the GEOS 1 satellite of whistler mode signals transmitted by the Omega Navigation System transmitter in northern Norway, J. Geophys. Res., 88, 4015-4025, 1983b.
Nunn, D., A theory of VLF emissions, Planet. Space Sci., 19, 1141-1167, 1971.
Nunn, D., A theoretical investigation of banded chorus, J. Plasma Phys., 11, 189-212, 1974a.
Nunn, D., A self-consistent theory of triggered VLF emissions, Planet. Space Sci., 22, 349-378, 1974b.
Nunn, D., Comment on 'A feedback model of cyclotron interaction between whistler mode waves and energetic electrons in the magnetosphere' by R. A. Helliwell and T. L. Crystal, J. Geophys. Res., 80, 4397-4398, 1975.
Nunn, D., The quasistatic theory of triggered VLF emissions, Planet. Space Sci., 32, 325-350, 1984.
Oliven, M. N., and D. A. Gurnett, Microburst phenomena. 3. An association between microbursts and VLF chorus, J. Geophys. Res., 73, 2355-2362, 1968.
Olsen, R. C., S. D. Shawhan, D. L. Gallagher, J. L. Green, C. R. Chappel, and R. R. Anderson, Plasma observations at the Earth's magnetic equator, J. Geophys. Res., 92, 2385-2407, 1987.
Omura, Y., and H. Matsumoto, Computer simulations of basic processes of coherent whistler wave-particle interactions in the magnetosphere, J. Geophys. Res., 87, 4435-4444, 1982.
Omura, Y., and H. Matsumoto, Competing processes of whistler and electrostatic instabilities in the magnetosphere, J. Geophys. Res., 92, 8649-8659, 1987.
Ondoh, T., and S. Hashizume, The effect of proton gyration in the outer atmosphere represented on the dispersion curve of whistler, J. Geomag. Geoelec., 12, 32-37, 1960.
O'Neil, T. M., J. H. Winfrey, and J. H. Malmberg, Nonlinear interaction of a small cold beam and a plasma, Phys. Fluids, 14, 1204-1212, 1971.
Ossakow, S. L., I. Haber, and R. N. Sudan, Computer studies of collisionless damping of a large amplitude whistler wave, Phys. Fluids, 15, 935-937, 1972a.
Ossakow, S. L., I. Haber, and E. Ott, Simulation of whistler instabilities in anisotropic plasmas, Phys. Fluids, 15, 1538-1540, 1972b.
Ossakow, S. L., E. Ott, and I. Haber, Nonlinear evolution of whistler instabilities, Phys. Fluids, 15, 2314-2326, 1972c.
Ossakow, S. L., E. Ott, and I. Haber, Theory and computer simulation of whistler turbulence and velocity space diffusion in the magnetospheric plasma, J. Geophys. Res., 78, 2945-2958, 1973a.
Ossakow, S. L., E. Ott, and I. Haber, Simulation of gyroresonant electron-whistler interactions in the outer radiation belts, J. Geophys. Res., 78, 3970-3975, 1973b.
Oya, H., Conversion of electrostatic plasma waves into electromagnetic waves: Numerical calculation of the dispersion relation for all wavelengths, Radio Sci., 6, 1131, 1971.
Oya, H., and A. Morioka, Plasma wave phenomena observed by EXOS-B and EXOS-C satellites, Proceedings of Chapman Conference on Plasma Waves and Instabilities in Magnetospheres and at Comets, October 12-16, 1987, Sendai/Mt. Zao, Japan, (this issue), 1989.
Palmadesso, P. J., and K. Papadopoulous, Wave Instabilities in Space Plasmas, D. Reidel, Hingham, Mass., 1979.
Park, C. G., Generation of whistler-mode sidebands in the magnetosphere, J. Geophys. Res., 86, 2286-2294, 1981.
Perraut, S., R. Gendrin, P. Robert, A. Roux, and C. De Villedary, ULF waves observed with magnetic and electric sensors on GEOS-1, Space Science Reviews, 22, 347-369, 1978.

Perraut, S., A. Roux, P. Robert, R. Gendrin, J.-A. Sauvaud, J.-M. Bosqued, G. Kremser, and A. Korth, A systematic study of ULF waves above F_{H+} from GEOS 1 and 2 measurements and their relationships with proton ring distributions, J. Geophys. Res., 87, 6219-6236, 1982.

Pope, J. H., Diurnal variation in the occurrence of "dawn chorus", Nature, 180, 433, 1957.

Pope, J. H., Effect of latitude on the diurnal maximum of "dawn chorus", Nature, 185, 87-88, 1960.

Pope, J. H., A high-latitude investigation of the natural very-low-frequency electromagnetic radiation known as chorus, J. Geophys. Res., 68(1), 83-99, 1963.

Pope, J. H., and W. H. Campbell, Observation of a unique VLF emission, J. Geophys. Res., 65(8), 2543-2544, 1960.

Poulsen, W., and U. S. Inan, Satellite observations of discrete VLF emissions at L>4, J. Geophys. Res. 93, 1817-1838, 1988.

Preece, W. H., Earth currents, Nature, 49(1276), 554, 1894.

Raghuram, R., T. F. Bell, R. A. Helliwell, J. P. Katsufrakis, Quiet band produced by VLF transmitter signals in the magnetosphere, Geophys. Res. Lett., 4, 199, 1977.

Reeves, G. D., P. M. Banks, A. C. Fraser-Smith, T. Neubert, R. I. Bush, D. A. Gurnett, and W. J. Raitt, VLF wave stimulation by pulsed electron beams injected from the space shuttle, J. Geophys. Res., 93, 162-174, 1988.

Reinleitner, L. A., W. S. Kurth, and D. A. Gurnett, Chorus-related electrostatic bursts at Jupiter and Saturn, J. Geophys. Res., 89, 75, 1984.

Rodriguez, P., Long duration lion roars associated with quasi-perpendicular bow shocks, J. Geophys. Res., 90, 241-248, 1985.

Rosenberg, T. J., R. A. Helliwell, and J. P. Katsufrakis, Electron precipitation associated with discrete very low frequency emissions, J. Geophys. Res., 76, 8445,-8452 1971.

Rosenberg, T. J., J. C. Sieren, D. L. Matthews, K. Marthinsen, J. A. Holtet, A. Egeland, D. L. Carpenter, and R. A. Helliwell, Conjugacy of electron microbursts and VLF Chrous, J. Geophys. Res., 86, 5819-5832, 1981.

Roux, A., and R. Pellat, A study of triggered emissions, in Magnetic Particles and Fields, ed. by B. M. McCormac, p. 209, D. Reidel, Hingham, Mass., 1976.

Roux, A., and R. Pellat, A theory of triggered emissions, J. Geophys. Res., 83, 1433-1441, 1978.

Roux, A., and J. Solomon, Self-consistent solution of the quasi-linear theory: Application to the spectral shape and intensity of VLF waves in the magnetosphere, J. Atmos. Terr. Phys., 33, 1457-1471, 1971.

Roux, A., S. Perrout, J. L. Rauch, C. deVilledary G. Kremser, A. Knorth, and D. T. Young, Wave particle interactions near Ω_{He+} onboard GEOS 1 and 2: Generation of ion cyclotron waves and heating of He^{+} ions, J. Geophys. Res., 87, 8174-8190, 1982.

Russell, C. T., R. E. Holzer, and E. J. Smith, OGO 3 observations of ELF noise in the magnetosphere: 1. Spatial extent and frequency of occurrence, J. Geophys. Res., 74, 755-777, 1969.

Russell, C. T., R. E. Holzer, and E. J. Smith, OGO 3 observations of ELF noise in the magnetosphere: 2. The nature of the equatorial noise, J. Geophys. Res., 75, 755-768, 1970.

Rycroft, M. J., VLF emissions in the magnetosphere, Radio Science, 7, 811-830, 1972.

Rydbeck, O. E., and J. Askne, Whistler-mode and ionized stream interactions, in Wave Interaction and Dynamic Nonlinear Phenomena in Plasmas, Engineering Proceedings P-42, Penn. State Univ., Univ. Park, PA, 1963.

Sagdeev, R. A., ed., Nonlinear and Turbulent Processes in Physics, Vol. 1, 2, and 3, Harwood Academic Publishers, Char-London-Paris-New York, 1983.

Sato, N., Short-period magnetic pulsations associated with periodic VLF emissions (T ~ 5.6s), J. Geophys. Res., 89, 2781-2787, 1984.

Sato, N. and H. Fukunishi, Interaction between ELF-VLF emissions and magnetic pulsations: Classification of quasi-periodic ELF-VLF emissions based on frequency-time spectra, J. Geophys. Res., 86, 19-29, 1981.

Sato, N., and S. Kokubun, Interaction between ELF-VLF emissions and magnetic pulsations: Quasi-periodic ELF-VLF emissions associated with Pc3-4 magnetic pulsations and their geomagnetic conjugacy, J. Geophys. Res., 85, 101-113, 1980.

Sato, N., and S. Kokubun, Interaction between ELF-VLF emissions and magnetic pulsations: Regular period ELF-VLF pulsations and their geomagnetic conjugacy, J. Geophys. Res., 86, 9-18, 1981.

Sato, N., H. Fukunishi, T. Ozaki, and T. Yoshino, Simultaneous ground-satellite observations of quasi-periodic (QP) ELF-VLF emissions near L=6, J. Geophys. Res., 86, 9953-9960, 1981.

Sato, N., K. Hayashi, S. Kokubun, T. Oguti, and H. Fukunishi, Relationships between quasi-periodic VLF emission and geomagnetic pulsations, J. Atmos. Terr. Phys., 36, 1515, 1974.

Sazhin, S. S., A model of quasiperiodic VLF emissions, Planet. Space Sci., 26, 399-401, 1978.

Sazhin, S. S., M. Hayakawa, and Y. Tanaka, On the fine structure of the ground-based VLF chorus as an indicator of the wave-particle interaction processes in the magnetospheric plasma, Planet. Space Sci., 33, 385-386, 1985.

Scarf, F. L., D. A. Gurnett, and W. S. Kurth, Jupiter plasma wave observations: An initial Voyager 1 overview, Science, 204, 991, 1979.

Scarf, F. L., D. A. Gurnett, W. S. Kurth, R. R. Anderson, and R. R. Shaw, An upper bound to the lightning flash rate in Jupiter's atmosphere, Science, 213, 684, 1981.

Scarf, F. L., D. A. Gurnett, W. S. Kurth, F. V. Coroniti, C. F. Kennel, and R. L. Poynter, Plasma wave measurements in the magnetosphere of Uranus, J. Geophys. Res., 92, 15,217-15,224, 1987.

Shaw, R. R., and D. A. Gurnett, Whistlers with harmonic bands caused by multiple stroke lightning, J. Geophys. Res., 76, 1851-1854, 1971.

Shawhan, S. D., Magnetospheric plasma waves, in Solar System Plasma Physics, Vol. III, ed. by L. J. Lanzerotti, C. F. Kennel, and E. H. Parker, 211, 1979a.

Shawhan, S. D., Magnetospheric plasma wave research 1975-1978, Rev. of Geophys. and Space Phys., 17, 705, 1979b.

Shawhan, S. D., G. B. Murphy, P. M. Banks, P. R. Williamson, W. J. Raitt, Wave emissions from DC and modulated electron beams on STS 3, Radio Sci., 19, 471-486, 1984.

Sieren, J. C., Fast hisslers in substorms, J. Geophys. Res., 80, 93-97, 1975.

Smirnova, N. A., Fine structure of the ground-observed VLF chorus as an indicator of the wave-particle interaction processes in the magnetospheric plasma, Planet. Space Sci., 32, 425-438, 1984.

Smith, E. J., and B. T. Tsurutani, Magnetosheath lion roars, J. Geophys. Res., 81, 2261-2266, 1976.

Smith, E. J., R. E. Holzer, and C. T. Russell, Magnetic emissions in the magnetosheath at frequencies near 100 Hz, J. Geophys. Res., 74, 3027-3036, 1969.

Stiles, G. S., Comment on 'Fast time resolved spectral analysis of VLF banded emissions' by F. V. Coroniti, R. W. Fredricks, C. F. Kennel, and F. L. Scarf, J. Geophys. Res., 80, 4401-4403, 1975.

Stiles, G. S., and R. A. Helliwell, Frequency-time behavior of artificially stimulated VLF emissions, J. Geophys. Res., 80, 608-618, 1975.

Stiles, G. S., and R. A. Helliwell, Stimulated growth of coherent VLF waves in the magnetosphere, J. Geophys. Res., 82, 523-530, 1977.

Stix, T. H., The Theory of Plasma Waves, McGraw-Hill, New York, 1962.

Storey, L. R. O., An investigation of whistling atmospherics, Phil. Trans. Roy. Sci., London, A246, 113-141, 1953.

Sudan, R. N. and J. Denavit, VLF emissions from the magnetosphere, Phys. Today, 26, 32-39, 1973.

Sudan, R. N., and E. Ott, Theory of triggered VLF emissions, J. Geophys. Res., 76, 4463-4476, 1971.

Taylor, W. W. L., and D. A. Gurnett, Morphology of VLF emissions observed with the Injun 3 satellite, J. Geophys. Res., 73, 5615-5626, 1968.

Taylor, W. W. L., and S. D. Shawhan, A test of incoherent Čerenkov radiation for VLF hiss and other magnetospheric emissions, J. Geophys. Res., 79, 105, 1974.

Temerin, M., and R. L. Lysak, Electromagnetic ion cyclotron mode (ELF) waves generated by auroral electron precipitation, J. Geophys. Res., 89, 2849-2859, 1984.

Thide, B., H. Derblom, A. Hedberg, H. Kopka, and P. Stubbe, Observations of stimulated electromagnetic emissions in ionospheric heating experiments, Radio Sci., 18, 851, 1983.

Thide, B., H. Kopka, and P. Stubbe, Observations of stimulated scattering of a strong high-frequency radio wave in the ionosphere, Phys. Rev. Lett., 49, 1561, 1982.

Thomas, J. O., and B. J. Landmark, eds., Plasma Waves in Space and in the Laboratory, Vol. 1 and 2, American Elsevier Publ. Co., New York, 1970.

Thorne, R. M., and B. T. Tsurutani, The generation mechanism for magnetosheath lion roars, Nature, 293, 384-386, 1981.

Thorne, R. M., S. R. Church, W. J. Malloy, and B. T. Tsurutani, The local time variation of ELF emissions during periods of substorm activity, J. Geophys. Res., 82, 1585-1590, 1977.

Tkalcevic, S., U. S. Inan, and R. A. Helliwell, Nonlinear pitch angle scattering and trapping of energetic particles during landau resonance interactions with whistler mode waves, J. Geophys. Res., 89, 10,813-10,826, 1984.

Tokar, R. L., D. A. Gurnett, F. Bagenal, and R. R. Shaw, Light ion concentration in Jupiter's inner magnetosphere, J. Geophys. Res., 87, 2241, 1982a.

Tokar, R. L., D. A. Gurnett, and F. Bagenal, The proton concentration in the vicinity of the Io plasma torus, J. Geophys. Res., 87, 10395, 1982b.

Tokuda, H., VLF emissions and geomagnetic disturbances at the auroral zone: Part I. Chorus bursts and preceding geomagnetic disturbances, J. Geomag. Geoelec., 14, 33-40, 1962.

Tsurutani, B. T., and E. J. Smith, Postmidnight chorus: A substorm phenomenon, J. Geophys. Res., 79, 118-127, 1974.

Tsurutani, B. T., and E. J. Smith, Two types of magnetospheric ELF chorus and their substorm dependences, J. Geophys. Res., 82, 5112-5128, 1977.

Tsurutani, B. T., E. J. Smith, R. R. Anderson, K. W. Ogilvie, J. D. Scudder, D. N. Baker, and S. J. Bame, Lion roars and non-oscillatory drift mirror waves in the magnetosheath, J. Geophys. Res., 87, 6060-6072, 1982a.

Tsurutani, B. T., R. M. Thorne, E. J. Smith, R. R. Anderson, K. W. Ogilvie, J. D. Scudder, and D. N. Baker, Observations of instabilities in a high beta plasma: Electron cyclotron and drift-mirror instabilities in the terrestrial magnetosheath, Proceedings 1982 International Conference on Plasma Physics, eds. H. Wilhelmsson and J. Weiland, Chalmers University of Technology, Goteborg, Sweden, 36, 1982b.

Ungstrup, E., Narrow band VLF electromagnetic signals generated in the auroral ionosphere by the high-frequency two-stream instability, J. Geophys. Res., 80, 4272-4278, 1975.

Ungstrup, E., and I. M. Jackerott, Observations of chorus below 1500 cycles per second at Godhavn, Greenland, from July 1957 to December 1961, J. Geophys. Res., 68(8), 2141-2146, 1963.

Ungstrup, E., T. Neubert, and A. Bahnsen, Observation on GEOS-1 of 10.2 to 13.6 kHz ground based transmitter signals, Space Sci. Rev., 22, 453-464, 1978.

Vomvoridis, J. L., and J. Denavit, Test particle correlation by a whistler wave in a nonuniform magnetic field, Phys. Fluids, 22, 367-377, 1979.

Vomvoridis, J. L., and J. Denavit, Nonlinear evolution of a monochromatic whistler wave in a nonuniform magnetic field, Phys. Fluids, 23, 174-183, 1980.

Vomvoridis, J. L., T. L. Crystal, and J. Denavit, Theory and computer simulations of magnetospheric very low frequency emissions, J. Geophys. Res., 87, 1473-1489, 1982.

Walker, A. D. M., The theory of whistler propagation, Rev. Geophys. Space Phys., 14, 629-638, 1976.

Warwick, J. W., Radiophysics of Jupiter, Space Sci. Rev., 6, 841, 1967.

Watts, J. M., An observation of audio-frequency electromagnetic noise during a period of solar disturbance, J. Geophys. Res., 62(2), 199-206, 1957a.

Watts, J. M., Audio frequency electromagnetic hiss recorded at Boulder in 1956, Geophys. Pura e Appl., 37, 169-173, 1957b.

Watts, J. M., J. A. Koch, and R. M. Gallet, Observations and results from the "hiss recorder," an instrument to continuously observe the VLF emissions, J. Res. NBS., 67D(5), 569-579, 1963.

Winckler, J. R., An investigation of wave-particle interactions and particle dynamics using electron beams injected from sounding rockets, Space Sci. Rev., 15, 751-780, 1974.

Yamamoto, T., Nonlinear development of the plasma-beam instability in a magnetized plasma, J. Plasma Physics, 15, 357-370, 1976.

Yoshida, S., and T. Hatanaka, The disturbances of exosphere as seen from the VLF emission, J. Phys. Soc. Japan, 17 (Supplement A-II), 78-83, 1962a.

Yoshida, S., and T. Hatanaka, Variations in the VLF emissions with reference to the exosphere, Rept. Iono. Space Res., Japan, 16, 387-409, 1962b.

Young, D. T., S. Perraut, A. Roux, C. De Villedary, R. Gendrin, A. Korth, G. Kremser, and D. Jones, Wave-particle interactions near Ω_{He^+} Observed on GEOS 1 and 2: 1. Propagation of ion cyclotron waves in He^+-Rich Plasma, J. Geophys. Res., 86, 6755-6772, 1981.

GENERATION OF ELF ELECTROMAGNETIC WAVES AND DIFFUSION OF ENERGETIC ELECTRONS IN STEADY AND NON-STEADY STATE SITUATIONS IN THE EARTH'S MAGNETOSPHERE

J. Solomon and N. Cornilleau-Wehrlin

Centre de Recherches en Physique de l'Environnement
Terrestre et Planétaire, CRPE/CNET, 92131 Issy-les-Moulineaux, France

A. Korth and G. Kremser

Max Planck Institut für Aeronomie, 3411 Katlenburg-Lindau, FRG

Abstract. Interaction of energetic electrons (16 - 300 keV) and ELF waves (100 Hz - 3 kHz) has been studied by using data of the GEOS-1 and -2 satellites. In steady-state, it is demonstrated that this interaction enters quite well into the framework of the quasi-linear theory of pitch-angle diffusion. By calculating the wave growth rate at different magnetic latitudes, it is shown that inside the plasmasphere the path integrated gain is large enough for explaining generation of hiss, in agreement with theoretical and experimental results on wave propagation. Outside the plasmasphere hiss generation deserves further study. At the time of storm sudden commencements, non-steady state behaviour is observed that results in antiphase oscillations of the maximum of the wave spectrum and of the corresponding growth rate and anisotropy.

1. Introduction

It is well known that electromagnetic electron-cyclotron waves in the ELF/VLF frequency range (100 Hz - 3 kHz) are amplified in the presence of an anisotropic distribution of energetic electrons (10 keV-200 keV). Such anisotropies build up in some regions of the magnetosphere, particularly by convection processes [Ashour-Abdalla and Cowley, 1974; Solomon and Pellat, 1978]. Consequences of the interaction between waves and particles have been studied extensively. Cornwall [1964] has stressed the importance of it with respect to the lifetime of trapped electrons. Kennel and Petschek [1966] have explained the limitation of the intensity of the flux of energetic electrons in the frame of the quasi-linear diffusion in pitch-angle. Etcheto et al. [1973] have extended it, in a self-consistent way, to the calculation of the equilibrium wave spectrum. They have shown that in a steady state situation the maximum amplitude of the wave spectrum can be linked in a simple way to both the cold plasma density n_0 and the intensity of the source of interacting particles dn_2/dt :

$$B^2_{fmax} \propto n_o \frac{dn_2}{dt}$$

Concerning the influence of the cold plasma, some studies have been carried out [Chan et al., 1974 ; Cornilleau-Wehrlin et al., 1978b] that have confirmed the dependence of $B^2{}_{fmax}$ on n_o. The factor dn_2/dt can be obtained theoretically from a knowledge of the precipitated flux. This problem has been discussed in connection with ground data [Cornilleau-Wehrlin et al., 1978a]. In particular, Kremser et al. [1986] have analysed simultaneous particle and field measurements on GEOS 2 and X-ray observations from balloons and evaluated the importance of WPI on the precipitation of energetic electrons.

Etcheto et al. [1973] have also emphasized that, in a real physical situation, the temperature anisotropy A of the energetic electrons depends on the frequency of the waves interacting with them. This point is of great importance since the anisotropy A must be greater than a critical anisotropy A_c, which is frequency dependent, for the instability to occur [Kennel and Petchek, 1966]. Maeda and Lin [1981], using energetic electron fluxes measured onboard EXPLORER 45, have computed the growth rate of the electron-cyclotron instability and its evolution with time in order to explain the frequency band broadening of magnetospheric VLF emissions observed on the same satellite. They have particularly examined the dependence of the electron anisotropy with energy.

In the present paper we review the results obtained either onboard GEOS 1 or onboard GEOS 2, using simultaneous detailed measurements of wave spectra and of the energy and pitch-angle dependence of the electron flux. These experimental results are discussed in the light of the quasi-linear theory and compared with the theoretical predictions or modelisations.

In part 2 we recall the theoretical results of Kennel and Petschek [1966] concerning the anisotropy and wave growth rate, then we summarize the calculation method used in Cornilleau-Wehrlin et al. (quoted C-W et al. from now on) [1985]. In part 3 we summarize the results of C-W et al. [1985], dealing with steady state situations. This last paper examines the fact that, in an equilibrium situation with continuous emission of waves, the anisotropy A of the energetic electrons must approach the critical anisotropy A_c in the frequency range where pitch-angle diffusion occurs. We then inspect the fact that waves are not emitted, either when $A < A_c$, or when the integrated growth rate along a given line of force is not large enough to fulfill some specific condition. In part 4, in the opposite situation of large growth rate, we discuss the problem of hiss generation inside and outside the plasmasphere. Both wave propagation and amplification are examined [Church and Thorne, 1983 ; Huang et al., 1983 ; Parrot and Lefeuvre, 1986]. Measurements of waves and electron fluxes at different magnetic latitude show that, at least inside the plasmasphere, the total spatial amplification of the waves is large enough for explaining hiss generation [Solomon et al., 1988]. Part 5 deals with a non-stationnary problem: at the time of Storm Sudden commencements, dynamical situation is observed which has been studied in detail by C-W et al. [1988].

2. Electron Anisotropy, Wave Growth Rate, Diffusion

Right-hand circularly polarized waves can interact with hot electrons by gyroresonance if the resonance condition

$$v_{//} = V_R = (\omega-\omega_{ce})/k$$

is satisfied, where $v_{//}$ is the velocity of the electrons in the direction parallel to the static magnetic field and ω_{ce}, ω and k are the angular gyrofrequency, the angular wave frequency and the parallel wave number, respectively (in what follows // and ⊥ always refer to the static magnetic field direction).

The waves are amplified if the temporal growth rate γ is positive ($B_f^2 = B_o^2\, e^{2\gamma t}$). γ can be expressed as follows [see Kennel and Petschek, 1966] :

$$\gamma = \pi\omega_{ce}\,(1 - \omega/\omega_{ce})^2\, \eta(E_R)\,(A(E_R) - \omega/(\omega_{ce} - \omega)) \qquad (1)$$

where $A(E_R)$ is the temperature anisotropy of energetic electrons, $\eta(E_R)$ represents the relative number of resonating particles, and $E_R = (1/2)m_e V_R^2$ is the resonant energy, m_e the electron mass.

The condition $\gamma > 0$ means that the last term of expression (1) must be positive. As we are concerned with right-hand circularly polarized waves for which $\omega < \omega_{ce}$, the anisotropy must be greater than a minimum value, called critical anisotropy A_c, which is frequency dependent :

$$A_c = \omega/(\omega_{ce} - \omega) \qquad (2)$$

For the computation of A and γ from the measurements we describe briefly the method used by C-W et al [1985]. They have expressed (1) as a function of measured quantities or parameters, namely the differential fluxes $j(\alpha,E)$, the pitch-angle α and the total energy E, the reduced frequency $x = \omega/\omega_{ce}$, the electron cyclotron frequency $f_{ce} = \omega_{ce}/2\pi$ and the plasma frequency f_{pe}.

The resonant energy can be written :

$$E_R(\mathrm{keV}) = 255(f_{ce}/f_{pe})^2\,(1-x)^3/x \qquad (3)$$

and the anisotropy

$$A(E_R) = \frac{\displaystyle\int_0^{\pi/2} \tan^2\alpha\, \frac{\partial j(\alpha,E)}{\partial\alpha}\Big|_E\, d\alpha}{\displaystyle 2\int_0^{\pi/2} \tan\alpha\; j(\alpha,E)\, d\alpha} \qquad (4)$$

The main problem consists in evaluating $(\partial j/\partial\alpha)_E$, i.e. for E = constant. In C-W et al's method it is assumed that locally, for a given α, the flux dependence on α and E is separable. The differential flux can then be written in the following way :

$$j(\alpha,E) = \mathrm{const}\,.\,k(E)\,.\,(\sin\alpha)^{2m(\alpha_i)} \qquad (5)$$

where $m(\alpha_i)$ has a constant value on each interval of width $\delta\alpha$ centered at $\alpha = \alpha_i$.In C.W. et al. $\delta\alpha = 10^o$ and $\alpha_i = 10^o, 20^o,...$

This gives :

$$A(E_R) = \frac{\displaystyle\int_0^{\pi/2} \tan\alpha\; j(\alpha,E)\; 2m(\alpha_i)\, d\alpha}{\displaystyle 2\int_0^{\pi/2} \tan\alpha\; j(\alpha,E)\, d\alpha} \qquad (6)$$

where $E = E_R/\cos^2\alpha$ and the integrations are performed with E_R = constant. The choice of (5) is somewhat arbitrary but convenient. Once $A(E_R)$ is calculated, it is easy to compute the corresponding growth rate :

$$\gamma(E_R)/\omega_{ce} = \frac{1.7\;10^{-6}}{f_{pe}^2}\,(E_R)^{1/2}\,(1-x)^2\Big[A(E_R) - \frac{x}{1-x}\Big]\int_0^{\pi/2} j(\alpha,E)\tan\alpha\, d\alpha \qquad (7)$$

the numerical coefficient is valid for E_R expressed in keV, f_{pe} in kHz and j in $(\mathrm{cm}^2\ \mathrm{sr\ s\ keV})^{-1}$.

Other quantities that are relevant to the discussion are the pitch-angle diffusion coefficient D(x) and the associated diffusion strength, described by the parameter y_o^2. Following Kennel and Petschek [1966] and Roux and Solomon [1970, 1971], the diffusion coefficient is

$$D(x) = \frac{e^2\, B_f^2(x)\, V_g(x)}{2m_e^2 |V_R(x) - V_g(x)|} = \frac{e^2}{m_e^2}\,\frac{x}{1+2x}\,B_f^2(x) \qquad (8)$$

V_g being the group velocity of the wave. The diffusion is

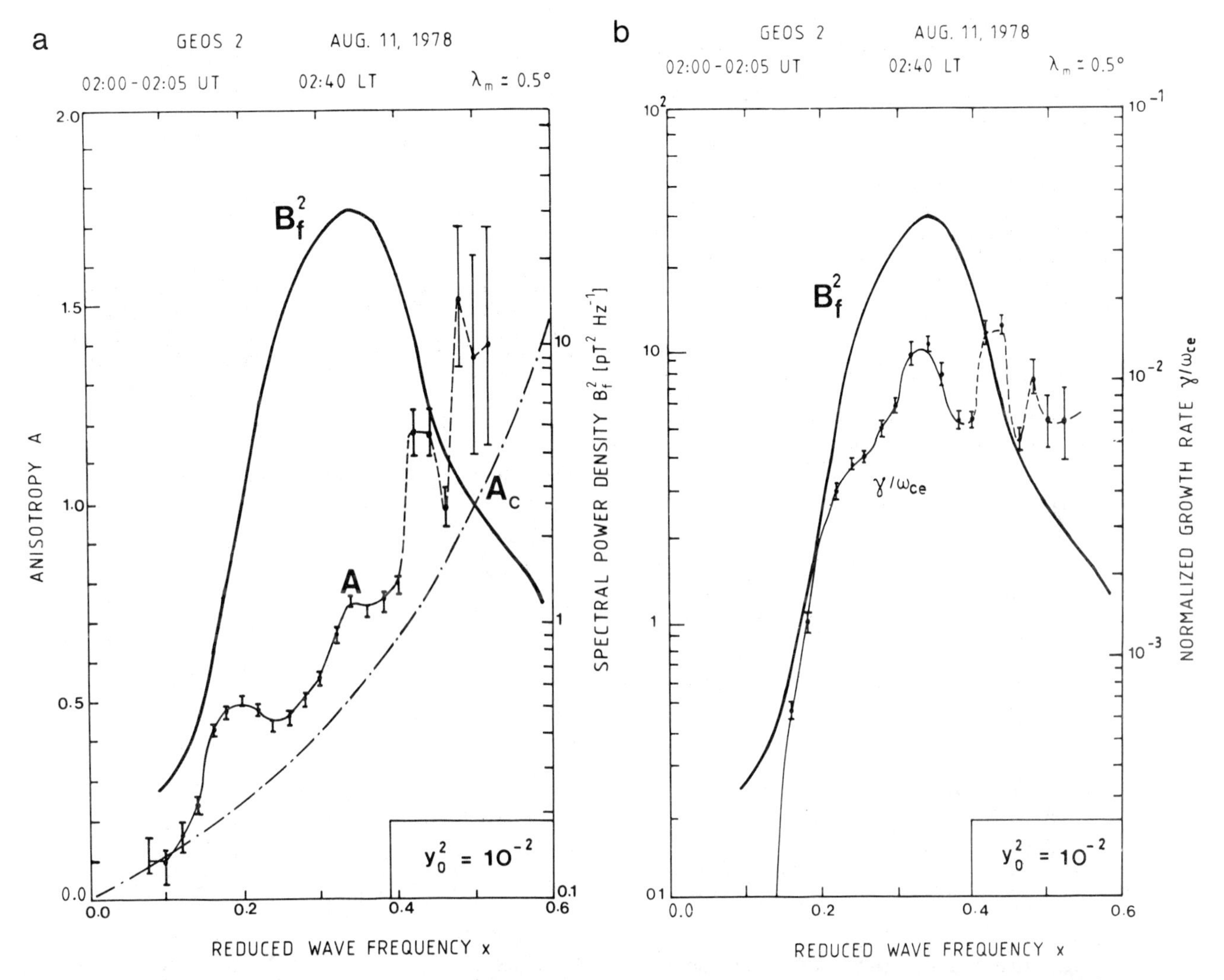

Fig. 1. (a) The wave spectrum (B_f^2) averaged over 5 minutes and the electron anisotropy A for the same time interval are both plotted versus the reduced wave frequency $x = f/f_{ce}$. The critical anisotropy A_c is plotted for comparison. The error bars for the anisotropy are estimated from the statistical error on the differential flux and from the number N of pitch angle ranges used. A is drawn with dotted lines when obtained for $N < 4$. The diffusion strength parameter y_o^2 is given at the maximum intensity of the wave (30 $pT^2\, Hz^{-1}$). This is a case of strong diffusion in the wave generation region ($\lambda_m = 0.5°$) with $f_{ce} = 2.3$ kHz and $f_{pe} = 9$ kHz. (b) For the same time interval, the wave spectrum (B_f^2) and the reduced temporal growth rate γ/ω_{ce} are plotted versus the reduced wave frequency x. The error bars on γ/ω_{ce} are estimated as for A [after C-W et al., 1985].

considered to be strong or weak, depending on whether the y_o^2 value is small or large with respect to 1 [Kennel and Petschek, 1966]. y_o^2 is obtained from

$$y_o^2 = \alpha_o^2/D(x)\, T_E \tag{9}$$

where α_o is the equatorial half loss cone angle and T_E a quarter of the bounce period of an electron, given by $T_E \approx LR_E/V_R$.

3. Steady State Situation

C-W et al. [1985] consider events for which waves are continuously emitted during some tens of minutes with a relatively constant intensity level.

In Figure 1 and 2 are displayed two events obtained onboard GEOS 2 in the vicinity of the geomagnetic equator (magnetic latitude $\lambda_m = 0.5°$). The critical anisotropy A_c, the experimentally determined anisotropy $A(E_R) = A(x)$, the

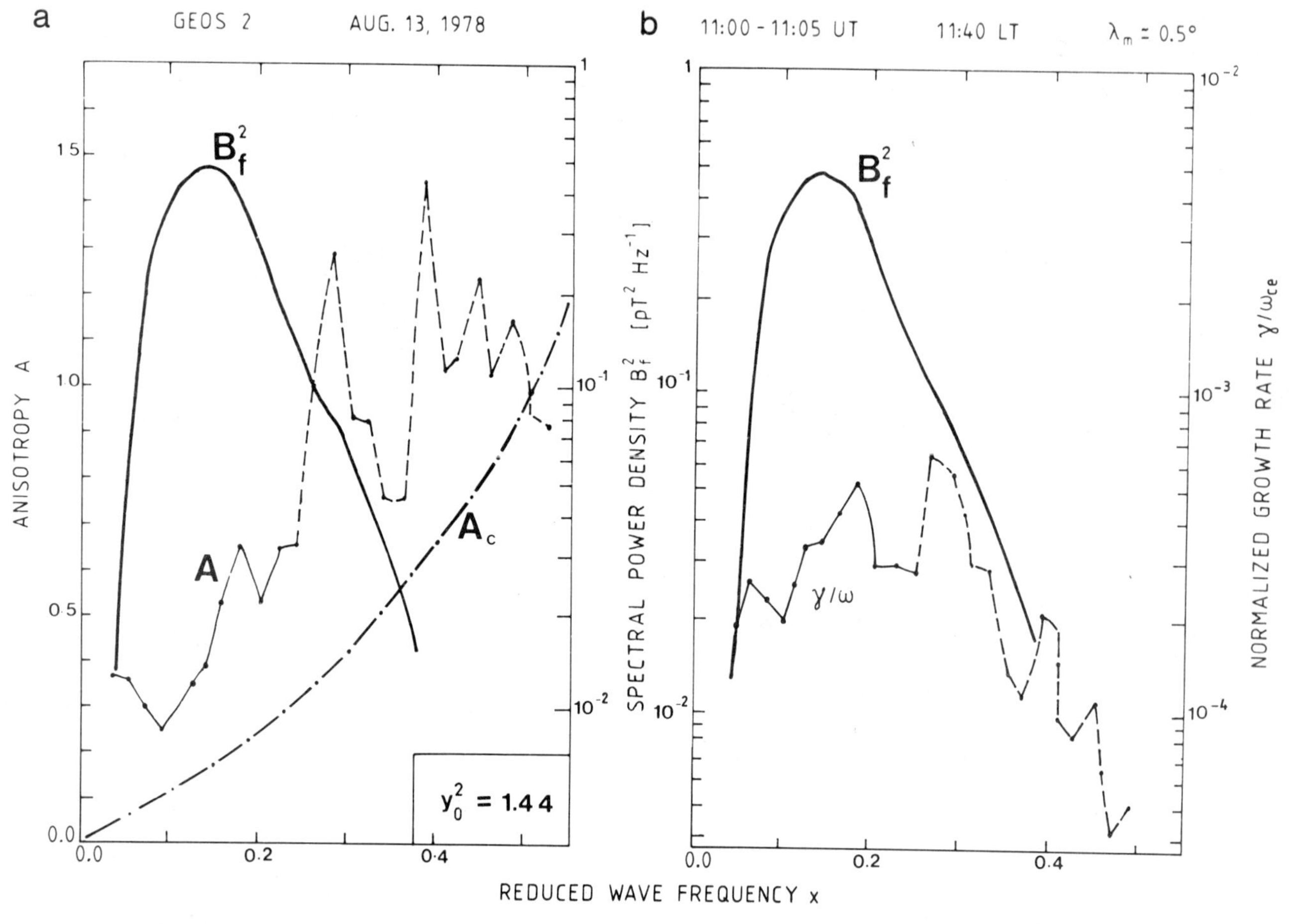

Fig. 2. Same as Figure 1, but in weak diffusion. f_{ce} = 2.8 kHz and f_{pe} =24 kHz [C-W et al., 1985].

corresponding normalized growth rate γ/ω_{ce}, and the wave spectrum B_f^2 are plotted versus the reduced wave frequency x. Note that A, γ and B_f^2 are obtained from 5-min. averaged data. The main difference between the two events stems from their respective wave intensity level resulting in strong pitch-angle diffusion ($y_o^2 = 10^{-2}$) or weak pitch-angle diffusion ($y_o \approx$ 1.44). In Figures 3 and 4 are plotted the same quantities but for events which have been obtained onboard GEOS-1 at higher latitudes ($\lambda_m > 20°$).

3.1. The Behavior of the Anisotropy in the Presence of Waves

When analysing the results obtained for the anisotropy the most striking one is the obvious frequency dependence (i.e. resonant energy dependence) of this quantity in the presence of waves. Let us recall that if in (5) m does not depend upon α and E, or if one takes a bimaxwellian distribution function, A is no longer a function of x and one gets A=m or $A=T_{\perp}/T_{//}$ -1, respectively. This means that such modellings of the particle distribution function are rough approximations, giving an indication of a global behavior, but would by far not have been sufficiently accurate for a comprehensive study of WPI. Furthermore, the frequency dependence of A resembles that of A_c in the presence of waves. Conversely, in the absence of waves the functions A_c and A(x) show no similarities (see figure 6 of C-W et al.).

Let us now discuss in more detail the difference (A(x) - A_c) with respect to the diffusion strength and the geomagnetic latitude. We first consider the GEOS 2 events. One can see that the stronger the diffusion, the smaller (A(x) - A_c). The event of August 11, 1978 (Figure 1) is a case of strong diffusion ($y_o^2 \approx 10^{-2}$). (A - A_c) is smaller on the average for 0.25 < x < 0.4, the reduced wave frequency range corresponding to the maximum of the wave spectrum, than outside this frequency range. This can be expected from quasi-linear diffusion : the stronger the wave amplitude, the quicker the waves diffuse the resonating particles, thus decreasing the anisotropy of the distribution function on a time within the minimum lifetime of the electrons. This means that a continuous source of resonating electrons with A > A_c is needed for a stationary process to exist. This is clearly illustrated by the "no waves/waves" event displayed in Figure 4. The anisotropy then just has to be large enough to produce a growth rate which allows to compensate for the wave losses. The growth rate γ/ω_{ce} is mainly the product of two terms, $\eta(E_R)$ and $(A(E_R) - A_c)$ (see formula (1)).

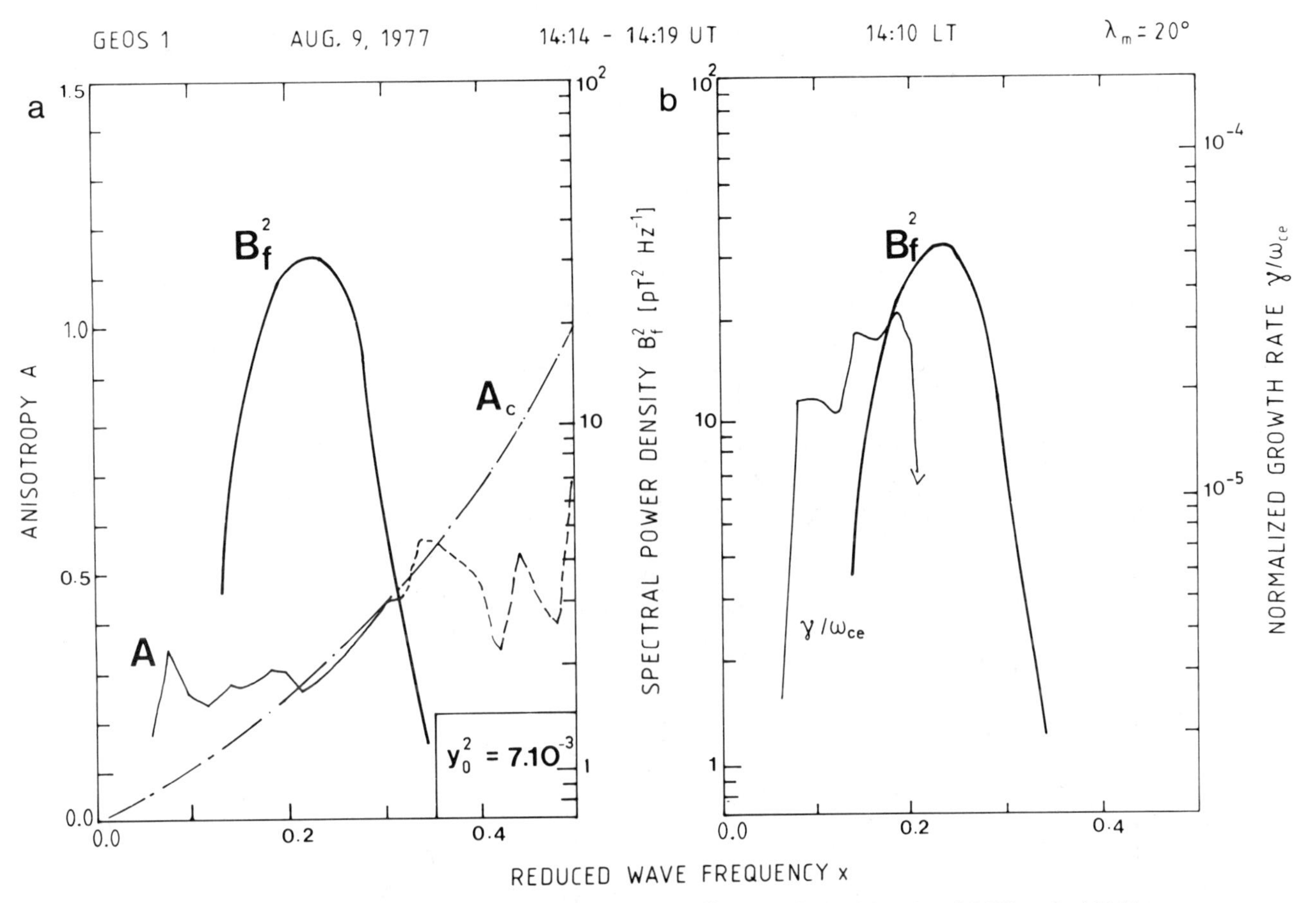

Fig. 3. Same as Figure 1, but for a magnetic latitude $\lambda_m \approx 20°$, GEOS-1 data [C-W et al., 1985].

Therefore in order to maintain a steady state if the source is weaker (i.e. the relative number of resonating particles η) the anisotropy term $(A(E_R) - A_c)$ has to be larger. Then it is easy to understand the August 13, 1978 event (Figure 2) which is a weak diffusion case. $(A - A_c)$ is much larger than in the case of Figure 1 and has no minimum at the frequency of the maximum of the wave power density. As the diffusion is weak, the waves affect the particle distribution function only slightly.

Looking at GEOS 1 data (Figures 3 and 4), we find at $\lambda_m \sim 20°$ the same kind of dependence of $(A(x) - A_c)$ on the diffusion strength, but on the whole $(A(x) - A_c)$ is smaller than at the equator apart from the lower frequency range ($x < 0.1$). Particularly, in strong diffusion one event has $A < A_c$ for the upper frequency range ($x > 0.2$) comprising part of the wave spectrum. This appears for the corresponding growth rates as a frequency shift with respect to the wave spectra (Figure 3).

3.2. The Temporal Growth Rate

Considering now the temporal growth rate, we notice that its behavior is highly dependent upon the region of observation. The GEOS 2 ($\lambda_m \sim 0°$) and GEOS 1 ($\lambda_m \sim 20°$) results are different by several aspects : the value and the variability of the growth rate and the frequency dependence of the growth rate as compared to that of the wave spectrum.

From Figures 1 and 2, one recognizes the high variability of the values for $(\gamma/\omega_{ce})_{max}$, ranging from 10^{-4} to 10^{-2} in the generation region close to the geomagnetic equator, giving rise to different wave amplitudes ($B^2{}_{fmax} \approx 0.4$ to 32 pT^2 Hz^{-1}, respectively) and diffusion strengths. The higher the growth rate, the more intense the waves. Conversely, for GEOS 1 measurements ($\lambda_m \sim 20°$), the normalized growth rate $(\gamma/\omega_{ce})_{max}$ has an almost constant value of the order of 10^{-5}, whatever the wave amplitude ($B^2{}_{fmax} = 8.10^{-3}$ to 32 pT^2 Hz^{-1}) or the diffusion strength. This constant value indicates that the diffusion is most efficient at latitudes below $\lambda_m \sim 20°$.

As far as the frequency dependence of the temporal growth rate is concerned, we note (Figures 1 and 2) that the maximum of the growth rate and of the wave spectrum are approximately reached for the same x values at the equator. We emphasize that these maxima belong to the most reliable points (continuous lines), i.e. obtained for values of the resonant energy for which the differential flux is known on a large pitch-angle range (see part 2, method to evaluate γ/ω_{ce} and A). Another remarkable feature is the clear lower and upper cut-off of the normalized growth rate that are similar to that of the wave spectrum (especially in Figure 2). The experimental result that the peak values of $\gamma(x)/\omega_{ce}$ and $B_f{}^2(x)$ occur for the same x-value is not obvious from a theoretical point of view. In fact, during the linear phase of the interaction at the very start of an event, on a time scale $t_{lin} \sim 1/\gamma \sim 100$ ms, the

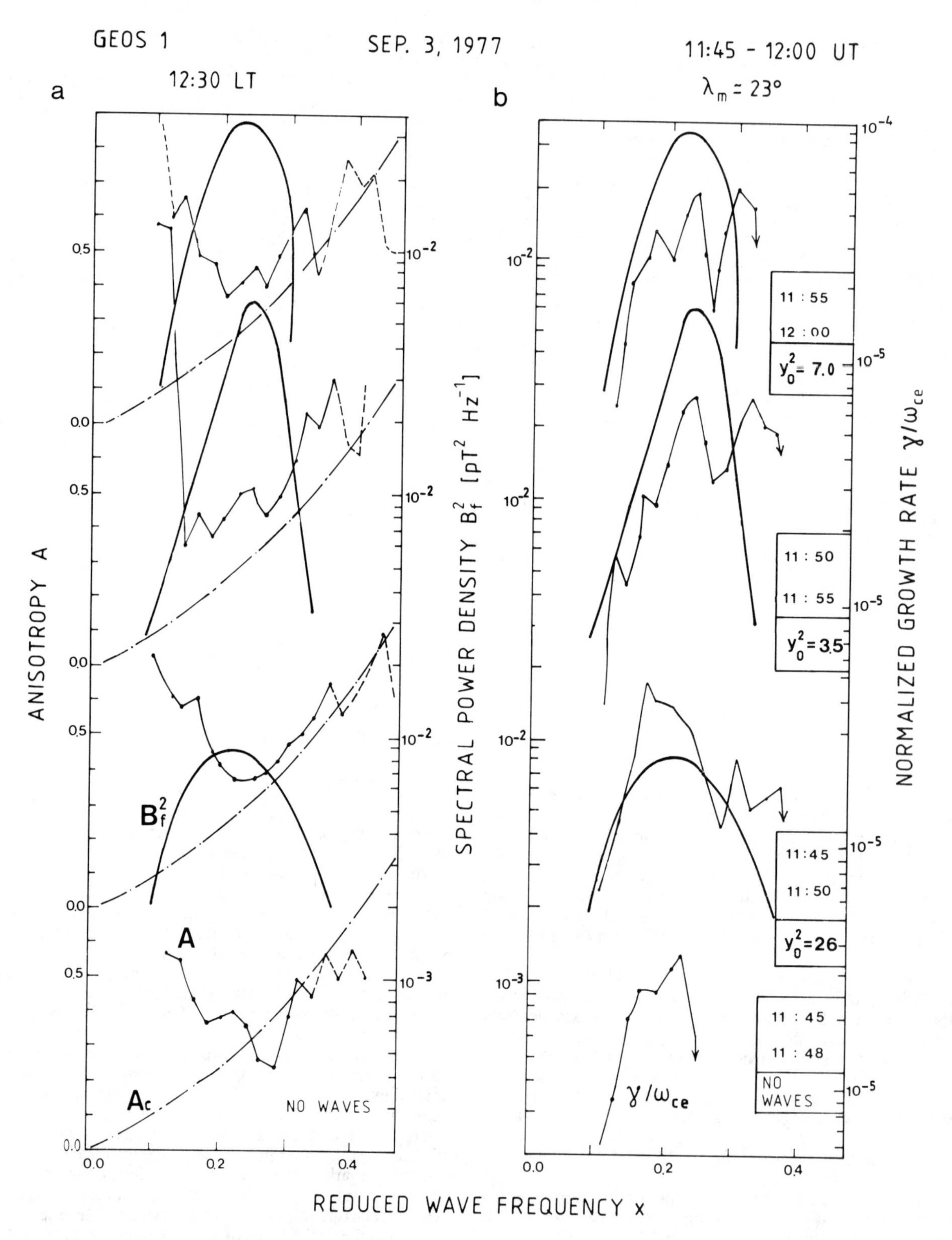

Fig. 4. Same as Figure 1. Time evolution for a GEOS 1 event : from bottom to top, four successive time intervals are represented. Note the no waves event when $A < A_c$ (bottom left). f_{ce} = 3.3 kHz and f_{pe} = 17.5 kHz [C-W et al., 1985].

maximum with respect to x of γ/ω_{ce} and B_f^2 should coincide ($B_f^2 \sim \exp(\gamma t)$). This linear phase is clearly not observable in our case because we have use 5-min. averaged data. For a time $t > t_{lin}$, non-linear effects (pitch-angle diffusion) take place which modify the electron distribution function and consequently the wave growth rate. The persistence of the coincidence with respect to x of the maximum of B_f^2 and γ/ω_{ce}, which is observed in steady-state situations is a first clue of a local wave generation (see part 4).

The same behavior is seen for the GEOS 1 events when the diffusion is weak or medium (Figure 4). The lower cut-off of γ/ω_{ce} is due to a lack of resonating particles rather than to the anisotropy (A(x) can be much larger than A_c for $x < 0.1$: see particularly Figure 4). This is quite understandable as the smallest x values correspond to the highest resonant energies (see equation (3)) for which the electron flux is much smaller than for medium energies ($E_R(x_{max}) \approx 20$ keV). The upper cut-off mainly stems from the fact that $(A(x) - A_c)$ tends towards zero in the upper frequency range.

On the contrary we have noticed for strong diffusion the frequency shift towards smaller x values of γ/ω_{ce} with respect to B_f^2 (Figure 3). Particularly the maximum of the wave spectrum is reached for x values corresponding to a negative growth rate. This can be considered as an effect of the pitch angle diffusion along a given line of force: the reduced frequency x_{max} corresponding to $B^2 f_{max}$ sweeps through a range of decreasing x values as f_{max} is constant while f_{ce} increases.

3.3. The Spatial Growth Rate

According to Kennel and Petschek [1966] and Etcheto et al. [1973], an equilibrium situation implies that in order to compensate the wave losses at the equator the spatial growth rate integrated along the field line Γ satisfies the following relation :

$$\Gamma \equiv 2 \int_0^{\lambda_1} \frac{\gamma}{V_g} \frac{ds}{d\lambda} d\lambda \sim \text{Log} \frac{1}{R} \qquad (10)$$

where R is the reflexion coefficient at the ionosphere, $s(\lambda)$ the path length measured along the field line at the magnetic latitude λ, λ_1 the latitude where the growth rate vanishes, and $ds/d\lambda \approx LR_E$ for $\lambda < 20°$ [Roederer, 1970]. From the events that we have studied, we know that there is a decrease of γ/V_g at a fixed frequency by at least one order of magnitude between the magnetic equator and the latitude $\lambda_m \sim 20°$. An estimate of (10) in a weak diffusion case gives $\Gamma \approx 3$ [Γ(dB) =10 Log_{10} (exp Γ) $\cong$ 13]. In agreement with this result, we have not observed waves in a steady state at the equator when the spatial growth rate γ/V_g does not exceed a certain critical value (Figure 5). But in general we will show in the next chapter that larger values of Γ are obtained both inside and outside the plasmasphere.

4. Generation Mechanism Inside and Outside the Plasmasphere

ELF/VLF hiss is a very common phenomenon observed inside and outside the plasmasphere of the earth. Up to now we have shown that this broad-band whistler mode emission enters well the theoretical frame of the quasi-linear diffusion in pitch-angle of the energetic electrons, but we have not yet explained the wave intensity itself.

Several recent papers have been concerned with the origin of ELF hiss inside the plasmasphere [Thorne et al., 1979 ; Church and Thorne, 1983 ; Huang and Goertz, 1983 ; Huang et al., 1983]. These papers differ in several respects, particularly in their choices for a model of the distribution function of the electrons. As a result, the values that they obtained for the total gain Γ differ notably. Nevertheless, concerning the wave propagation itself all these authors arrived at the same conclusion. Using three dimensional ray-tracing calculations they showed that for a wave launched with $\theta = 0°$ at the magnetic equator (where θ is the angle between the wave normal and the static field B_0) θ increases progressively, reaching large values at magnetic latitudes $\lambda_m \sim 30°$-$40°$. Then the waves can be reflected in different ways but in general do not return in the equatorial region with small values of θ (i.e. $\theta < 30°$). These features appear in good agreement with several recent experimental studies on the propagation characteristics of ELF hiss [Hayakawa et al., 1986b; Parrot and Lefeuvre, 1986]. An immediate consequence of the above facts is that taking into account both the rapid reduction of the wave growth rate and the appearance of Landau absorption for

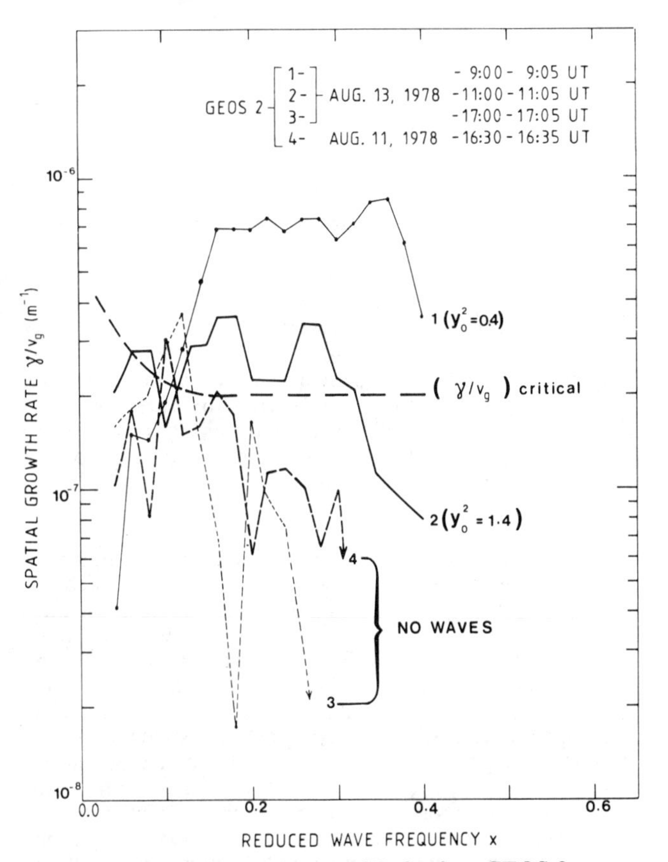

Fig. 5. The spatial growth rates γ/V_g for four GEOS 2 events are compared to the approximate critical limit above which the wave energy can be restored. Case 2 correspond to wave event already studied (Figure 2) whereas no waves are observed for cases 3 and 4. In these last two cases f_{ce} = 2 kHz, f_{pe} = 33 kHz and 20 kHz, respectively [C-W et al., 1985].

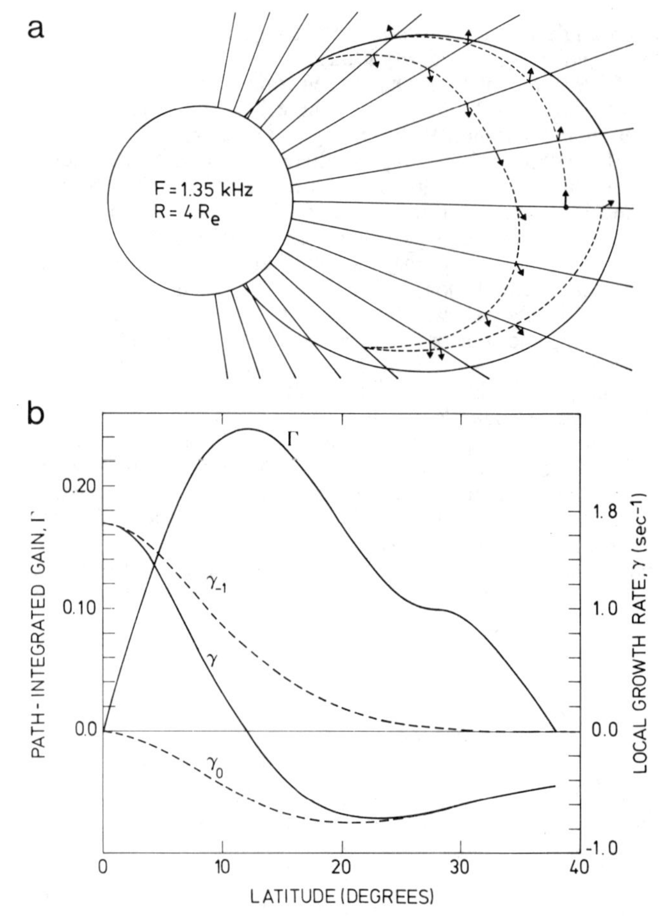

Fig. 6. (a) Ray path of a 1.35 kHz ray, launched with zero initial wave normal angle at 4 R_E at the equator. The latitude is indicated at 10° intervals by radial lines from the earth. The plasmapause at L = 4.6 is shown. The wave normal angle subsequent to launch is indicated by arrows. (b) Corresponding local and path-integrated gains. γ is the sum of the Landau term γ_0 and of the anomalous electron cyclotron term γ_{-1} [after Huang et al., 1983].

increasing values of θ, the concept of waves being amplified up to their observed level while propagating parallel to B_0 back and forth in the equatorial region seems obsolete. In this context even the largest values of the spatial integrated gain Γ found by Huang et al. [1983] (Γ< 2) and by Church and Thorne [1983] (Γ < 4) are quite insufficient to explain, starting from the magnetic thermal level, the generation of hiss up to its typical spectral intensity $B^2{}_f \sim 1pT^2Hz^{-1}$ in a single transit of the waves through the equatorial region. As pointed out in the above papers, there could exist situations more favorable to feed back by wave propagation, either through the existence of cyclic waves or because that the ray-tracing models based on smooth density variations are invalid. Nevertheless their general conclusions about the possibility of local hiss generation are mostly negative.

An example of a typical ray path obtained by Huang et al. [1983] is shown on top of Figure 6. Clearly, the wave normal angle θ exhibits large values (θ > 30°) for a magnetic latitude λ > 10°. The corresponding local and path-integrated gains are displayed at the bottom of Figure 6. Growth rate calculations depend sensitively on the form chosen for the distribution function F. Huang et al. [1983] have taken $F = C\,(\sin\alpha)^{2m}/E^n$ where C is a normalization constant and with, in general, m = 1 and n = 2. Such a choice gives some weight to low energy electrons, which results in an increasing Landau damping (γ_o) when θ increases with latitude.But before everything, values obtained for Γ depend largely on the choice of C, i.e. on the intensity of the energetic electron flux that is supposed to be present in the plasmasphere at the time of hiss generation. For similar calculations, Church and Thorne [1983] have used a lorentzian form for the distribution function $F = C\,(\sin\alpha)^{2m}/(E+E_{th})^n$ where E_{th} is a measure of the average energy in the distribution and with mostly m = 1 and n = 3. One typical result is displayed in Figure 7. First, Landau attenuation is rather weak. Moreover, because of the larger intensity that they have assumed for the flux of the energetic electron, they have obtained larger values of Γ than those of Huang et al. [1983]. Nevertheless even the largest values

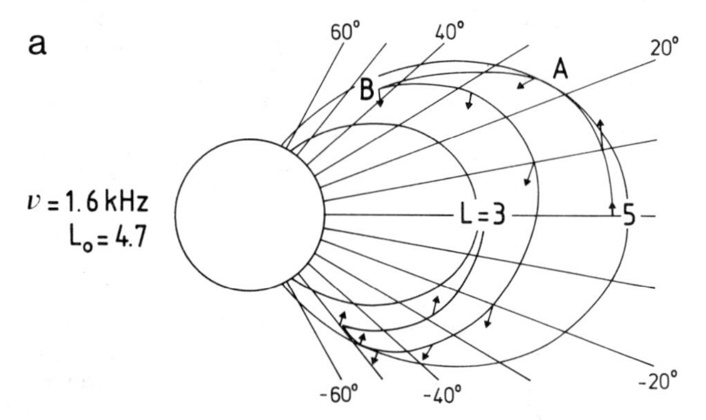

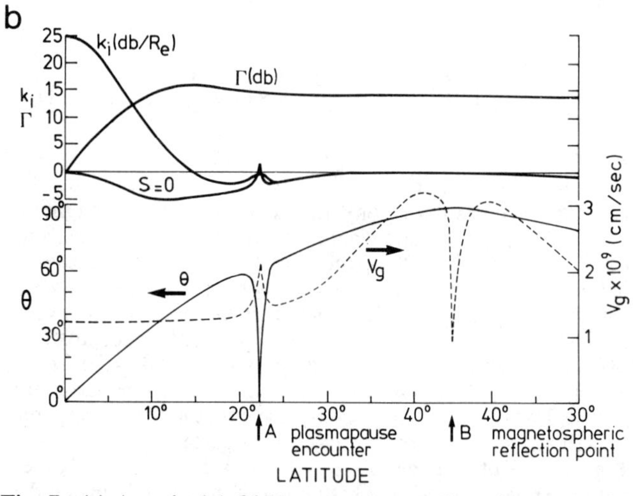

Fig. 7. (a) A typical 1.6 kHz ray trajectory for whistler mode waves originating in the outer plasmasphere at L_0= 4.7 at the equator and with zero initial wave normal angle. (b) The local convective amplification $k_i = \gamma/V_g$ and net integrated gain Γ, assuming a model 2-keV (= E_{th}) Lorentzian energetic electron distribution and a $\sin^2\alpha$ pitch angle dependence. The major contribution to a wave growth is confined to the equatorial region ; Landau attenuation (s=0 curve) at high latitude is relatively insignificant [after Church and Thorne, 1983].

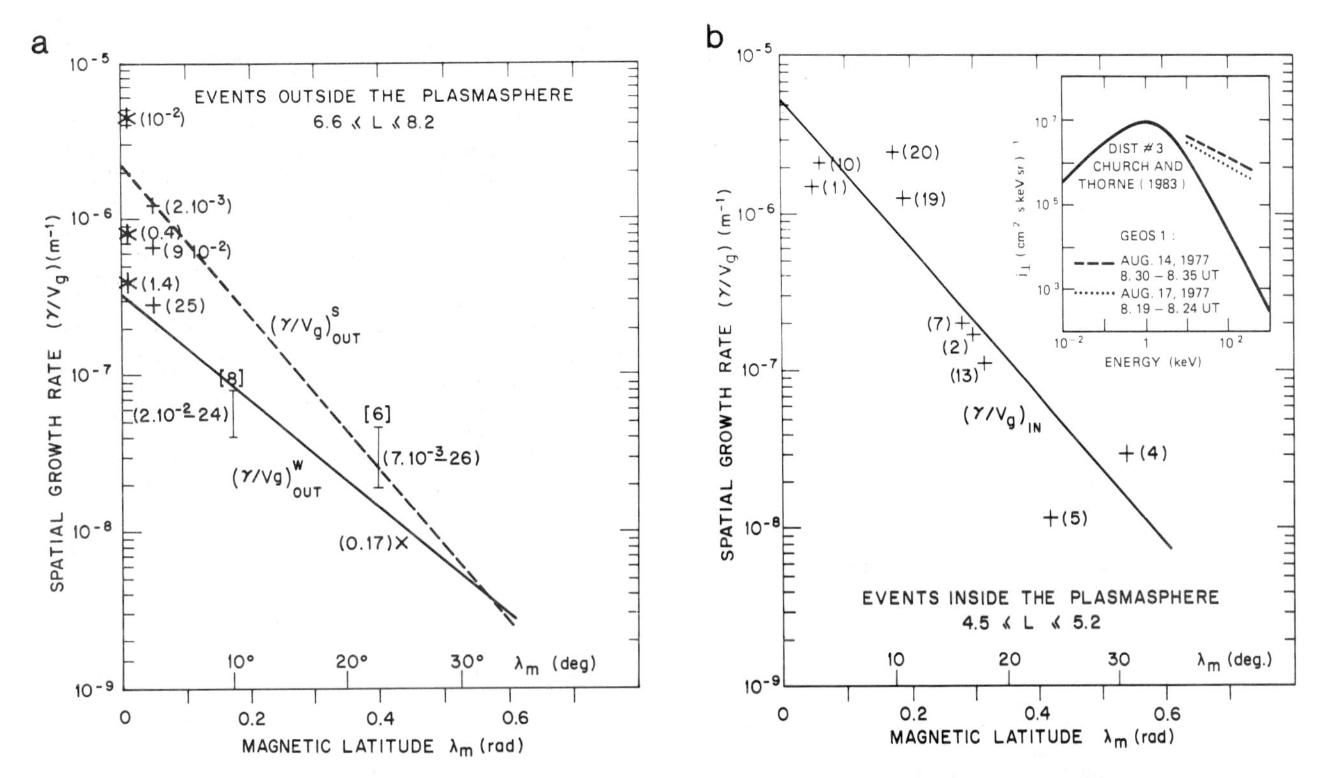

Fig. 8. (a) The maximum spatial growth rates γ/V_g of events observed outside the plasmasphere is displayed versus magnetic latitude. Stars and straight crosses correspond to events obtained at latitudes $\lambda_m = 0.5°$ and 3.5° respectively, the cross to an event obtained at $\lambda_m = 25°$. At latitudes $\lambda_m = 10°$ and $\lambda_m = 23°$ the range of values of γ/V_g is indicated by vertical bars with the corresponding number of events given above these bars. The numbers between parentheses indicate the value or the range of values of y^2_0 for the corresponding events. The two exponential fits to the data corresponding roughly to strong diffusion (Superscript S ; dashed line) or weak diffusion (Superscript W ; solid line) are also shown. Fig. 8. (b) Same as (a) for events observed just inside the plasmasphere. Because only weak diffusion events were observed, a single exponential fit to the data (solid line) was sufficient. In the right upper part of the figure is displayed the perpendicular differential flux of the electrons versus energy. DIST # 3 is the favoured model of Church and Thorne [1983]. It has been compared to GEOS flux data in the energy range 15-300 keV. These flux were fitted by using a power law $(E)^{-n}$. One finds $n \approx 0.5$ (correlation coefficient $r^2 \approx 0.9$). GEOS fluxes are much stronger than the flux given by DIST # 3 [after Solomon et al., 1988].

obtained by Church and Thorne [1983] (~ 30 dB) appear insufficient to explain plasmaspheric Hiss Generation.

In a recent paper [Solomon et al., 1988], this problem of hiss generation has been tackled in a different way. The authors have taken advantage of simultaneous measurements of energetic electron fluxes and of ELF waves obtained onboard the GEOS-1 or the GEOS-2 satellite. The detailed flux data in pitch-angle and energy allow them to compute following the method of C-W et al. [1985], the anisotropy and wave growth rate at different magnetic latitudes ($0 < \lambda < 30°$) and L-values ($4.5 < L$).

In Figure 8 are shown the results obtained both outside the plasmasphere ($L \sim 7.4$) and just inside the plasmasphere ($L \sim 4.8$). The experimental spatial growth rate is plotted versus magnetic latitude. The experimental results have been fitted by using exponential laws. Outside the plasmasphere two different laws have been obtained corresponding respectively to the strongest and to the weakest possible spatial amplification (in m^{-1}) :

$$(\gamma/V_g)^S_{OUT} \approx 2.2 \times 10^{-6} \exp(-11.1\, \lambda_m) \quad (11)$$

(correlation coefficient $r^2 = 0.80$)

and

$$(\gamma/V_g)^W_{OUT} = 3.4 \times 10^{-7} \exp(-7.9\, \lambda_m) \quad (12)$$

(correlation coefficient $r^2 = 0.88$)

Inside the plasmasphere strong diffusion events seem rare and consequently only one law has been obtained :

$$(\gamma/V_g)_{IN} \approx 5.3 \times 10^{-6} \exp(-10.8\, \lambda_m) \quad (13)$$

(correlation coefficient $r^2 = 0.81$)

From formula (11), (12), or (13) inserted in (10), the total gain in power intensity for a wave launched at the magnetic equator and propagating parallel to the magnetic field B_0 up to

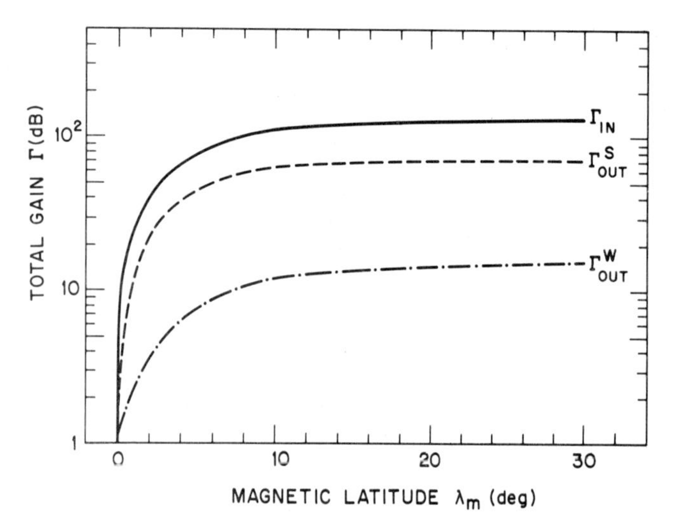

Fig. 9. The spatial integrated amplification versus magnetic latitude both inside (Γ_{IN}) and outside the plasmasphere (Γ^S_{OUT} and Γ^W_{OUT}) computed from the fits shown in Fig. 8 [Solomon et al, 1988].

a latitude $\lambda_m < 30°$, has been estimated. Results of the calculations of Γ are displayed in Figure 9.

As already stressed, there are clear indications, both theoretical and experimental, that in general field-aligned waves launched from the equator propagate to higher latitude exhibiting a rapid increase in the wave normal angle θ, and that after reflection they do not return in the equatorial region with small θ values (i.e. $\theta < 30°$). Bearing this last constraint in mind, the most simple scenario for explaining hiss generation would be that field-aligned waves, starting from noise level at the equator, were amplified to their observed level during their first journey to higher latitudes. The plausibility of such a scenario depends both on the observed propagation characteristics of ELF hiss and on the estimated values for the integrated spatial amplification of the waves.

Before discussing these two points, one can note that there exists an initial clue in favor of a local hiss generation. It is that, outside the plasmasphere, around the equator ($\lambda_m < 3°$) in strong diffusion and at any observed magnetic latitude in weak diffusion ($\lambda_m < 30°$), the peak values of the wave spectrum and of the associated growth rate occur at much the same frequency [see C-W et al., 1985]. This is also observed inside the plasmasphere at the expense of a choice of values of the cold plasma density n_o, which are in reasonable agreement with partial density measurements [Solomon et al, 1988]. It would seem difficult to explain satisfactorily such a close correlation between $B^2{}_f$ and γ/ω_{ce} if the waves were being not amplified in the region where we observe them.

Let us now consider the problem of the observed intensity level of the waves inside the plasmasphere. Church and Thorne [1983] estimated a value of the magnetic thermal noise $B^2{}_{fthermal} \sim 10^{-11} pT^2Hz^{-1}$ at a frequency f ~ 500 Hz. Inside the plasmasphere (Figure 9) the total spatial gain for the waves in $\Gamma_{IN} \sim 130$ dB, which results for the wave amplification in a maximum intensity $B^2{}_{fmax} \sim 10^2 pT^2Hz^{-1}$. Non-parallel propagation of the waves certainly reduces this last value somewhat both because of the growth rate decrease and through absorption of the waves due to the Landau effect. Ray-tracing calculations by Huang et al [1983] and Church and Thorne [1983] have shown that the wave normal angle θ of a wave launched at the equator with $\theta = 0°$ increases progressively, reaching $\theta \sim 30°$ at a latitude $\lambda_m = 10°$ (Figures 6 and 7). At that latitude their calculations give a reduction of a factor 2-3 of the spatial growth rate γ/V_g and a Landau attenuation $\Gamma_{landau} < 5$ dB. Nevertheless, those effects do not seem large enough to prevent plasmaspheric hiss from reaching, with a total gain $\Gamma > 100$ dB, a typical intensity $B^2{}_f \sim 1$ pT^{2}Hz^{-1}.

One might worry about such differences existing between values of Γ obtained by Church and Thorne ($\Gamma < 30$ dB) and ours ($\Gamma > 100$ dB) inside the plasmasphere. The reason for it stems chiefly from the much larger differential fluxes of energetic electrons observed on board GEOS 1 as compared to the flux models used by Church and Thorne [1983] (right upper part of Figure 8b). Particularly, the experimental electron energy spectra are much harder ($j \sim E^{-0.5}$) in the measured energy range than was generally assumed in previous works [Kennel, 1966 ; Liemohn, 1967 ; Kennel and Thorne, 1967]. The possibility of wave amplification up to the observed intensity in a single transit of the equatorial region with small θ values is also supported by a recent experimental study on the propagation characteristics of ELF hiss observed onboard GEOS 1 [Parrot and Lefeuvre, 1986]. This study shows that, just inside the plasmasphere, for magnetic latitudes $\lambda_m < 10°$ there is a tendency for downcoming waves (i.e. waves propagating away from the equator) to exhibit small θ values while upgoing waves present an almost flat distribution in θ values (Figure 10). For latitudes $\lambda_m > 10°$ both downcoming and upgoing waves exhibit large θ values. For latitudes $\lambda_m < 10°$ the ratio of downcoming to upgoing waves found by Parrot and Lefeuvre [1986] is ~ 1.8. Therefore one can suppose that, once amplified, parts of the waves are reflected at higher latitudes and propagate back and forth several times in the plasmasphere, exhibiting large wave normal angles, without being much attenuated because of moderate Landau damping. Ray tracing calculations show that the ray pathes migrate towards lower L and the waves can populate most of the plasmasphere [Thorne et al., 1979 ; Church and Thorne, 1983] resulting in parasitic diffusion of energetic electrons [Lyons et al., 1972].

At this point we have not yet considered the situation outside the plasmasphere ($L \geq 6.6$), which appears different in several respects from the one inside the plasmasphere. For the total spatial gain Γ_{OUT} one sees (Figure 9) that 16 dB < Γ_{OUT} < 70 dB, whereas larger wave intensities ($B^2{}_f > 10 pT^2 Hz^{-1}$) than inside the plasmasphere are often observed, resulting in a strong diffusion regime [C-W et al., 1985, 1987 ; Parrot and Lefeuvre, 1986]. These last authors have found that within $\lambda_m < 10°$ small-θ waves dominate for $x < 0.25$ but that on the contrary large-θ waves ($30° < \theta < 60°$) are slightly more numerous for $x > 0.25$ (Figure 11). Moreover they have found that the ratio of downgoing waves to upgoing waves is ~ 0.5, a value that is quite different from the factor 1.8 obtained inside the plasmasphere. An interesting additional fact is that numerous hiss events observed outside the plasmasphere are associated with regions of irregularities in electron density possibly corresponding to detached plasma regions [Cornilleau-Wehrlin et al., 1978; Hayakawa et al., 1986a; Parrot and Lefeuvre, 1986]. Apart from the effect of a larger amplification due to cold plasma density enhancement, these irregularities could also serve as waveguides, explaining the small θ-values observed within $\lambda_m < 10°$ for $x < 0.25$ in spite

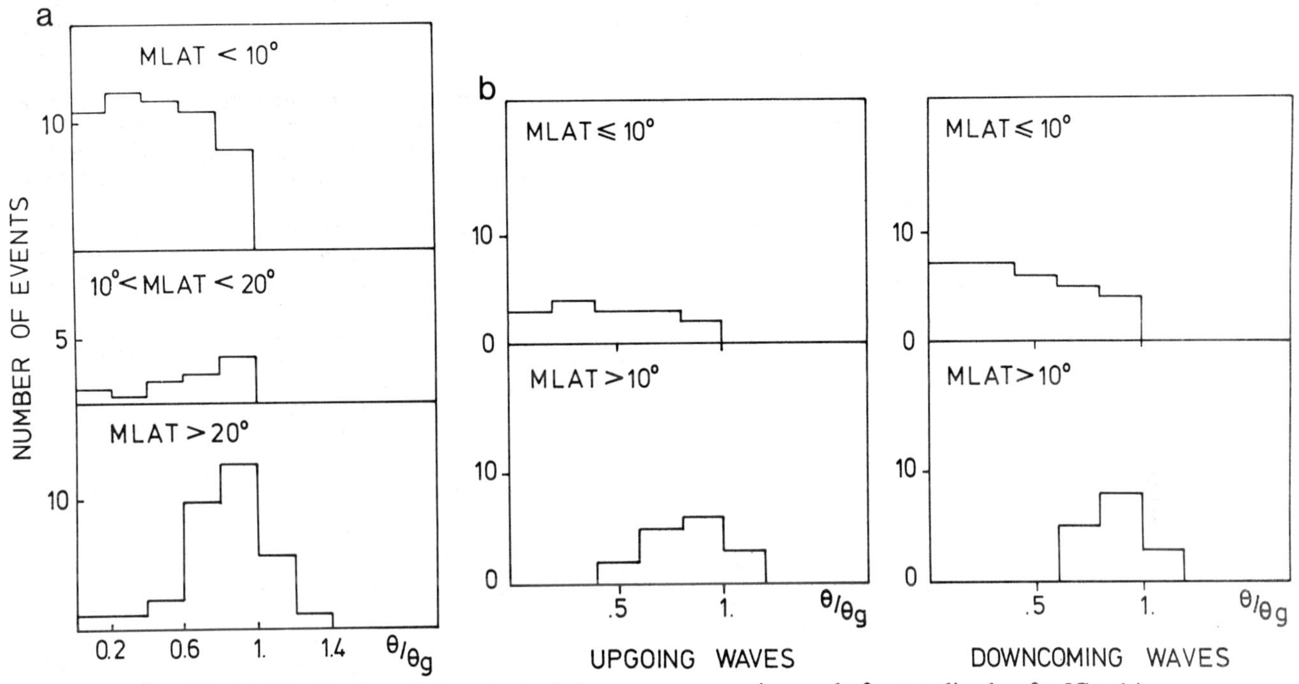

Fig. 10. (a) Distribution of the values of the wave propagation angle θ normalized to θ_g [Gendrin (1961) angle : cos $\theta_g = 2x$] within the plasmasphere. (b) Same as (a) but with a distinction between upgoing waves and down coming waves (after Parrot and Lefeuvre, 1986).

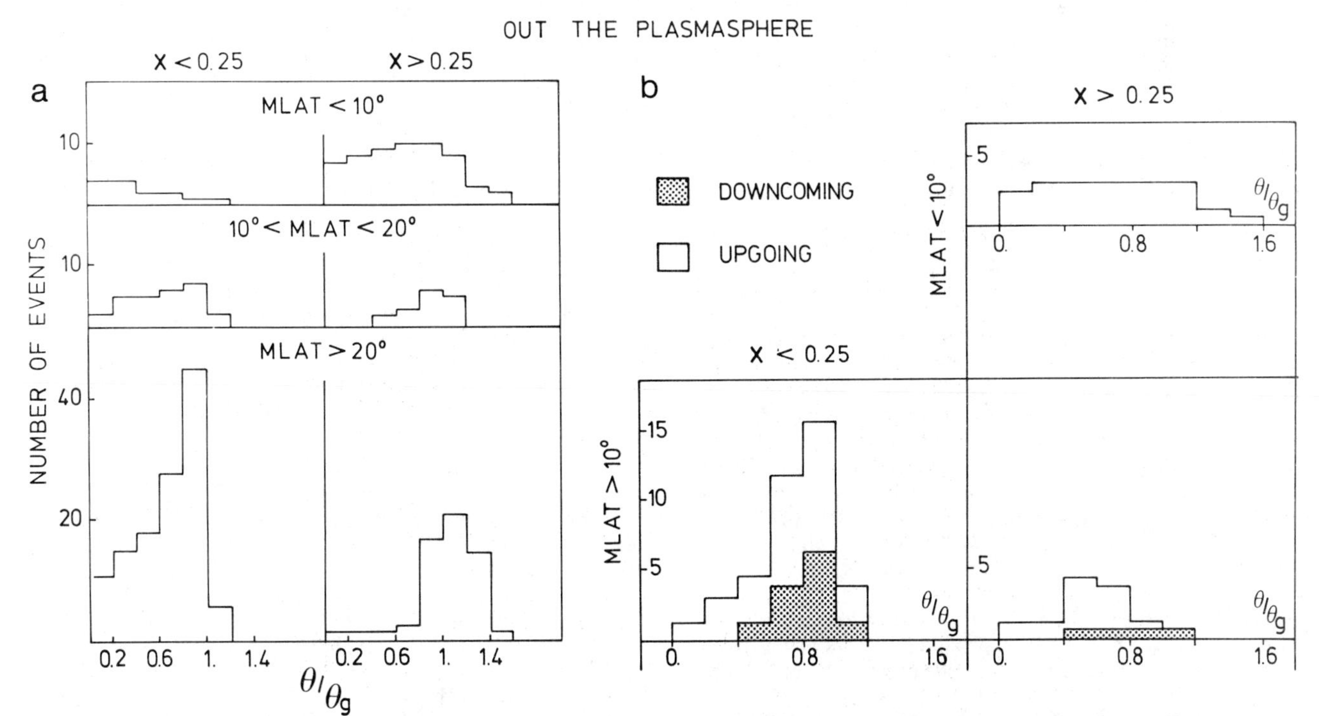

Fig. 11. (a) Same as Figure 10a, but out the plasmasphere and with a distinction between the small normalized frequencies ($x < 0.25$) and the large ones ($x > 0.25$). (b) Same as Figure 10b, but out the plasmasphere and with a distinction between $x < 0.25$ and $x > 0.25$. There is no significant data for $x < 0.25$ and MLAT < 10° [after Parrot and Lefeuvre, 1986].

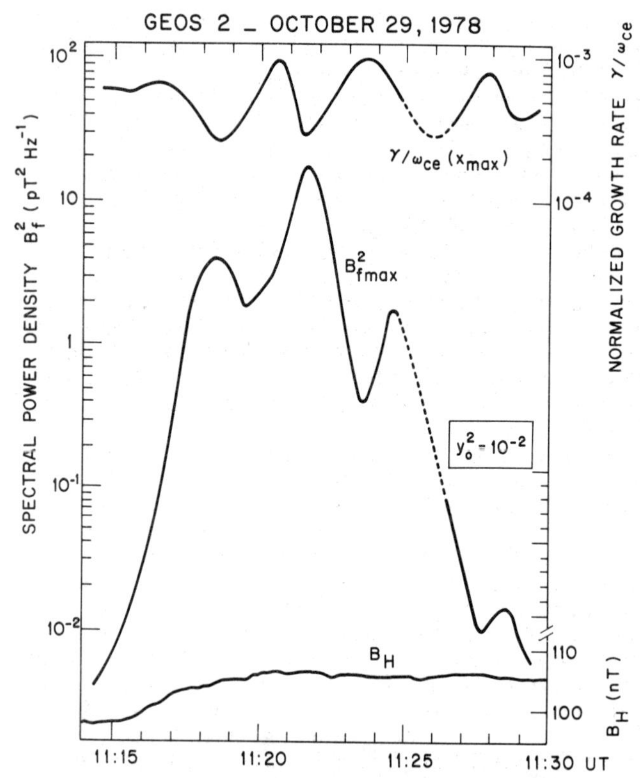

Fig. 12. Temporal evolution of the maximum of the wave spectral density $B^2{}_{fmax}$ and of the corresponding growth rate γ/ω_{ce} (x_{max}). The static magnetic field B_H is also plotted in the figure [after C-W et al., 1988].

of the relative dominance of upgoing over downgoing waves. In this case 2-3 transits of the waves through the equatorial region, starting from thermal level and taking $\Gamma \sim 50$ dB (Figure 9), would explain the wave intensity observed outside the plasmasphere. Moreover one must not exclude a leakage of waves from inside the plasmasphere through the plasmapause at high latitudes constituting what was called exohiss [Muzzio and Angerami, 1972 ; Thorne et al, 1973]. This exohiss propagating from high latitudes towards the equatorial region and subject to a possible further amplification could also contribute to the observed upgoing waves in the whole frequency range. Concerning waves propagating at large θ-values, it is not a simple matter to evaluate whether growth or damping will occur because of finite Larmor radius effects. Both right- and left-hand circularly polarized waves and also higher-order cyclotron resonance will contribute to the overall growth rate. In general, tedious numerical integrations are required. Thorne and Sommers [1986] have sought for an analytical solution of this problem in the case of different models of the particle distribution functions (bi-maxwellian, bi-lorenzian and loss-cone distributions). Although all the consequences of this new approach have not yet been developed, it should greatly reduce the effort to establish the frequency- and θ-range of instability.

The arguments used to explain the observed intensity level of the waves, particularly just inside the plasmasphere, are deceptively simple. One must bear in mind that by using the experimental values of the growth rate γ/ω_{ce} at different magnetic latitudes one has the enormous advantage of taking implicitly into account the intricate continuous process of pitch-angle diffusion of the energetic electrons by growing waves. That such a process is at work along the lines of force is obvious when one notices that the values of the anisotropy obtained at different latitudes lie quite close to the critical anisotropy Ac (see Figures 1, 2, 3, 4, and Solomon et al.[1988].

5. Non Stationary Effects

The event displayed in Figure 4 demonstrates that it takes 5 to 10 minutes to achieve steady state in a weak diffusion case. At the time of storm sudden commencements (SSC), rapid magnetospheric compression increases the temperature anisotropy A of the energetic electrons on much a shorter time scale. In this context, recent studies have been made by using GEOS-2 data [Korth et al., 1986 ; C-W et al., 1988]. Following the method of C-W et al. [1985] calculations of the anisotropy A(x) (formula 6) and of the wave growth rate $\gamma(x)$ (formula 7) have been performed by using 1- min. averaged data. The temporal evolution of these quantities and of $B^2{}_f(x)$ have been studied for the reduced frequency $x = x_{max}$ of the maximum of the wave spectrum $B^2{}_{fmax}$ for each one-minute interval.

Figure 12 gives the results for the temporal evolution of $B^2{}_{fmax}$ and γ/ω_{ce} (x_{max}) for the 29th of October. B_H, the compressional component of the static magnetic field (in the V, D, H system) is also given at the bottom of the figure.

Clearly, $B^2{}_{fmax}$ and γ/ω_{ce} (x_{max}) exhibit antiphase oscillations, not seen in B_H. How can these antiphase oscillations be interpreted ? We consider here those particles for the calculation of γ/ω_{ce} (x_{max}) which resonate with that part of the wave spectrum that corresponds to its maximum. These particles are most efficiently diffused in pitch-angle by the waves. Once the growth rate is strong enough to produce high amplitude waves, the waves in turn diffuse particles and reduce the growth rate. When γ/ω_{ce} has become low enough, as it is the case at 1118:30 in Figure 12, $B^2{}_f$ itself decreases. As $B^2{}_f$ decreases, the particle diffusion also decreases and γ/ω_{ce} starts to grow again, followed by an increase of $B^2{}_{fmax}$. The same process starts again, inducing the observed antiphase oscillations. This explanation implies that one admits the existence of a steady source of particles at the time scale of the oscillations (~ 3 min).

Let us now examine which parameter contributes to the oscillations of the growth rate by looking at Figure 13. In this figure, $\gamma/\omega_{ce}(x_{max})$ is plotted, together with 3 other quantities: $(A - A_c)(x_{max})$, $\eta'(x_{max})$ and x_{max} itself. $\eta'(x_{max})$ is defined by $\gamma/\omega_{ce} = \eta'(A-A_c)$ and is proportional to the relative number of resonating particles (see formula (1)). Let us consider the time interval during which x_{max} is constant, i.e. from 1114 until 1123 UT: $(A-A_c)$ (x_{max}) oscillates in phase with $\gamma/\omega_{ce}(x_{max})$ and both quantities have the same modulation rate. At the same time, $\eta'(x_{max})$ remains constant. Thus the oscillations of γ/ω_{ce} in fact stem from the anisotropy oscillations, which is in agreement with the hypothesis of a quasi-linear diffusion in pitch-angle. It can be inferred from the values of $\eta'(x_{max})$ that the particle source was constant. For this strong diffusion event ($y^2{}_o = 10^{-2}$), the modulation rate of γ/ω_{ce}, taken as $\gamma_{max}/\gamma_{min}$, is equal to 5. After 11.23 UT, oscillations appear in $\eta'(x_{max})$. Yet, this effect is chiefly a

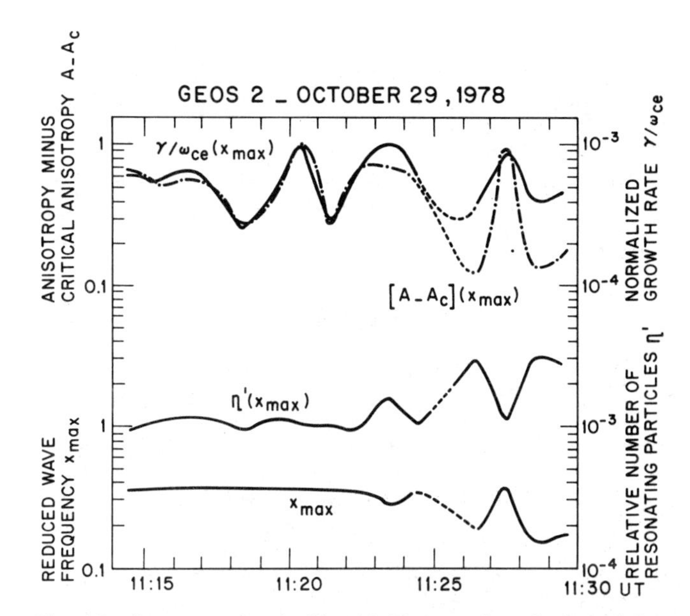

Fig. 13. Same event as in Fig. 12. Temporal evolution of the two factors that contribute to the growth rate γ/ω_{ce} : $\eta'(x_{max})$, the relative number of resonating electrons and $(A-A_c)(x_{max})$. $\gamma/\omega_{ce}(x_{max})$ is also plotted for comparison. x_{max} is also displayed [C-W et al., 1988].

mathematical artefact due to fluctuations in x_{max} itself and to the lower energy measurement threshold (16 keV) : as $E_R \propto 1/x$, any variation in x results in a change in the pitch-angle range over which integration to compute η' can be made. This last effect is counteracted by an opposite effect in the anisotropy calculation, so that on the whole, the growth rate itself is not changed considerably [for further details, see C-W et al., 1985].

To summarize the above observations, the antiphase oscillations of B^2_{fmax} on the one hand, of (γ/ω_{ce}) (x_{max}) and $(A - A_c)(x_{max})$ on the other hand, seem to be due to quasi-linear pitch-angle diffusion of the electrons which resonate with the highest amplitude wave, this diffusion leading to a periodic relaxation of the anisotropy.

One more astonishing point is that the same kind of oscillations, with a period of 3-4 minutes, is observed for weak diffusion events, although the low intensity waves are expected to be less efficient for diffusing resonating particles in pitch-angle. Nevertheless it must be noticed that the modulation rate of γ/ω_{ce} (x_{max}) (~ 2) is less than for the strong diffusion event (~ 5).

The periodic oscillations of the anisotropy could be due to compressional pulsations in the static magnetic field. Such causes of the modification of the anisotropy have been invoked first by Coroniti and Kennel [1970] to explain modulations in precipitated electron fluxes and also to explain quasi-periodic emissions [Sazhin, 1978 ; Tixier and Cornilleau-Wehrlin, 1986]. It is clear from the B_H data displayed in Figures 12 that locally there are no compressional ULF pulsations of 2-4 min periods. We have selected one SSC after which GEOS-2 measurements show a modulation in the B_H component, whereas the VLF wave intensity increases (Figure 14). We again observe the antiphase position of B^2_f on the one hand, and γ/ω_{ce} (x_{max}) and $(A - A_c)(x_{max})$ on the other hand, just after the SSC at about 0809 UT. When a large modulation of B_H appears at 0814 UT, all parameters oscillate approximatively in phase (B_H, B^2_{fmax}, γ/ω_{ce}, $(A - A_c)$). Of course, both the effects due to the continuous increase in B_H and to the pulsations are not completely separable and the experimental result is not so pure as in figures 12 and 13. Therefore the presence of pulsations does not seem to be the cause of the antiphase modulation between B^2_f and $(A - A_c)$, but rather to bring all these quantities in phase.

Thus we can interpret the clear antiphase oscillations of the maximum wave intensity B^2_{fmax} with respect to the growth rate and anisotropy of the resonating particles γ/ω_{ce} (x_{max}) and $(A - A_c)(x_{max})$ as a quasi-linear effect, as mentioned above. This effect is very local in the phase space velocity. This happens for weak diffusion as well for strong diffusion, yet the modulation rate is different, increasing with the diffusion strength.

Now two questions remain unsolved : first, why is the periodicity of the order of 3 min in each case ? No link has been found, either with the steepness of the SSC (variation in the B_H field), or with particle life time (more than 5 min) [see C-W et al., 1985]. Second, why does this happen in wave events occurring after an SSC ? One can think that the anisotropy increase is faster than in other events (drift, injection) and that the quasi-linear reaction is also faster, resulting in some phase space oscillations. Recent work on auroral pulsations [Davidson and Chiu, 1986], based on an extension of a phenomenological treatment of WPI due to Schulz [1974] could be of some interest for the present problem. Nevertheless, the fact that details of the distribution

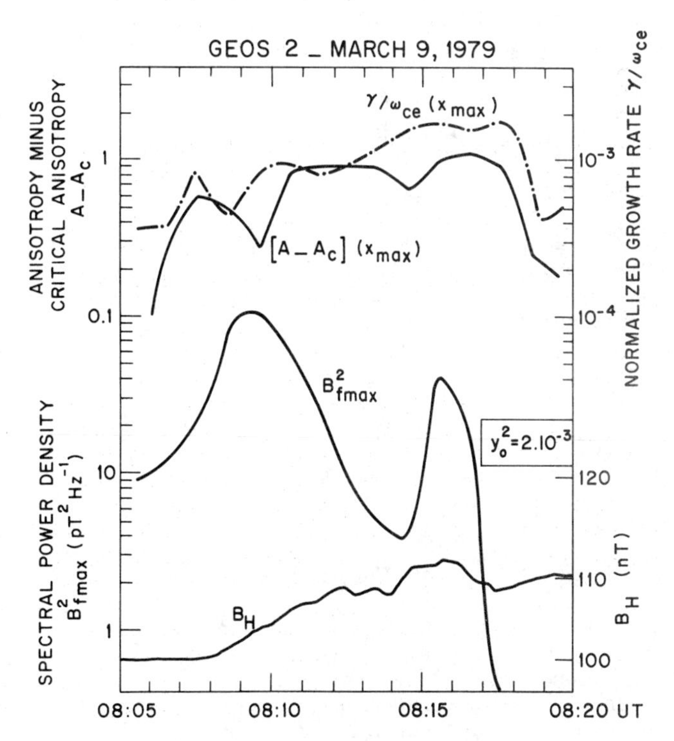

Fig. 14. Same as Fig. 12. In this case, the compressional component B_H of the magnetic field oscillates [C-W et al., 1988].

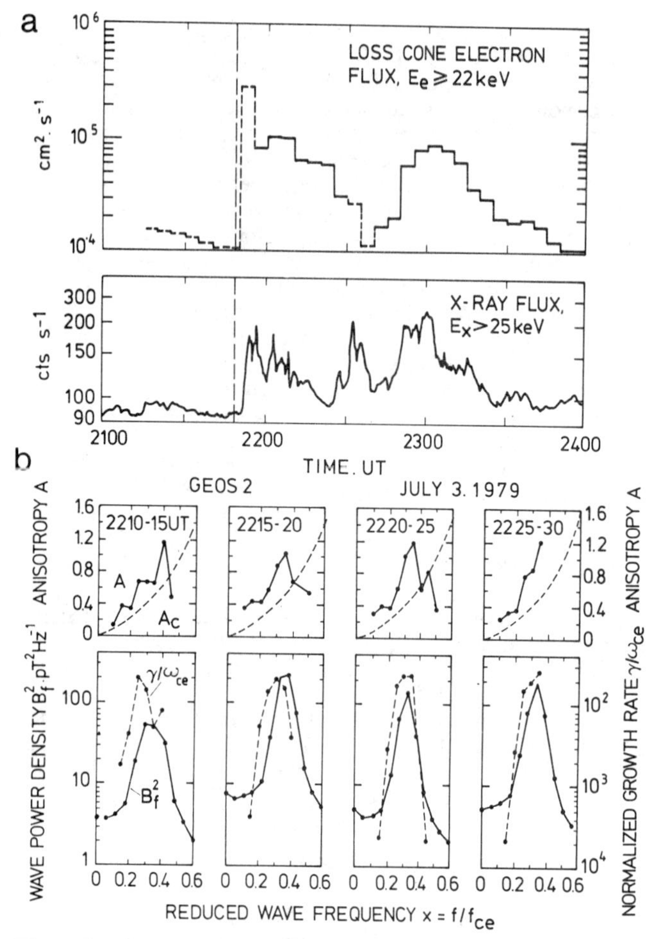

Fig. 15. (a) Electron flux measured in the loss cone at GEOS-2 (geostationary orbit) and the corresponding X-ray measurements obtained on a balloon during a precipitation event - the dashed vertical line indicates the substorm onset. (b) Calculations of the anisotropy A and of the normalized wave growth rate γ/ω_{ce} and their comparison to the observed wave power density B_f^2 for the time interval 22 10 - 22 30 UT. [after Kremser et al., 1986].

function $F(\alpha,E)$ in pitch-angle and energy cannot be smeared out without altering the results of the calculations (e.g. the proper behaviour of $A(x)$ with respect to A_c is obtained only when using detailed measurements of $F(\alpha,E)$) arises the question concerning the validity of such a treatment in the present case.

6. Conclusion

In this review, we have discussed different aspects of the interaction between energetic electrons ($E > 15$ keV) and whistler mode ELF/VLF waves ($f \sim 0.2$ - 2 kHz) in the earth's magnetosphere, particularly by using detailed measurements of waves and particles obtained onboard the GEOS-1 and -2 satellites. It appears that when an anisotropic source of energetic electrons exists, an equilibrium situation for WPI can be set up. The most striking result is that, due to quasi-linear diffusion in pitch-angle of the energetic electrons, the temperature anisotropy A is maintained slightly above the critical anisotropy A_c in the wave frequency range, A_c being the threshold above which instability occurs. More precisely, close to the magnetic equator, the larger the wave intensity, the larger the growth rate γ and the smaller the positive difference $(A-A_c)$ in the wave frequency range. When magnetic latitude λ_m increases, there is a large decrease of the normalized growth rate (γ/ω_{ce}), by at least one order of magnitude when λ_m varies from 0° to 20°. This fact confirms a preferential amplification of the waves in the equatorial regions.

Inside the plasmasphere, the integrated spatial growth rate Γ(> 100 dB) is large enough to account for the observed wave intensity (~ 1 $pT^2 Hz^{-1}$) in a single transit of the wave through the equatorial region ($\lambda_m < 20°$), what appears compatible with theoretical and experimental results on wave propagation. Outside the plasmasphere the situation is more intricate, both because of the smaller obtained Γ values and of more complicated results about wave propagation.

Non steady-state situations are found during Storm Sudden Commencements. Oscillations with a period of 2-4 minutes are observed in the maximum intensity of the waves that are in antiphase both with the corresponding growth rate and anisotropy. This effect is interpreted as the result of the quasi-linear diffusion in pitch-angle of the resonant electrons.

Other important effects of gyroresonant interactions have not been mentioned as the precipitations of particles in the atmosphere by the waves. Kremser et al. [1986] using both GEOS-2 data and X-ray measurements from balloons have evidenced the link between ELF electro-magnetic waves and precipitated electrons $E > 20$ keV (Figure 15). Torkar et al. [1987] have studied, using similar data set, the relationship between equatorial energetic electrons drifting eastwards at $L = 6.6$, ELF waves and auroral X-rays. Gurnett et al. [1979] and Tsurutani et al. [1981] have studied plasma wave turbulence at the magnetopause by using data of the satellites ISEE-1 and -2. Particularly Tsurutani et al. [1981] have shown that intense broad-band whistler mode noise are well correlated with enhanced fluxes of electrons (in the energy range 1 - 6 keV) and of protons (in the energy range 5.4 - 6.6 keV), resulting in a possible strong pitch-angle diffusion regime and particle precipitations which could contribute substantially to the day side aurora. Intense electrostatic waves measured at the same time could strongly interact with low energy electrons (below 1 keV).

References

Ashour-Abdalla, M., and S.W.H. Cowley, Wave-particle interactions near the geostationary orbit, in Magnetospheric Physics, edited by B.M. Mc Cormac, pp. 241-270, D. Reidel, Hingham, Mass., 1974.

Chan, K.W., R.E. Holzer, and E.J. Smith, A relation between ELF hiss amplitude and plasma density in the outer plasmasphere, J. Geophys. Res., 79, 1989-1993, 1974.

Church, S.R., and R.M. Thorne, On the origin of plasmaspheric hiss : Ray path integrated amplification, J. Geophys. Res., 88, 7941-7957, 1983.

Cornilleau-Wehrlin, N., R. Gendrin, and M. Tixier, VLF wave : conjugated ground-satellite relationships, Space Sci. Rev., 22, 419-431, 1978a.

Cornilleau-Wehrlin, N., R. Gendrin, F. Lefeuvre, A. Bahnsen, E. Ungstrup, R.J.L. Grard, D. Jones, and W.

Gibbons, VLF electromagnetic waves observed onboard GEOS-1, Space Sci. Rev., 22, 371-382, 1978b.

Cornilleau-Wehrlin, N., J. Solomon, A. Korth, and G. Kremser, Experimental study of the relationship between energetic electrons and ELF waves observed onboard GEOS: a support to quasi-linear theory, J. Geophys. Res., 90, 4141-4154, 1985.

Cornilleau-Wehrlin, N., J. Solomon, A. Korth and G. Kremser, Non stationary effects in wave-particle interactions during storm sudden commencements (SSC), Physica Scripta, 37, 437-442, 1988.

Cornwall, J.M., Scattering of energetic trapped electrons by very-low frequency waves, J. Geophys. Res., 69, 1251-1258, 1964.

Coroniti, F.V., and C.F. Kennel, Electron precipitation pulsations, J. Geophys.Res., 75 1279-1289,1970.

Davidson G.T., and Y.T. Chiu, A closed nonlinear model of wave-particle interactions in the outer trapping and morningside auroral regions, J. Geophys. Res., 91, 13,705-13,710, 1986.

Etcheto, J., R. Gendrin, J. Solomon, and A. Roux, A self consistent theory of magnetospheric ELF hiss, J. Geophys. Res., 78, 8150-8166, 1973.

Gurnett, D.A., R.R. Anderson, B.T. Tsurutani, E.J. Smith, G. Paschmann, G. Haerendel, S.J. Bame, and C.T. Russel, Plasma wave turbulence at the magnetopause : Observations from ISEE 1 and 2, J. Geophys. Res., 84, 7043-7058, 1979.

Hayakawa M., N. Ohmi, M. Parrot, F. Lefeuvre, Direction finding of ELF hiss emissions in a detached plasma region of the magnetosphere, J. Geophys. Res., 91, 135-141, 1986a.

Hayakawa M., M. Parrot, and F. Lefeuvre, The wave normals of ELF hiss emissions observed onboard GEOS-1 at the equatorial and off equatorial regions of the plasmasphere, J. Geophys. Res., 91, 7989-7999, 1986b.

Huang, C.Y., and C.K. Goertz, Ray-tracing studies and path integrated gains of ELF unducted whistler mode waves in the earth's magnetosphere, J. Geophys. Res., 88, 6181-6187, 1983.

Huang, C.Y., C.K. Goertz, and R.R. Anderson, A theoretical study of plasmaspheric hiss generation, J. Geophys. Res., 88, 7927-7940, 1983.

Kennel, C.F., Low-frequency whistler mode, Phys. Fluids, 9, 2190-2202, 1966.

Kennel, C. F., and H.E. Petschek, Limit on stably trapped particle fluxes, J. Geophys. Res., 71, 1-28, 1966.

Kennel, C.F., and R.M. Thorne, Unstable growth of unducted whistler propagating at an angle to the geomagnetic field, J. Geophys. Res., 72, 871-878, 1967.

Korth, A., G. Kremser, N. Cornilleau-Wehrlin and J. Solomon, Observations of energetic electrons and VLF waves at geostationary orbit during storm sudden commencements (SSC), in Solar wind-magnetosphere coupling, edited by Y. Kamide and J.A. Slavin, pp 391-399, Terra Scientific Pub. Cy (TerraPub), Tokyo, 1986.

Kremser, G., A. Korth, S. Ullaland, J. Stadsnes, W. Baumjohann, L. Block, K.M. Torkar, W. Riedler, B. Aparicio, P. Tanskanen, I.B. Iversen, N. Cornilleau-Wehrlin, J. Solomon, E. Amata, Energetic electron precipitation during a magnetospheric substorm and its relationship to wave particle interaction, J. Geophys. Res., 91, 5711-5718, 1986.

Liemohn, H.B., Cyclotron-resonance amplification of VLF and ULF whistlers, J. Geophys. Res., 72, 39-55, 1967.

Lyons, L.R., R.M. Thorne, and C.F. Kennel, Pitch-angle diffusion of radiation belt electrons within the plasmasphere, J. Geophys. Res., 77, 3455-3474, 1972.

Maeda, K., and C.S. Lin, Frequency band broadening of magnetospheric VLF emissions near the equator, J. Geophys. Res., 86, 3635-3639, 1981.

Muzzio, J.L.R., and J.J. Angerami, OGO-4 observations of extremely low frequency hiss, J. Geophys. Res., 77, 1157-1173, 1972.

Parrot, M., and F. Lefeuvre, Statistical study of the propagation characteristics of ELF hiss observed on GEOS-1 inside and outside the plasmasphere, Ann. Geophys., 4A (5), 363-383, 1986.

Roederer, J.G., Dynamics of geomagnetically trapped radiation, Ed. by J.G. Roederer and J. Zähringer, Springer-Verlag Pub., Berlin, 1970.

Roux, A., and J. Solomon, Mécanismes non-linéaires associés aux interactions ondes-particules dans la magnétosphère, Annls Géophys., 26, 279-297,1970.

Roux, A., and J. Solomon, Self-consistent solution of the quasi-linear theory: application to the spectral shape and intensity of VLF waves in the magnetosphere, J. Atmos.Terr. Phys., 33, 1457-1471, 1971.

Sazhin, S.S., A model of quasi-periodic VLF emissions, Planet.Space Sci. , 26, 399-401, 1978.

Schulz, M., Particle saturation of the outer zone : a non-linear model, Astrophys. and Space Sci, 29, 233-242, 1974.

Solomon, J., and R. Pellat, Convection and wave-particle interactions, J. Atmos. Terr. Phys., 40, 373-378, 1978.

Solomon, J., N. Cornilleau-Wehrlin, A. Korth and G. Kremser, An experimental study of ELF/VLF hiss generation in the earth's magnetosphere, J. Geophys. Res., 93, 1839-1847, 1988.

Thorne, R.M., E.J. Smith, R.K. Burton, and R.E. Holzer, Plasmaspheric hiss, J. Geophys. Res, 78,1581-1596,1973.

Thorne, R.M., S.R. Church and D.J. Gorney, On the origin of plasmaspheric hiss : the importance of wave propagation and the plasmapause, J. Geophys. Res., 84, 5241-5247, 1979.

Thorne, R.M., and D. Sommers, Analytical solutions to the general problem of oblique wave growth and damping, Phys. Fluids, 29, 4091-4102, 1986.

Tixier, M., and N. Cornilleau-Wehrlin, How are VLF quasiperiodic emissions controlled by harmonics of field line oscillations ? The result of a comparison between ground and GEOS satellites measurements, J. Geophys. Res., 91, 6899-6919, 1986.

Torkar, K.M., W. Riedler, G. Kremser, A. Korth, S. Ullaland, J. Stadsnes, L.P. Block, I.B. Iversen, P. Tanskanen, J. Kangas, N. Cornilleau-Wehrlin, J. Solomon, A study of the interaction of VLF waves with equatorial electrons and its relationship to auroral X-rays in the morning sector, Planet. Space Sci., 35 , 1231-1253, 1987.

Tsurutani, B.T., E.J. Smith, R.M. Thorne, R.R. Anderson, D.A. Gurnett, G.K. Parks, C.S. Lin, and C.T. Russell, Wave-particle interactions at the magnetopause : Contributions to the dayside aurora, Geophys. Res. Lett., 8, 183-186, 1981.

ACCELERATION OF THERMAL PLASMA IN THE MAGNETOSPHERE

Maha Ashour-Abdalla

Institute of Geophysics and Planetary Physics and Department of Physics,
University of California, Los Angeles, CA 90024

David Schriver

Department of Physics, University of California, Los Angeles, CA 90024

Hideo Okuda

Plasma Physics Laboratory, Princeton University, Princeton, NJ 08544

Abstract. Recent observations revealing the composition of magnetospheric plasmas have demonstrated the importance of mixed plasmas in magnetospheric plasma physics. In this paper we use analytic theory and numerical simulations to investigate the physics of two types of mixed plasmas. We begin our study by considering the transverse acceleration of ions on auroral field lines in order to determine the effects of multi-ion species. In the auroral zone the components of a multi-ion plasma, including hydrogen and oxygen, interact with each other as well as with a two-component electron plasma composed of both a magnetospheric beam and background ionospheric components. This interaction occurs as a mixed ion-ion hybrid mode. In the second part of our study we look at how an electron plasma, with both hot and cold components as well as ion beams, affects the plasma sheet boundary layer. We find that in the presence of this mixed electron plasma, warm ion beams can drive the electron acoustic instability; this phenomenon may be responsible for broadband electrostatic noise in the boundary layer.

I. Introduction

For many years it has been generally accepted that the magnetosphere is a mixture of plasmas of solar wind and ionospheric origins [Shelley et al., 1972]. Recently, however, observations disclosing the composition of magnetospheric plasmas have become available, and as a result we have begun to recognize the importance of mixed plasmas in the dynamics of the magnetospheric system [Horwitz, 1987 and references therein]. First, there is species mixing with multiple species of heavy ions [see Chappell et al., 1987]. The second type is thermal mixing with both hot and cold electrons which will not attain equilibrium unless there are collective processes that equalize the temperatures [Schield and Frank, 1970]. In both cases, the mixed nature of the plasma adds an additional degree of freedom in kinetic theory and provides new modes which the traditional two-species dispersion relation does not have.

In this paper we will give examples of these mixed plasma modes and use both analytic theory and numerical simulations to study the nonlinear physics associated with these modes. As a first example we will consider the heating of ions on auroral field lines. Evidence for transverse heating of auroral ions has been been found both on field lines that carry the primary magnetospheric field aligned current and on field lines that carry the ionospheric return current. In this paper, we will concentrate on the primary current region where the physics of multispecies plasmas is important. We will consider the heating of a hydrogen-oxygen plasma in the presence of a mixed electron population that includes both magnetospheric electrons and ionospheric electrons. In such a plasma it is possible to excite a mixed ion-ion hybrid mode: the Buchsbaum resonance [Buchsbaum, 1960].

Next we will consider heating in the plasma sheet boundary layer and the generation of broadband electrostatic noise. In this situation two-temperature plasma effects may be important. The boundary layer is characterized by beams in the ion distribution and by a two-temperature electron distribution with a hot plasma sheet component and a cold ionospheric component. In this two-temperature electron plasma, the warm ion beams can generate electron acoustic and beam resonant waves.

2. Species Mixing - Transverse Heating of Ionospheric Ions in the Auroral Zone

One of the most interesting regions of the magnetosphere is the auroral zone in the northern and southern polar regions of the earth from which plasma energy stored in the tail region is dissipated into the ionosphere. Some of the prominent features of the auroral zone are field-aligned currents flowing into and out of the ionosphere, double layers, ion beams and transverse ion acceleration (ion conics).

A number of different electrostatic waves can contribute to and be collectively responsible for the transverse acceleration of the ions in the auroral zone. The theoretical studies of wave-particle interactions aimed at gaining an understanding of the acceleration of ions on auroral field lines began with the work of Kindel and Kennel [1971]; they showed that of the various instabilities driven by a drifting electron Maxwellian distribution, the ion cyclotron waves are the most easily destabilized. Indeed, the currents observed by satellite were well above the threshold required for destabilization [Klumpar and Heikkila, 1982]. Following the pioneering work of Kindel and Kennel [1971], various authors considered the acceleration of ionospheric ions [Palmadesso et al., 1974; Ungstrup et al., 1979; Lysak et al., 1980; Dusenbery and Lyons, 1981; Okuda and Ashour-Abdalla, 1983; Ashour-Abdalla and Okuda, 1984]. In particular, Ashour-Abdalla and Okuda [1984] considered the heating of oxygen ions in the presence of a hydrogen-oxygen plasma. Their studies showed that in such a plasma, both electrostatic hydrogen cyclotron waves and oxygen cyclotron waves are excited. Depending on the plasma parameters (e.g. oxygen density concentration, thermal electron drifts, etc.) the hydrogen-cyclotron waves were found to heat the hydrogen ions, whereas the oxygen ions are heated by oxygen-cyclotron waves.

Palmadesso et al. [1974], Chang and Coppi [1981] and Singh and Schunk [1982] suggested that in the primary current region, the lower hybrid waves might transfer energy from the keV auroral electron beam to the ionospheric ions thereby heating the ions in the transverse direction. The problem with this theory is that, although lower hybrid waves may be able to heat hydrogen ions, it is difficult to see how such high frequency waves could heat thermal oxygen ions. It would be interesting to find an analogous wave for the oxygen ions.

This paper considers plasma modes that can occur only in a multicomponent plasma, in particular an O^+-H^+ plasma, and the effect of these plasma modes on heavy ion acceleration. In such a plasma new instabilities arise from the combined effects of the multiple components of ions and electrons, effects which are absent in a pure electron-ion, two-component plasma. It is found that the addition of oxygen ions to a multi-component plasma creates a new instability, the ion-ion hybrid mode, whose frequency is adequate for the wave to interact with oxygen ions [Buchsbaum, 1960; Ashour-Abdalla et al., 1987; Schriver and Ashour-Abdalla, 1988].

2.1 Linear Theory

To understand the heating of ions in various regions of the auroral zone we first use linear theory to determine which electrostatic waves can act as a medium for transferring free energy from the field-aligned electron current to the perpendicular energy of ionospheric ions. Once these waves are identified from the dispersion relation, the actual details of the energy transfer will be studied using computer simulations (section 2.2).

Here we briefly review plasma linear theory for a three-component plasma that includes electrons, hydrogen and oxygen. When in the presence of heavy ions in the ionosphere, such as oxygen ions at low altitudes, it is possible to excite the ion-ion hybrid mode by bombarding the ionosphere with auroral electrons. The cold plasma dispersion relation for waves propagating nearly perpendicularly to the magnetic field, $k_\parallel/k_\perp \ll 1$, can be written as

$$\varepsilon(\omega,k_\perp) = 1 - \frac{k_\parallel^2}{k^2}\frac{\omega_{pe}^2}{\omega^2} + \frac{\omega_{pH}^2}{\Omega_H^2 - \omega^2} + \frac{\omega_{po}^2}{\Omega_o^2 - \omega^2} - \frac{k_\parallel^2}{k^2}\frac{\omega_{pb}^2}{(\omega - k_\parallel v_o)^2} = 0 \qquad (1)$$

where we assume $\omega \ll \Omega_e$, $\omega_{pe}^2 \ll \Omega_e^2$, $k_\perp\rho_H \ll 1$ and $k_\perp\rho_o \ll 1$. Here ω_{pe}, ω_{pH} and ω_{po} are the electron, hydrogen and oxygen plasma frequencies, respectively, and Ω_e is the electron gyrofrequency. The Larmor radius of hydrogen and oxygen are given by ρ_H and ρ_o defined as the species thermal velocity divided by the gyrofrequency. It is clear that for $k_\parallel = 0$, the two resonant frequencies are given by solution of

$$\omega^4 - (\omega_{pH}^2 + \omega_{po}^2 + \Omega_H + \Omega_o^2)\,\omega^2 + (\omega_{pH}^2\Omega_o^2 + \omega_{po}^2\Omega_H^2 + \Omega_H^2\Omega_o^2) = 0 \qquad (2)$$

When $\omega_{pe}^2/\Omega_e^2 \ll 1$, $\omega_{pH}^2/\Omega_H^2 > 1$ and $\omega_{po}^2/\Omega_o^2 > 1$, the two resonant frequencies are given by

$$\omega_{LH}^2 = \omega_{pH}^2 + \omega_{po}^2 \qquad (3)$$

$$\frac{\omega_{IH}^2}{\Omega_H \Omega_o} = \frac{n_H \Omega_o + n_o \Omega_H}{n_H \Omega_H + n_o \Omega_o} \tag{4}$$

Equation (3) represents the lower hybrid resonance frequency in the presence of hydrogen and oxygen ions, while equation (4) represents the ion-ion hybrid or Buchsbaum resonance in the presence of two-ion species (Buchsbaum, 1960; Stix, 1962). When an electron beam, such as a beam of auroral electrons bombarding the ionosphere is present it is possible to excite the waves by wave-particle interactions, thereby transferring the electron beam energy to the ions. The instability is driven by the drifting auroral electrons given by the last term of equation (1). Equation (1) clearly indicates that the ion-ion hybrid mode becomes unstable for $\Omega_o < k_\parallel v_o < \Omega_H$, while the lower hybrid mode is destabilized for $k_\parallel v_o > \Omega_H$.

Since a thermal plasma exists in the auroral region, we must numerically solve the warm plasma dispersion relation to accurately determine the frequency and growth rate of the unstable waves. The electrostatic warm plasma dispersion relation in the presence of an external magnetic field can be written as

$$0 = 1 + \sum_\alpha \frac{1}{k^2 \lambda_\alpha^2} \left[1 + \frac{\omega}{\sqrt{2} k_\parallel v_{t\alpha}} e^{-y_\alpha^2} \sum_{n=-\infty}^{+\infty} Z\left(\frac{\omega + n\Omega_\alpha}{\sqrt{2} k_\parallel v_{t\alpha}} \right) I_n(y_\alpha^2) \right]$$
$$+ \frac{1}{k^2 \lambda_b^2} \left[1 + \frac{\omega - k_\parallel V_d}{\sqrt{2} k_\parallel v_{tb}} e^{-y_b^2} \sum_{n=-\infty}^{+\infty} Z\left(\frac{\omega + n\Omega_e - k_\parallel V_d}{\sqrt{2} k_\parallel v_{tb}} \right) I_n(y_b^2) \right] \tag{5}$$

The sum α is the background plasma which includes oxygen, hydrogen and ambient electrons, and the third term with the subscripts b is the auroral beam term. The wavenumber is given by k, with $k_\perp$ and $k_\parallel$ the perpendicular and parallel wavenumbers, λ_α is the Debye length, $v_{t\alpha}$ is the thermal velocity of the species α, and V_d is the beam drift speed. The modified Bessel functon is given by $I_n(y_s)$ with the argument defined as $y_s = k_\perp v_{ts}/\Omega_s$, and Z(x) is the plasma Z function defined by Fried and Conte (1961).

In a warm plasma, ion cyclotron waves can exist in addition to hybrid modes. For a simple two component ion-electron plasma, the lower hybrid frequency is a crossover: the modes above it have frequencies essentially constant over k near each multiple of the gyrofrequency (with cyclotron damping), whereas below ω_{LH} the mode structure changes such that at small k the frequency starts at $(n+1)\Omega_i$ and drops to $n\Omega_i$ with increasing k [Bernstein, 1958]. A beam can destabilize the hybrid mode and the harmonics below it because the cyclotron damping is small over a wide range of k for these modes; above the hybrid frequency the harmonics are cyclotron damped, and wave growth (if any) is small. Thus the value of the lower hybrid frequency determines how many cyclotron harmonics can be excited below it. If $\omega_{LH} < 2\Omega_H$, only the lower hybrid frequency determines how many cyclotron harmonics can be excited below it. If $\omega_{LH} < 2\Omega_H$, only the lower hybrid mode could be unstable with any significant growth rate. Since $\omega_{LH}/\Omega_H \simeq \omega_{pH}/\Omega_H \simeq 42\, \omega_{pe}/\Omega_e (n_H/n_e)^{1/2}$, the hydrogen density and Ω_e/ω_{pe} are important parameters in determining how many hydrogen cyclotron harmonics can be unstable. For $\Omega_e/\omega_{pe} > 21$, only the lower hybrid wave is unstable, regardless of the hydrogen density.

The ion-ion hybrid mode is the heavy ion analog to the lower hybrid mode. Above the ion-ion hybrid frequency, the normal modes occur near the multiples of the oxygen gyrofrequency for all k; below ω_{IH} the modes go from $(n+1)\Omega_o$ to $n\Omega_o$ with increasing k. In contrast to the lower hybrid frequency, however, the ion-ion hybrid frequency is somewhat insensitive to Ω_e/ω_{pe}, but highly dependent on the oxygen to hydrogen density ratio (see equation 4). Less oxygen lowers the ion-ion frequency, thereby suppressing oxygen cyclotron harmonic destabilization. When $n_o/n_H \leq 0.2$, $\omega_{IH} < 2\Omega_o$, and the ion-ion mode will be the primary unstable oxygen mode.

With an electron beam (current) present, the lower hybrid and ion-ion modes can be driven unstable at oblique wave propagation. The ratio of hydrogen to oxygen density concentration, n_H/n_o, as well as the ratio of the electron gyrofrequency to electron plasma frequency (Ω_e/ω_{pe}), plays an important role in determining which instability of the two has the larger growth rate [Schriver and Ashour-Abdalla, 1988]. This can be seen in Figure 1, with the left panel showing the real frequency of both the ion-ion and lower hybrid modes plotted versus the hydrogen to oxygen density ratio (n_H/n_o) for different values of Ω_e/ω_{pe}. Maximum growth rates are plotted for each instability on the right versus n_H/n_o, and in both figures dashed lines are used for the ion-ion hybrid mode and solid lines for the lower hybrid. Here, the propagation angle is $k_\parallel/k = .003$ with an electron beam at $V_d = 6v_{tb}$. The left panel shows that when hydrogen is dominant ($n_H/n_o > 1$), the ion-ion frequency approaches Ω_o, while the lower hybrid mode approaches the hydrogen plasma frequency. As Ω_e/ω_{pe} decreases, the frequency of the lower hybrid mode increases, and one would not expect much interaction with heavy ions. The ion-ion hybrid mode, however, is always confined to the region

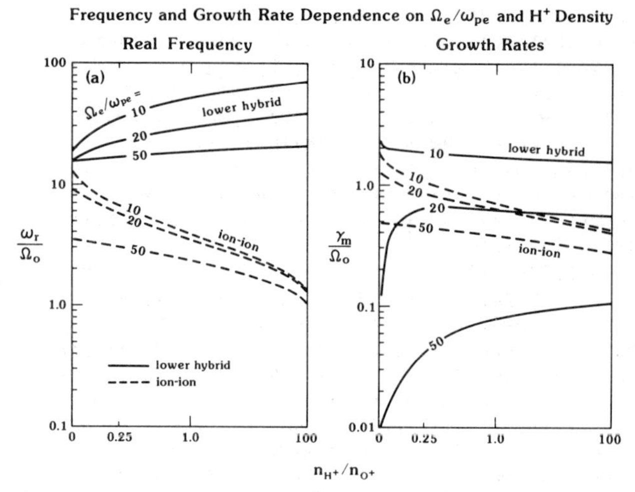

Fig. 1. Real frequency (left panel) and maximum growth rates (right panel) are shown versus the hydrogen to oxygen density ratio for both the ion-ion hybrid (dashed lines) and the lower hybrid (solid lines) instabilities. Both modes are shown for three values of Ω_e/ω_{pe}.

between Ω_H and Ω_o, and thus can interact with oxygen for a wide range of parameters. The growth rates (right panel) of both instabilities decrease with increasing Ω_e/ω_{pe}; for $\Omega_e/\omega_{pe} < 10$ the lower hybrid wave should dominate for all values of hydrogen concentration, but when $\Omega_e/\omega_{pe} > 25$, ion hybrid waves have a faster growth rate than the lower hybrid waves regardless of which ion species is more abundant.

2.2 Nonlinear Effects

In any particular region of the auroral field lines the dominant instability excited depends largely on the total plasma density, relative ion densities and magnetic field strength. Therefore a given auroral region would be expected to allow one instability to be stronger than another. To understand the heating at different points along the field lines we need to know the characteristic frequencies, ω_{pH}, ω_{po}, ω_{pe} and Ω_e, as functions of altitude. Figure 2 shows the characteristic frequencies of altitude in the auroral zone, based on quiet time satellite observations; f_{ce} is the electron gyrofrequency; f_{pe} is the electron plasma frequency; f_{po} is the oxygen plasma frequency; and f_{pH} is the hydrogen plasma frequency. To assist in determining the types of heating to be expected on the auroral field lines as functions of altitude we have defined three zones where we expect one instability to be more dominant than another. The parameters used to determine when a particular instability is dominant are Ω_e/ω_{pe} and n_H/n_o [Schriver and Ashour-Abdalla, 1988]. The characteristic frequencies shown on Figure 2 can vary as a function of magnetic activity, solar activity and seasonal variations, and thus the altitudes used to distinguish between zones are variably dependent on plasma conditions. The heating zones we will define will be determined by f_{ce}/f_{pe} and n_o/n_H, with Figure 2 used as a guide to establish the estimated altitude limits of each region. In zone 1, where the electron density is high ($f_{ce}/f_{pe} < 10$) and where oxygen is the dominant ion (roughly below 2000 kms), we expect the lower hybrid instability to have the largest growth rates. Zone 2 is defined as the region where $n_o \lesssim n_H$ and $f_{ce}/f_{pe} > 10$. This region ranges from about 2000 to 15,000 km and although not shown in Figure 2 includes the density cavities often observed in the nightside auroral zone [Calvert, 1981; Persoon et al., 1988] where $f_{ce}/f_{pe} > 10$ is usually satisfied. Here the ion-ion hybrid instability is dominant. The final zone, 3, lies above 15,000 kms where the magnetic field magnitude decreases, allowing f_{ce}/f_{pe} to fall back below 10; now hydrogen is clearly the dominant ion. The lower hybrid instability should dominate here as in zone 1, but the resultant transverse ion heating will be different from that in zone 1, since $n_H \gg n_o$.

To understand the heating in the different zones, each of the regions will be addressed separately, using linear theory and simulations as guides to distinguish the wave spectrum and ion transverse heating to be expected in each

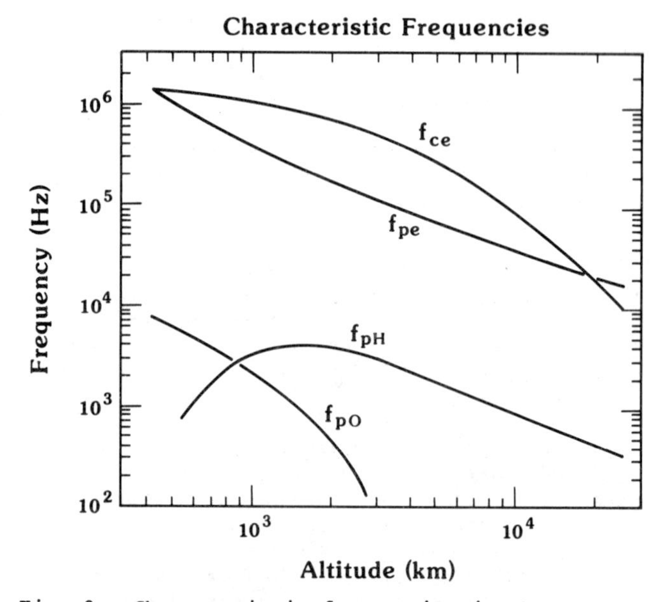

Fig. 2. Characteristic frequencies in the auroral zone as a function of altitude are shown based on quiet time observations. Each is given in Hz; f_{ce} is the electron cyclotron frequency and f_{pe}, f_{pH} and f_{po} and the electron, hydrogen and oxygen plasma frequencies, respectively.

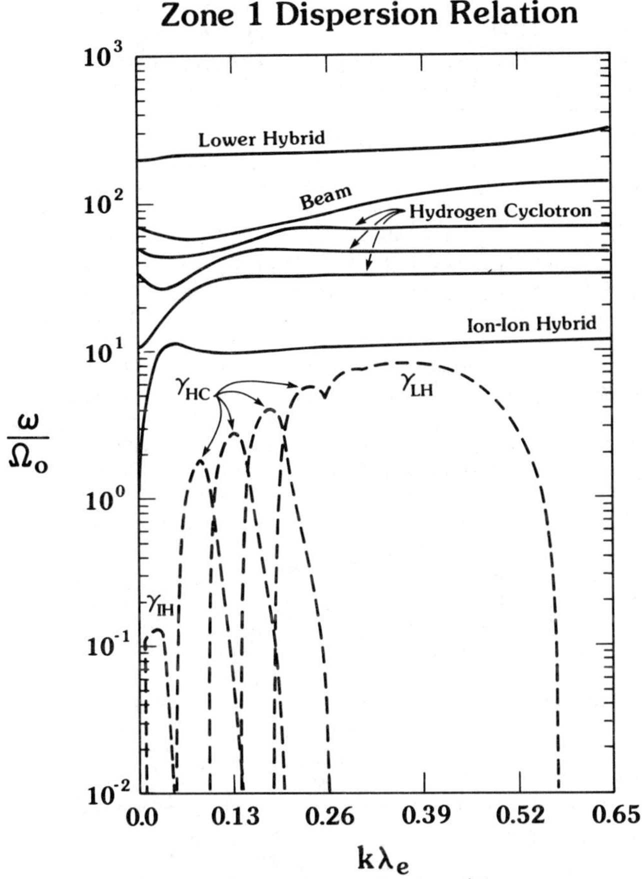

Fig. 3. Linear dispersion relation (frequency versus wavenumber) is shown for the parameters of zone 1 with $\Omega_e = \omega_{pe}$ and $n_o/n_H = 9$ (for other parameters, see text). Real frequency is indicated by solid lines and growth rates by dashed lines. When the beam mode couples with a normal mode of the system, i.e. a hydrogen cyclotron mode, that mode is driven unstable as seen in the figure.

zone as Ω_e/ω_{pe} and n_o/n_H vary. But even though each of the zones are addressed separately, the following examples all assume an $O^+ - H^+$ plasma and use a drifting electron Maxwellian as the free energy source.

Zone 1. Here we consider a region with a high total density and with oxygen dominant such that $\Omega_e/\omega_{pe} < 10$ and $n_o > n_H$. These conditions are generally found at low altitudes, below about 2000 km.

If the electron drift is large enough so that both the lower hybrid wave and ion-ion waves are unstable, then according to linear theory the lower hybrid instability will be the fastest growing wave, with frequency $\omega_{pe}{}^2 = (\omega_{po}{}^2 + \Omega_H{}^2)^{1/2}$ [Schriver and Ashour-Abdalla, 1988]. Since oxygen is the dominant ion, the ion-ion frequency will fall just below the hydrogen gyrofrequency $\omega_{IH} \lesssim \Omega_H$, and although unstable the growth rates will be much lower than those of the lower hybrid instability.

Because the frequency of the unstable modes is large compared to the oxygen gyrofrequency, we would expect the hydrogen to be heated more than oxygen. This is seen to be the case in a number of runs satisfying the parameter conditions for zone 1. We present the results here for a particular run with parameters $n_o = 9n_H$ and $\Omega_{ce} = \omega_{pe}$. The drift velocity, V_d, is 6 times the electron beam thermal velocity, v_{tb}, and the electron beam density, n_b, is one fourth the total electron density. This drift velocity is in line with observations and is above stability threshold for the wave modes discussed here [Klumpar and Heikkila, 1982; Frank, private communication]; other runs with different drifts have been used and the results are qualitatively the same. The warm plasma dispersion relation for these parameters is shown in Figure 3. The solid lines show the real frequency, and the dashed lines show the imaginary frequency versus $k\lambda_e$. The electron beam mode corresponds to $\omega_d \simeq kV \cos\theta_b - \omega_{pb}$, and is the dispersion branch that increases with $k\lambda_e$. When the negative energy beam couples with a natural mode of the system, that mode is driven unstable. The low frequency ion-ion hybrid mode ($\omega \sim 10\Omega_o$) and oxygen cyclotron harmonics below $9\Omega_o$ are unstable at very low k, $k\lambda_e < 0.04$, with growth of the order of $\gamma - .1\Omega_o$. We note that a series of growth rate peaks occur at higher $k\lambda_e$; these are the hydrogen cyclotron waves: each peak corresponds to where the beam mode couples with a cyclotron harmonic, and the largest peak represents the lower hybrid mode being driven unstable.

Now that we have identified the unstable wave modes we examine the resulting nonlinear ion heating. Figure 4 shows the parallel electron velocity distribution (top panel) and the hydrogen and oxygen perpendicular velocity distribution. The initial distribution at $t = 0$ is shown for comparison. The beam electrons at $\Omega_H t = 245$, have diffused out to form a monotonically decreasing function of energy as a result of wave growth. Both ion species show transverse heating, but the hydrogen ions have been bulk heated much more than have the oxgyen ions.

The physics of this heating can be understood if we recall (from linear theory) that the dominant mode is the lower hybrid wave. The lower hybrid phase velocity is low in this parameter regime mainly because oxygen is dominant. Since the phase velocity of the lower hybrid mode is $\omega/k_\perp \simeq .5v_{tH} = 2v_{to}$, this results in bulk heating of the entire hydrogen distribution, but since only higher energy oxygen particles resonantly interact with the waves, much less heating occurs in the oxygen velocity distribution.

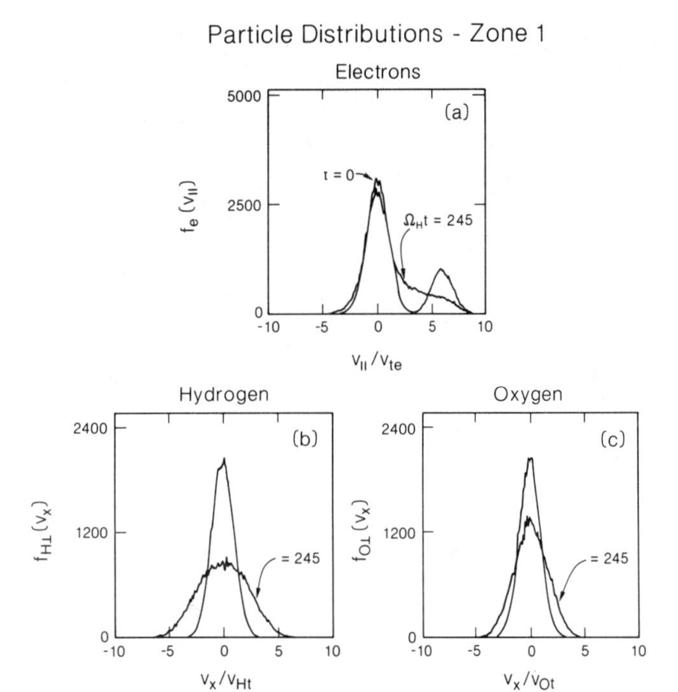

Fig. 4. Simulation results for the electron parallel velocity distribution (panel a), and the hydrogen and oxygen perpendicular velocity distributions (panels b and c, respectively) for zone 1 parameters. Superimposed on each distribution (shown at Ω_{Ht} = 245) is the initial distribution for comparison. For this case hydrogen is heated the most.

We have carried out a number of runs with different initial parameters that satisfy the criteria for zone 1, and the same results were seen. The distinguishing features of zone 1 are that the lower hybrid instability is dominant, hydrogen is bulk heated and oxygen is slightly bulk heated. In some cases, because of a higher wave phase velocity, oxygen formed more of a high energy tail (rather than the slight bulk heating seen in Figure 4), whereby most of the distribution was unaffected, but the higher energy particles in the tail of the distribution were able to interact with the waves and be accelerated creating a non-Maxwellian distribution like that seen in Figure 7c.

Zone 2. Now we turn to zone 2 and allow Ω_e/ω_{pe} to exceed 10 and the hydrogen density to be comparable to or greater than the oxygen density. Linear theory predicts that the ion-ion hybrid instability will be dominant and the frequency will fall about midway between the oxygen and hydrogen gyrofrequencies [Schriver and Ashour-Abdalla, 1988]. Since the frequency is low, the main effect will be oxygen transverse bulk heating; hydrogen will be unaffected.

At altitudes within zone 2, or more precisely between 4,000 and 15,000 kms, Persoon et al. [1988] have shown that in auroral zone density cavities, n_e can plunge below $1cm^{-3}$, and Ω_e/ω_{pe} is usually greater than 20. We have tried to model auroral cavities by looking at two cases. In both cases, Ω_e/ω_{pe} = 50, but in one case the density of hydrogen is equal to the oxygen density, and in the other case hydrogen is the dominant species with $n_H = 9n_O$. Figure 5 shows the phase space and perpendicular velocity distribution for oxygen near the end of the simulation run, at $\Omega_O t$ = 60. The top panels show the simulation results for equal densities of hydrogen and oxygen, and the lower panels show the case where hydrogen is the dominant species. The upper panel phase space indicates strong transverse heating, and this is confirmed in the particle distribution function; however, the velocity distribution in the lower panel shows a much wider thermal spread and a high energy tail. This illustrates the fact that oxygen heating is greater when oxygen is the minority species. While at first somewhat surprising, when taken with the facts that the ion-ion hybrid instability is the dominant wave mode and its frequency decreases with decreasing oxygen density, the result is easily understood.

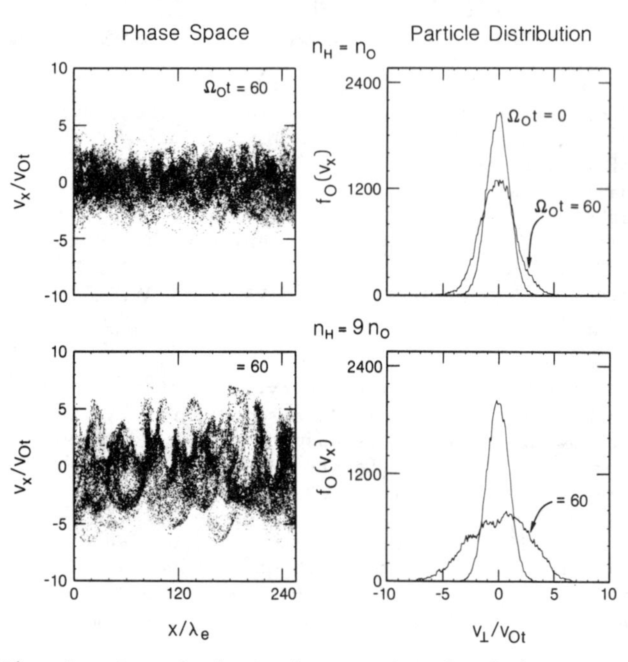

Fig. 5. Zone 2 simulation results for Ω_e/ω_{pe} = 50 and $n_H = n_O$ (upper panels) and $n_H = 9n_O$ (lower panels). Both phase space and perpendicular velocity distribution are shown for each case. When oxygen is the minority species (lower panel), more heating occurs since the ion-ion hybrid instability has a lower frequency.

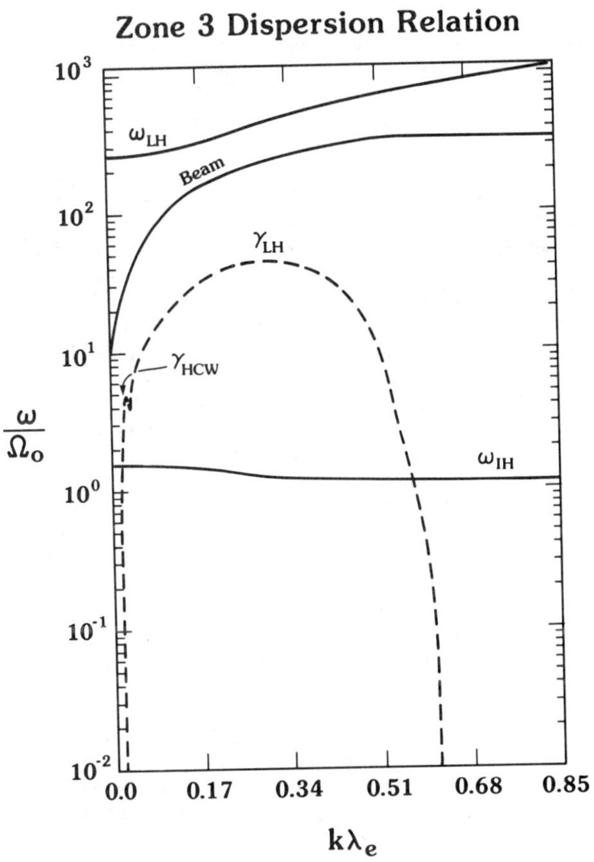

Fig. 6. Dispersion relation for zone 3 parameters with $\Omega_e/\omega_{pe} = 3$ and $n_H = 9n_o$, in the same format as Figure 3. Here, the lower hybrid instability is dominant, but has a higher frequency than in zone 1 (see Figure 3.)

In both simulation runs, the hydrogen species showed no sign of heating, indicating that all free energy is channelled into oxygen heating.

Zone 3. At higher altitudes (zone 3), we move back to our original parameter of $\Omega_e/\omega_{pe} < 10$, but in contrast to zone 1, hydrogen is now the more abundant ion. As in zones 1 and 2, we present results from a simulation with particle parameters which illustrate the types of wave growth and ion heating seen in a number of runs with similar parameters that satisfy the conditions of zone 3.

The case we have chosen to present has $n_H = 9n_o$, $\Omega_e = 3\omega_{pe}$, $V_d = 6v_{tb}$, $n_b = 0.25n_t$ and $k_\parallel/k = 0.0175$. Here we are at angles less oblique than in the previous case, because at higher θ no wave growth occurs in the simulation run.

The dispersion relation for these parameters is shown in Figure 6, following the format of Figure 3. Clearly the lower hybrid mode is most unstable with the hydrogen cyclotron waves accounting for the growth rate humps at lower k. The ion-ion hybrid frequency is very low ($\omega < 2\Omega_o$), and its growth rate is so small in comparison with the lower hybrid that it cannot be seen on this scale.

The simulation results are summarized in Figure 7. The top two panels show the wave data for the fastest growing and most powerful wave mode, mode 3. The frequency and wavenumber manifestly indicate that this mode is the lower hybrid instability. The lower two panels show the hydrogen and oxygen perpendicular velocity distribution rate in the run. The hydrogen distribution shows a high energy tail, but the oxygen is totally unaffected. This is to be expected since the lower hybrid frequency is very high, pushing the phase velocity up to $v_{phase} = 1.5v_{tH} = 6.0v_{to}$. Clearly, the hydrogen high energy particles are able to resonantly absorb energy from the waves, but the oxygen is moving too slowly for any kind of resonant heating.

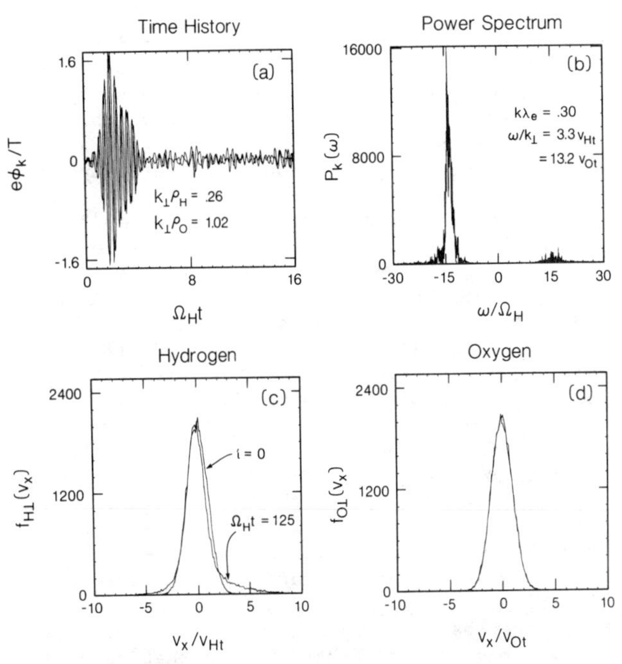

Fig. 7. Zone 3 simulation results showing the fastest growing mode time history (panel a) and power spectrum (panel b). These results agree with linear theory predictions that this mode is the lower hybrid instability. The lower two panels show the effects on the hydrogen (panel c) and oxygen (panel d) distributions; oxygen is not affected at all while hydrogen forms a high energy tail.

In this part of the paper, we have considered the heating of ionospheric ions due to waves driven unstable by field-aligned currents in a multispecies O^+-H^+ plasma. To understand the heating in different zones on auroral field lines we have carried out a series of simulations in different parameter regimes. At low altitudes the electron plasma density is high and oxygen ions constitute the dominant ion species; the lower hybrid instability is dominant resulting mostly in bulk heating of hydrogen although some heating can occur in the oxygen distribution. At mid-altitudes where the electron density drops and the relative hydrogen ion concentration increases, the ion-ion hybrid instability is excited which results in a bulk heating of the oxygen ions and hardly any effect on the hydrogen ions. At higher auroral altitudes, where the relative hydrogen density increases even further and Ω_e/ω_{pe} decreases, the lower hybrid instability is dominant with a higher frequency than at lower altitudes (zone 1), which results in a high energy tail heating of hydrogen and no affect whatsoever on oxygen.

3. Thermal Mixing - Excitation of Broadband Electrostatic Noise in the Plasma Sheet Boundary Layer

Continuing with our theme of the nonlinear development of modes in multicomponent plasmas, we now look at the plasma sheet boundary layer. In this case our starting point is wave observations which is unlike the auroral problem where we knew that ionospheric ions are heated and had little information about the waves.

Intense broadband electrostatic noise (BEN) has been observed in the boundary layer region of the earth's magnetosphere [Scarf et al., 1974; Gurnett et al., 1976]. The observations showed the noise to be present over a broad range of frequencies from about 10 Hz all the way to the electron plasma frequency, near 10 kHz. In association with these waves, ion beams, flowing sunward and/or anti-sunward along magnetic field lines, have been observed [Frank et al., 1976; DeCoster and Frank, 1979; Eastman et al., 1984]. In fact, the most intense waves occur in regions of very large flow velocities ($U \gtrsim 10^3$ km/s). In addition to single ion beams traveling in either the earthward or tailward direction, sometimes two counterstreaming ion beams are observed.

The conventional instabilities that have been suggested to explain the broadband electrostatic waves are the ion acoustic and the ion-ion instabilities [Grabbe and Eastman, 1984; Omidi, 1985; Dusenbery and Lyons, 1985; Akimoto and Omidi, 1986; Ashour-Abdalla and Okuda, 1986a, Schriver and Ashour-Abdala, 1987].

Let us see how this is set up in the plasma sheet boundary layer. We usually consider the situation where cold ionospheric ion beams interact with warm ions and electrons in the plasma sheet. The ions and electrons in the plasma sheet are considered to have an isotropic Maxwellian distribution with $T_e = T_i$ while ionospheric ion beams are drifting with temperatures T_b and drift speed U. Now if $U > C_s$, where C_s is the sound speed, the ion acoustic instability is driven at parallel propagation, which is a resonant type instability resulting from the interaction of the ion beam with the hot electrons. At propagation oblique to the magnetic field, for an angle given roughly by $\cos\theta = C_s/U$, the ion-ion instability is excited, which is a fluid type instability resulting from the net flow between the two ion species [Akimoto and Omidi, 1986]. When $U < C_s$, only the ion-ion instability is excited. These waves can extend in frequency from .001 to .1 ω_{pe} where ω_{pe} is the electron plasma frequency. Results from a simulation study of broadband electrostatic noise [Ashour-Abdalla and Okuda, 1986a] showed that the ion acoustic waves saturated by heating the plasma sheet electrons, whereas the plasma sheet ions and the beam ions were heated by the ion-ion mode. The difficulty with these instabilities though, is that although they result in waves with large amplituded with frequencies near ω_{pe}, they are only unstable when the beam temperature is much less than the plasma sheet electrons and ions. Neither the ion acoustic nor ion-ion mode can grow when the beam temperature becomes comparable to plasma sheet temperatures because of ion Landau damping.

Taking a closer look at the observed beam temperature, we find that the most intense waves occur in regions of very large flow which are characteristic of hot beams created in the tail flowing towards the earth [Grabbe and Eastman, 1984]. The typical temperatures are $T_b \geq 500$eV, and drift velocities as large as 3000 km/s. For these parameters the ion-acoustic and ion-ion mode will not be operative. It would be interesting to find analogous instabilities that would be unstable for the case of the hot ion beams observed in the plasma sheet boundary layer.

Prompted by a suggestion of the presence of a very cold population of electrons at temperatures near 1 eV (Frank, private communication) and observations of cold electrons in the plasma sheet boundary layer by Etcheto and Saint-Marc [1985] with temperatures below 30 eV, we propose a new mechanism for the excitation of broadband electrostatic noise in the plasma sheet boundary layer. In keeping with our theme of mixed plasmas we consider a two temperature electron distribution and show that in the presence of cold electrons, warm ion beams can drive the electron acoustic instability. This is one of the many mixed modes which may be responsible for the high frequency component of broadband electrostatic noise [Ashour-Abdalla and Okuda,

Initial Condition

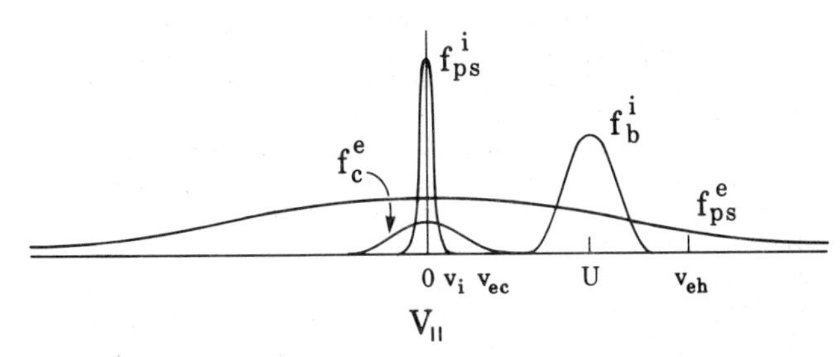

Fig. 8 Schematic model used for the plasma sheet boundary layer which shows velocity distributions for cold ionospheric electrons, warm plasma sheet ions and electrons, and an ion beam.

1986b; Grabbe, 1987; Schriver and Ashour-Abdalla, 1987].

3.1 Linear Theory

Consider an electrostatic oscillation propagating parallel to the ambient magnetic field. Assuming the presence of cold and hot electrons, thermal ions and a warm ion beam drifting along the magnetic field, the dispersion relation can be written as

$$1 = \frac{1}{2k^2\lambda_i^2} Z'\left(\frac{\omega}{\sqrt{2}\,kv_i}\right) + \frac{1}{2k^2\lambda_b^2} Z'\left(\frac{\omega - kU}{\sqrt{2}\,kv_b}\right) + \frac{1}{2k^2\lambda_h^2} Z'\left(\frac{\omega}{\sqrt{2}kv_h}\right) + \frac{1}{2k^2\lambda_c^2} Z'\left(\frac{\omega}{\sqrt{2}\,kv_c}\right) \quad (6)$$

where the subscripts h, c, i and b represent hot electrons, cold electrons, thermal ions and beam ions, respectively; Z' is the derivative of the plasma dispersion function [Fried and Conte, 1961]. All species can be approximated as being unmagnetized because of the low magnetic field found in the plasma sheet boundary layer [Omidi, 1985; Schriver and Ashour-Abdalla, 1987]. Both the ions and electrons at the plasma sheet are assumed to be isotropic Maxwellian distributions with temperatures T_h and T_i respectively. The ion beam is assumed to be a drifting Maxwellian with temperature T_b and drift speed U, whereas the cold electrons are modelled as an isotropic Maxwellian with temperature T_c. A schematic of the distribution functions is shown in Figure 8.

It is well known that an electron acoustic wave propagates at phase speed satisfying $v_c < \omega/k < v_h$, resulting in little damping even in the absence of ion beams [Watanabe and Taniuti, 1977]. In the presence of beam ions, an electron acoustic mode may be destabilized if the electron Landau damping is exceeded by inverse ion Landau damping from the beam ions.

Assuming $v_c < \omega/k < v_h$, $\omega/kv_i \gg 1$ and $(\omega - ku)/kvb < 1$, and writing $\omega = \omega_r + i\gamma$, we find that

$$\omega_r = \frac{\omega_{pc}}{(1/k^2\lambda_b^2 + 1/k^2\lambda_h^2)^{1/2}} = \frac{kv_h\sqrt{\dfrac{n_c}{n}}}{\left(1 + \dfrac{T_h}{T_b}\dfrac{nb}{nh}\right)^{1/2}} \quad (7)$$

and

$$\frac{\gamma}{\omega_r} = \frac{1}{2}\frac{\omega_r^2}{\omega_{pc}^2}\left(\frac{\pi}{2}\right)^{1/2} \cdot \left[\frac{kU}{k^2\lambda_b^2 kv_b} - \frac{\omega_r}{k^2\lambda_b^2}\left(\frac{1}{kv_b} + \frac{\lambda_b^2}{\lambda_h^2}\frac{1}{kv_h}\right)\right] \quad (8)$$

Equation (7) represents the electron acoustic wave in the presence of ion beams. Equation (8) indicates that the electron acoustic wave becomes unstable when inverse Landau damping of the beam ion exceeds the hot electron Landau damping. Note that we neglected the cold electron Landau damping as well as the thermal ion Landau damping, both of which remain small for small values of k.

3.2 Simulation Results

Numerical simulations have been carried out using a one-dimensional electrostatic particle model in the presence of a uniform external magnetic field. Four different components of the particles are considered: hot and cold electrons, thermal and beam ions. The simulation parameters are mass ratio, $m_i/m_e = 400$ (which

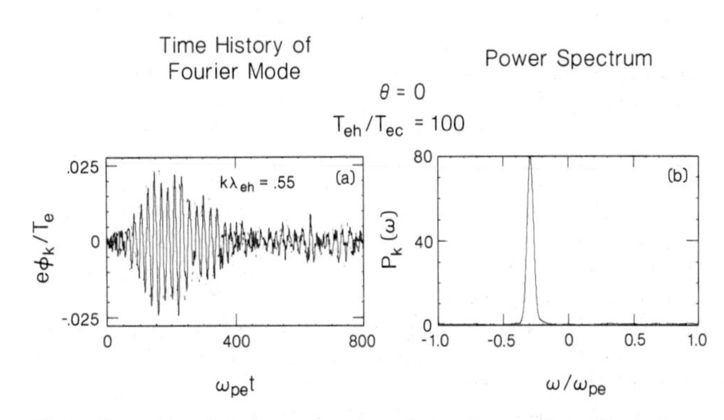

Fig. 9. Simulation results based on the distributions of Figure 8 (see text for exact parameters). Shown here are the time history and power spectrum of the fastest growing mode which corresponds with linear theory as being the electron acoustic instability.

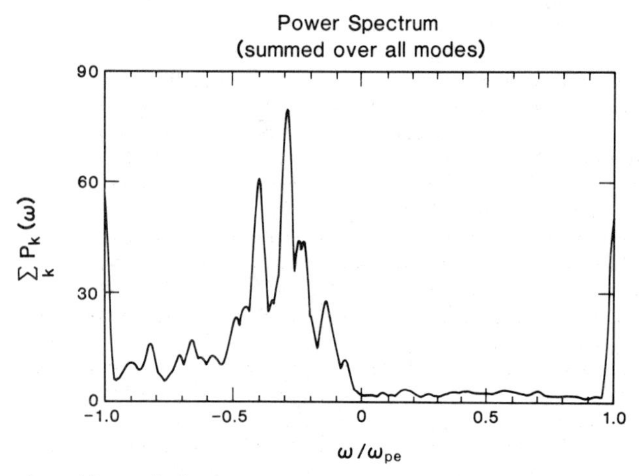

Fig. 10. Similar to the previous figure except here the power spectrum is summed over all k modes. Taking into account all Fourier modes illustrates the broadbanded nature of the instability.

is reduced for simulation tractability since the full dynamics of both ions and electrons must be followed), hot to cold electron temperature ratio, $T_h/T_c = 100$, hot electron to ion temperature ratio, $T_h/T_i = 1$ and hot electron to ion beam temperature ratio, $T_h/T_b = 1$. The choice of $T_h = T_i = T_b$ guarantees that the ion acoustic wave will be damped. The system length was chosen as $L = 256\Delta$, where Δ is the unit grid scaled to $\Delta = 1/8\ \lambda_h$. Other parameters used are $n_h = 0.8n_o$, $n_c = 0.2n_o$, $n_i = n_b$ and $U/v_b = 12.5$. We assume $\vec{k}_\parallel \vec{B}_o$. Figure 9 shows the time history, $e\phi_k/T_e$, and the frequency spectrum of the most unstable mode in the system, mode 9, satisfying $k\lambda_h = 0.55$. The exponentially growing waves saturate at $e\phi_k/T_e \sim 0.02$ near $\omega_{pe}t = 150$ and then decreases in amplitude thereafter. The corresponding power spectrum clearly indicates the presence of a peak around $\omega/\omega_{pe} \simeq 0.3$.

Now this was the power spectrum for an individual Fourier mode but to be able to compare the simulation results with satellite observations we must sum up all the wavelength modes of the simulation data. Figure 10 shows the result of this summing up; one aspect is a much broader distribution than in Figure 9. This is reminiscent of the broadband electrostatic waves observed on ISEE-1, and extends all the way to the cold electron plasma frequency. It should be noted that while the amplitude of the electrostatic potential is smaller than that for the ion-acoustic waves [Ashour-Abdalla and Okuda, 1986a], the fastest growing modes for the electron acoustic waves occur at large wavenumbers resulting in an electric field, $E \sim$ mV/m, in agreement with satellite observations.

What are the nonlinear saturation mechanisms? Figure 11 shows the phase space plots for the cold electrons at different times. This figure shows that the cold electrons are trapped first by the unstable waves which are forming holes in phase space at $\omega_{pe}t = 200$. The deterioration of this coherent structure seen at $\omega_{pe}t = 400$ results in the strong heating of cold electrons.

Figure 12 confirms this statement. In the upper panel we plot (a) the electric field energy, (b) the hot electron temperature, (c) the cold electron temperature and (d) the cold electron velocity distribution function. It is interesting to note that the hot electron temperature hardly changes, whereas the cold electron temperature increases significantly. This can also be seen in panel (d). In fact the hot electrons are heated during the unstable phase as shown for $\omega_{pe}t \lesssim 300$, but they lose energy afterwards. It is clear that during the unstable phase both cold and hot electrons are heated as they absorb energy from the drifting ion beam. When the instability saturates, cold electrons continue to be heated albeit, at a slower rate, whereas the hot electrons lose energy. This heat loss is caused partly by collisional coupling between hot and cold electrons, which transfers hot electon energy to

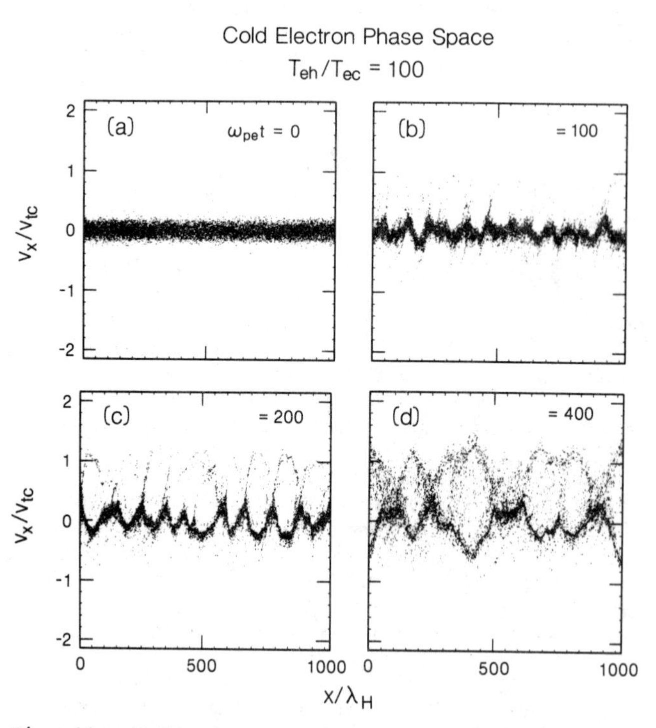

Fig. 11. Cold electron phase space evolution forfor the same simulation case discussed in the previous two figures. Electron phase trapping occurs in panel c, and as the wave amplitude decreases after saturation, the electrons begin to relax to a heated thermal distribution (panel d).

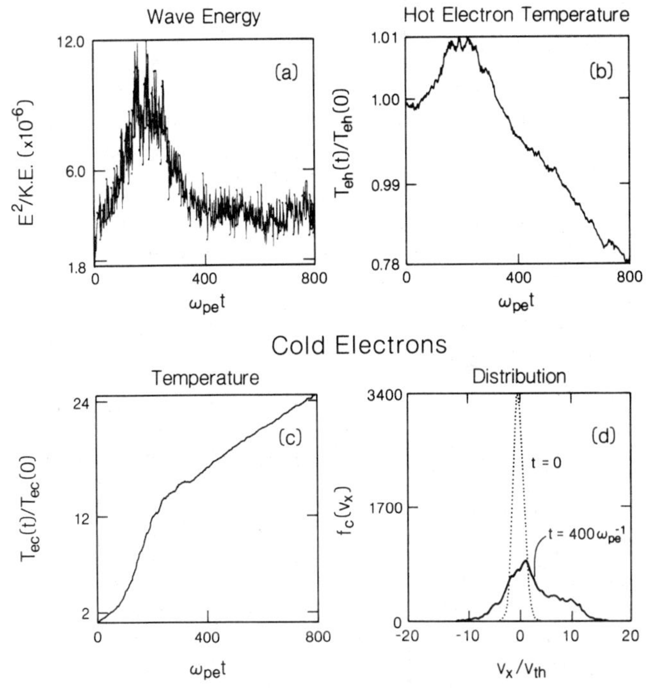

Fig. 12. Time histories of wave energy (panel a), hot electrons (panel b), and cold electrons (panel c) are shown for the simulation run described in the last three figures as well as the cold electron velocity distribution (panel d). After wave saturation, the cold electrons continue to heat while the hot electrons lose energy (see text for explanation).

cold electrons. However the energy gained by the cold electrons when $\omega_{pe}t \geq 300$ is more than the energy loss from the hot electrons. Obviously the additional energy gained by the cold electrons is taken from ion kinetic energy which is transferred in the form of marginally unstable electron acoustic waves. The linear theory and simulation results clearly show how the addition of cold electrons significantly enhances wave growth by a warm ion beam. Without cold electrons, no wave growth is possible. With cold electrons, a broadbanded, intense, wave spectrum is excited which in turn heats the cold electrons. The heating of the cold plasma may be indicative of the process by which ionospheric plasma gains energy to ultimately become part of the hot plasma sheet itself. The importance of thermal mixing in the earth's magnetotail is seen clearly in the results of this section and cannot be neglected in understanding energy transport in the plasma sheet.

4. Conclusions

We have presented two examples of how the effects of multicomponent plasmas differ significantly from those of a basic two-component ion-electron plasma. In both cases, magnetospheric plasma interacts with ionospheric plasma and by including all plasma species, the physics is enriched by allowing new wave modes to be unstable that in turn have different nonlinear effects from those of a two-component plasma.

Comparisons of theory with wave and particle measurements show good agreement in the plasma sheet boundary layer [Gurnett et al., 1976; Eastman et al., 1984]. In the auroral zone it is more difficult to compare theory and observations because wave data has been inconclusive in determining the existence of low frequency waves (below Ω_H) because of Doppler shifting effects [Kintner, 1986]; also, as far as we know there is no study done that correlates the presence of ion (hydrogen and oxygen) conics with Ω_e/ω_{pe}. Nevertheless, there are some wave observations at mid-altitudes (zone 2) that show waves with frequency less than Ω_H [Temerin and Lysak, 1984], which may be indicative of the ion-ion hybrid mode. As well, Gorney et al. [1981] have shown that ion conics are usually found above 2000 km. This is in general agreement with the theory of section 2. The main results are summarized as follows:

I. Species Mixing - Auroral Zone

Includes a heavy ion, oxygen, in a hydrogen-electron plasma driven unstable by an electron current.

1) Allows an electrostatic ion-ion hybrid mode, present only when two ions are included, to be driven unstable at low frequencies (i.e., $\Omega_o < \omega < \Omega_H$).
2) The ion-ion hybrid instability is dominant when $\Omega_e/\omega_{pe} > 10$, while the lower hybrid instability is more important for $\Omega_e/\omega_{pe} < 10$.
3) Using Ω_e/ω_{pe} and n_o/n_H variation with altitude in the auroral zone, regions of preferred ion heating, based on the dominant instability, can be determined as follows:
   i) Low altitudes ($\lesssim$ 2000 km): transverse hydrogen bulk heating and oxygen high energy tail formation by lower hybrid instability ($\omega \gtrsim \Omega_H$).
   ii) Mid-altitudes (< 15000 km and > 2000 km): transverse oxygen bulk heating by ion-ion hybrid instability ($\Omega_o < \omega_{IH} < \Omega_H$).
   iii) High altitudes (> 15000 km): hydrogen high energy tail created by lower hybrid instability ($\omega_{LH} \gg \Omega_H$).

II. Thermal Mixing - Plasma Sheet Boundary Layer

Includes cold ionospheric electrons in a hot (thermal) ion-electron plasma driven unstable by an ion beam.

1) Allows the electron acoustic and beam resonant modes, present only when cold electrons are included, to be driven unstable by a warm ion beam.
2) A broadbanded, high frequency wave spectrum is formed in good agreement with BEN observations.
3) Electron acoustic (and beam resonant) wave saturation is due to cold electron heating which brings ionospheric plasma to plasma sheet temperatures.

These results show that the effects of multicomponent plasmas are very significant indeed, with species mixing in the auroral zone helping to account for oxygen transverse acceleration and thermal mixing in the plasma sheet boundary layer explaining not only the high frequency portion of BEN, but how ionospheric electrons are heated in the plasma sheet boundary layer.

In order to completely understand magnetospheric solar wind interactions and energy transport in space plasmas, it is becoming more and more apparent that the consequences of multicomponent plasmas are extremely important and cannot be neglected. Future space plasma research must take into account species and thermal mixing of plasmas in order to explain the complex wave and particle behavior observed by satellites.

Acknowledgements. We wish to thank R.J. Walker for discussions and useful comments on this paper. This work has been supported by NASA STTP Grant NAGW-78, U.S. Air Force Contract F19628-85-K-0027 and National Science Foundation Grants ATM 85-03434, ATM 85-13215 and ATM 85-12512. Computing resources were provided by San Diego Supercomputer Center.

References

Akimoto, K. and N. Omidi, The generation of broadband electrostatic noise by an ion beam in the magnetotail, Geophys. Res. Lett., 13, 97, 1986.

Ashour-Abdalla, M. and H. Okuda, Turbulent heating of heavy ions on auroral field lines, J. Geophys. Res., 89, 2235, 1984.

Ashour-Abdalla, M. and H. Okuda, Theory and simulations of broadband electrostatic noise in the geomagnetic tail, J. Geophys. Res., 91, 6833, 1986a.

Ashour-Abdalla, M. and H. Okuda, Electron acoustic instabilities in the geomagnetic tail, Geophys. Res. Lett., 13, 366, 1986b.

Ashour-Abdalla, M., H. Okuda and S.Y. Kim, Transverse ion heating in multicomponent plasmas, Geophys. Res. Lett., 14, 375, 1987.

Bernstein, I.B., Waves in a plasma in a magnetic field, Phys. Rev., 109, 10, 1958.

Buchsbaum, S.J., Resonance in a plasma with two ion species, Phys. Fluids, 3, 418, 1960.

Calvert, W., The auroral plasma cavity, Geophys. Res. Lett., 8, 919, 1981.

Chang, T.S. and B. Coppi, Lower hybrid acceleration and ion evolution in the supra-auroral region, Geophys. Res. Lett., 8, 1253, 1981.

Chappell, C.R., T.E. Moore and J.H. Waite, The ionosphere as a fully adequate source of plasma for the Earth's magnetosphere, J. Geophys. Res., 92, 5896, 1987.

DeCoster, R.J. and L.A. Frank, Observations pertaining to the dynamics of the plasma sheet, J. Geophys. Res., 84, 5099, 1979.

Dusenbery, P.B. and L.R. Lyons, Generation of ion-conic distribution by upgoing ionospheric electrons, J. Geophys. Res., 86, 7627, 1981.

Dusenbery, P.B. and L.R. Lyons, The generation of electrostatic noise in the plasma sheet boundary layer, J. Geophys. Res., 90, 10935, 1985.

Eastman, T.E., L.A. Frank and W.K. Peterson, The plasma sheet boundary layer, J. Geophys. Res., 89, 1553, 1984.

Etcheto, J. and A. Saint-Marc, Anomalously high plasma densities in the plasma sheet boundary layer, J. Geophys. Res., 90, 5338, 1985.

Frank, L.A., K.L. Ackerson and R.P. Lepping, On hot tenuous plasmas, fireballs and boundary layers in the earth's magnetotail, J. Geophys. Res., 81, 5859, 1976.

Fried, B. and S. Conte, The Plasma Dispersion Function, Academic Press, New York, 1961.

Gorney, D.J., A. Clarke, D. Crowley, J. Fennel, J. Luhmann and P. Mizera, The distribution of ion beams and conics below 8000 km, J. Geophys. Res., 86, 83, 1981.

Grabbe, C.L., Numerical study of the spectrum of broadband electrostatic noise in the magnetotail, J. Geophys. Res., 92, 1185, 1987.

Grabbe, C.L. and T.E. Eastman, Generation of broadband electrostatic waves in the magnetotail, J. Geophys. Res., 89, 3865, 1984.

Gurnett, D.A., L.A. Frank and R.P. Lepping, Plasma waves in the distant magnetotail, J. Geophys. Res., 81, 6059, 1976.

Horwitz, J.L., Core plasma in the magnetosphere, Rev. Geophys., 25, 579, 1987.

Kindel, J.M. and C.F. Kennel, Topside current instabilities, J. Geophys. Res., 76, 3055, 1971.

Kintner, P.M., Experimental identification of electrostatic plasma waves with ion conic acceleration regions, Ion Acceleration in the Magnetosphere and Ionosphere, edited by T. Chang, Geophysical Monograph 38, Washington, 1986.

Klumpar, D.M. and W.J. Heikkila, Electrons in the ionospheric source cone: evidence for runaway electrons as carriers of downward Birkeland currents, Geophys. Res. Lett., 9, 873, 1982.

Lysak, R.L., M.K. Hudson and M. Temerin, Ion heating by strong electrostatic ion cyclotron turbulence, J. Geophys. Res., 85, 678, 1980.

Okuda, H. and M. Ashour-Abdalla, Acceleration of hydrogen ions and conic formation along auroral field lines, J. Geophys. Res., 88, 899 1983.

Omidi, N., Broadband electrostatic noise produced by ion beams in the earth's magnetotail, J. Geophys. Res., 90, 12330, 1985.

Palmadesso, P.J., T.P. Coffey, S.L. Ossakow and K. Papadopoulos, Topside ionosphere ion heating due to electrostatic ion cyclotron turbulence, Geophys. Res. Lett., 1, 105, 1974.

Persoon, A.M., D.A. Gurnett, W.K. Peterson, J.H. Waite, Jr., J.L. Burch and J.L. Green, Electron density depletions in the nightside auroral zone, J. Geophys. Res., 93, 1871, 1988.

Scarf, F., L.A. Frank, K.L. Ackerson and R.P. Lepping, Plasma wave turbulence at distant crossings of the plasma sheet boundaries and neutral sheets, Geophys. Res. Lett., 1, 189, 1974.

Schield, M.A. and L.A. Frank, Electron observations between the inner edge of the plasma sheet and the plasmapause, J. Geophys. Res., 75, 5401, 1970.

Schriver, D. and M. Ashour-Abdalla, Generation of high-frequency broadband electrostatic noise: the role of cold electrons, J. Geophys. Res., 92, 5807, 1987.

Schriver, D. and M. Ashour-Abdalla, Linear instabilities in multicomponent plasmas and their consequences on the auroral zone, J. Geophys. Res., 93, 2633, 1988.

Shelley, E.G., R.G. Johnson and R.D. Sharp, Satellite observations of energetic heavy ions during a geomagnetic storm, J. Geophys. Res., 77, 6104, 1972.

Singh, N. and R.W. Schunk, Numerical calculations relevant to the initial expansion of the polar wind, J. Geophys. Res., 87, 9154, 1982.

Stix, T.H., The Theory of Plasma Waves, McGraw-Hill Co., New York, pg. 44, 1962.

Temerin, M. and R.L. Lysak, Electromagnetic ion cyclotron mode (ELF) waves generated by auroral electron precipitation, J. Geophys. Res., 89, 2849, 1984.

Ungstrup, E., D.M. Klumpar and W.J. Heikkila, Heating of ions to superthermal energies in the topside ionosphere by electrostatic ion cyclotron waves, J. Geophys. Res., 84, 4289, 1979.

Watanabe, K. and T. Taniuti, Electron acoustic mode in a plasma of two temperature electrons, J. Phys. Soc. Japan, 43, 1819, 1977.

QUASINEUTRAL BEAM PROPAGATION IN SPACE

K. Papadopoulos,* A. Mankofsky and A. Drobot

Science Applications International Corporation, 1710 Goodridge Drive, McLean, Virginia 22102

Abstract. The propagation of dense energetic neutralized ion beams or plasmoids injected into a plasma across the ambient magnetic field under ionospheric conditions is considered. Using a simple physical model supported by two-dimensional hybrid simulations, it is shown that thin dense beams can propagate ballistically over lengths many times their gyroradius. This occurs when the following conditions are satisfied: $\Delta/R_o \ll 1$, where Δ is the cross field beam width and R_o the gyroradius of an ambient ion having a velocity equal to the beam velocity u_b, and $n_b M_b/n_o M_o \gg 1$, where $n_b(M_b)$, $n_o(M_o)$ are the density (mass) of beam and ambient ions. The scaling of the ballistic propagation lengths and times with beam and ambient parameters is presented along with comments on the applicability of the model to space and astrophysics.

Introduction

The propagation of high speed neutralized ion beams, often called plasmoids, across a magnetic field is among the oldest of problems in plasma physics. It first arose in investigations of the origin of magnetospheric storms and substorms [Chapman and Ferraro, 1931; Ferraro, 1952]. Despite the long history of investigation, a clear model has yet to emerge. Early theoretical models established by Chapman and Ferraro [1931], Ferraro [1952], Tuck [1959] and Chapman [1960], indicated that a neutralized beam with a large width Δ transverse to the ambient magnetic field $\mathbf{B}_o(\Delta \gg R_b$, where R_b is the gyroradius of the beam ions), will in general compress the magnetic field but will not propagate significantly. Propagation can potentially occur in the diamagnetic regime when $\beta_b \equiv 4\pi n_b M_b u_b^2/B_o^2 \gg 1$, where n_b, M_b, u_b are the density, mass and cross field velocity of the beam ions. This propagation mode can be properly described by MHD and is equivalent to the propagation of a solid conductor moving across $\mathbf{B}_o$. During propagation the beam picks up and carries along the ambient plasma and magnetic field, in a fashion similar to the pick-up of cometary ions by the solar wind [Cargill et al., 1988]. The mass loading, along with the various pick-up ring instabilities, soon destroys the beam coherence. Another propagation mode proposed by Schmidt [1960] addressed the narrow beam regime, $R_e < \Delta \ll R_b$ where R_e is the electron gyroradius. This propagation mode occurs in the non-diamagnetic $\beta_b \ll 1$ regime (often called the electrostatic regime). In this case the flow energy is not sufficient to alter the magnetic field configuration; therefore, the ambient magnetic field controls the electron and ion dynamics. A polarization electric field develops by the differential motion of the magnetized electrons ($R_e \ll \Delta$) and the unmagnetized ions ($R_b > \Delta$). The polarization field $\mathbf{E}$, coupled with the ambient field $\mathbf{B}_o$, allows the neutralized beam to move by an $\mathbf{E} \times \mathbf{B}_o$ drift [Schmidt, 1960]. This mode of propagation was experimentally observed by Baker and Hammel [1965]. Peter and Rostoker [1982] noted that dielectric shielding due to the presence of an ambient plasma does not affect the beam propagation as long as $V_{Ap}/V_{Ab} < 1$, where V_{Ab}, V_{Ap} are the beam and plasma Alfvén speeds. Scholer's [1970] model of artificial propagation of ion clouds in the magnetosphere belongs to this class of low kinetic β, subalfvénic propagation modes.

In this paper we examine plasmoid propagation in the narrow beam regime discussed by Schmidt [1960] but for the high $\beta_b(\beta_b \gg 1)$ strongly diamagnetic case. It will be shown first by a simple analytic model, and then by a set of 2-D computer simulations using a hybrid code [Mankofsky et al., 1987], that extremely long range propagation across the magnetic field is possible in the strongly diamagnetic $\beta_b \gg 1$ regime if $n_b \gg n_o$ and $R_e \ll \Delta < u_b/\Omega_o \equiv R_o$, where Ω_o is the cyclotron frequency of the ambient plasma ions. The propagation physics and the

* Permanent Address: Department of Physics, University of Maryland, College Park, Maryland 20742.

self-similar stationary field configurations are completely different from the high β_b MHD regime. Preliminary results on this propagation mode appeared recently [Papadopoulos et al., 1988].

In the MHD regime where the beam width is much larger than the beam ion gyroradius, R_b, we expect the background field to be excluded by ion diamagnetic effects. As a consequence the background plasma and magnetic field will be swept by the beam resulting in strong coupling at the beam front. In the low β_b situation described by Schmidt where the beam propagates because of the presence of a polarization electric field the interpenetration of the beam by the background plasma is also likely to lead to strong coupling and hence poor propagation. However, for a high β ion beam which is narrower than its ion gyroradius, the possibility exists for diverting the background plasma and magnetic field asymmetrically around the beam with minimum coupling or beam-background interpenetration. In this situation the high β_b ions have sufficient energy to displace the background plasma at the beam head. Intuitively we expect the beam front to erode at a small rate while the beam body remains undisturbed.

Propagation Model - Physical Description

Consider a cold, dense ($n_b \gg n_o$), high kinetic beta ($\beta_b \gg 1$) ion beam interacting with an ambient magnetoplasma such as shown in Fig. 1. It is convenient to perform our analysis in the beam reference frame. In this frame for $\mathbf{B}_o = B_o\hat{e}_y$, the ambient magnetized plasma flows with a cross field velocity $\mathbf{u}_o = -u_b\hat{e}_z$, with the aid of a motional electric field $\mathbf{E} = -(\mathbf{u}_e \times \mathbf{B}_o)/c = -\hat{e}_x(u_bB_o)/c$, where $\mathbf{u}_e$ is the fluid velocity of the plasma electrons and of the magnetic flux. The equations of motion of the background plasma ions (charge e, mass M_o) are

$$\frac{du_x}{dt} = \frac{e}{M_o}\left[E_x - \frac{u_zB_y}{c}\right] , \tag{1a}$$

$$\frac{du_z}{dt} = \frac{e}{M_o}\left[E_z + \frac{u_zB_y}{c}\right] . \tag{1b}$$

The value of the motional electric field $E_x(z)$ is given by

$$\mathbf{E}(z) = -\frac{\mathbf{u}_e(z) \times \hat{e}_yB(z)}{c} = \hat{e}_x\frac{u_e(z)B(z)}{c} . \tag{2}$$

In the region $z > 0$ ahead of the beam-plasma interface (Fig. 1a), $u_e(z) = -u_b$ and $u_z = -u_b$ so that the r.h.s. of (1a) is zero. Namely, the ions, the electrons and the flux follow straight ballistic orbits. At the plasma interface,

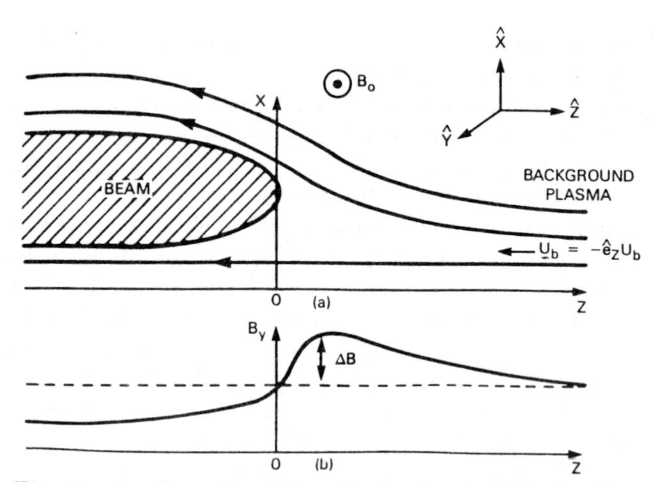

Fig. 1. A schematic of the physics involved during the interaction of the ion beam with the background plasma.

$z \approx 0$, the fluid velocity u_e reduces to

$$u_e(z) = \frac{n_ou_b}{n_o + n_b(z)} \tag{3}$$

to maintain charge and current neutrality. This is accompanied by a diamagnetic current at the front and a field compression (Fig. 1b) such that $B(z)/B_o \approx 1+n_b(z)/n_o$. From Eq. (1a) the reduction in $u_e(z)$ produces a net force $(e\Delta B(z)/M_oc)u_b$ in the positive x-direction which diverts the plasma ions upwards (Fig. 1a). In a high β flow the electrons and the flux follow the path of the ions. For $\Delta \ll R_o$ a change of the plasma frame speed $u_e(z)$ of the order of $\Delta u_e/u_b \approx \Delta/R_o$ is sufficient to establish a quasistationary state in which the background magnetoplasma is diverted to one side of the beam. In the laboratory frame the above interaction corresponds to the beam front diverting the ambient plasma and magnetic field sideways and propagating freely. The beam front suffers an erosion due to the energy required to divert the plasma sideways. The resulting loss is, however, minimal for $\beta_b \gg 1$, allowing for long range beam propagation. An important ingredient of this propagation mode is the highly asymmetric steady state. This is contrary to the one expected by fluid or MHD models. It is dominated by kinetic ion effects. The validity of MHD models can only be expected for scale lengths, Δ, larger than the beam or background ion gyroradius. Furthermore, the MHD equations which neglect ion inertial effects yield only symmetric solutions. A fluid treatment which includes ion inertia will allow asymmetric solutions but will miss the highly nonlinear and essentially kinetic ion beam behavior at the beam front. Notice that in the electrostatic (or nondiamagnetic) case [Schmidt, 1960] where $\Delta B \cong 0$ and $\Delta u_e = 0$ flow diversion does not occur and the beam and ambient plasmas will interpenetrate. Interpenetration can cause instabilities that destroy the coherence of the propagation mode. Such beam dispersal was observed in the laboratory experiments of Birko and Kirchenko [1978] in which an ion acoustic wave was driven unstable on the surface of a low-energy beam in the $\beta_b < 1$ regime, apparently due to the relative drift between heavy and light ions. Similar dispersal was also noted in 2-D particle simulations of cross field ion beam propagation [Shanahan, 1984]. In this case, the wave excitation was attributed to the Buneman instability. The excited surface wave penetrated throughout the bulk of the plasma, was accompanied by strong electron heating and resulted in dispersal of the beam.

In the next section we describe representative results from a series of computer simulations which elucidate the physical description presented above, and help clarify the resulting scaling laws. Of utmost importance in the study is the time over which the beam propagates ballistically. The definition of ballistic propagation depends on the particular application or measurement resolution. For example, if one is concerned with absence of magnetic effects, ballistic propagation will occur over a time scale τ such that the lateral beam displacement $\Delta x_b \ll R_b$. For applications concerned with delivering the beam energy flux at a detector located on a ballistic path at a distance $z_o = u_b t$, the criterion will be $\Delta x_b/\Delta \ll 1$ at $t = z_o/u_b$ (Δ is the lateral beam width). These issues are discussed in Section 5.

Representative Simulation Results

Since the problem is inhomogeneous, two-dimensional, and requires kinetic ion treatment, analytic progress past minor improvements on the above simple and intuitive picture is not possible. We therefore use numerical simulations. A two-dimensional hybrid simulation code is the appropriate tool. The details of the code applied to the problem can be found in Mankofsky et al. [1987]. Briefly, the ions are treated as discrete particles, using standard particle-in-cell techniques to follow their motion in the electromagnetic (EM) fields. Summing over the particles provides the ion charge and current density. The electrons are treated as a massless fluid, described via the momentum and energy equations. These equations, along with the ion equations of motion, are solved self-consistently on a uniform two-dimensional grid for the ion velocity vectors, the EM fields, and the electron pressure, in the nonradiative limit (i.e., Darwin Hamiltonian).

The equations solved in the simulation are:

$$\frac{1}{c}\frac{\partial}{\partial t}\mathbf{B} = -\nabla \times \mathbf{E}$$

and

$$\nabla \times \mathbf{B} = \frac{4\pi}{c}(\mathbf{J}_e + \mathbf{J}_i)$$

The first equation, Faraday's Law, is used to determine the change in magnetic field by inductive electric fields. In the second equation, Ampère's Law, the displacement current has been neglected implying the condition $\nabla \cdot (\mathbf{J}_e + \mathbf{J}_i) = 0$. This in conjunction with quasineutrality, $n_e \simeq n_i$, permits us to determine the velocity, $\mathbf{u}_e$, of the electron fluid as a function of magnetic field and ion current density. Further neglecting electron inertia, but retaining electron pressure and collisionality the electric field is found from force balance for the electrons using a generalized Ohm's Law,

$$\mathbf{E} = -\frac{\mathbf{u}_e}{c} \times \mathbf{B} - \frac{\nabla P_e}{n_i \mid e \mid} - \frac{m_e}{\mid e \mid}\sum_s \nu_{es}(\mathbf{u}_{es} - \mathbf{u}_{es}) \ .$$

P_e is the electron pressure determined from an appropriate electron energy equation, and ν_{es} is the effective collision frequency with specie "s". The collision frequency can be based on classical Coulomb interaction or used to account for nonlinear plasma coupling through the use of an anomalous prescription. The electric field contains both an electrostatic as well as an inductive component. This model can account for an extremely large range of physical phenomena. It includes ambipolar expansion, magnetic field convection, and magnetic field diffusion. The model resolves Alfv́en waves, whistlers, and kinetic ion effects, and because it is explicit leads to the limitation $\Delta t V^* < \Delta x$, where Δt is the timestep, Δx the smallest cell size, while V^* is the maximum of the Alfv́en and whistler wave phase velocities or the fastest ion velocity in a problem. The omission of electron inertia does not allow a proper treatment of the electron skin depth so we limit our results to scale lengths much greater than c/ω_{pe}.

The system used in the ion beam simulations has periodic boundary conditions in the direction transverse to the flow (i.e., x-axis). In the flow direction (z-axis) plasma is injected from the right boundary at a rate $n_o u_b$ and permitted to leave on the left at the local flux rate. The magnetic field is allowed to float at these boundaries. We describe below simulation results designed to illustrate quantitatively aspects of the high β propagation model.

In the simulations presented here, the ambient plasma was composed of O^+ with density $n_o = 10^5\#/\text{cm}^3$ and temperature 0.25 eV, and was embedded in a magnetic field $B_o = 0.3$ Gauss. These are parameters typical of the ionospheric F-region. The beam was composed of protons and was given a Gaussian profile in the x and z directions. Beam velocities $u_b = 10^8$, 2×10^8 and 4×10^8 cm/sec were studied, while the total number of beam particles varied

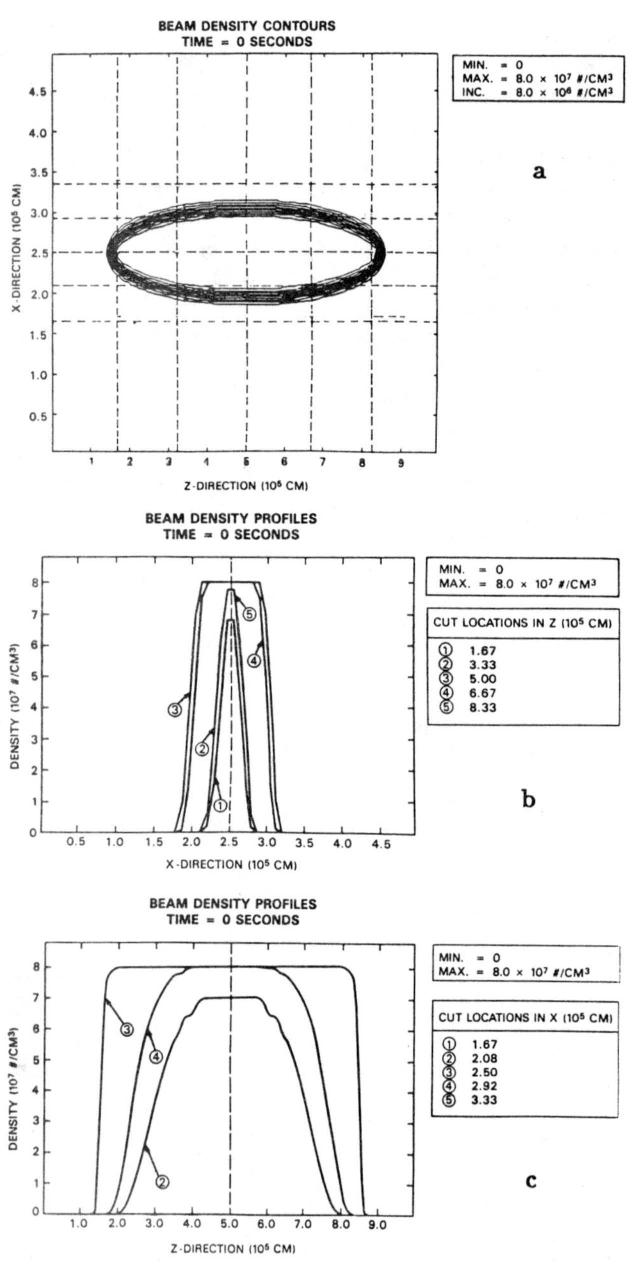

Fig. 2. Initial beam profiles for all runs. The parameters correspond to run #1. For the other runs they can be scaled according to the value of the peak beam density n_b. (a) Beam isodensity contours. This figure shows also the locations of the vertical and horizontal diagnostic cuts in the $x-z$ plane. (b) Beam density profiles along vertical cuts 1-5. (c) Beam density profiles along horizontal cuts 1-5.

between $10^{17} - 10^{19}$, corresponding to peak beam densities in the range of $n_b \approx 5 \times 10^6 - 8 \times 10^7 \#/\text{cm}^3$. Table I lists the parameters of the simulation runs in real (dimensional) and dimensionless units. The runs described here were performed with the magnetic field $\mathbf{B}_o$ out of the plane of the simulation. (Runs were also performed with the magnetic field in the plane of the simulation. We will comment on these later.) All the simulations were performed in the beam reference frame. In this frame, the beam particles are initially stationary while the magnetoplasma flows with $\mathbf{u} = -u_b\hat{e}_z$, with the aid of an appropriate motional electric field. The geometry and the initial conditions of the beam in the simulations are shown in Figs. 2a-c. Figure 2a shows the beam isodensity contours at $t = 0$. To facilitate the understanding of the physics, diagnostic cuts at the positions labeled 1-5 were taken along the x and z axes. Figures 2b, c show the initial beam profiles for cuts 1-5 along the x and z axes. The physics of the interaction described in Section 2 becomes clear by referring to typical simulation results.

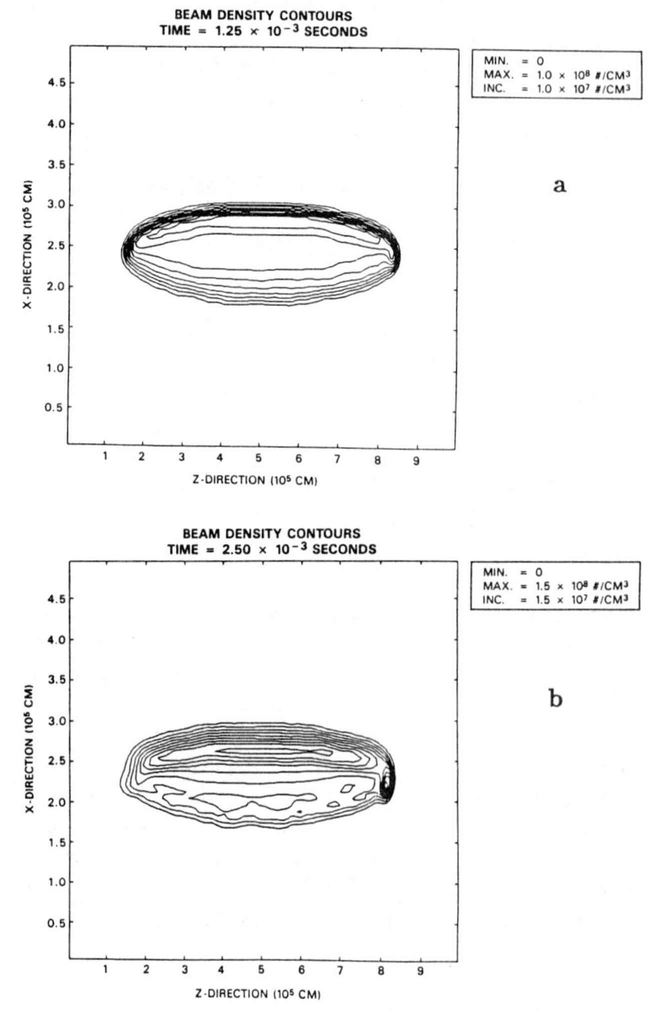

Fig. 3. Isodensity beam contours for run #1. (a) At $t = 1.25$ msec or (equivalently $\Omega_b t = 4$ or propagation distance $4R_b$). (b)At $t = 2.5$ msec.

Run #1 of Table I is examined first. Computer time constraints limited the runs to times $t \leq 2.5$ msec, which corresponds to $\Omega_b t \approx 8$ or propagation distances of 8 R_b ($\Omega_b = eB_o/M_b c = 3 \times 10^3$ sec^{-1} is the proton cyclotron frequency).

The evolution of the beam for the parameters of run #1 can be seen from Figs. 3a, b and 4a, b. Figures 3a, b show the beam isodensity contours at times $t = 1.25$ and 2.5 msec, corresponding to $\Omega_b t = 4$ and 8 and equivalent beam propagation distances of 4 and 8 R_b. It is clear that the beam has maintained its macroscopic integrity as a plasmoid and followed a ballistic trajectory. Similar conclusions are derived from Figs. 4a, b, which show the beam profiles as a function of x at the diagnostic cuts. Figures 4a, b should be compared with Fig. 2b at $t = 0$. Notice that both the isodensity contours and the density profiles show density compression at the center of the beam by almost a factor of two (the vertical scale in the figures changes according to the peak value). Detailed examination of Fig. 4b shows the presence of a secondary density peak at the front of the beam (position #5) which is displaced downwards (i.e. in the negative x-direction). This corresponds to the front erosion discussed in the simplified model of Section 2 and is also apparent at the front in Fig. 3b. The erosion of the beam front provides the momentum that balances the diversion of the flow of the ambient magnetoplasma in the positive x-direction. The erosion rate can be computed quantitatively by resorting to Figs. 2b and 4a, b and examining the temporal evolution of the density of the beam front at diagnostic location 5. The stationary magnetic field structure, which causes the flow diversion, is shown in the form of isomagnetic contours $\Delta\mathbf{B}(x, z)$, i.e. the difference between $\mathbf{B}(x, z)$ and the initial homogeneous $\mathbf{B}_o$, in Figs. 5a, b. Notice the compression of the magnetic field at the front followed by a diamagnetic cavity at the back. As described in Section 2, for $\Delta < R_o$ the

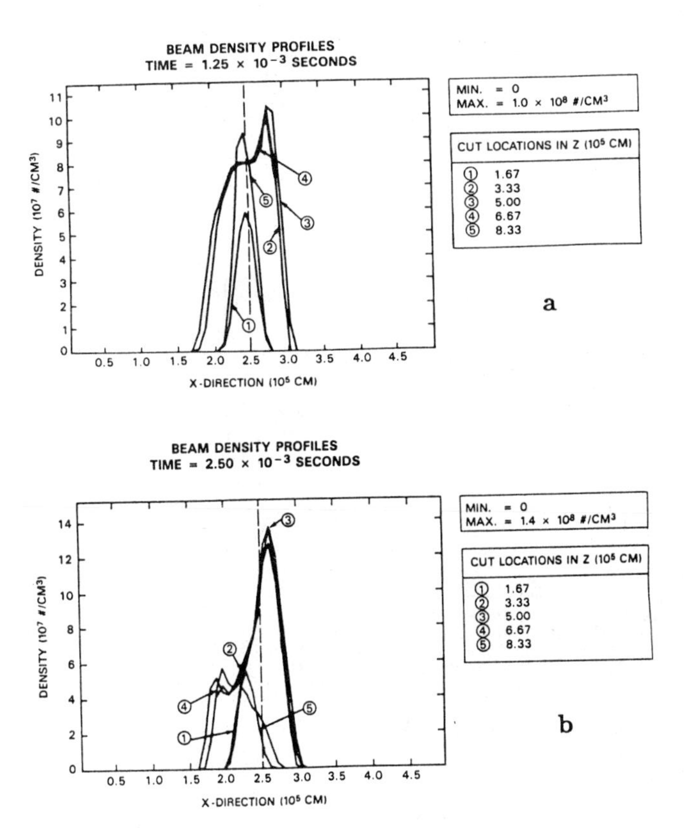

Fig. 4. Vertical beam density profiles at the diagnostic locations for run #1. (a) At $t = 1.25$ msec. (b) At $t = 2.5$ msec.

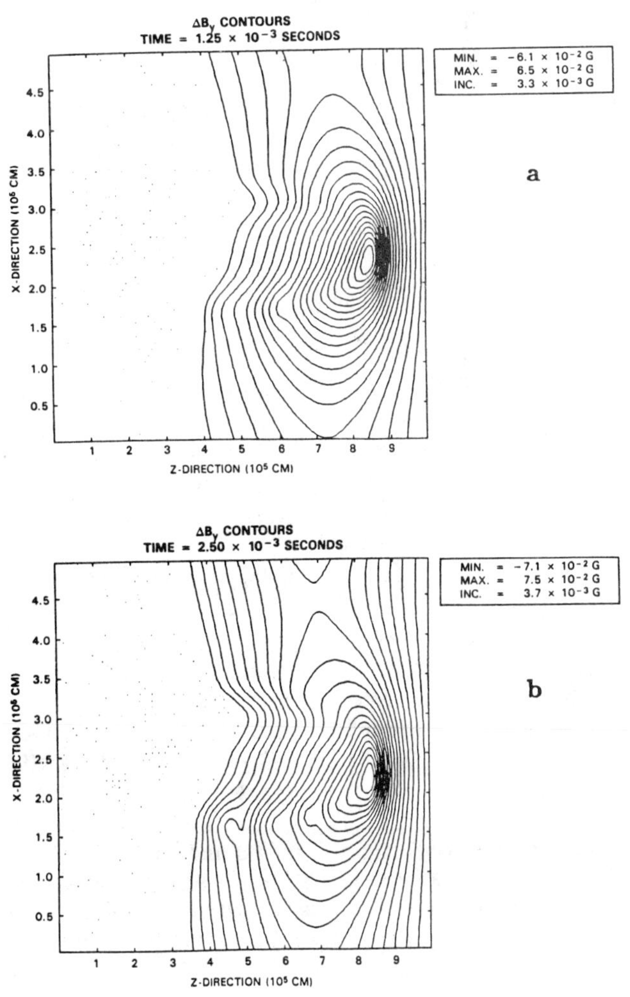

Fig. 5. Isomagnetic contours ΔB for run #1. (Solid lines represent compression of the magnetic field over the ambient, while dotted lines represent magnetic field depression). (a) At $t = 1.25$ msec. (b) At $t = 2.5$ msec.

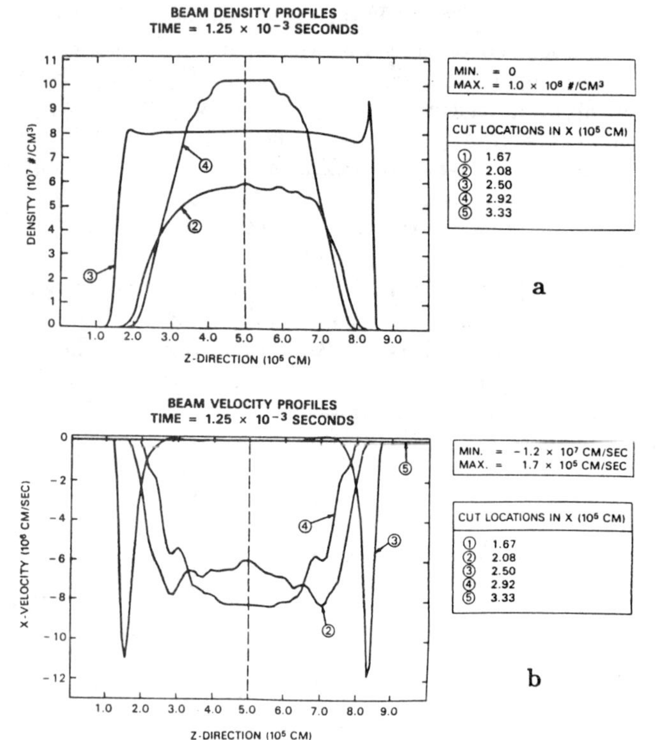

Fig. 6. Horizontal profiles of the beam density (a) and lateral velocity (b) for run #1 at $t = 1.25$ msec.

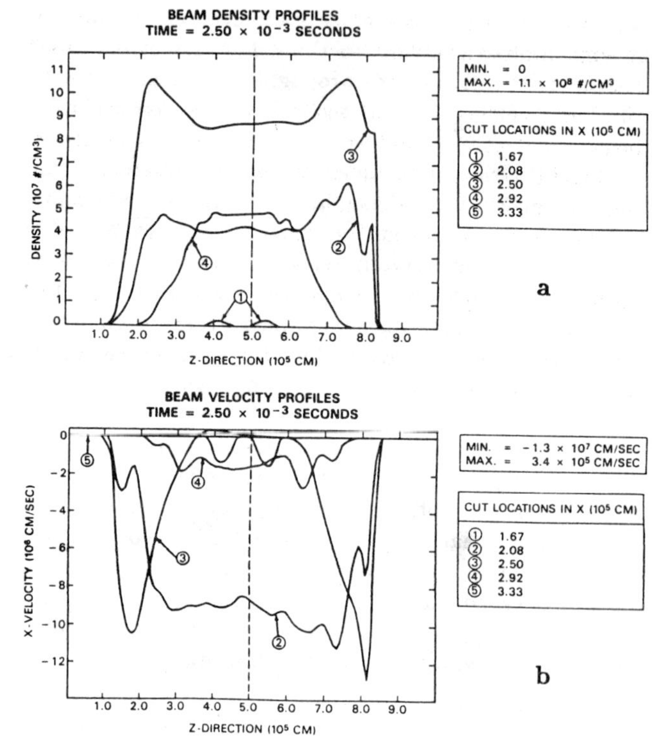

Fig. 7. Same as Fig. 6 at $t = 2.5$ msec.

compression level saturates at the value of $\Delta B/B_o$ necessary to divert the incoming plasma ions sideways to the point that they do not penetrate into the beam. It should be noted that the magnetic field structure, which was set up as early as $\frac{1}{3}\Omega_b^{-1}$, remained essentially stationary over the length of the simulations.

To understand the phenomenology controlling the scaling of the ballistic propagation time scale, we examine the evolution of the horizontal beam density profile (Figs. 6a, 7a) and lateral drift velocity (Figs. 6b, 7b) at the center cut (location #3, representing the maximum of the beam density and momentum flux) at times 1.25 and 2.5 msec. A comparison of Figs. 3a, 6a and 7a in conjunction with Figs. 4a, b shows that the beam density has been compressed at the center point, while the longitudinal length is preserved. Furthermore Figs. 6b and 7b demonstrate that the center of the beam follows a ballistic trajectory (i.e. $u_x = 0$) except at the front and the rear. The front of the beam is eroding at a downward velocity which has saturated at a value $u_x \approx 10^7$ cm/sec. As can be seen by examining the horizontal downward velocity profile at location #3 in Figs. 6b and 7b, the erosion is penetrating backwards towards the beam center.

Let us finally examine the flow behavior of the ambient plasma. Figures 8a, b show the flow speed at $t = 1.25$ and

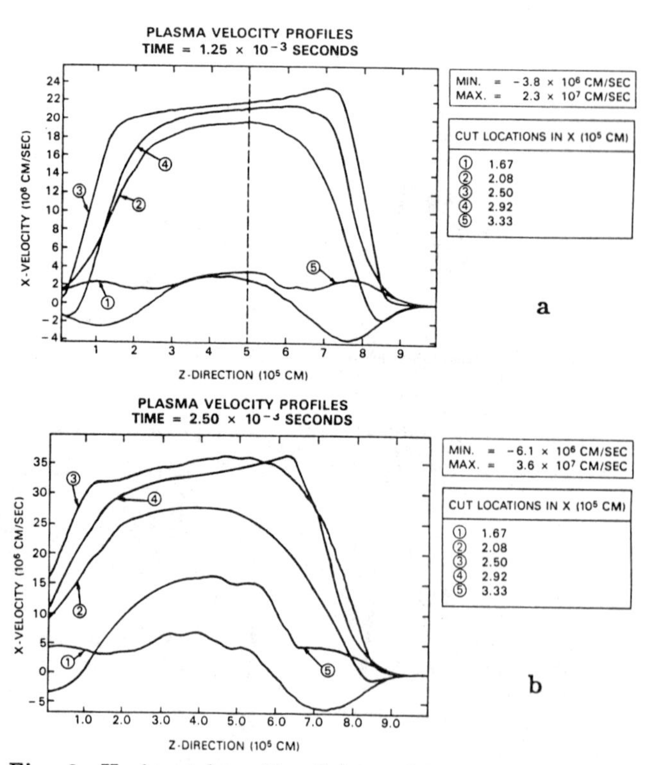

Fig. 8. Horizontal profile of the ambient plasma lateral velocity profiles for run #1. (a) At $t = 1.25$ msec. (b) At $t = 2.5$ msec.

2.5 msec at the horizontal cuts. It can be seen that the plasma is diverted upwards at the front with an increasing speed, reaching a value of about 2×10^7 cm/sec towards the beam center. This implies that if we started with a beam penetrated by ambient plasma, the ambient plasma ions and their neutralizing electrons will move outside the beam. Since, as can be seen in Fig. 8, there is no upward plasma flow from below (i.e. cut #1) toward the beam center, the plasma density inside the beam center will be reduced until there is no more plasma in the beam. This of course will occur only if the front erosion time scale is longer than the plasma evacuation time. We will return to this point later.

Scaling Considerations

The previous discussion identified two critical issues which control the scaling of ballistic beam propagation. The first refers to the scaling of the downward drift of the beam center (i.e. position at the intersection of horizontal and vertical cuts #3) and the upward drift of the plasma. The second refers to the rate of front erosion. As long as the front has not eroded, the beam center will propagate along a trajectory determined by the self-consistent electromagnetic fields at the center of the beam or plasmoid.

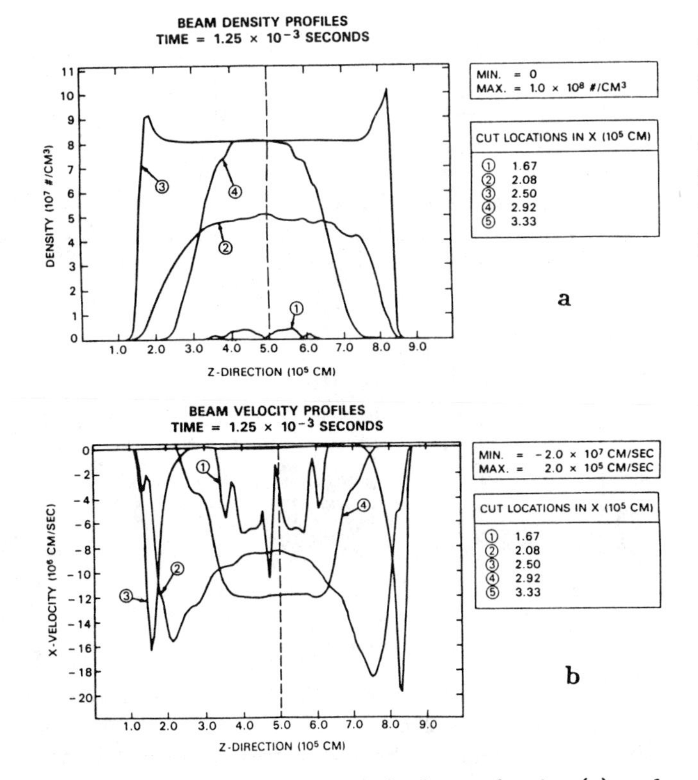

Fig. 9. Horizontal profiles of the beam density (a) and lateral velocity (b) for run #2 at $t = 1.25$ msec.

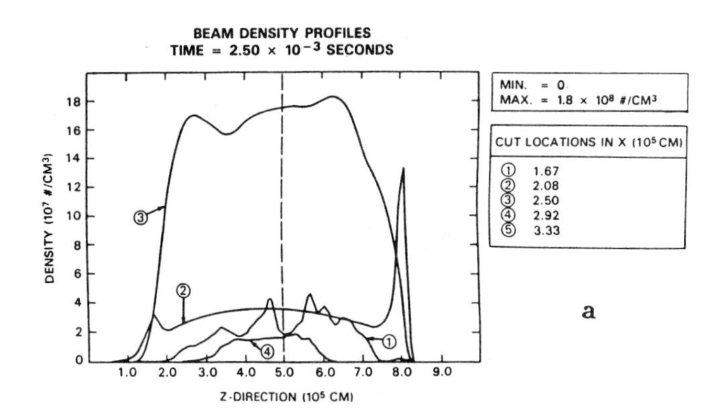

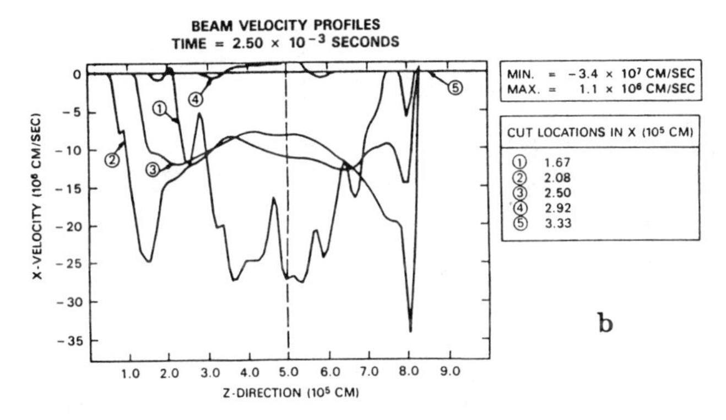

Fig. 10. Same as Fig. 9 at $t = 2.5$ msec.

We examine separately the scaling with beam velocity and with beam density.

Runs #1-3 all have the same number of beam particles (4.8×10^{18}) or an equivalent $n_b/n_o = 800$, but the flow velocity corresponds to 10^8, 2×10^8, and 4×10^8 cm/sec respectively. Figures 9a, b show the horizontal profiles of the beam density and downward displacement at time t=1.25 msec for run #2. The center density profile and center downward velocity are essentially similar to run #1. However, the erosion speed at the front of cut #3 (Fig. 9b) is almost a factor of two faster than in run #1 (i.e. 2×10^7 cm/sec vs 1.1×10^7 cm/sec). This is confirmed by referring to the same profiles at time $t = 2.5$ msec (Figs. 10a, b). While there is still substantial beam density along the center cut #3 (Fig. 10a), the front has eroded to such an extent that the beam center is now drifting downwards at a speed of 10^7 cm/sec. In the above two runs the erosion rate scales almost linearly with u_b. The same trend is evident from an examination of run #3 ($u_b = 4 \times 10^8$ cm/sec). Figure 11 shows the density (a) and downward beam velocity (b) at an earlier time ($t = 0.75$ msec). While the density profile is similar to Figs. 6a and 10a at $t = 1.25$ msec, the downward erosion speed is now 4×10^7 cm/sec, again revealing linear scaling with beam velocity. As a result of the faster erosion rate the profiles at $t = 1.25$ msec (Fig. 12) are

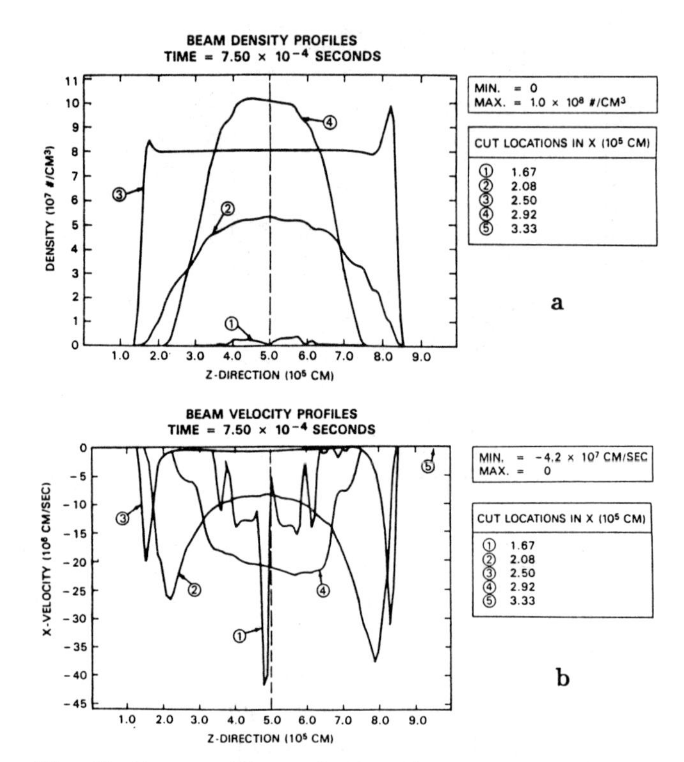

Fig. 11. Same as Fig. 10 for run #1 at $t = 0.75$ msec.

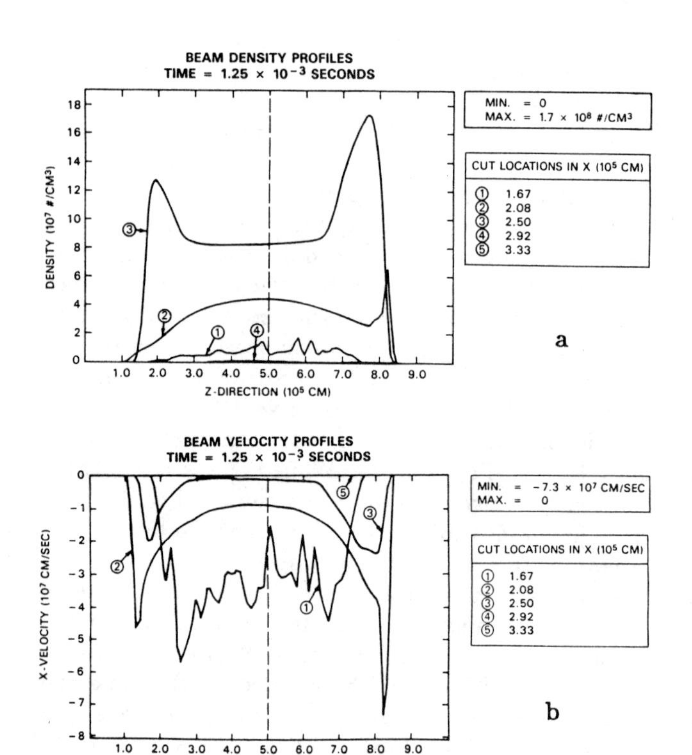

Fig. 12. Same as Fig. 11 at $t = 1.25$ msec.

similar to the ones of run #2 at 2.5 msec (Fig. 11). By time $t = 2.5$ msec the beam has been displaced and modified substantially. However, it has still maintained its plasmoid-like entity, although it is no longer following a ballistic propagation path (Fig. 13a). Figure 13b shows $\Delta \mathbf{B}$ contours at the same time. They should be compared with the contours of the ballistic propagation mode (Fig. 5). Finally, Figs. 14a, b show the plasma flow profiles at $t = 1.25$ msec for runs #2 and #3. It can be seen from these and Fig. 9a that, as expected from the previous results and momentum conservation, the diversion velocity of the plasma at the front as well as the outflow velocity of the plasma from the beam center scale almost linearly with u_b. It should be noted that for the above runs the

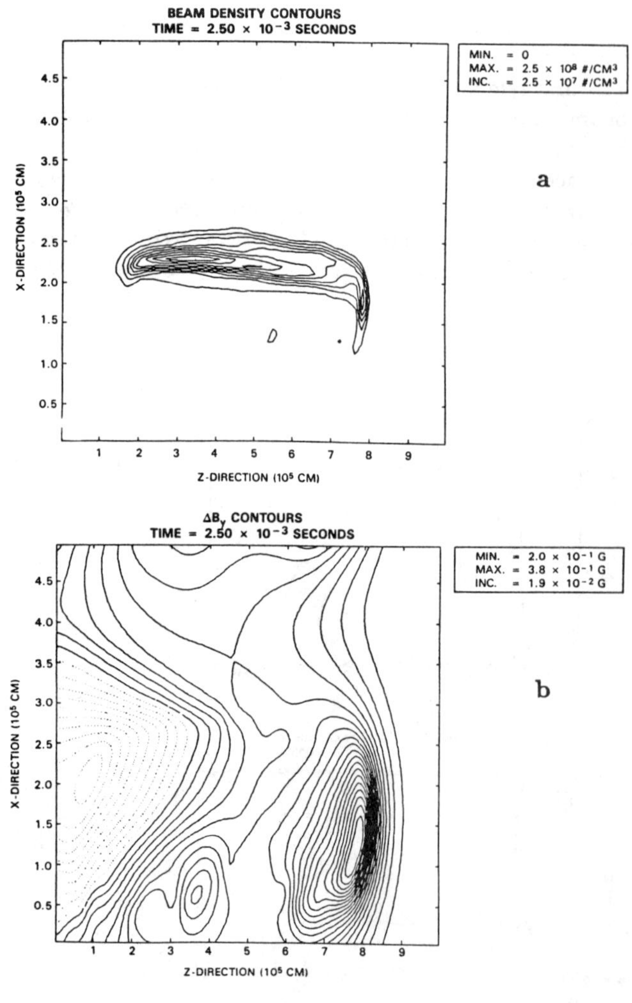

Fig. 13. Isodensity (a) and isomagnetic ΔB (b) contours for run #3 at $t = 2.5$ msec (solid lines represent compression while dashed lines represent depression of the ambient magnetic field).

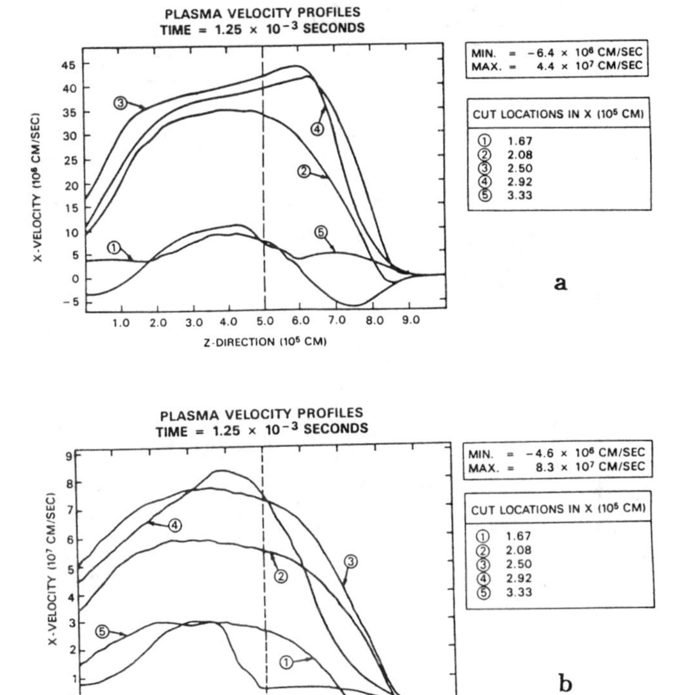

Fig. 14. Horizontal profile of the ambient plasma lateral velocity at $t = 1.25$ msec. (a) For run #2. (b) For run #3.

value of the magnetic field compression at the front also scales linearly with u_b.

We next address scaling issues related to beam-to-plasma density ratios. Runs #1, 4 and 5 all have the same velocity $u_b = 10^8$ cm/sec, but the number of beam particles is 4.8×10^{18}, 1.2×10^{18} and 3×10^{17}, corresponding to $n_b/n_o \approx 800$, 200, and 50. Figures 15a, b and 16a, b show horizontal profiles of beam density and lateral velocity for run #4 at $t = 1.25$ and 2.5 msec. A comparison of Fig. 15 with Fig. 6 shows that over the time scale of $t = 1.25$ msec ($\Omega_b t = 4$) the beam density profiles are basically self-similar for runs #1 and 4. However, in the lower density case (run #4), the center of the beam drifts with $u_x = 2 \times 10^6$ cm/sec while the front erosion speed is 1.7×10^7 cm/sec (Fig 15b). In comparing this with runs #1-3 (Figs. 6b, 9b, 11b) we note that for the high density cases there was essentially no drift of the center of the beam before erosion. The front erosion speeds, however, were $10^7, 2 \times 10^7$ and 4×10^7 cm/sec. Namely, a change in density by a factor of four resulted in a 70% change in the front erosion rate. Referring to Fig. 16 we note that while most of the energy density still remains at the beam center, a combination of a faster erosion rate and a

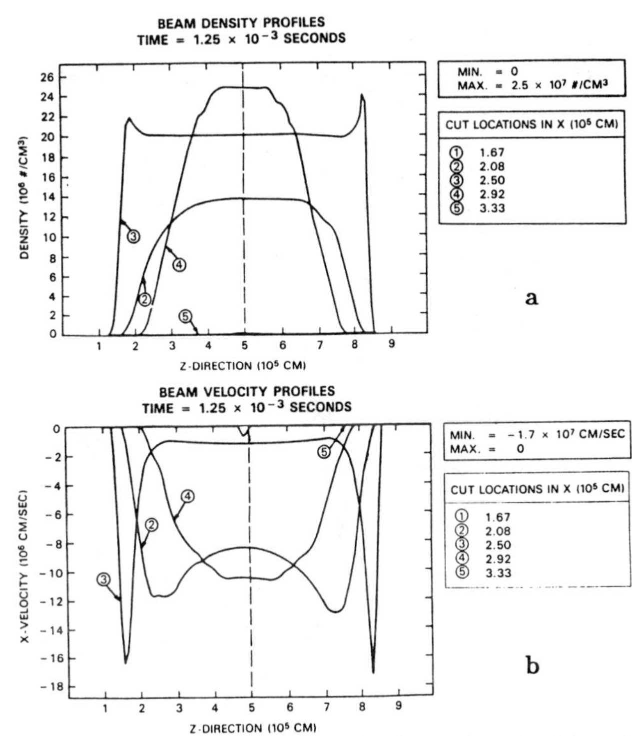

Fig. 15. Horizontal profiles of the beam density (a) and lateral velocity (b) for run #4 at $t = 1.25$ msec.

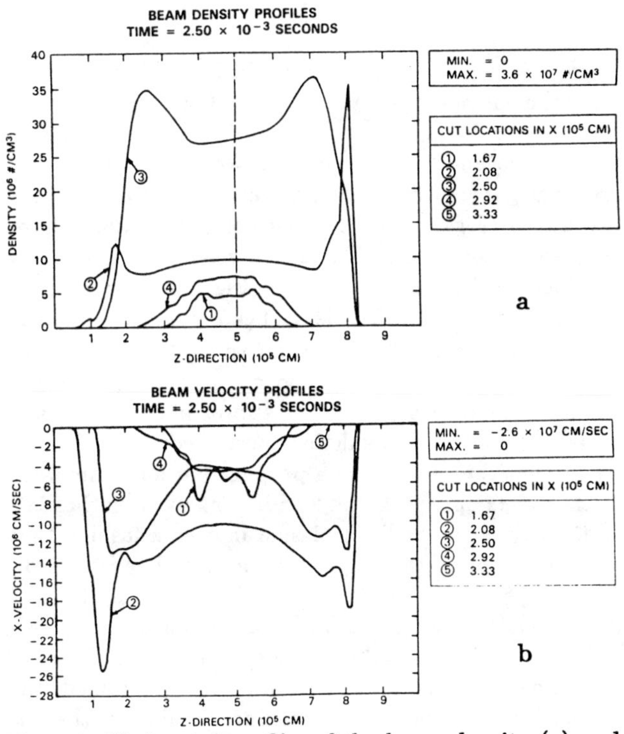

Fig. 16. Horizontal profiles of the beam density (a) and lateral velocity (b) for run #4 at $t = 2.5$ msec.

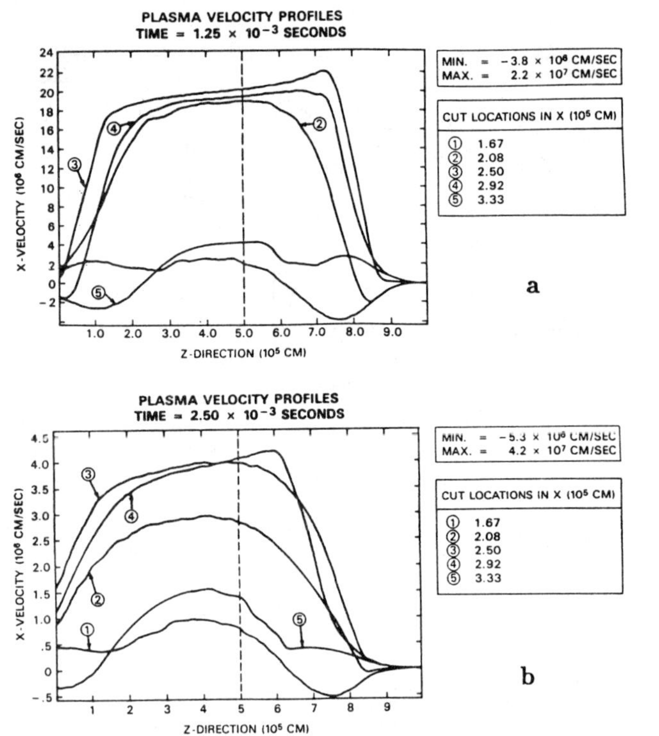

Fig. 17. Horizontal profiles of the ambient plasma lateral velocity for run #4 (a) $t = 1.25$ msec (b) $t = 2.5$ msec.

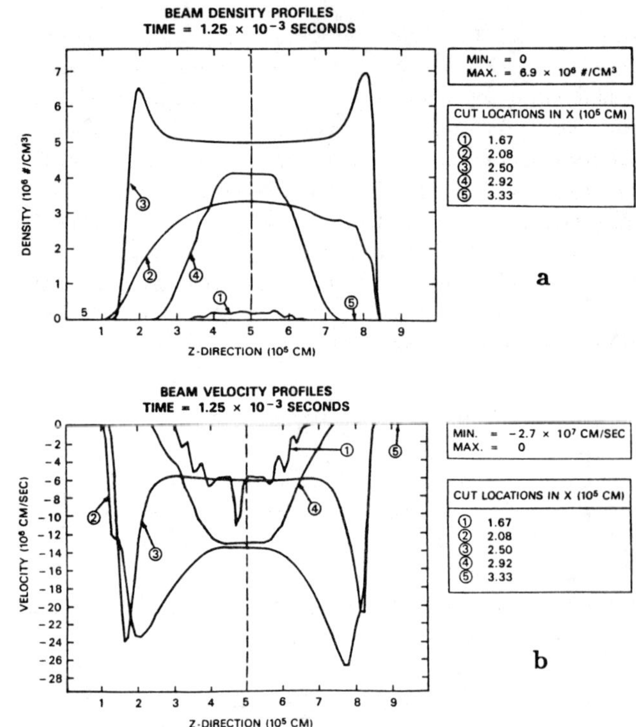

Fig. 18. Horizontal profiles of the beam density (a) and lateral velocity (b) for run #5 at $t = 1.25$ msec.

larger downward displacement of the beam center will destroy the ballistic propagation mode. Figures 17a, b show the velocity displacement profiles of the ambient plasma at $t = 1.25$ and 2.5 msec for run #4. Notice that they are qualitatively and quantitatively similar to the ones for run #1 (Figs. 8a, b). Furthermore, the outflow shows constant acceleration. The above scalings with density continue for the case of run #5 (Figs. 18a, b; 19a, b), which has a factor of four fewer beam particles than run #4 and sixteen times fewer than run #1. It should finally be noted that for the above runs the value of the maximum field compression was approximately the same.

In summary, the simulations described above demonstrate that the presence of a propagating ion beam with $n_b \gg n_o$ sets up an electrodynamic configuration that laterally diverts the ambient plasma in such a fashion that no penetration of the main beam occurs. The lateral diversion speed scales linearly with the beam velocity and is independent of the beam-to-plasma density ratio. Although specific simulation studies with various values of the ambient magnetic field $\mathbf{B}_o$ were not performed, the physical understanding dictates linear scaling of the lateral speed with $\mathbf{B}_o$. The beam responds to this configuration in a manner that is consistent with conservation of momentum in the plasma frame. As a consequence of

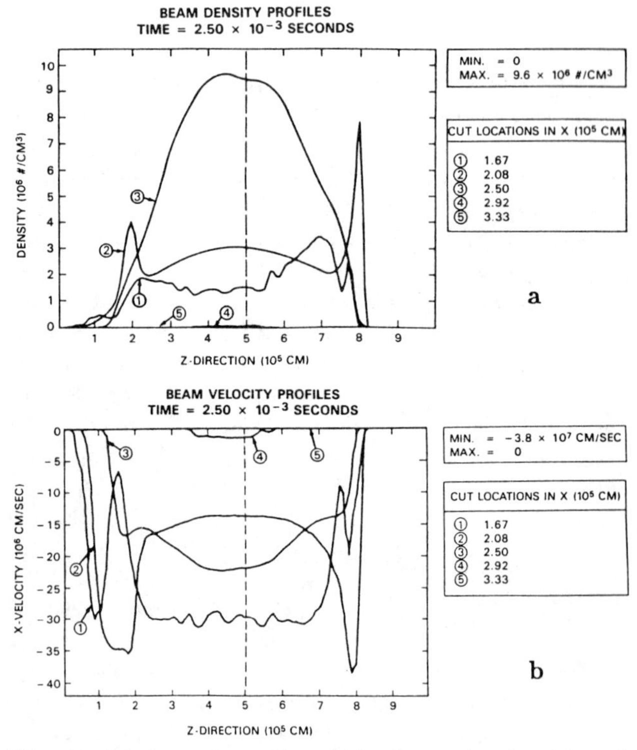

Fig. 19. Horizontal profiles of the beam density (a) and lateral velocity (b) for run #5 at $t = 2.5$ msec.

this, for time scales that are shorter than the front erosion time (i.e. for long thin beams) and for $n_b M_b \gg n_o M_o$, the plasma is evacuated from the central beam region with minimum beam displacement. In fact, the simulations show that the dynamics of the system are such that focusing is produced at the center while momentum is balanced by shedding of surface particles from the beam. The beam will subsequently follow a ballistic path until erosion of the front destroys it. The erosion rate was found to scale linearly with the beam velocity and by the previous argument with $\mathbf{B}_o$, while it is inversely proportional to (n_b/n_o) at the front.

Time Scale for Ballistic Propagation

From the previous analysis we can derive the following general conclusions concerning the time scale for cross field ballistic propagation of a neutralized ion beam of width Δ and length L with density n_b and ion mass M_b.

(i) The plasma will be evacuated from the beam center at an average rate given by

$$\frac{d^2 x_p}{dt^2} = \frac{eE_x}{M_o} \quad , \tag{4}$$

where

$$E_x = u_b \frac{\Delta B}{c} \quad . \tag{5}$$

The evacuation time scale τ_1 for a beam of lateral width Δ is easily computed from Eqs. (4) and (5) as

$$\Omega_b \tau_1 < (2\frac{\Delta}{R_o})^{1/2} \frac{M_o}{M_b} \quad . \tag{6}$$

During this time, the beam displacement Δx_b is given by

$$\frac{\Delta x_b}{\Delta} < \frac{n_b M_b}{n_o M_o} \quad . \tag{7}$$

For times greater than τ_1, no further displacement occurs until the beam front is eroded.

(ii) The front of the beam acts as a shield that allows the bulk of the beam to propagate as described above. The rate of beam erosion can be estimated by simple energy and momentum considerations. The beam erosion rate dz/dt is given by balancing momentum in the z-direction, i.e.

$$n_b M_b \frac{dz}{dt} = n_o M_o (u_b - u_b') \quad , \tag{8}$$

where u_b' is the ambient plasma speed in the z-direction after it has been diverted by the magnetic compression at the front. From conservation of energy,

$$u_b'^2 + u_x^2 = u_b^2 \quad , \tag{9}$$

where u_x is given from Eq. 1a as

$$u_x^2 = 2\frac{e\Delta B}{M_o c} u_b \Delta \tag{10}$$

(i.e. the transverse energy equals the potential drop). For $u_x^2 \ll u_b^2$, i.e. $\Delta/R_o \ll 1$, Eq. (9) becomes

$$u_x^2 = (u_b - u_b')(u_b + u_b') \simeq 2u_b(u_b - u_b') \quad ,$$

so that

$$u_b - u_b' = \Omega_o \frac{\Delta B}{B_o} \Delta \quad . \tag{11}$$

The erosion rate is then found from Eqs. (8) and (11) as

$$\frac{1}{u_b}\frac{dz}{dt} = \frac{n_o M_o}{n_b M_b} \frac{\Delta}{R_o} \frac{\Delta B}{B} < \frac{n_o M_o}{n_b M_b} \frac{\Delta}{R_o} \quad . \tag{12}$$

Notice that for $\Delta/R_o \ll 1$ and $n_o M_o \ll n_b M_b$ the erosion rate is a very small fraction of the beam speed. For a beam of length L we can define a beam erosion time as the time to penetrate to $z = L/2$ from the front. Then Eq. (12) gives an erosion time scale τ_2 as

$$\Omega_b \tau_2 > \frac{1}{2} \frac{n_b}{n_o} \frac{L}{\Delta} \quad . \tag{13}$$

An alternative interpretation of Eq. (12) applies to the case of constant beam injection. Under conditions such that $\Delta z_b/\Delta \ll 1$ as given by Eq. (8), Eq. (12) states that a beam injected into a magnetoplasma with an injection rate faster than the erosion rate will propagate ballistically at all times.

Summary and Conclusions

We presented above a physical model, supported by computer simulation studies, which demonstrated that dense ($n_b M_b \gg n_o M_o$), thin ($\Delta/R_o \ll 1, \Delta/L \ll 1$) neutralized ion beams or plasmoids can propagate ballistically if injected in the ionosphere or magnetosphere across the ambient magnetic field. The results presented were based on two-dimensional hybrid simulations performed in the plane perpendicular to the magnetic field. We are in the process of performing three-dimensional simulations as well as extending the simulations of the scaling discussed above to other M_b/M_o ratios, different values of

$\mathbf{B}_o$ and other values of Δ/R_o. These, along with specific applications to active experiments, plasmoid propagation and astrophysical jets, will be published elsewhere.

Before closing we should remark that two-dimensional simulations were also performed with the magnetic field in the simulation plane. Lateral diversion consistent with the earlier simulation results was observed. The only additional feature was that the field lines were slightly draped around the beam as they drifted upwards. As expected, the mode of beam propagation was not affected to any observable degree.

Acknowledgements. The authors are grateful to Drs. R. Sudan, T. Antonsen, P. Wheeler, and J. Denavit for many critical comments and suggestions during the course of this research. This work was supported by the U.S. Department of Energy under contract number DE-AC03-85SF15935.

References

Baker, D.A. and J.E. Hammel, Experimental Studies of the Penetration of a Plasma Stream into a Transverse Magnetic Field, *Phys. Fluids 8*, 713-722 (1965).

Birko, V.F. and G.S. Kirchenko, *Sov. Phys. Tech. Phys. 23*, 202 (1978).

Cargill, P., S. Sarma and K. Papadopoulos, Lower Hybrid Waves and Their Consequences for Cometary Bowshocks, *J. Geophys. Res.* (communicated), 1988.

Chapman, S., Idealized Problems of Plasma Dynamics Relating to Geomagnetic Storms, *Rev. Mod. Phys. 32*, 919-933 (1960).

Chapman, S. and V.C.A. Ferraro, A New Theory of Magnetic Storms, *Terrestrial Magnetism and Atmospheric Electricity 37*, 147-156 (1931).

Ferraro, V.C.A., On the Theory of the First Phase of a Geomagnetic Storm: A New Illustrative Calculation Based on an Idealized (Plane not Cylindrical) Model Field Distribution, *J. Geophys. Res. 57*, 15-49 (1952).

Mankofsky, A., R.N. Sudan, and J. Denavit, *J. Comp. Phys. 70*, 89 (1987).

Papadopoulos, K., A. Mankofsky and A. Drobot, *Phys. Rev. Lett. 61*, 94 (1988).

Peter, W. and N. Rostoker, *Phys. Fluids 25*, 730 (1982).

Schmidt, G., Plasma Motion Across Magnetic Fields, *Phys. Fluids 3*, 961-965 (1960).

Scholer M., On the Motion of Artificial Ion Clouds in the Magnetosphere, *Planet. Space Sc. 18*, 977, 1984.

Shanahan, W. (unpublished), 1984.

Tuck, J.L., Plasma Jet Piercing of Magnetic Fields and Entropy Trapping in a Conservative System, *Phys. Rev. Lett. 3*, 313-315 (1959).

RAY PATHS OF ELECTROMAGNETIC AND ELECTROSTATIC WAVES IN THE EARTH AND PLANETARY MAGNETOSPHERES

Iwane Kimura*

Department of Electrical Engineering II, Kyoto University, Kyoto 606, Japan

Abstract. Various wave phenomena whose ray path analyses are essential for the interpretation of the phenomena are briefly reviewed, namely the whistler- mode, ion cyclotron mode, free space mode, Z mode, and the upper hybrid or electrostatic electron cyclotron harmonic mode. Then three-dimensional ray tracing techniques are summarized in both cold and hot plasmas. Some applications of ray paths calculated by the latest computer programs are also shown.

Introduction

In the present paper, studies concerned with wave phenomena in the Earth and planetary magnetospheres are reviewed with emphasis on their ray paths and the importance of the ray tracing techniques to trace their paths for the interpretation of the phenomena is also stated.

The modes of waves concerned are whistler mode, ion cyclotron mode, free space modes (R-X and L-O modes), Z mode, and electrostatic electron cyclotron harmonic wave (ECHW) mode. For the last ECHW mode, the effect of plasma temperature of plasma distribution function is essential. For other modes, the effects of energetic particles in addition to a cold plasma have also been taken into account in order to investigate growth or damping of the waves along their ray paths. Such considerations are sometimes necessary to interpret a phenomenon completely with its generation mechanism.

In what follows, typical phenomena, for which investigations of their ray paths are essential for interpretation, are briefly reviewed for the above mentioned modes. Then, ray tracing techniques for general use are introduced. Relatively new techniques are discussed for ray tracing in a nondipolar magnetic field model of the earth magnetosphere, and for ray tracing in a hot magnetospheric plasma, the latter being utilized to trace the paths of ECHW and of any other hot plasma waves if the damping is sufficiently small. Some results of these new ray tracing techniques are finally presented.

Whistler Mode

Whistler Mode Phenomena in the Earth Magnetosphere

Several whistler mode phenomena interpreted by the results of ray tracing have been reviewed by the present author [Kimura, 1985], such as subprotonospheric whistlers, magnetospherically reflected (MR) whistlers, walking trace (WT) and etc. Some of these phenomena can be explained by taking account of ion effects in the ionosphere and magnetosphere. The effect of ions on the whistler mode propagation is that (1) below the lower hybrid frequency, the group velocity or wave energy can cross the external magnetic field lines, (2) the polarization of the whistler mode switches from right-handed to left-handed below the crossover frequency, and (3) the whistler mode waves are reflected at the cutoff frequency. In order for the whistler waves to penetrate the ionosphere completely with the right-handed polarization, mode coupling must take place at the crossover frequencies, which depend on the species and the number densities of ions.

Ray paths of ground-based VLF signals propagating in the magnetosphere have been found to be delicately dependent on the plasma distribution and their gradients in the magnetosphere. It turns out that multiple ray paths can reach a satellite with completely different wave normal directions, the fact being detected as several different or widely spread doppler shifts at the satellite [e.g. Edgar, 1976;Matsuo et al., 1985]. By the measurement of the doppler frequency shift, a direction of the wave normal can be estimated. In a recent study of Aldra Omega signals observed by ISIS-2 tracked at Syowa Station, Antarctica [Matsuo et al., 1985], the Omega signals accompanied by triggered emissions were often found and the original Omega signal was observed generally with a slight positive doppler shift and sometimes with a large negative doppler shift for the orbits from north to south. A wave with a slight positive doppler shift has a wave normal directed toward the satellite, which orbited southwards. By ray tracing from the source hemisphere, this ray path has a small wave normal angle in the interaction region around the equatorial plane. On the other hand, the waves with a large negative doppler shift are found to have a large wave normal angle at the equatorial plane. These facts have very important information in considering the mechanism of triggered emissions or wave-particle interaction phenomena.

Referring to the wave-particle interaction, a ray tracing study of Siple triggered emissions observed by Jikiken(EXOS-B) satellite indicated that the paths directly reached the satellite were non-ducted although the triggering took place in a ducted path [Bell et al., 1983]. Bell [1986] has also utilized ray tracing to establish the wave magntic field amplitude threshold for nonlinear trapping of energetic gyroresonant and Laudau resonant electrons by nonducted VLF waves in the magnetosphere. His paper confirms an importance of Landau type wave-particle interaction for a large wave normal angle for generation of triggered emissions, which might be related with Omega triggered emissions mentioned previously.

As to the mechanism of plasmaspheric hiss generation, ray tracing has been made including growth or damping rate along the propagation paths [Huang et al., 1983;Church and Thorne, 1983;Solomon et

*Also at Institute of Space and Astronautical Science, Sagamihara City, Kanagawa Prefecture 229, Japan

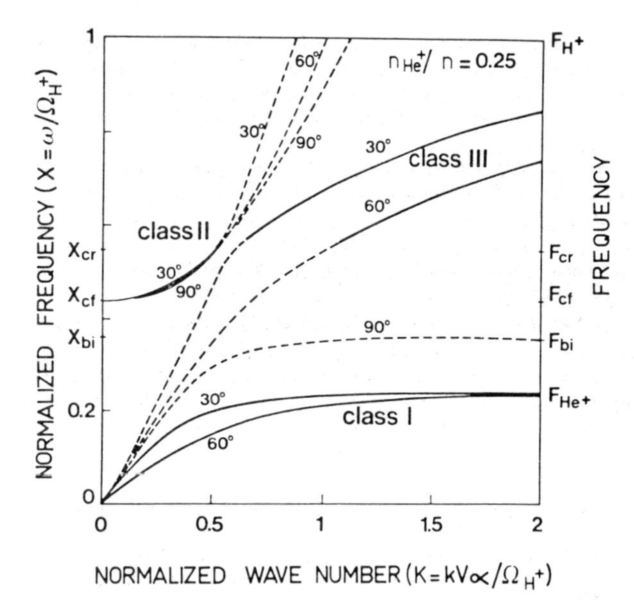

Fig. 1. Dispersion relation of ULF waves in the presence of He^+ ions. The solid curves denote left-handed waves, while the dashed curves denote right-handed waves [Rauch and Roux, 1982].

al., 1987]. Huang et al. [1983] concluded that integrated growth for multiply reflected waves was not enough to explain the intensity of the plasmaspheric hiss. Church and Thorne [1983] on the other hand drew a conclusion that the observed broadband hiss spectrum can be maintained by the modest net gains anticipated along ray trajectories that are recycled back to the favorable equatorial growth region. Recently, Solomon et al. [1987] obtained much larger gains than Church and Thorne using a realistic storm-time electron density distribution observed by GEOS 1 and 2, and found that one traverse through an unstable equatorial region results in a sufficient amplification gain for plasmaspheric hiss generation.

Another utilization of ray tracing is the backward ray tracing to locate the source region based on the wave normal direction deduced from the observed wave distribution function (WDF). Several such works have been done, such as Lefeuvre and Helliwell [1985], and references therein. Cairo and Lefeuvre [1986] made three-dimensional forward and inverse ray tracing to explain the WDF found for ELF/VLF hiss detected on GEOS satellite.

For duct propagations, ray tracing has also been utilized to study the conditions of trapping of whistler mode waves into the ducts [Strangeways and Rycroft, 1980; Strangeways, 1981] and the conditions of wave leakage from narrow ducts [Strangeways, 1986]. The trapping and interhemispheric propagation of whistlers in ducts has also been treated by Gorney and Thorne [1980]. They found that the conventional bell-shaped duct is not the most efficient means of inter-hemispheric gradient; instead one sided mini-ledges provide a more viable mechanism.

Whistler Mode Phenomena in the Planetary Magnetospheres

Some wave phenomena judged as whistler mode waves have been observed in the vicinity of Venus [Ksanformaliti et al., 1979], in the vicinity of Io torus of Jupiter [Gurnett et al., 1979;Menietti and Gurnett, 1980] and during the encounters with Saturn [Gurnett el al., 1981; Scarf et al., 1983] and Uranus [Kurth et al., 1986]. Ray tracing has been used for whistlers observed in the Io torus around Jupiter in order to quantitatively confirm the dispersion of the observed whistlers in the estimated electron density distribution in the Io torus [Menietti and Gurnett, 1980]. Ray paths and path integrated gain were calculated by Thorne and Moses [1985] for unstable whistler mode waves in the frequency range near the local crossover frequency (F_{cr}) propagating in the torus near the equator, while the waves experience polarization reversal at the condition of $f = F_{cr}$.

Ion Cyclotron Waves

Ray tracing studies on the ion cyclotron mode in an electron–proton cold plasma was made by Kitamura and Jacobs [1968], in order to explain how Pc 1 waves could bounce back and forth between geomagnetically conjugate points in the Earth magnetosphere. They concluded that the left-hand waves could be guided along the field lines, but the waves could not bounce more than several times due to damping.

Ray paths of the ion cyclotron modes in an He^+ rich cold plasma were studied by Rauch and Roux [1982], for interpretation of intense ULF waves around the helium ion cyclotron frequency (F_{He^+}), which were commonly observed on GEOS 1 and 2 spacecrafts. Dispersion relation of ULF waves in the presence of He^+ ions in addition to protons is illustrated in Figure 1 (Rauch and Roux, 1982). For frequencies less than the proton cyclotron frequency (F_{H^+}), the dispersion relation exhibits three branches, each of which shows the following characteristics. i) Below F_{He^+}, class I waves are left-handed and propagate along magnetic field lines. ii) Class II waves, left-handed for frequencies between the cutoff frequency (F_{co}) and the crossover frequency(F_{cr}) and right-handed above F_{cr}, are not guided along the field lines. iii) Class III waves are right-handed below F_{cr} and left-handed above it. These waves are guided above the bi-ion hybrid frequency (F_{bi}), and are not guided below F_{bi}.

From these results, Rauch and Roux [1982] concluded that class III waves which are unstable in the equatorial region for $f > F_{cr}$ and are guided along a field line for $f > F_{bi}$, are amplified owing to their multiple crossing of the equatorial (unstable) region, at the expense of energetic protons.

A ray tracing study was made by Thorne and Moses [1985] on ion cyclotron waves near the two-ion crossover frequency in the Io plasma torus, which consists of protons and O_2^+ ions. It was found that the propagation characteristics of ion cyclotron waves for frequencies below F_{H^+} are strongly influenced by changing ion composition in the Io torus, and obliquely propagating waves experience a natural reversal of polarization when their frequency becomes equal to F_{cr}. Path-integrated gain for unducted ion cyclotron waves in the torus was calculated.

Free Space Modes in the Earth and Planetary Magnetospheres

The terms "free space modes" are used here to specify R-X and L-O modes whose velocity becomes the light velocity in vacuum, when the plasma frequency is much less than the wave frequency. HF wave propagation in the ionosphere is also in this category. The ray paths of these high frequency signals have been an important subject from the view point of communication, so that a great number of ray tracing works have so far been made. In the present review, one of the reports [Jones and Stephenson, 1975] explaining a computer program of 3-D ray tracing in the ionosphere is referred here.

Auroral Kilometric Radiation

The first topics of the free space modes in the earth magnetosphere is auroral kilometric radiation (AKR). Many theoretical studies have been made since the first observational result on AKR was published [Gurnett, 1974]. The initial identification of the mode of AKR was R-X mode. However, Oya and Morioka [1983] could identify from the cutoff frequency measurements by Jikiken satellite that their observed AKR was in the L-O mode. Benson [1982;1984] also confirmed the existence of L-O mode AKR on ISIS-1 ionograms. These two results of completely opposite sense of polarization was found to be reconciled by Hashimoto [1984] by 3-D ray tracing of AKR waves, if we assume that at the source both R-X and L-O modes are present but the R-X mode component is much more intense than the L-O mode component. He calculated ray paths using an electron density model with an auroral plasma cavity and with the plasmapause, and using an initial condition that the AKR is generated in the cavity along a nightside field line at 70° invariant latitude at a frequency slightly above the R-X mode cutoff frequency. According to the ray tracing, it turns out that the R-X mode component is dominant at high latitudes and at lower latitudes, only the L-O mode component can be received, because the R-X mode cannot reach there by refraction through the high density region in the plasmasphere. The above assumption that the R-X mode is stronger than the L-O mode at the source is consistent with the cyclotron maser mechanism by Wu and Lee [1979] and Lee and Wu [1980], if the plasma frequency is much less than the cyclotron frequency.

Omidi and Gurnett [1984] calculated the total growth of AKR integrated along its ray path, using a particle distribution function obtained by S3-3 satellite. They concluded that the electron distribution actually used, is not capable of amplifying cosmic noise background to the observed AKR intensify and that much steeper slopes ($\partial F/\partial v$) at the edges of the loss cone are required. The presence of such distribution functions in the auroral zone is plausible if one assumes that backscattered electrons in this region have energies less than a few hundred eV.

Radiation from the Jupiter

It is well known that the Jovian Decametric (DAM) radiation consists of two major components; Io-phase dependent component and independent component. The Io-dependent source is considered to be located in the Jovian ionosphere near the foot point of a magnetic field line passing through the location of the satellite Io, because the Io dynamo effect can be an energy source of the emission. To confirm this model, 3-D ray tracing was made by Hashimoto and Goldstein [1982] based on the assumptions that the emission in R-X mode is generated from the above mentioned location in the Jovian ionosphere at a frequency slightly above the R-mode cutoff frequency f_R. By taking account of a proper ionospheric refraction effect the Io-asymmetry characteristics can be satisfactorily explained. Menietti et al.. [1984a,b] made 3-D ray tracing of DAM emission by taking account of Io torus effects but without ionospheric effect to explain a characteristic spectral arc phenomenon of DAM emission, so-called "nested arcs", and the results suggest that much of the variation in the nested arcs is a result of propagation effects rather than emission process differences. In their latest paper [Menietti et al., 1987] they made a ray tracing to investigate whether the DAM source is located in the southern and/or the northern hemisphere.

Z mode in the Earth and Planetary Magnetospheres

DE-1 observations of high latitude auroral phenomena in the radial distance range between 2 and 5 Re have revealed that in addition to strong AKR at frequencies above the electron gyrofrequency, auroral hiss in the whistler mode and Z mode radiation are observed essentially every pass over the auroral zone [Gurnett et al., 1983]. This Z mode emission, which is especially observed in the low-density region, has a sharply defined upper cutoff near the electron gyrofrequency and extends downward in frequency to the L mode cutoff frequency. The Z mode radiation is thought to correspond to a part of "continuum radiation" found in the Hawkeye polar region data [Gurnett el al., 1983].

According to the ray paths and the integrated growth along the paths calculated by Menietti and Lin [1985, 1986] using the starting condition that a wave is excited at frequencies slightly lower than the local gyrofrequency, they concluded that the cyclotron maser mechanism alone is not enough to explain the observed bandwidth and a long-distance propagation of the Z mode emission.

Hashimoto and Calvert [1987] made a similar ray tracing but they have found that for the wave frequency near the plasma frequency at the source region and for the initial wave normal off from the meridian plane, the Z mode wave can propagate over a long distance and result in the spatial distribution of the waves consistent with the observation.

Upper Hybrid Mode and Electrostatic Cyclotron Harmonic Mode

Terrestrial nonthermal radiation which is also named as myriametric radiation by Lembege and Jones [1982] was first detected in the frequency range 30 – 110 kHz by Brown [1973] and in the range 5 – 20 kHz by Gurnett and Shaw [1973]. When the radiation is generated as a free space mode in the magnetospheric cavity, it is predominantly in the L-O mode [Gurnett and Shaw, 1973], although some R-X mode components are also present [Shaw and Gurnett, 1980] which possibly produced after multiple reflections at the cavity walls [Jones, 1976a,b].

Gurnett [1975] noted that the continuum radiation seemed to be closely associated with intense bands of electrostatic noise near the electron plasma frequency at the plasmapause. Jones [1976a] first provided a theory to relate the continuum radiation to upper hybrid waves. Barbosa [1980, 1985] and Barbosa and Kurth [1980] have shown the characteristics of the ray paths of electrostatic cyclotron harmonic wave (ECHW) close to the magnetic equator in order to show their confinement near the magnetic equatorial plane, and to check a possibility of their convective growth.

Lembege and Jones [1982] made ray tracing using Poeverlein's construction method to confirm an idea; a possibility of a mode conversion from ECHW to L-O mode through Z mode. This idea of mode conversion was based on Oya [1971]. In their work, the ray tracing for ECHW was made under the electrostatic approximation, and the results were used as the initial condition for cold plasma ray tracing of Z mode.

Engel and Kennel [1984] calculated the growth and damping along ray paths of electrostatic mode (ECHW).

A complete computer program for ray tracing in a hot plasma, which can deal with electrostatic as well as electromagnetic modes in the magnetospheric plasma, has been developed in Kyoto University [Hashimoto et al., 1987;Yamaashi et al., 1987]. The mode conversion of electrostatic electron cyclotron wave to Z mode electromagnetic waves has been satisfactorily confirmed by the works made by this computer program. Detailed description of the techniques and the results will be reviewed in the following section.

Ray Tracing Techniques

Basic Equations for Ray Tracing

Basic equations of ray tracing governing ray path vector $\boldsymbol{r}$ and wave normal vector $\boldsymbol{k}$ are [Weinberg, 1962]

$$\frac{d\boldsymbol{r}}{d\tau} = \frac{\partial D}{\partial \boldsymbol{k}} \quad (1)$$

$$\frac{d\boldsymbol{k}}{d\tau} = -\frac{\partial D}{\partial \boldsymbol{r}} \quad (2)$$

where $D(\boldsymbol{r}, \boldsymbol{k}, \omega)$ = constant is the dispersion relation; ω is the wave frequency and τ is a quantity proportional to the phase propagation time. A ray path is obtained by simultaneously integrating the right-hand side of the above equations (1) and (2) by τ. In a hot plasma or by the effect of particle collision, the dispersion relation generally provide a complex $\boldsymbol{k}$ or $\boldsymbol{\rho}(= c\boldsymbol{k}/\omega)$. However, if the imaginary part of $\boldsymbol{k}$ or the imaginary part (χ) of the refractive index $(\boldsymbol{\rho} = \mu - i\chi)$ is relatively large, the ray path thus obtained becomes meaningless. Therefore we have to check at every step of integration if the quantity χ is much smaller than the real part μ of ρ. The real propagation time t can be calculated by integrating the following differential equation.

$$\frac{dt}{d\tau} = -\frac{\partial D}{\partial \omega} \quad (3)$$

Refractive Index or Dispersion Function

Cold plasma. The refractive index ρ of an electromagnetic wave in cold plasma is represented by [Stix, 1962]

$$A\rho^4 - B\rho^2 + C = 0 \quad (4)$$

in which A, B and C are functions of plasma and cyclotron frequencies of electron and ions, and of the angle ψ between the geomagnetic field direction and the refractive index vector $\boldsymbol{\rho}$. The angle ψ is given by

$$\psi = \cos^{-1}\{(B_r\rho_r + B_\theta\rho_\theta + B_\phi\rho_\phi)/\boldsymbol{B}\rho\} \quad (5)$$

where B_r, B_θ, B_ϕ are r, θ, ϕ components of the geomagnetic field vectors $\boldsymbol{B}_T$, and $\rho_r, \rho_\theta, \rho_\phi$ are r, θ, ϕ components of $\boldsymbol{\rho}$. In the dipole model, B_r, B_θ are represented by a simple function of θ only and $B_\phi = 0$.

Electrostatic approximation in a hot plasma. In the following we neglect the effect of ions and assume the isotropic Maxwellian distribution with no drift velocity then the dispersion relation in the electrostatic approximation, which is derived from a scalar potential, is given by [Miyamoto, 1980]

$$D = (k\rho_L)^2 + \left(\frac{\Pi}{\Omega}\right)^2 \left\{1 + \sum_{n=-\infty}^{\infty} \left(1 - \frac{n\Omega}{\omega + n\Omega}\right) I_n(b)e^{-b} \cdot \zeta_n Z(\zeta_n)\right\} = 0 \quad (6)$$

where

$$b = (k\rho_L \sin\psi)^2, \quad \zeta_n = \frac{\omega + n\Omega}{\sqrt{2}k\rho_L\Omega\cos\psi} \quad (7)$$

In the above equations, $\Pi(= 2\pi f_p)$ and $\Omega(= 2\pi f_c)$ are the angular electron plasma and cyclotron frequencies, ρ_L is the Larmor radius defined by v_T/Ω where $v_T = \sqrt{T/m}$, and $Z(\zeta_n)$ is the so-called plasma dispersion function and I_n(b) is the modified Bessel function. This approximated dispersion relation is much simpler than the following hot plasma dispersion relation and can save CPU time in ray tracing, although its application is limited.

Full dispersion in an isotropic Maxwellian hot plasma. The dispersion relation in a hot plasma without any electrostatic approximation is given by [Miyamoto, 1980]

$$D = \begin{vmatrix} H_{xx} - k^2 + (\frac{\omega}{c})^2 & H_{xy} & H_{xz} \\ -H_{xy} & H_{yy} - k^2\cos^2\psi + (\frac{\omega}{c})^2 & H_{yz} + k^2\sin\psi\cos\psi \\ -H_{xz} & H_{yz} + k^2\sin\psi\cos\psi & H_{zz} - k^2\sin^2\psi + (\frac{\omega}{c})^2 \end{vmatrix} = 0 \quad (8)$$

$$\boldsymbol{H} = \left(\frac{\Pi}{c}\right)^2 \sum_n (\zeta_0 Z(\zeta_n))\, e^{-b}\boldsymbol{X}_n + 2\zeta_0^2\boldsymbol{L} \quad (9)$$

where $\boldsymbol{L}$ is a tensor, in which all elements except L_{zz} are zero,

$$\alpha = k\rho_L\sin\psi, \quad b = \alpha^2, \quad \rho_L = v_T/\Omega \quad (10)$$

and $\boldsymbol{X}_n$ is a tensor whose elements are functions of modified Bessel function of order n with the argument b.

For $|\psi| \sim 90°$, the asymptotic expansion of the plasma dispersion can be used [Hashimoto et al., 1987].

Derivatives of the Refractive Index or Dispersion Function

In order to proceed the ray tracing, the right-hand sides of (1) and (2), namely the derivatives of refractive index or dispersion function with respect to the space coordinates and to the components of the wave number vector must be calculated, at every step of the ray paths. The former depends both on the distribution of plasma parameters and the model of the background magnetic field. If these quantities are given by functions of space coordinates, the derivatives can be calculated analytically. The latter are also calculated analytically from the dispersion relation.

Background Magnetic Field Models

For ray tracing of whistler mode waves in the earth magnetosphere, the simplest model of the geomagnetic field is a dipole model. Most of the ray tracing techniques treating wave phenomena in the earth magnetosphere have been based on the dipole model. Actually, however, real geomagnetic field lines are known to be fairly deformed from dipole field lines, so that for whistler mode waves, whose propagation are strongly governed by the geomagnetic field direction, ray paths calculated using the dipole field model will differ considerably from those in the actual magnetosphere.

In order to represent the magnetic field intensity and direction in the non-dipolar model, the so-called spherical harmonic expansion model, such as IGRF (International Geomagnetic Reference Field) model, can be used [Kimura et al., 1985]. In the Jovian magnetosphere, such a non-dipolar model has been used [Menietti et al., 1984a,b;1987].

Plasma Density Models

For ray tracing calculation, background plasma densities have to be defined as functions of space coordinates. In the earth magnetosphere, a diffusive equilibrum model, a collisionless model and a combination of these two, taking account of the plasmapause [e.g. Aikyo and Ondoh, 1971] are typical models. In these models, however, the plasma densities are defined along geomagnetic field lines, not simply by space coordinates. Therefore to know the densities at a certain point, the field line that passes through the point has to be known. In case of the dipole field model, each field line passing through a point of interest is represented by a simple function of the coordinates of the point. In the non-dipolar model mentioned in the previous section, each field line can not be represented by a simple function of the point coordinates, but must be determined by field line tracing, that requires additional numerical integration. Such a time consuming process can be hastened by performing multiple field line tracing in advance in the region concerned before the ray tracing calculation [Kimura et al., 1985].

Ray tracing in the polar region, especially for AKR, is made by using a special electron density profile in which a density cavity exists along an auroral magnetic field line, in addition to the magnetospheric plasma density model [e.g. Hashimoto, 1984].

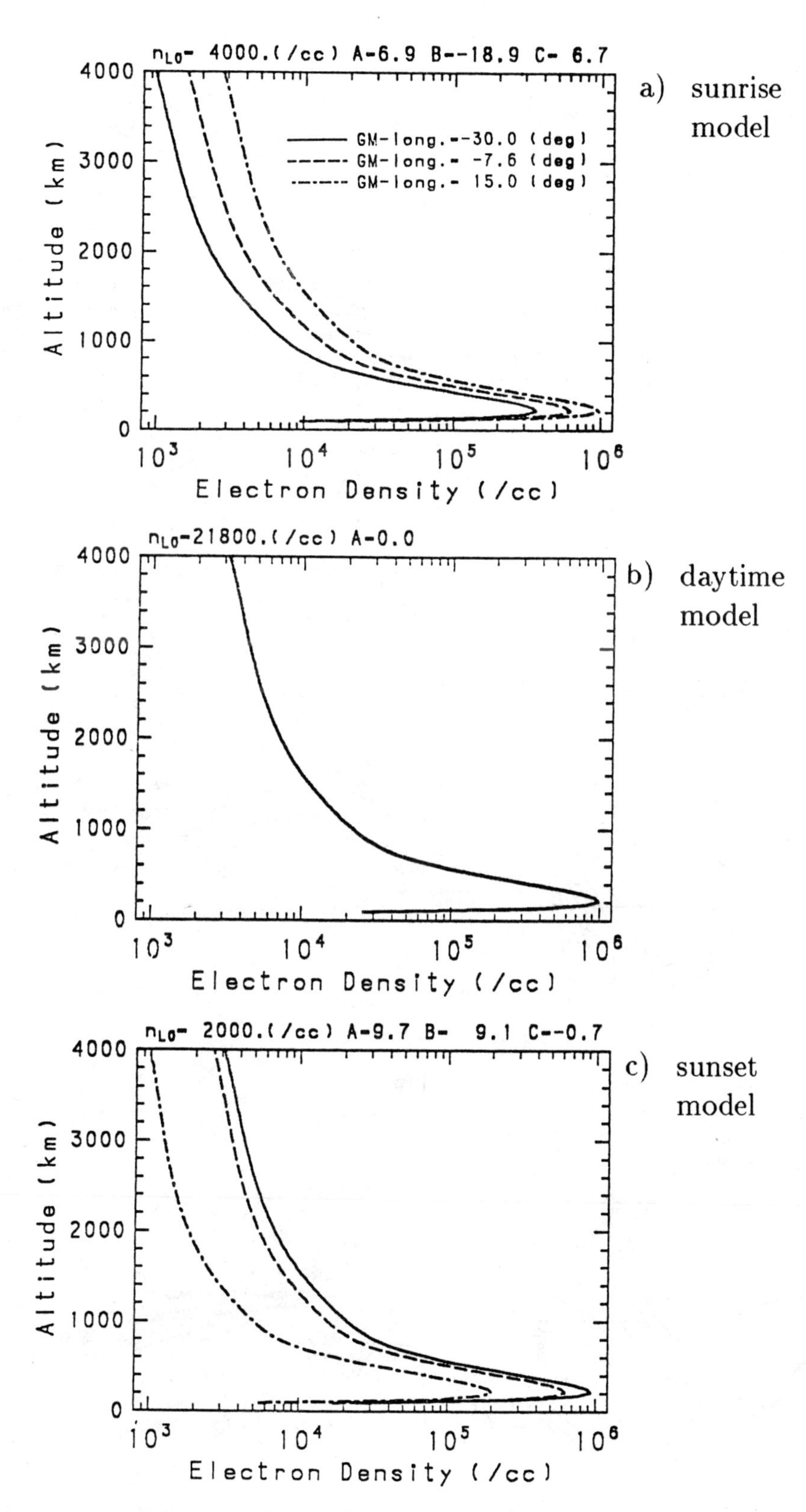

Fig. 2. Electron density profiles to be used for ray tracing, taking account of longitudinal gradients of electron density. (a) sunrise model, (b) daytime model, (c) sunset model.

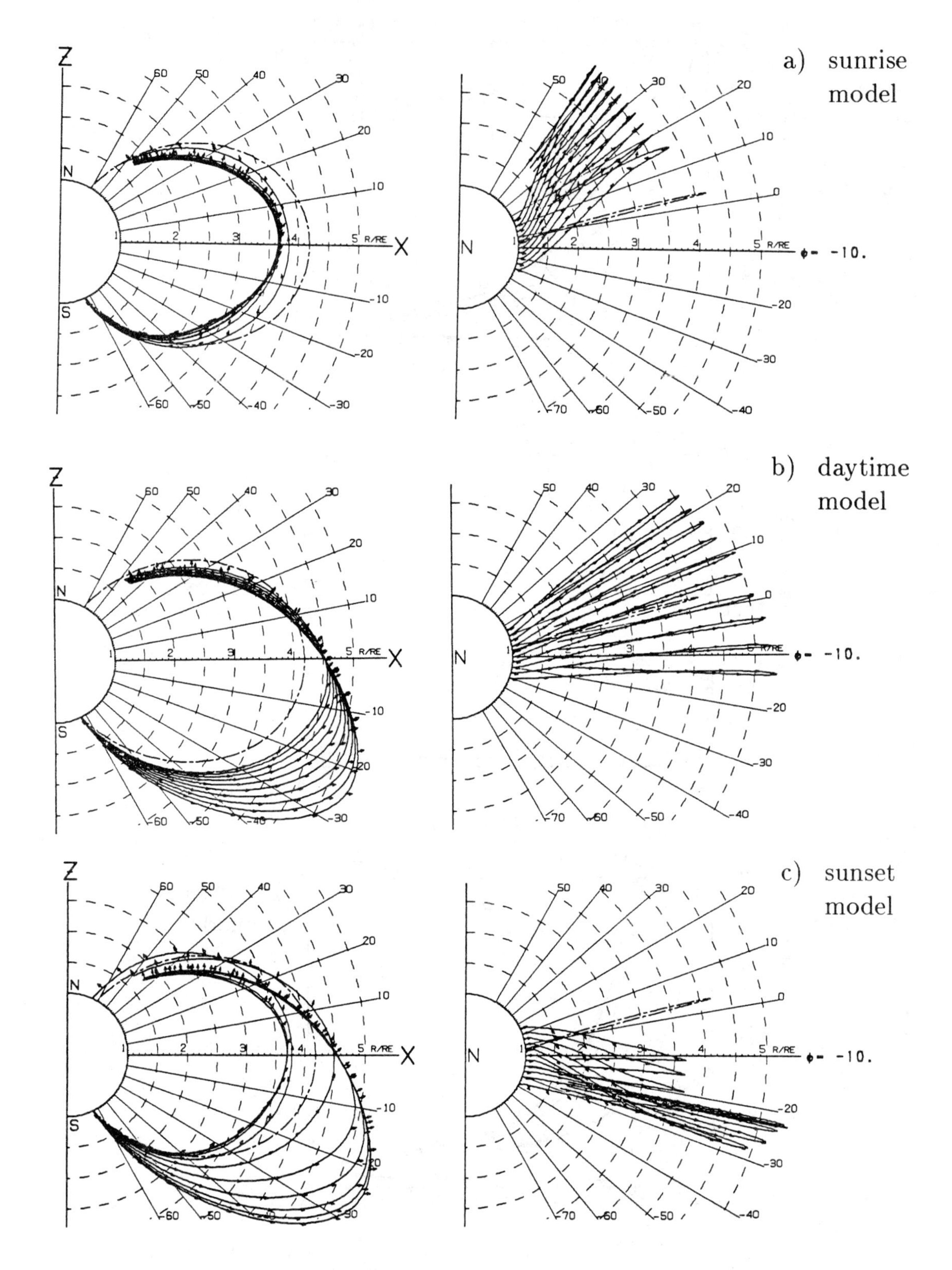

Fig. 3. Ray paths of a 5 kHz signal transmitted from Siple transmitter, in the non-dipolar magnetic field. (a), (b) and (c) correspond to those calculated using the density profiles in the three conditions shown in Figure 2.

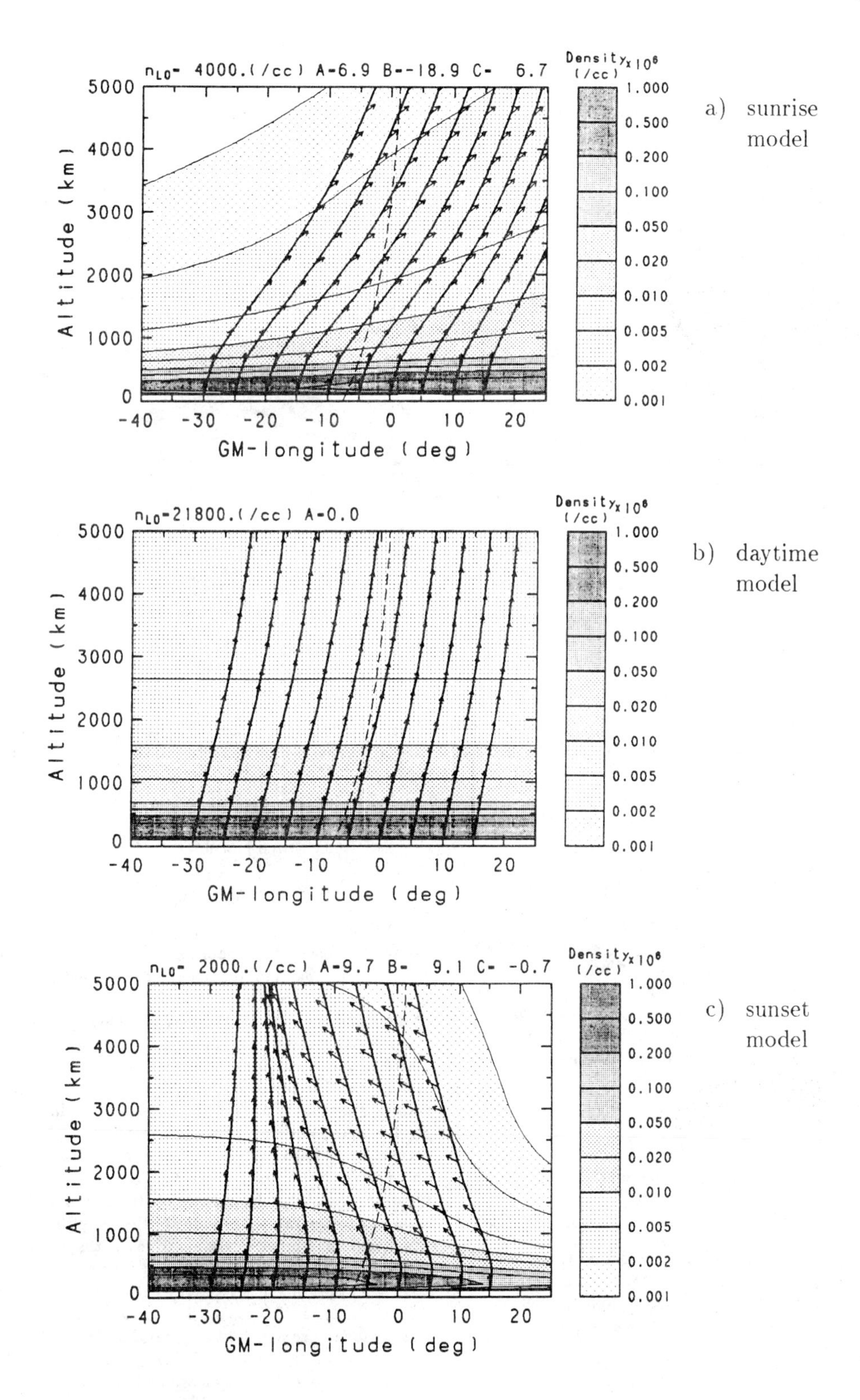

Fig. 4. Ray paths in lower altitudes, shown together with contour maps of electron density. (a), (b) and (c) correspond to the same conditions as in Figure 2. Wave normal directions are denoted by arrows on the ray paths. A magnetic field line passing through Siple station is drawn by a broken line in each panel.

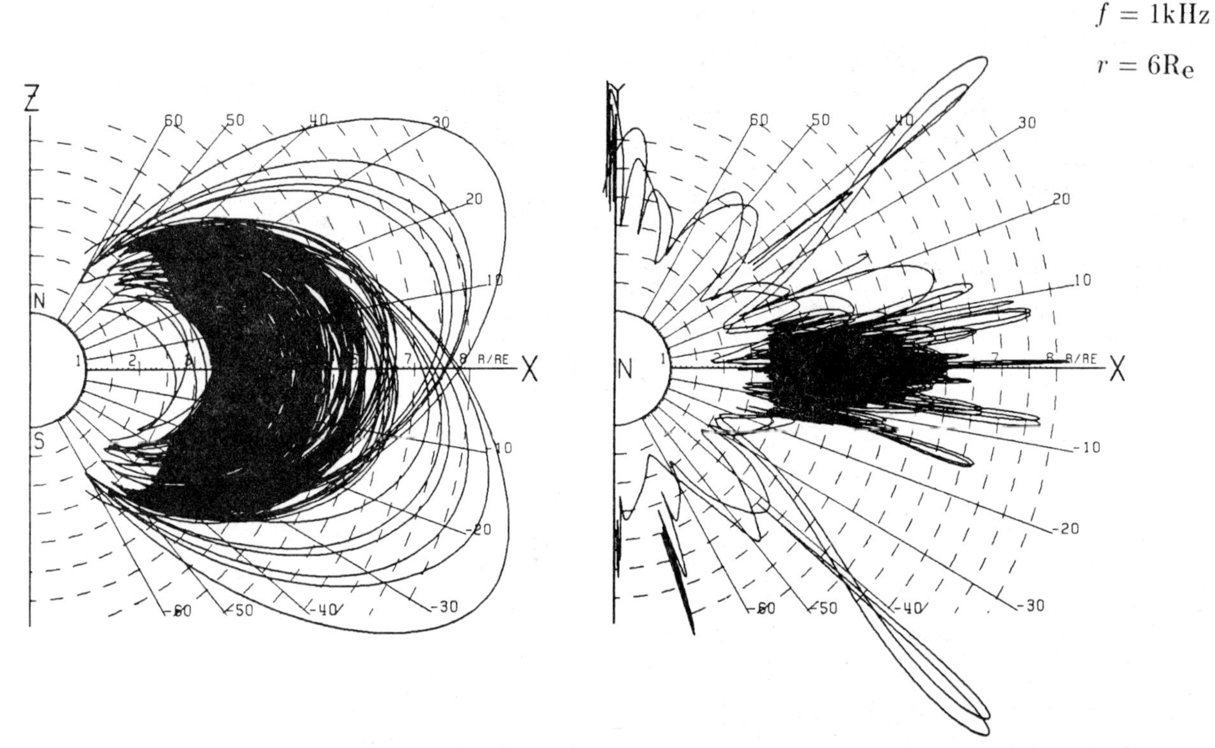

Fig. 5. One kilohertz ray paths starting from six earth radii at the equatorial plane. The left panel represents the projection of the ray paths to a magnetic meridian plane and the right panel represents those to the magnetic equatorial plane.

Some Applications of the New Ray Tracing Techniques

3-D Ray Tracing in a Cold Plasma in a Nondipolar Magnetic Field Model

Hereafter we will show a typical example of 3-D ray tracing in a cold plasma in a nondipolar geomagnetic field. Namely, ray paths of Siple transmitter signal in the magnetosphere starting from the sunrise and sunset regions of the ionosphere are calculated. A longitudinal gradient is, then, taken into account by multiplying the following longitude (ϕ) dependent function with constants A, B, and C:

$$\frac{A}{1+\exp\left(\frac{\phi-C}{B}\right)} \tag{11}$$

to the ordinarily adopted plasma density profile function

$$N_E = N_{DE} N_L \tag{12}$$

where N_{DE} represents diffusive equilibrium model with a plasmapause (Aikyo-Ondoh model) and N_L represents lower ionosphere model

For sunrise time, medium daytime, and sunset time the following parameters are used for N_{L0}, A, B, and C in the above (11).

(A-1) sunrise time model

$$N_{L0} = 4000(/cc),\ \ A = 6.9,\ \ B = -18.9,\ \ C = 6.7$$

(A-2) daytime model

$$N_{L0} = 21800(/cc),\ \ A = 0$$

(A-3) sunset time model

$$N_{L0} = 2000(/cc),\ \ A = 9.7,\ \ B = 9.1,\ \ C = -0.7$$

Figures 2 a), b), and c) show electron density profiles corresponding to the above three conditions (sunrise, daytime, and sunset time models), and the solid, broken and dot-dashed line correspond to geomagnetic longitude of −30°, −7.6°, and 15.0°. Siple station is located at a geomagnetic longitude of −7.7°. Three panels in Figures 3 illustrate, for the above three conditions, the ray paths of a 5kHz signal starting from an altitude of 120km at the Siple latitude 64.8°S and at longitudes from 30°W to 15°E with a 5° interval. The left panels show the ray paths projected on a same meridian plane, whereas the right panels show the ray paths projected on the geomagnetic equatorial plane. Outstanding deviation of the ray paths at the sunrise and sunset time from those in the daytime is quite obvious. Figure 4 illustrate the ray paths in lower altitudes in more detail. The arrows attached to the ray paths indicate the direction of the wave normal. The above mentioned deviation of ray paths due to longitudinal gradients of electron density is reduced in the dipole model. It is, then, concluded that as a whole the ray paths from Siple station in the nondipolar model deviate estward from the Siple dipole meridian plane. This tendency is seen on the data of Siple VLF signals observed by EXOS-B/Siple experiments [Bell et al., 1983].

Another example of application of the 3-D ray tracing in a nondipolar model is to find a condition on how far in longitude ray paths starting from a deep equatorial region can propagate away. The initial motivation of such a ray tracing was to study whether or not ELF emission generated by storm sudden commencement (SC) can propagate to the dawnside or duskside of the ionosphere. The ray tracing was made by using a plasma profile with the plasmapause at $L = 4$. One kHz signal is emitted from $L = 6$ on the equatorial plane with initial wave normal directions isotropically distributed. As the result, several ray paths can reach meridian plane more than 90° in longitude away from the source meridian, as shown in Figures 5. This phenomenon is a result of 3-D ray tracing and is attributed to LHR reflections, and the non-dipolar magnetic field is not always essential for this phenomenon.

3-D Ray Tracing of ECHW and Z Mode in a Hot Plasma

The following example of ray tracing is made in a dipole geomagnetic model. In order to confirm Oya and Jones' mechanism of mode conversion from electrostatic electron cyclotron harmonic mode waves (ECHW) to a free space L-O mode through R-X-Z and L-X-Z mode, ray paths starting in the electromagnetic L mode from the radio window (represented by a circle in Figure 6) are traced inversely, propagating first in L-X-Z then in R-X-Z, and finally passing over the f_q resonance.

Some examples of ray paths are shown in Figures 7. In these figures the starting locations are $L = 3.2$ on the magnetic equatorial plane with initial wave normal angle of 180° in a), $L = 3.2$ with the angle of 174° in b), and $L = 3.5$ with the angle of 180° in c). In all the three cases, the starting condition is chosen to satisfy the radio window condition, namely the wave frequency is the local plasma frequency there. In a) and b), Z mode region is from $L = 3.20$ to 3.24, where the upper hybrid frequency is equal to the wave frequency. The sharp reflection point around $L = 3.30$ indicates for the wave to pass over the f_q resonance condition. After the reflection the ray path going down to L less than 3.15 corresponds to the large wave number portion of ECHW mode. b) clearly indicates the the trapping of ECHW in the vicinity of the magnetic equatorial plane, as indicated by Barbosa [1980], and Barbosa and Kurth [1980].

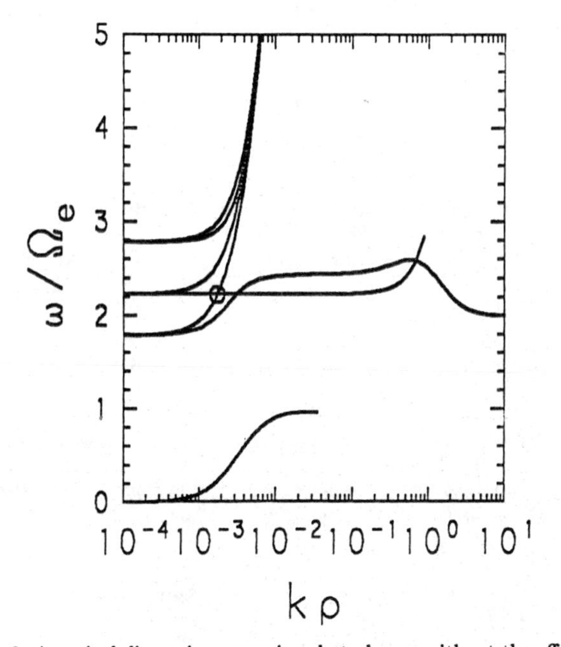

Fig. 6. A typical dispersion curve in a hot plasma without the effect of ions. The small wave number region corresponds to electromagnetic modes and the large wave number region corresponds to electrostatic modes, such as the electrostatic electron cyclotron harmonic wave (ECHW) mode. A circle represents the radio window.

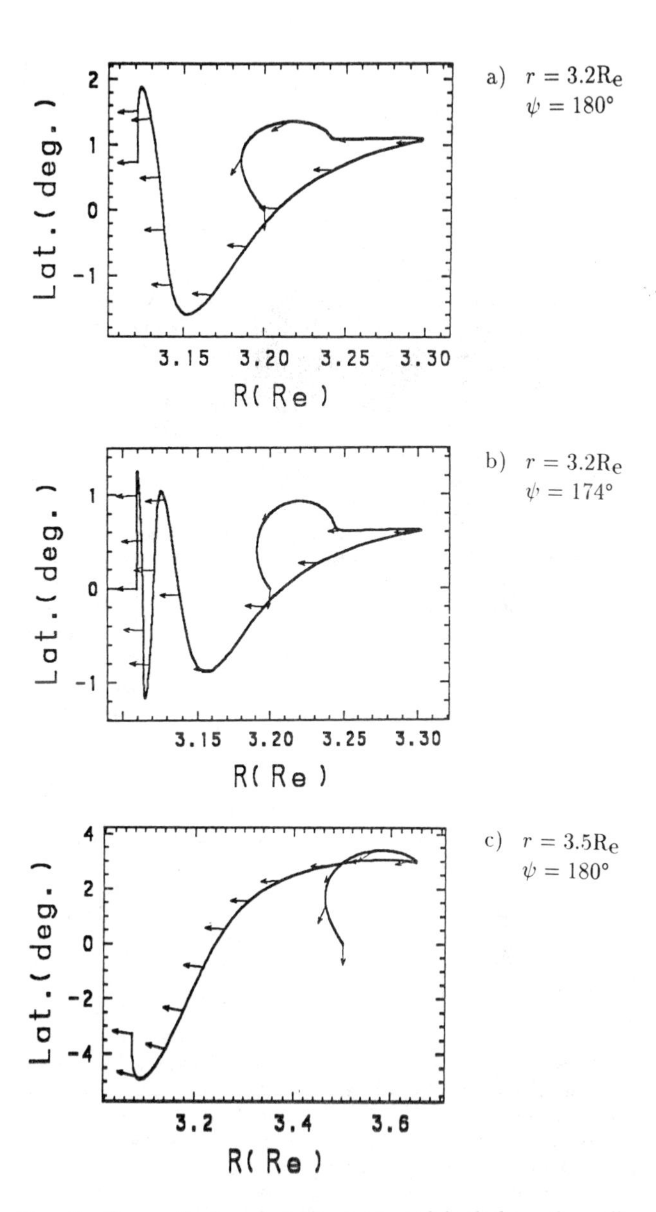

Fig. 7. Ray paths in a hot plasma, traced back from the radio window. (a), (b) and (c) correspond to the paths calculated with different initial conditions (see text in detail).

Figure 7 c) corresponds to a case where the ray path does not encounter the f_q resonance condition, so that the reflection point at $L = 3.66$ corresponds to the conditon that the wave frequency is equal to the upper hybrid frequency. In these ray path calculations, the path are terminated when the imaginary part of the wave number exceed one hundredth of the real part.

From these ray paths, it is clearly shown that a mode conversion smoothly takes place between Z mode electromagnetic mode and pure electrostatic cyclotron harmonic mode(ECHW). In these ray tracing, path integrated damping can also be calculated. In the above cases, however, the damping effect was not found to be serious.

In our program, we can also take account of temperature anisotropy of plasma, and non-thermal components, which cause growth as well as damping on the ray path.

Discussion and Conclusions

By the recent remarkable development of computational techniques, most of the requirements for tracing the ray paths of plasma waves in space plasmas have been satisfied. Namely three-dimensional cold plasma ray tracing programs for the waves in the ionosphere and magnetosphere in a dipole maganetic field model have been widely utilized by many scientists in the world. The 3-D cold plasma ray tracing in a non-dipolar model has also become available, which can be used to investigate an expected ray path distortion due to the difference of the magnetic field model from the dipole model. As an example of such trials, the effect of longitudinal gradients of ionospheric electron density in the dawn and dusk regions on the ray paths of a ground based VLF transmitter signal was introduced here.

In a hot plasma, so-called dispersion solver has been developed successfully to be able to calculate and display $\omega - k$ diagram, though no description is given on the dispersion solver in the present paper. The ray tracing in the hot plasma is an extension of this technique combined with the cold plasma ray tracing technique. For plasma distribution functions, a variety of functions can be involved besides the simple Maxwellian distribution with an isotropic temperature. Inclusion of the effect of plasma temperature and/or velocity distributions of particles generally gives rise to a imaginary part of refractive index for both cold and hot plasma modes, which causes growth or damping of the wave. Path integrated growth or damping has been treated in many papers, as previously reviewed. In such a case, it must be reminded that the calculated ray path is correct only when the imaginary part of the refractive index is sufficiently small.

We also need to remember that some part of the ray paths of electrostatic modes can be calculated much faster by using the electrostatic approximation than that by using the full hot dispersion relation. However, for the problems like a mode conversion between the electrostatic and the electromagnetic modes, the full hot plasma ray tracing program plays an essential role.

By many elaborate observational instruments to be installed on the future satellites, further abundant and high quality data of plasma waves in the space around the earth and around the other planets will be accumulated, by which necessary information on the amplitude, the polarization, and the mode will be determined. Ray tracing techniques will become more and more necessary for a variety of problems as reviewed in the first part of the present paper.

Acknowledgments. I am grateful to K. Hashimoto for his assistance in preparing this paper, and to H. Matsumoto and Y. Omura for their valuable discussion and comments. The computer programs for 3-D cold plasma and hot plasma ray tracing have been developed by K. Hashimoto, T. Matsuo, and many graduate and undergraduate students who belong to our laboratory. Ray paths illustrated in the present paper were calculated by A. Sawada, a master course student. The works by these staff and students are highly acknowledged. All computations were made at the Data Processing Center of Kyoto University.

Finally, I appreciate R. Thorne and S. Kokubun for their valuable comments and suggestions as the reviewers of this paper.

References

Aikyo, K., and T. Ondoh, Propagation of nonducted VLF waves in the vicinity of the plasmapause, *J. Radio Res. Labs.*, *14*, 153, 1971.

Barbosa, D. D., On the convective properties of magnetospheric Bernstein waves, *J. Geophys. Res.*, *85*, 2341, 1980.

Barbosa, D. D., and W. S. Kurth, Superthermal electrons and Bernstein waves in Jupiter's inner magnetosphere, *J. Geophys. Res.*, *85*, 6729, 1980.

Barbosa, D. D., Electrostatic wave propagation and trapping near the magnetic equator, *Ann. Geophysicae.*, *3*, 63, 1985.

Bell, T. F., The wave magnetic field amplitude threshold for nonlinear trapping of energetic gyroresonant and Landau resonant electrons by nonducted VLF waves in the magnetophere, *J. Geophys. Res.*, *91*, 4365, 1986.

Bell, T. F., U. S. Inan, I. Kimura, H. Matsumoto, T. Mukai, and K. Hashiomoto, EXOS-B/Siple station VLF wave-particle interaction experiments. 2. Transmitter signals and associated emissions, *J. Geophys. Res.*, *88*, 295, 1983.

Benson, R. F., Harmonic auroral kilometric radiation of natural origin, *Geophys. Res. Lett.*, *9*, 1120, 1982.

Benson, R. F., Ordinary mode auroral kilometric radiation, with harmonics , observed by ISIS-1, *Radio Sci.*, *19*, 543, 1984.

Brown, L. W., The galactic radio spectrum between 130 kHz and 2600 kHz, *Astrophys. J.*, *180*, 359, 1973.

Cairo, L., and F. Lefeuvre, Localization of ELF/VLF hiss observed in the magnetosphere: Three-dimensional ray tracing, *J. Geophys. Res.*, *91*, 4352, 1986.

Church, S. R., and R. M. Thorne, On the origin of plasmaspheric hiss: Ray path integrated amplification, *J. Geophys. Res.*, *88*, 7941, 1983.

Edger, B. C., Theory of VLF doppler signatures and their relation to magnetospheric density structure, *J. Geophys. Res.*, *81*, 3327, 1976.

Engel, J., and Kennel, C. F., Effect of parallel refraction on magnetospheric upper hybrid waves, *Geophys. Res. Lett.*, *11*, 865, 1984.

Gorney, D. J. and R. M. Thorne, A comparative ray-trace study of whistler ducting processes in the Earth's plasmasphere, *Geophys. Res. Lett.*, *7*, 133, 1980.

Gurnett, D. A., and R. R. Shaw, Electromagnetic radiation trapped in the magnetosphere above the plasma frequency, *J. Geophys. Res.*, *78*, 8136, 1973.

Gurnett, D. A., The earth as a radio source: terrestrial kilometric radiation, *J. Geophys. Res.*, *79*, 4227, 1974.

Gurnett, D. A., The earth as a radio source: The nonthermal continuum, *J. Geophys. Res.*, *80*, 2751, 1975.

Gurnett, D. A., R. R. Shaw, R. R. Anderson, W. S. Kurth, and F. L. Scarf, Whistlers observed by Voyager 1: Detection of lightning on Jupiter, *Geophys. Res. Lett.*, *6*, 511, 1979.

Gurnett, D. A., W. S. Kurth, and F. L. Scarf, Narrowband electromagnetic emissions from Saturn's magnetosphere, *Nature*, *292*, 733, 1981.

Gurnett, D. A., S. D. Shawhan, and R. R. Shaw, Auroral hiss, Z mode radiation, and auroral kilometric radiation in the polar magnetosphere, *J. Geophys. Res.*, *88*, 329, 1983.

Hashimoto, K., and M. L. Goldstein, A theory of the Io phase asymmetry of the Jovian decametric radiation, *J. Geophys. Res.*, *88*, 2010, 1983.

Hashimoto, K., A reconciliation of propagation modes of auroral kilometric radiation, *J. Geophys. Res.*, *89*, 7459, 1984.

Hashimoto, K., K. Yamaashi, and I. Kimura, Three-dimensional ray tracing of electrostatic cyclotron harmonic waves and Z mode electromagnetic waves in the magnetosphere, *Radio Sci.*, *22*, 579, 1987.

Hashimoto, K., and W. Calvert, Z mode observations with DE 1 and analysis by 3-D ray tracing, *Chapman Conference on Plasma Waves and Instabilities and Comets, Sendai*, 1987.

Huang, C. Y., C. K. Goertz, and R. R. Anderson, A theoretical study of plasmaspheric hiss generation, *J. Geophys. Res.*, *88*, 7927, 1983.

Jones, D., and J. Stephenson, A versatile three-dimensional ray tracing computer program for radio waves in the ionosphere, *OT Report, 75-76*, 1975.

Jones, D., Mode coupling of a Cerenkov radiation as a source of noise above the plasma frequency, *Proc. the 10th Eslab. Symp., Vienna (1975)*, p.281, D. Reidel .Hingham Mass 1976a

Jones, D., Source of terrestrial non-thermal radiation, *Nature*, *260*, 686, 1976b.

Kimura, I., Whistler mode propagation in the earth and planetary magnetospheres and ray tracing technique, *Space Sci. Rev.*, *42*, 449, 1985.

Kimura, I., T. Matsuo, M. Tsuda, and K. Yamauchi, Three dimensional ray tracing of whistler mode wave in a non-dipolar magnetoshere, *J. Geomagn. Geoelectr.*, *37*, 945, 1985.

Kitamura, T. and J. A. Jacobs, Ray paths of Pc 1 waves in the magnetosphere, *Planet. Space Sci.*, *16*, 863, 1968.

Kurth, W. S., D. A. Gurnett, and F. L. Scarf, Sporadic narrowband radio emissions from Uranus, *J. Geophys. Res.*, *91*, 11958, 1986.

Ksanfomaliti, L. V., N. M. Vasil'chikov, O. F. Ganpntserova, E. V. Petrova, A. O. Sowvorov, G. F. Filippov, O. V. Vablonskaya, and L. V. Yabrova, Electrical discharges in the atmosphere of Venus, *Pisma Astrom. Zh.*, *5*, 229, 1979.

Kurth, W.S., D. A. Gurnett and F. L. Scarf, Sporadic narrow band radio emissions from Uranus, *J. Geophys. Res.*, *91*, 11958, 1986.

Lee, L. C., and C. S. Wu, Amplification of radiation near cyclotron frequency due to electron population inversion, *Phys. Fluids*, *23*, 1348, 1980.

Lefeuvre, F., and Helliwell, R. A., Characterization of the sources of VLF hiss and chorus observed on GEOS 1, *J. Geophys. Res.*, *90*, 6419, 1985.

Lembege, B., and D. Jones, Propagation of electrostatic upper-hybrid emission and Z mode waves at the geomagnetic equatorial plasmapause, *J. Geophys. Res.*, *87*, 6187, 1982.

Matsuo, T., I. Kimura, and H. Yamagishi, ISIS-I and ISIS-II observation of emissions triggered by doppler shifted Norway Omega signals, *Mem. Natl. Inst. Polar Res. Special Issue*, *36*, 165, 1985.

Menietti, J. D., and D. A. Gurnett, Whistler propagation in the Jovian magnetosphere, *Geophys. Res. Lett.*, 7, 49, 1980.

Menietti, J. D., J. L. Green, S. Gulkis, and N. F. Six, Three-dimensional ray tracing of the Jovian magnetosphere in the low-frequency range, *J. Geophys. Res.*, *89*, 1489, 1984a.

Menietti, J. D., J. L. Green, S. Gulkis, and N. F. Six, Jovian decametric arcs: An estimate of the required wave normal angles from three-dimensional ray tracing, *J. Geophys. Res.*, *89*, 8089, 1984b.

Menietti, J. D., and C. S. Lin, Ray tracing of Z-mode emissions from source regions in the high altitude auroral zone, *Geophys. Res. Lett.*, *12*, 385, 1985.

Menietti, J. D., and C. S. Lin, Ray tracing survey of Z mode emissions from source regions in the high-altitude auroral zone, *J. Geophys. Res.*, *91*, 13559, 1986.

Menietti, J. D., J. L. Green, N. F. Six, and S. Gulkis, Ray tracing of Jovian decametric radiation from southern and northern hemisphere sources: Comprarison with Voyager observations, *J. Geophys. Res.*, *92*, 27, 1987.

Miyamoto, K., *Plasma Physics for Nuclear Fusion*, MIT Press, Cambridge.Mass.1980

Omidi, N. C., C. S. Wu, and D. A. Gurnett, Generation of auroral kilometric and Z mode radiation by the cyclotron maser mechanism, *J. Geophys. Res.*, *89*, 883, 1984.

Omidi, N. C., and D. A. Gurnett, Path integrated growth of auroral kilometric radiation, *J. Geophys. Res.*, *89*, 10801, 1984.

Oya, H., Conversion of electrostatic plasma waves into electrostatic waves:numerical calculation of the dispersion relation for all wavelength, *Radio Sci.*, *6*, 1113, 1971.

Oya, H., and A. Morioka, Observationnal evidence of Z and L-O mode waves as the origin of auroral kilometric radiation from the Jikiken (EXOS-B) satelliate, *J. Geophys. Res.*, *88*, 6189, 1983.

Rauch, J. L. and A. Roux, Ray tracing of ULF waves in a multicomponent magnetospheric plasma: Consequences for the generation mechanism of ion cyclotron waves, *J. Geophys. Res.*, *87*, 8191, 1982.

Sawada, A and I. Kimura, Sunrise and sunset ionosphere effects on the ray paths from Siple station in the magnetosphere, *unpublished materials*, 1987.

Sawada, A., and I. Kimura, Ray tracing survey for widely longitudinal propagation of whistler-mode waves in the magnetosphere, *Chapman Conference on Plasma waves and instabilities in magnetosheres and Comets, Sendai*, 1987.

Scarf, F. L., D. A. Gurnett, W. S. Kurth, and R. L. Poynter, Voyager plasma wave measurements at Saturn, *J. Geophys. Res.*, *88*, 8971, 1983.

Shaw, R. R., and D. A. Gurnett, A test of two theories for the low-frequency cutoffs of nonthermal continuum radiation, *J. Geophys. Res.*, *85*, 4571, 1980.

Solomon, J. N., N. Cornilleau-Wehrlin, A. Korath and G. Kremser, Generation of ELF electromagnetic waves and diffusion of energetic electrons in steady and non-steady state situations in the earth's magnetosphere, *Chapman Conference on Plasma waves and instabilities in magnetosheres and Comets, Sendai*, 1987, and to be published in the present AGU Monograph.

Stix, T. H., *The Theory of Plasma Waves*, McGraw-Hill Book Co. Inc, 1962.

Strangeways, H. J. and M. J. Rycroft, Trapping of whistler-waves through the side of ducts, *J. Atmos. Terr. Phys.*, *42*, 983, 1980.

Strangeways, H. J., Trapping of whistler-mode waves in ducts with tapered ends, *J. Atmos. Terr. Phys.*, *43*, 1071, 1981.

Strangeways, H. J., Whistler leakage from narrow ducts, *J. Atmos. Terr. Phys.*, *48*, 455, 1986.

Thorne, R. M. and J. J. Moses, Resonant instability near the two-ion crossover frequency in the Io plasma torus, *J. Geophys. Res.*, *90*, 6311, 1985.

Weinberg, S., Eikonal method in magnetohydrodynamics, *Phys. Rev.*, *126*, 1899, 1962.

Wu, C. S., and L. C. Lee, A theory of the terrestrial kilometric radiation, *Astrophys. J.*, *230*, 621, 1979.

Yamaashi, K., K. Hashimoto, and I. Kimura, 3-D electrostatic and electromagnetic ray tracing in the magnetosphere, *Mem. Natl. Inst. Polar Res. Special Issue*, *47*, 192, 1987.

THEORY OF THE DRIFT MIRROR INSTABILITY

Akira Hasegawa

AT&T Bell Laboratories, Murray Hill, New Jersey 07974

Liu Chen

Plasma Physics Laboratory, Princeton University, Princeton, New Jersey 08540

Abstract. We present a new comprehensive theory of the drift mirror instability which takes into account the inhomogeneity in the direction parallel as well as perpendicular to the ambient magnetic field, the effect of finite Larmor radius and the coupling to the shear Alfveń wave.

I. Introduction

The drift mirror instability (Hasegawa 1969) is an instability in a high beta plasma which is driven by an anisotropic plasma pressure with its frequency given by the diamagnetic drift frequency. The threshold condition is given by the classic mirror instability condition [Chandrasekhar et al, 1957], $\beta_\perp^{-1} < P_\perp / P_\parallel - 1$, where $\beta = nT/(B^2/2\mu_o)$ is the ratio of the plasma energy density to the magnetic energy density, P is the plasma pressure and the subscripts $\perp$ and $\parallel$ indicate the component perpendicular and parallel to the background magnetic field. However the real part of the frequency is given by the diamagnetic drift frequency due to the coupling to the drift wave with the perpendicular wavelengths comparable to the ion gyroradius. Southwood [1976] treated the drift-mirror instability taking into account of the dipole geometry of the earth magnetic field.

Recently the drift mirror instability is attracting much interest in relation to an increased number of observations of "compressional waves" in various space plasma locations such as in the magnetosphere [Kremser et al., 1981; Wedeken et al., 1984; Takahashi et al., 1985a; Takahashi et al., 1985b, Lin and Barfield, 1985; Higuchi et al., 1986;Greenstadt et al., 1986; Takahashi and Higbie 1986; Baumjohann et al., 1987, Takahashi et al., 1987], in the magnetosheaths of different planets [Tsurutani et al. 1982; Tsurutani et al., 1984] or near the comet Halley [Tsurutani et al. 1987]. Furthermore, multi-satellite observations have revealed strong evidences of short perpendicular wavelengths [Takahashi et al., 1985b].

However simultaneous observations on the background plasma data often fail to satisfy the instability condition and the observed frequencies are also sometimes off from the predicted diamagnetic drift frequency.

In order to account for the observed frequency and an existence of shear component of the observed waves, Walker et al., [1982] considered the coupling between the drift mirror mode and the shear Alfveń wave. Although the idea is right, the theory has a major flaw because the authors erroneously replaced ω simply by $\omega - k_\perp v_d$, where v_d is the drift velocity of particles, in deriving the dispersion relation. Attempts of correcting the error have been made by many authors [Lin and Parks, 1982; Migliuolo, 1983, Ng and Patel, 1983]. In particular Pokhotelov et al. [1986] have pointed out that the curvature driven mode (ballooning mode) can reduce the threshold condition when the coupling to the shear Alfveń mode (field line oscillation) is taken into account. The theory however does not take into account the inhomogeneity along the field line hence fails to explain the mode structure. In addition since the threshold condition should be averaged over along the field line, to neglect this effect is critical in evaluating the significance of the claim. Most recently Cheng and Lee [1987] have analyzed the drift mirror mode by taking into account the inhomogeneity and the eigenmode structure along the field line. However they failed to include the coupling with the Alfveń wave as well as the full ion Larmor radius effect in their analysis.

In this manuscript, we present a comprehensive theory of the coupling between the drift mirror instability and the shear Alfveń wave in an inhomogeneous axisymmetric plasma which includes the effect of a radial as well as parallel pressure and magnetic field gradients, anisotropic pressure and curvature of the magnetic field.

We first present in Section II the hydromagnetic calculation based on an adiabatic response of the plasma. There we elucidate the coupling mechanism which occurs due to the gradient of the perpendicular component of the plasma pressure in the direction of $\nabla\psi$ (where ψ is the flux function) and the instability mechanisms; the anisotropic pressure for the mirror mode and the combination of curvature and pressure gradients for the Alfveń mode. The coupled equations illuminate the mechanism whereby the instability conditions for either the mirror mode or the Alfveń mode can be met at a condition which is less stringent than the case of each mode uncoupled.

In Section III, we present results of the kinetic theory which includes the wave particle interactions through the bounce and the drift resonances and the finite Larmor radius correction. The coupled equations have a structure identical to the hydromagnetic calculation of Section II in the absence of these effects.

II. Hydromagnetic Theory

The study of plasma instability in the hydromagnetic frequency range has a long history. Although to derive the full nature of the drift mirror instability, the kinetic theory which includes the finite Larmor radius is required, the condition of instability can be derived by the energy principle. In fact there are a large number of papers published for various range of parameters on the related subjects [Van Dam et al. 1982].

In this section we derive the coupled equation between the mirror and the Alfveń mode based on hydromagnetic equations. The derivation is significantly simplified under the assumption of almost perpendicular propagation in the azimuthal direction whereby the magnetic field perturbation in the azimuthal direction is negligible. This assumption is justified by many observations of primarily compressional modes with an meridian polarization in the magnetosphere. Under this assumption, both of the equilibrium and the wave fields are expressed by two field variables ψ and B where ψ is the flux function and $B(\psi)$ is the magnitude of the magnetic flux density. The vector magnetic field $\mathbf{B}$ is related to the flux function ψ through,

$$\mathbf{B} = \nabla\psi \times \nabla\phi \tag{1}$$

where ϕ is the azimuthal angle. We note that for the spherical coordinate

$$\nabla\phi = \frac{\hat{\boldsymbol{\phi}}}{r\sin\theta}. \tag{2}$$

Because of the anisotropic pressure the equilibrium plasma pressure becomes a function of both ψ and B. The equilibrium condition is given by

$$\nabla\cdot\overleftrightarrow{\mathbf{P}} + \mathbf{J}\times\mathbf{B} = 0 \tag{3}$$

where the pressure tenser $\overleftrightarrow{\mathbf{P}}$, which is a function of ψ and B, is given by

$$\overleftrightarrow{\mathbf{P}} = P_\perp\overleftrightarrow{\mathbf{I}} + (P_\parallel - P_\perp)\hat{\mathbf{b}}\hat{\mathbf{b}}, \tag{4}$$

$\hat{\mathbf{b}} = \mathbf{B}/B$ is the unit vector in the direction of the magnetic field, $\overleftrightarrow{\mathbf{I}}$ is the unit tenser, and $\mathbf{J}$ is the current density given by

$$\mathbf{J} = \frac{1}{\mu_o}\nabla\times\mathbf{B}. \tag{5}$$

The parallel and perpendicular components of the equilibrium relation (3) can be written with (4)

$$\hat{\mathbf{b}}(\hat{\mathbf{b}}\cdot\nabla)P_\parallel + (P_\parallel - P_\perp)\hat{\mathbf{b}}(\nabla\cdot\hat{\mathbf{b}}) = 0 \tag{6}$$

or

$$B\frac{\partial P_\parallel}{\partial B} + P_\perp - P_\parallel = 0 \tag{6'}$$

and

$$\nabla_\perp P_\perp + (P_\parallel - P_\perp)(\hat{\mathbf{b}}\cdot\nabla)\hat{\mathbf{b}} + \frac{1}{\mu_o}\mathbf{B}\times(\nabla\times\mathbf{B}) = 0 \tag{7}$$

or

$$\tau\nabla_\perp \ell n B + \frac{\mu_o}{B^2}\frac{\partial P_\perp}{\partial\psi}\nabla_\perp\psi = \sigma\kappa, \tag{7'}$$

where

$$\tau = 1 + \frac{\mu_o}{B}\frac{\partial P_\perp}{\partial B} \tag{8}$$

$$\sigma = 1 + \frac{\mu_o}{B^2}(P_\perp - P_\parallel) \tag{9}$$

$$\boldsymbol{\kappa} = (\hat{\mathbf{b}}\cdot\nabla)\hat{\mathbf{b}} \tag{10}$$

and

$$\nabla = \nabla B\frac{\partial}{\partial B} + \nabla\psi\frac{\partial}{\partial\psi}. \tag{11}$$

We note that $\boldsymbol{\kappa}$ represents the curvature and τ represents the anisotropic pressure parameter which drives the mirror mode. For example, a bi-Maxwellian distribution given by

$$f(v_\perp, v_\parallel) = A(\psi)\exp\left(-\frac{v_\parallel^2}{2v_{T\parallel}^2} - \frac{v_\perp^2}{v_{T\perp}^2}\right)$$

can be expressed in terms of the magnetic moment μ and the energy ϵ as

$$f(\epsilon, \mu, \psi, B) = A(\psi)\exp\left[-\frac{\epsilon}{T_\parallel} - \mu B\left(\frac{1}{T_\perp} - \frac{1}{T_\parallel}\right)\right].$$

Thus

$$\frac{\partial f}{\partial B} = -\mu\left(\frac{1}{T_\perp} - \frac{1}{T_\parallel}\right)f$$

and

$$\tau = 1 + \frac{\mu_o}{B}\frac{\partial P_\perp}{\partial B} = 1 + \beta_\perp\left(1 - \frac{T_\perp}{T_\parallel}\right). \tag{8'}$$

We now consider a wave which propagates in the azimuthal as well as in the parallel direction. We describe the small amplitude wave field by δ. $\nabla\cdot\delta\mathbf{B} = 0$ requires an existence of an azimuthal component of the wave magnetic field. However for a almost perpendicular propagation, $k_\perp >> k_\parallel$, the azimuthal component δB_ϕ may be negligible. Then the wave magnetic field may be represented by two field variables, δB and $\delta\psi$ which represent the compressional component

$$\delta\mathbf{B}_\parallel = \delta B\hat{\mathbf{b}} \tag{12}$$

and the ψ (radially perpendicular to $\mathbf{B}$) component

$$\delta \mathbf{B}_\psi = \nabla \delta\psi \times \nabla \phi = |\nabla \phi| \hat{\mathbf{b}} \cdot \nabla \delta\psi \hat{\boldsymbol{\psi}}. \tag{13}$$

δB and δB_ψ are related to the current density perturbation δJ through the Maxwell equation,

$$ik_\phi \delta B = \mu_o \delta J_\psi \tag{14}$$

$$ik_\phi \delta B_\psi = -\mu_o \delta J_\parallel, \tag{15}$$

when k_ϕ is the azimuthal wavenumber. Equation (15) combined with the quasi neutrality condition, $\nabla \cdot \delta \mathbf{J} = 0$ gives,

$$\hat{\mathbf{b}} \cdot \nabla [ik_\phi |\nabla \phi| \hat{\mathbf{b}} \cdot \nabla \delta\psi] = \mu_o \nabla \cdot \delta \mathbf{J}_\perp. \tag{18}$$

The perturbed current density $\delta \mathbf{J}_\perp$ can be obtained from the linearized hydromagnetic equation of motion for an anisotropic pressure,

$$\delta \mathbf{J}_\perp = \frac{\hat{\mathbf{b}} \times \nabla \cdot \delta \overleftrightarrow{\mathbf{P}}}{B} + \frac{\delta \hat{\mathbf{b}} \times (\nabla \cdot \overleftrightarrow{\mathbf{P}})_\parallel}{B} - \frac{\delta B (\hat{\mathbf{b}} \times \nabla \cdot \overleftrightarrow{\mathbf{P}})}{B^2} + \delta \mathbf{J} p. \tag{17}$$

Here the perturbed pressure tenser $\overleftrightarrow{\mathbf{P}}$ is given from Eq. (4),

$$\nabla \cdot \delta \overleftrightarrow{\mathbf{P}} = \nabla \delta P_\perp + \delta[(P_\parallel - P_\perp)(\hat{\mathbf{b}} \cdot \nabla) \hat{\mathbf{b}}] + \delta[(P_\parallel - P_\perp) \hat{\mathbf{b}} \nabla \cdot \hat{\mathbf{b}}] + \delta[\hat{\mathbf{b}} \hat{\mathbf{b}} \cdot \nabla (P_\parallel - P_\perp)], \tag{18}$$

$$\delta \hat{\mathbf{b}} = \frac{B_\psi}{B} \hat{\boldsymbol{\psi}} \tag{19}$$

and $\delta[\ \ ]$ indicates the perturbed portion of $[\ \ ]$, for example

$$\begin{aligned} \delta[(P_\parallel - P_\perp)(\hat{\mathbf{b}} \cdot \nabla) \hat{\mathbf{b}}] &= (\delta P_\parallel - \delta P_\perp)(\hat{\mathbf{b}} \cdot \nabla) \hat{\mathbf{b}} \\ &+ (P_\parallel - P_\perp)(\delta \hat{\mathbf{b}} \cdot \nabla) \hat{\mathbf{b}} \\ &+ (P_\parallel - P_\perp)(\hat{\mathbf{b}} \cdot \nabla) \delta \hat{\mathbf{b}} \end{aligned} \tag{20}$$

Also from the equilibrium conditions (6) and (7),

$$(\nabla \cdot \overleftrightarrow{\mathbf{P}})_\parallel = 0, \quad (\nabla \cdot \overleftrightarrow{\mathbf{P}})_\perp = \frac{1}{\mu_o} (\nabla \times \mathbf{B}) \times \mathbf{B}.$$

The polarization current density $\delta \mathbf{J} p$ is obtained from

$$\delta \mathbf{J}_p = -\frac{\omega^2 \rho_m}{B} |\nabla \phi| \delta\psi \hat{\boldsymbol{\phi}} \tag{21}$$

where $\rho_m = m_i n$ is the mass density and ω is the angular frequency. The pressure perturbation $\delta P_\perp$ and $\delta P_\parallel$ are obtained using the fact that $P_\perp$ and $P_\parallel$ are the functions of B and ψ,

$$\delta P_\perp = \frac{\partial P_\perp}{\partial B} \delta B + \frac{\partial P_\perp}{\partial \psi} \delta\psi \tag{22}$$

$$\delta P_\parallel = \frac{\partial P_\parallel}{\partial B} \delta B + \frac{\partial P_\parallel}{\partial \psi} \delta\psi, \tag{23}$$

where $\partial P / \partial B$ and $\partial P / \partial \psi$ are the partial derivatives of P with respect to B and ψ for a fixed value of ψ and B respectively. We derive the wave equation assuming that these derivatives are given. Equations (16) and (17) together with (18), (19), (21), (22) and (23) are necessary and sufficient to derive the coupled wave equations we are seeking.

Let us first derive the mirror mode. The mirror mode is associated with the compressional component of the magnetic field, δB.

The relevant Maxwell's equation is (14) where the ψ component of the perturbed current density is obtained from Eq. (17) by constructing $\hat{\boldsymbol{\psi}} \cdot \delta \mathbf{J}_\perp$,

$$\begin{aligned} \delta J_\psi = \hat{\boldsymbol{\psi}} \cdot \delta \mathbf{J}_\perp &= -ik_\phi \frac{\delta P_\perp}{B} \\ &= -\frac{ik_\phi}{B} \left(\frac{\partial P_\perp}{\partial B} \delta B + \frac{\partial P_\perp}{\partial \psi} \delta\psi \right). \end{aligned}$$

Thus the wave equation reads,

$$\tau \delta B = \frac{\mu_o}{B^2} k_\phi (\hat{\boldsymbol{\phi}} \times \hat{\mathbf{b}}) \cdot \nabla \psi \frac{\partial P_\perp}{\partial \psi} \delta \tilde{\psi}, \tag{24}$$

where

$$\delta \tilde{\psi} = -|\nabla \phi| \delta\psi / k_\phi \tag{25}$$

We note here that 1. from Eq. (8′), $\tau \equiv 1 + \mu_o / B(\partial P_\perp / \partial B) = 1 - \beta_\perp (1 - T_\perp / T_\parallel)$ for a bi-Maxwellian distribution, thus giving the mirror instability condition, 2. the polarization current in ψ direction is ignored hence the magnetosonic wave (which is a wave at a frequency much higher than the shear and/or the drift wave considered here) is eliminated, 3. the wave particle interaction is expected to produce the drift mirror mode, and 4. the coupling to the shear mode (wave with $\delta\psi$ component of the field) occurs through the gradient of $P_\perp$ in the ψ direction.

The wave equation for the Alfveǹ mode is more complicated because there are many terms which contribute to $\nabla \cdot \delta \mathbf{J}_\perp$ in Eq. (17). After a tedious calculation of $\nabla \cdot \delta \mathbf{J}$ using Eq. (7), (18), (19), (21), (22) and (23), the resulting wave equation can be casted into the following form,

$$B \frac{\partial}{\partial \ell} \left[\frac{k_\phi^2}{B} \{1 + \frac{1}{2} (\beta_\perp - \beta_\parallel)\} \frac{\partial}{\partial \ell} \delta \tilde{\psi} \right]$$

$$+\frac{\omega^2 k_\phi^2}{v_A^2}\tilde{\delta\psi} = -\frac{\mu_o}{B^2}k_\phi(\hat{\phi}\times\hat{b})\cdot\nabla\psi\frac{\partial P_\perp}{\partial\psi}(\delta B+\Omega_B\tilde{\delta\psi})$$

$$-\frac{\mu_o}{B^2}k_\phi(\hat{\phi}\times\hat{b})\cdot\nabla\psi\frac{\partial P_\parallel}{\partial\psi}\Omega_\kappa\tilde{\delta\psi}, \qquad (26)$$

where

$$\Omega_B = k_\phi(\hat{\phi}\times\hat{b})\cdot\nabla \ell n B \qquad (27)$$

represents ∇B effect while

$$\Omega_\kappa = k_\phi(\hat{\phi}\times\hat{b})\cdot\boldsymbol{\kappa} \qquad (28)$$

represents the field line curvature effect (see Eq. (10)),

$$\frac{\partial}{\partial\ell} = \hat{b}\cdot\nabla \qquad (29)$$

and $v_A = B/\sqrt{\mu_o\rho_m}$ is the Alfveń speed.

In Eq. (26), the left hand side shows the shear Alfveń wave equation with the fire hose effect, $(\beta_\perp-\beta_\parallel)/2$. δB term in the right hand side shows the coupling to the compressional mode δB, which as can be seen from Eq. (24) is symmetric. $\Omega_B\delta\psi$ and $\Omega_\kappa\delta\psi$ terms represent the ballooning effect driven by $\nabla_\psi P_\perp$ and ∇B, as well as $\nabla_\psi P_\parallel$ and the curvature. The coupled equations (24) and (26) are the desired hydromagnetic wave equations which represent the mirror mode and the shear Alfveń mode. As can be seen, at near the threshold of the mirror instability $\tau \simeq 0$, the ballooning effect on the shear Alfveń wave is enhanced in the presence of $\partial P_\perp/\partial\psi$.

III. Kinetic Theory

To study the wave-particle interaction effects, we should use the Vlasov equation. For low frequency waves such as considered here, the gyrokinetic equation is a powerful tool because it can contain the finite Larmor radius effects. Because of the page limitation, we only present the result of the derivation. The details will be published elsewhere. The coupled wave equations (24) and (26) are modified to

$$\tau\delta B = \frac{\mu_o}{B^2}k_\phi\hat{\phi}\times\hat{b}\cdot\nabla\psi\frac{\partial P_\perp}{\partial\psi}\tilde{\delta\psi} + \mu_o\{\sum_{j=e,i} m_j < \frac{QF_0\mu}{\omega-\bar{\omega}_d}\overline{\Omega_d\delta\psi+\mu\delta B} >_j$$

$$-i\pi\sum_{j=e,i} < m_j\sum_{k\neq o} QF_0\mu\delta(\omega-\bar{\omega}_d-k\omega_b)\frac{2J_1(\lambda)}{\lambda}\cos I_{\ell_1}^{\ell}$$

$$\times\overline{\cos I_{\ell_1}^{\ell}(\Omega_d J_o(\lambda)\delta\psi+2\mu J_1(\lambda)\delta B/\lambda)} - \overline{|v_\parallel|\sin I_{\ell_1}^{\ell}\delta\psi\partial J_o(\lambda)/\partial\ell} >_j. \qquad (24')$$

$$B\frac{\partial}{\partial\ell}\left[\frac{k_\phi^2}{B}\{1+\frac{1}{2}(\beta_\perp-\beta_\parallel)\}\frac{\partial\tilde{\delta\psi}}{\partial\ell}\right]+\frac{\omega^2k_\phi^2\tilde{\delta\psi}}{v_A^2}$$

$$= -\frac{\mu_o}{B^2}k_\phi(\hat{\phi}\times\hat{b})\cdot\nabla\psi\frac{\partial P_\perp}{\partial\psi}(\delta B+\Omega_B\tilde{\delta\psi})$$

$$-\frac{\mu_o}{B^2}|k_\phi(\hat{\phi}\times\hat{b})\cdot\nabla\psi\frac{\partial P_\parallel}{\partial\psi}\Omega_\kappa\tilde{\delta\psi}$$

$$-\mu_o\{\sum_{j=e,i} m_j < \frac{QF_0}{\omega-\bar{\omega}_d}\Omega_d\overline{\Omega_d\delta\psi+\mu B} >_j$$

$$-i\pi\sum_{j=e,i} m_j < QF_0\sum_{k\neq o}\delta(\omega-\bar{\omega}_d-k\omega_b)(J_o(\lambda)\Omega_d\cos I_{\ell_1}^{\ell}$$

$$-|v_\parallel|\sin I_{\ell_1}^{\ell}\partial J_o(\lambda)/\partial\ell)$$

$$\times\overline{\cos I_{\ell_1}^{\ell}(\Omega_d J_o(\lambda)\delta\psi+2\mu J_1(\lambda)\delta B/\lambda)} - \overline{|v_\parallel|\sin I_{\ell_1}^{\ell}\delta\psi\partial J_o(\lambda)/\partial\ell} >_j. \qquad (26')$$

where

$$Q\,F_0 \equiv (\omega\partial_\epsilon+\hat{\omega}_*)F_0, \quad \Omega_d = \mu\Omega_B + v_\parallel^2\Omega_\kappa/B,$$

$$<\ldots> = \int d^3v(\ldots), \quad \bar{A} = (\oint\frac{d\ell}{|v_\parallel|}A)/\tau_b,$$

$$\lambda = k_\perp\rho, \quad I_a^b = \int_a^b\frac{d\ell}{|v_\parallel|}(\omega-\omega_d),$$

F_0 is the unperturbed distribution function, ρ is the Larmor radius a and b are bounce positions and ω_d and ω_b are the drift and bounce frequencies.

IV. Conclusion

We have derived the coupled wave equations for the drift mirror and the shear Alfveń modes based on the hydromagnetic theory as well as on the kinetic theory. The coupling occurs due to the gradient of perpendicular pressure in the direction perpendicular to the flux surface. By the coupling, the ballooning effect (the pressure gradient driven mode) and the mirror effect (the pressure anisotropy driven mode) helps each other to lower the threshold conditions.

The work done at Princeton University is supported by NSF grant ATM86-09585. The travel expense for A. Hasegawa to attend the conference is supported by NASA grant NAGW-894.

References

Baumjohann, W., N. Sckopke, J. Labelle, B. Klecker, H. Lühr and K. H. Glassmeier, Plasma and field observations of a compressional Pc5 event, J. Geophys. Res. (submitted), 1987.

Chandrasekhar, S., A. N. Kaufman, and K. M. Watson, The Stability of the pinch, Proc. Roy. Soc. A, 245, 435, 1958.

Cheng, C. Z., and C. S. Lin, Analysis of Compressible waves in the magnetosphere, Geophys. Res. Lett., 14, 884, 1987.

Greenstadt, E. W., R. L. McPherson, R. R. Anderson and F. L. Scarf, A storm time Pc5 event observed in the outer magnetosphere by ISEE1 and 2 wave properties, J. Geophys. Res., 91, 13, 398, 1986.

Hasegawa, A., Drift mirror instability in the magnetosphere, Phys. Fluids, 12, 2642, 1969.

Higuchi, T., S. Kokubun and S. Ohtani, Harmonic structure of compressional Pc5 pulsations at synchronous orbit, Geophys. Res. Lett., 13, 1101, 1986.

Kremser, G. A. Korth, J. A. Fejer, B. Wilken, A. V. Gurevich and E. Amata, Observation of quasi-periodic flux variations of energetic ions and elections associated with Pc5 geomagnetic pulsations, J. Geophys. Res., 86, 3345, 1981.

Lin, C. S. and J. N. Barfield, Azimuthal propagation of storm time Pc5 waves observed simultaneously by geostationary satellites GOES2 and GOES3, J. Geophys. Res., 90, 11, 075 1985.

Migliuolo, S., High-β theory of low frequency magnetic pulsations, J. Geophys. Res., 88, 2065 1983.

Ng, P. H. and V. L. Patel, The coupling of shear Alfveń and congressional waves in high-β magnetospheric plasma, J. Geophys. Res., 88, 10,035, 1983.

Pokhotelov, O. A. V. A. Pilipenko, Yu, M. Nezlina, J. Woch, G. Kremser, A. Korth and E. Amata, Excitations of high-β plasma instabilities at the geostationary orbit; theory and observations, Planet. Space Sci., 34, 695, 1986.

Southwood, D. J., A General Approach to Low-Frequency Instability in the Ring Current Plasma, J. Geophys. Res., 81, 3340, 1976.

Takahashi, R., P. R. Higbie and D. N. Baker, Azimuthal propagation and frequency characteristic of compressional Pc5 waves observed at geostationary orbit, J. Geophys. Res., 90, 1473, 1985a.

Takahashi, K., C. T. Russell and R. A. Anderson, ISEE1 and 2 observation of the spatial structure of compressional Pc5 wave, Geophys. Res. Lett., 12, 613, 1985b.

Takahashi, K. and P. Higbie, Antisymmetric standing wave structure associated with the compressional Pc5 pulsation of November 14, 1979, J. Geophys. Res., 91, 11, 163, 1986.

Takahashi, K., J. F. Fennell, E. Amata and P. R. Higbie, Field-aligned structure of the storm time Pc5 wave of November 14-15, 1979. J. Geophys. Res., 1987, to be published.

Tsurutani, B. T., E. J. Smith, R. R. Anderson, K. W. Ogilvie, J. D. Scudder, D. N. Baker and S. J. Bame, Lion roars and nonoscillatory drift mirror waves in the magnetosheath, J. Geophysics Res., 87, 6060, 1982.

Tsurutani, B. T., I. G. Richardson, R. P. Lepping, R. D. Zwickl, D. E. Jones, E. J. Smith and S. J. Bame, Geophys. Res. Lett., 11, 1102, 1984.

Tsurutani, B. T., A. L. Brinca, E. J. Smith, R. M. Thorne, F. L. Scarf, J. T. Gosling and F. M. Ipavich, MHD waves detected by ICE at distances $\geq 28\times 10^6$ km from comet Halley: cometary or solar wind origin? Astronomy and Astrophys., 1987, to be published.

Van Dam, J. W., M. N. Rosenbluth and Y. C. Lee "A Generalized Kinetic Energy Principle", Phys. Fluids, 25, 1349, 1982 and the references therein.

Walker, A. D. M., R. A. Greenwald, A. Korth and G. Kremser, Stare and GOES2 observations of a storm time Pc5 ULF pulsation, J. Geophys. Res., 87, 9135, 1982.

Wedeken, V., B. Inhester, A. Korth, K.-H. Glassmeier, R. Gendrin, L. J. Lanzerotti, H. Gough, C. A. Green, E. Amata, A Pedersen and G. Rostoker, Ground-satellite coordinated study of the April 5, 1979 events: flux variations of energetic particles and associated magnetic pulsations, J. Geophys. Res., 55, 120, 1984.

HYDROMAGNETIC WAVES IN THE DAYSIDE CUSP REGION AND GROUND SIGNATURES OF FLUX TRANSFER EVENTS

Hiroshi Fukunishi

Upper Atmosphere and Space Research Laboratory, Tohoku University, Sendai 980, Japan

L. J. Lanzerotti

AT&T Bell Laboratories, Murray Hill, New Jersey, U.S.A.

Abstract. The dayside cusp region is a window on solar wind-magnetosphere coupling processes. Low frequency (f $\lesssim$ 1Hz) magnetic field variations measured in this region provide key information on generation and transmission processes of hydromagnetic waves at the magnetopause. Recent observational and theoretical works have demonstrated that flux transfer events (FTE's) are an important energy source for hydromagnetic waves at the magnetopause in addition to the mechanisms normally associated with this boundary; upstream waves in the solar wind and surface waves excited by the Kelvin-Helmholtz instability. In the present paper, general features of long-period and short-period magnetic variations (pulsations) at cusp latitudes, based on recent ground-based observations, are first reviewed. Then, recent studies on possible ground signatures of FTE's are briefly presented, and the relationships between magnetic pulsation activity and FTE's are discussed. The present understanding of the characteristics of cusp magnetic pulsations and of ground signatures of FTE's can be summarized as follows: (1) hydromagnetic wave activity at cusp latitudes is characterized by the continuous occurrence of broadband, irregular pulsations in the daytime hours, with power enhancements tending to occur in the Pc 3 (f=22-100mHz) and Pc 5 (f=1.8-6.6mHz) period bands;(2) spectral peaks in these two frequency bands are highly correlated with similar peaks in auroral zone magnetic field data, suggesting that energy sources at cusp latitudes are readily transferred to the auroral zone across L shells; (3) FTE-type magnetic impulses occur simultaneously in conjugate regions, often accompanying damped-type Pc 5-band pulsations. These observations suggest that a sporadic reconnection process associated with FTE's excites standing Alfven oscillations on closed magnetic field lines ; (4) Magnetic pulsation activity in the Pc 1-2 period range (f=0.1-1Hz) at cusp latitudes is characterized by continuous occurrences of Pc 1-2 band activity and sporadic occurrences of IPRP (Intervals of Pulsations with Rising Periods) and Pc 1b (Pc 1 burst) events. The source region of the Pc 1-2 band appears to be the dayside cusp region, while the source region of the IPRP and Pc 1b events is likely to be the low-latitude boundary layer.

Introduction

Recent observational and theoretical works have demonstrated that there are two dominant energy sources for the excitation of shear Alfven mode oscillations of individual magnetic field lines in the dayside magnetosphere. One of these is surface waves on the magnetopause produced by a shear flow (Kelvin-Helmholtz type) instability [Southwood, 1968 ; Lee and Olson, 1980 ; Miura, 1984; Yumoto, 1984]. The other is the direct transmission through the magnetopause of hydromagnetic waves generated by upstream ions reflected and/or accelerated by the bow shock [Greenstadt et al., 1983 ; Russell and Hoppe, 1983 ; Watanabe and Terasawa, 1984]. From a comparison of power spectra of the ISEE 1 and 2 magnetic field data recorded simultaneously on both sides of the magnetopause, the transmission coefficient of upstream waves through magnetopause has been estimated to be about one percent [Greenstadt et al., 1983]. This is consistent with the theoretical and observational studies by Wolfe and Kaufmann [1975] of the transmission of individual wave events through a magnetopause tangential discontinuity. Observationally, both the surface waves excited on the magnetopause and the upstream waves transmitted across the magnetopause appear to be able to penetrate deeply into the magnetosphere as fast mode hydromagnetic waves and to excite shear Alfven mode oscillactions of local magnetic field lines at those locations where the frequencies of the source waves coincide with the eigen frequencies of the field lines [Southwood, 1984; Chen and Hasegawa, 1974]. However, the mechanism(s) for maintaining such large amplitudes as are seen deep into the magnetosphere are not well understood [e.g., Lanzerotti et al., 1981; Wolfe et al., 1985; Yumoto, 1986; Vero 1986]. Studies of magnetic pulsations in the dayside cusp region will provide the justification for the predicted energy sources since this region is connected to the dayside magnetopause by the high latitude magnetic field lines. However, compared with observations made in the auroral, sub-auroral, and low latitude regions, there have been only a limited number of studies of magnetic pulsations in the dayside cusp region. The main reason is that it is very difficult to establish a ground station network at cusp latitudes since the dayside cusp region is largely located in the Arctic Ocean in the northern hemisphere and in inland areas of the Antarctic Continent in the southern hemisphere for the greater part of a day.

The dayside cusp is also important as a window on solar wind-magnetosphere coupling processes. The most important coupling process is magnetic field reconnection at the dayside magnetopause. The reconnection model as first proposed by Dungey [1961] is a steady state model. Reconnection rates at the nose of the magnetosphere and in the magnetotail balance on the average. The recent concept of patchy dayside reconnection was proposed by Haerendel et al. [1978] and by Russell and Elphic [1978, 1979]. The Heos 2 spacecraft observed short duration increases in the magnetospheric field strength in the high latitude boundary layer. Haerendel et al. [1978] attributed these increases to temporally and spatially limited reconnection in the polar cusp. On the other hand, in the low latitude magnetosheath adjacent to the magnetopause the ISEE spacecraft observed "spikes" in the magnetic field magnitude [Russell and Elphic, 1978]. These "spikes" were accompanied by enhancements of energetic electrons, as well as by protons streaming out of the magnetosphere. Therefore Russell and Elphic suggested that spatially and temporally limited reconnection occurs in the near-equatorial region adjacent to the magnetopause and that the connected flux tubes, containing both high-density magnetosheath plasma and high-energy magnetospheric plasma, are convected over the spacecraft. These patchy reconnection events were termed "flux transfer events" (FTE's) by Russell and Elphic [1978]. FTE's were also reported inside the magnetosphere [Paschmann et al. 1982].

The spatial properties of FTE's were investigated by Saunders et al. [1984] and by Rijnbeek et al. [1984] in examinations of data from the ISEE 1 and 2 satellite pair. A twisting of magnetic field lines within FTE flux tubes was found by Saunders et al. [1984]. The twisting is well explained by a core field-aligned current with a magnitude of a few x 10^5 A. The MHD simulation study of FTE's by Fu and Lee [1985] suggested that helical flux tubes are formed at the magnetopause through a multiple X-line reconnection (MXR) process and that the flux tubes embedded in the magnetopause have field-aligned currents captured from the Chapman-Ferraro magnetopause currents.

Lanzerotti et al. [1986a] suggested that impulsive magnetic variations with durations of a few minutes observed at South Pole station are a possible ground signature of FTE's and that these impulses are produced by a moving circular Hall current loop in the ionosphere. These impulsive variations are often accompanied by damped-type Pc 5-band pulsations. Lee et al. [1988] pointed out that the impulsive magnetic variations are induced by the symmetric field-aligned currents associated with the magnetic islands formed in the MXR process, and that damped-type Pc 5-band pulsations are excited on adjacent closed field lines by the poleward convection of magnetic islands. On the other hand, Southwood [1985] pointed out that the solar wind provides momentum to move a flux tube foot through the ionosphere. The momentum transfer is achieved by Alfven waves which propagate along the flux tube and induce a current flow along the flanks of the flux tube. The current flow along the flanks connects to the ionospheric Pedersen currents, inducing twin-vortex type Hall current pattern. Southwood [1987] also pointed out that part of the incident field-aligned current surge will be carried on the adjacent closed field lines and that if reconnection has ceased by the time when the surge returns to the magnetopause vicinity, the surge will simply bounce back and forth on the closed tube, evolving rapidly into a damped standing Alfven wave. Therefore, it is likely that FTE's are also an important energy source for long-period magnetic pulsations near the magnetopause and inside the magnetosphere.

In the present paper, general features of long-period and short-period magnetic pulsations observed in the dayside cusp region are reviewed first. Then, some relationships between magnetic pulsations observed at cusp latitudes and pulsations observed in the auroral zone are discussed. Recent studies on possible ground signatures of FTE's, particularly magnetic and auroral signatures, are presented. Possible relationships between the occurrences of FTE's and magnetic pulsation activity are discussed. It is pointed out that FTE's seem to play an important role in the generation of magnetic pulsations at cusp latitudes.

Magnetic Pulsation Activity at Cusp Latitudes

Long-Period Magnetic Pulsations Observed at Cusp Latitudes

Magnetic pulsations observed in the dayside cusp region are important for studying the sources of energy for the excitation of hydromagnetic waves in the magnetosphere. However, as

TABLE 1. Locations of magnetometer stations at cusp latitudes. The magnetic local time of each station is the time at UT=00h00m. The invariant latitude and longitude and magnetic local time are calculated at the sea level for the epoch of 1985.0 using the IGRF 1980 model.

Station Name and Code		Geographic Lat(deg)	Geographic Long(deg)	Invariant Lat(deg)	Invariant Long(deg)	Magnetic LT(h:m)	L
Northern Hemisphere							
New Alsund	NAL	78.9	11.9	75.6	114.0	02:29	16.3
Hiss Island	HIS	80.6	58.1	74.6	145.0	04:33	14.1
Inuic	INK	68.3	226.7	70.9	270.8	12:56	9.3
Sacks Harbour	SAH	72.0	235.0	76.1	273.1	13:05	17.3
Cape Parry	CPY	70.2	235.3	74.5	276.5	13:19	14.1
Mould Bay	MBC	76.3	240.6	80.7	267.2	12:41	38.5
Cambridge Bay	CBB	69.1	255.0	77.6	303.9	15:08	21.6
Baker Lake	BLC	64.3	264.0	74.6	323.9	16:28	14.2
Rankin Inlet	RIT	62.8	267.9	73.6	331.6	16:59	12.5
Frobisher Bay	FBB	63.8	291.5	74.0	14.7	19:51	13.1
Godhavn	GDH	69.2	306.5	76.5	42.3	21:42	18.4
Sondre Stromfjord	STF	67.0	309.1	74.0	43.0	21:44	13.1
Southern Hemisphere							
South Pole	SPA	-99.0	0.0	-74.8	18.0	20:05	14.6
Davis	DVS	-68.6	78.0	-74.6	98.8	01:28	14.1
Mirny	MIR	-66.6	93.0	-77.3	121.6	02:59	20.6
Casey	CSY	-66.2	110.3	-80.8	154.8	05:12	39.2

noted above, because of logistical problems there have been only a limited number of studies of magnetic pulsations in the dayside cusp region. Magnetometer stations which have been used for cusp magnetic pulsation studies at various times are summarized in Table 1. The locations of these stations are shown in invariant coordinates in Figure 1 [courtesy of T. Araki and T. Iemori], where the Antarctic continent (heavier continental outline) has been mapped to the northern hemisphere and the northern conjugate points of southern magnetometer stations are marked by the solid dots. The locations of the conjugate points have been calculated for the epoch of 1985.0 using the IGRF 1980 model. Therefore, it is expected that the real conjugate points at these high latitudes move considerably, being dependent upon magnetic activity, local time, and season.

The most distinct feature of magnetic pulsation activity at cusp latitudes is the continuous occurrences of long-period irregular pulsations in the daytime hours [Rostoker et al., 1972 ; Troitskaya et al., 1972, 1980; Heacock and Hunsuker, 1977; Heacock and Chao, 1980]. These pulsations were termed ipcl (irregular period continuous) pulsations by Troitskaya et al. [1972]. Their amplitudes have the order of tens of gammas and the period range is 3-10 min, with dominant periods of 5-6 min [Troitskaya et al., 1980]. On the other hand, Heacock and Chao [1980] classified these dayside irregular pulsations as type Pi pulsations, based on magnetic field measurements made at Sachs Harbour (76.1 invariant latitude). The power spectra of dayside irregular pulsations are characterized by broadband enhancements reaching from a few millihertz to above 1 Hz [Heacock and Chao, 1980; Olson, 1986].

From spectral analysis using induction magnetometer data from Cape Parry (invariant

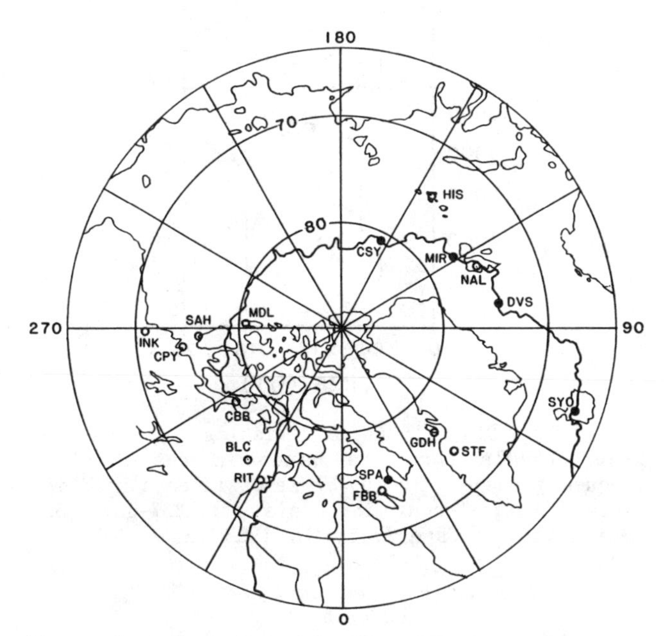

Figure 1. Locations of magnetometer stations at cusp latitudes in invariant coordinates. The Antarctic Continent and magnetometer stations are mapped to the northern hemisphere. The southern hemisphere magnetometer stations are marked by black dots.

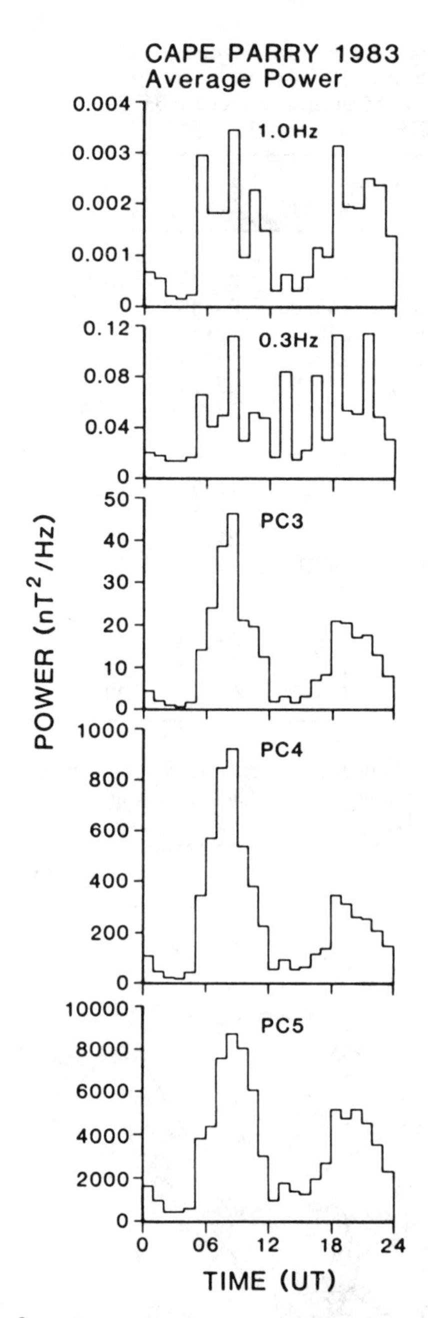

Figure 2. Diurnal characteristics of average power levels for five bands in the Pc 1-5 frequency range. The power level in each band is average over 1-hour intervals for 237-days of data taken at Cape Parry in 1983 (after Olson, 1986).

latitude 74.5; LT = UT + 13.3 hours), Olson [1986] showed that, on average, the power of dayside irregular magnetic pulsations decreases with frequency as $f^{-2.6}$. As shown in Figure 2, the diurnal variations of spectral power are quite similar from the Pc 1 range to the Pc 5 range [Olson, 1986]. The dayside enhancements occurred between 18 UT (∿07 MLT) and 02 UT (∿ 15 MLT). The power enhancements between 06 UT (∿19 MLT) and 12 UT (∿01 MLT) in Figure 2 is caused by Pi bursts associated with substorm activity in the night side at high latitudes. The diurnal variation in the Pc 1 period band in Figure 2 is more irregular than the Pc 3, 4 and 5 period bands. Olson [1986] suggested that this difference is due to the sporadic appearance of strong Pc 1 emissions.

Shown in Figure 3 are examples of magnetic variations acquired at South Pole station in the Antarctic (invariant lat. -74.8, long. 18.0) and its conjugate station, Frobisher Bay in Canada (invariant lat. 74.0, long. 14.7). The magnetic

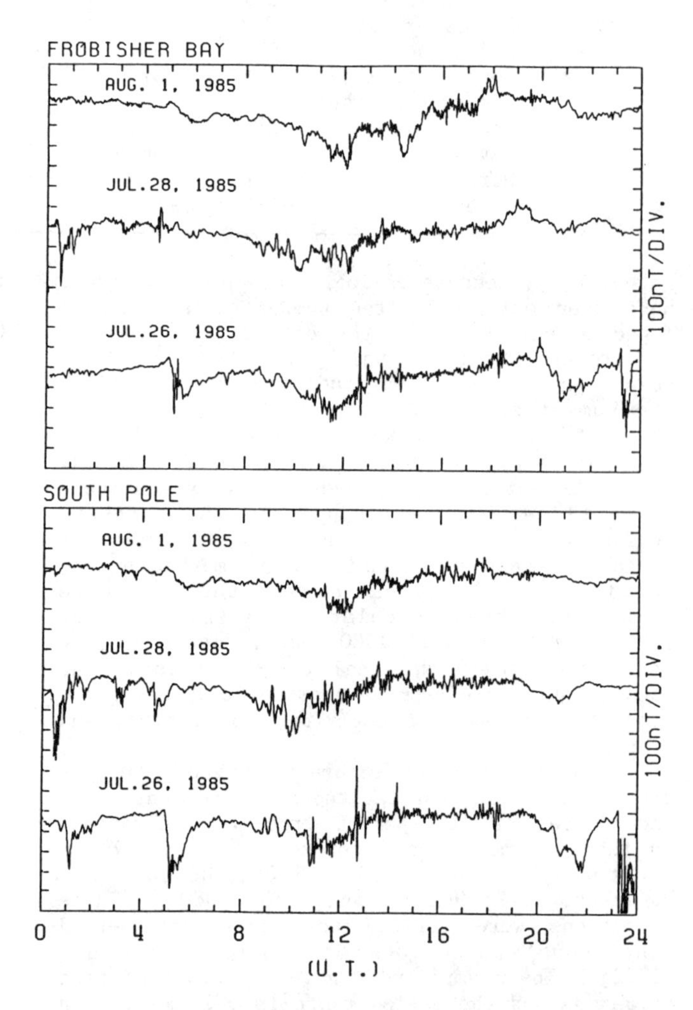

Figure 3. Examples of magnetic variations observed at cusp latitudes. The magnetic field data are the H component data acquired at the conjugate-pair stations South Pole/Frobisher Bay. The magnetic local noon of these stations is ∿16 UT.

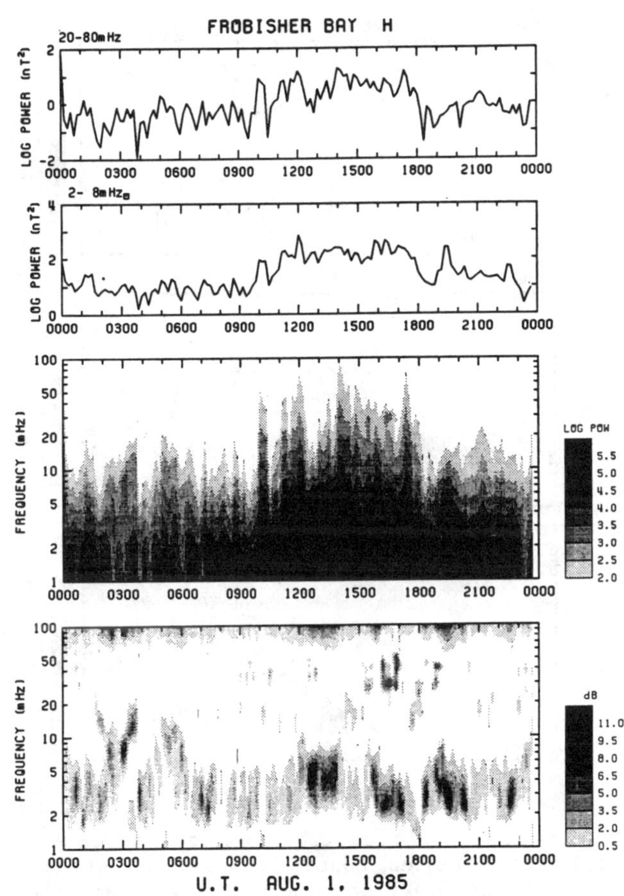

Figure 4a. Diurnal characteristics of magnetic pulsations observed at Frobisher Bay station on August 1, 1985. Top two panels show power levels integrated in the 20-80 mHz and 2-8 mHz bands, respectively. The third panel shows the dynamic spectrum of the H component variations without subtracting the broadband background level, while the bottom panel shows the background-subtracted dynamic spectrum. Magnetic local noon is ∿ 16 UT.

local time of these stations is appmximately UT - 4.0 hours (see Table 1). The magnetometer data obtained from a conjugate pair of stations at cusp latitudes are extremely useful for studying characteristics of cusp-associated magnetic pulsations [Lanzerotti et al., 1987b]. It is apparent in Figure 3 that irregular magnetic pulsations occur continuously in the daytime hours from ∿11 UT (∿07 MLT) to ∿19 UT (∿15 MLT). It is also found that large-amplitude, spike-like magnetic variations (up to 500 nT peak to peak) occurred intermittently in the daytime hours, particularly on July 26, 1985. These spike-like magnetic variations are discussed in the next section as a possible ground signature of FTE's.

Examples of dynamic spectra of the conjugate data are shown in Figures 4a, b. The dynamic spectra are characterized by broadband enhancements, as is apparent in the third panels in Figures 4a, b. In order to emphasize the spectral peaks in these broadband enhancements, background-subtracted dynamic spectra are displayed in the bottom pannels in Figures 4a, b. The background power is subtracted by fitting a second-order polynomial to the slope of the power spectrum, and then the power relative to the background level is displayed with an eight-step gray scale. It is apparent that the spectral peaks appear in two dominant frequency bands; one of these is the 20-80 mHz band and the other is the 2-8 mHz band. The former band roughly corresponds to Pc 3 band (defined as 22-100mHz), while the latter band roughly corresponds to the Pc 5 band (defined as 1.6-6.6 mHz). (Since it is extremely difficult to decide where the "boundaries" should be drawn for real hydromagnetic waves, in the present paper we use the terms Pc 3, Pc 4 (6.6-22 mHz) and Pc 5 for the frequency ranges in which the dominant energies

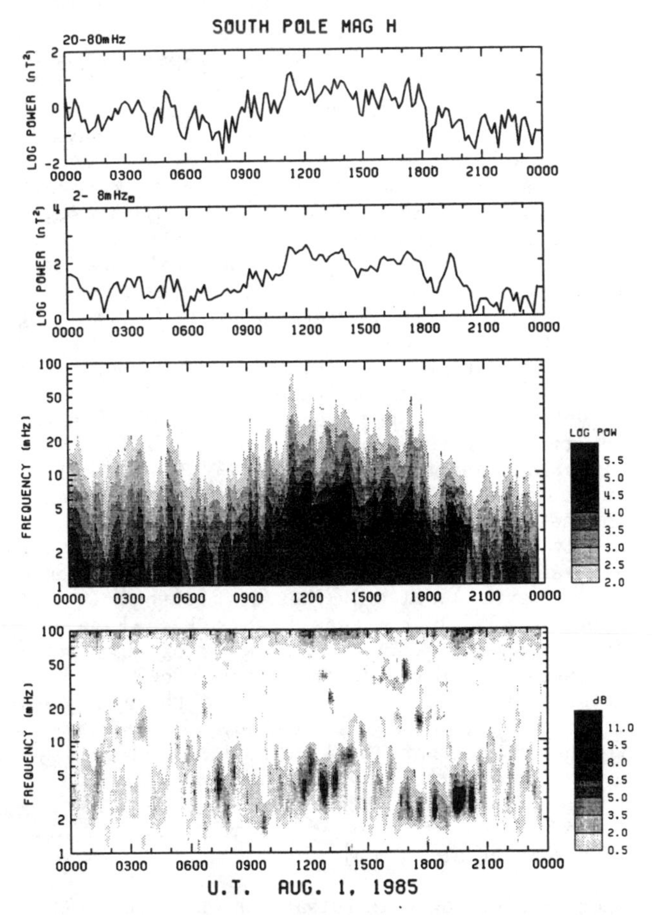

Figure 4b. Diurnal characteristics of magnetic pulsations observed at South Pole station on August 1, 1985. The notation is the same as that in Figure 4a.

are included.) The top two panels in Figures 4a, b show diurnal variations of total powers integrated in the 20-80 mHz and 2-8 mHz bands, respectively. Both bands show large enhancements of total power in the daytime hours.

Enhancements in the power of magnetic field fluctuations in the Pc 3 and Pc 5 bands have been reported by Kato et al. [1985], Olson [1986], and Lanzerotti and Maclennan [1987]. Olson [1986] pointed out that there are often regions of enhanced power found near 40 mHz and near 5 mHz. Engebretson et al. [1986], using search coil magnetometer data from South Pole station, have made a distinction between narrow-band Pc 3 pulsations and broadband irregular pulsations in the period range 0.5-40 sec (Pi 1). The occurrence of Pc 3 depended on the IMF cone angle and the frequency of the observed Pc 3 depended on the IMF magnitude. Therefore, Engebretson et al. [1986] suggested that upstream wave energy in the Pc 3 frequency range can reach low altitudes on cusp field lines and that this energy can be transfered across L shells to lower latitudes. The same conclusion has been obtained by Lanzerotti et al. [1986b] based on the observations of Pc 3 pulsations accompanied by quasi-periodic VLF emissions. Bolshakova and Troitskaya [1984] showed statistically that the maximum of Pc 3 amplitude follows the position of the dayside cusp to within 5-8 degrees.

The spectral structures of the background-subtracted dynamic spectra shown in Figures 4a, b are quite similar to those of Pc 3 and Pc 5 magnetic pulsations observed at auroral-zone stations [Tonegawa et al., 1984b]. Therefore, it is suggested that the HM energy sources for exciting Pc 3 and Pc 5 pulsations at cusp latitudes are transferred to L shells in the auroral zone. The cross spectral analysis using the simultaneous magnetometer data acquired at South Pole station in the cusp region and at Syowa station (invariant latitude -66.1; MLT = UT -0.4 hr) in the auroral zone have shown that spectral peaks in the Pc 3 and Pc 5 bands at cusp latitudes are highly correlated with those in the auroral zone. An example is given in Figure 5. The difference in magnetic local time between South Pole and Syowa is 3.5 hours (see Figure 1). Figure 5 demonstrates clearly that in spite of this large difference in magnetic local time, Pc 3 and Pc 5 pulsations in the auroral zone and those at cusp latitudes are highly correlated. However it is worth noting that Pc 3 pulsations observed at auroral-zone stations often have harmonic structures [Tonegawa and Fukunishi, 1984a], while those at cusp stations are usually monochromatic [Tonegawa et al., 1985].

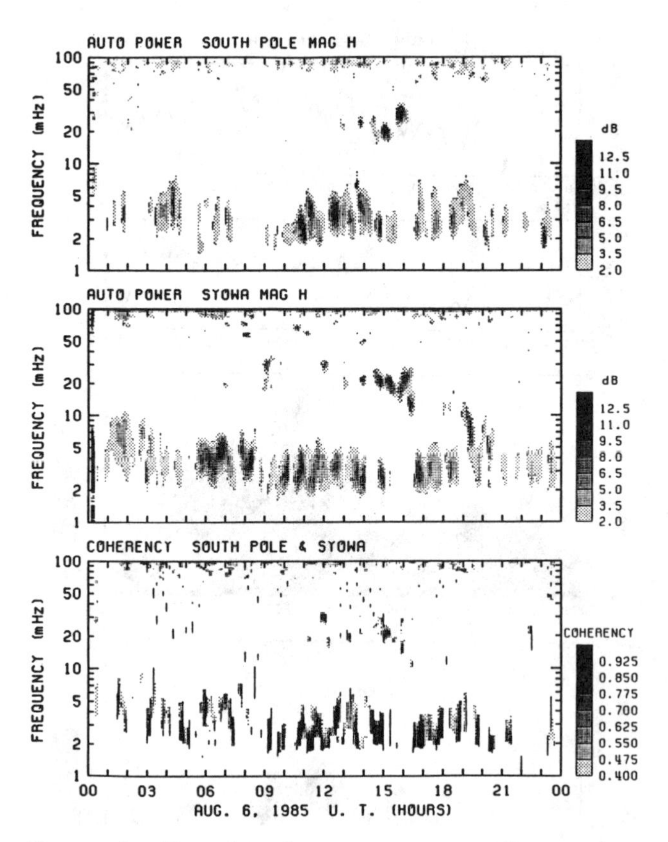

Figure 5. Example of cross-spectra of magnetic variations between South Pole station at cusp latitude and Syowa station in the auroral zone. The magnetic local time of South Pole is UT-3.9 hours, while that of Syowa is UT-0.4 hour.

Short-Period Magnetic Pulsations Observed at Cusp Latitudes

In addition to long-period magnetic pulsations, specific short-period magnetic pulsations occur in the dayside cusp region [Fraser-Smith, 1982; Fukunishi, 1984]. The frequency range of short-period magnetic pulsations is usually between 0.1 Hz and 1 Hz, which corresponds to the Pc 1-2 frequency range. Typical types of short-period pulsations are the Pc 1-2 band [Heacock, 1974], intervals of pulsations with rising periods [IPRP; Bolshakova et al., 1980], and Pc 1b [Pc 1 burst; Troitskaya et al., 1980]. Pulsations in the Pc 1-2 band are found to be continuous pulsations observed around magnetic local noon, while IPRP and Pc 1b are pulsations of discrete type occurring sporadically. These three types of magnetic pulsations have characteristic spectral structures, as shown in Figure 6.

The f-t spectra of the Pc 1-2 band are mainly unstructured and, occasionally, semistructured. The frequency range is 0.1-0.4 Hz, and the duration is 2-3 hours on average [Troitskaya et al., 1980]. IPRP events are observed as falling tone emissions, while Pc 1b events are seen as dots in the frequency-time display. Therefore, Hayashi et al. [1984] used the term "dots" instead of Pc 1b.

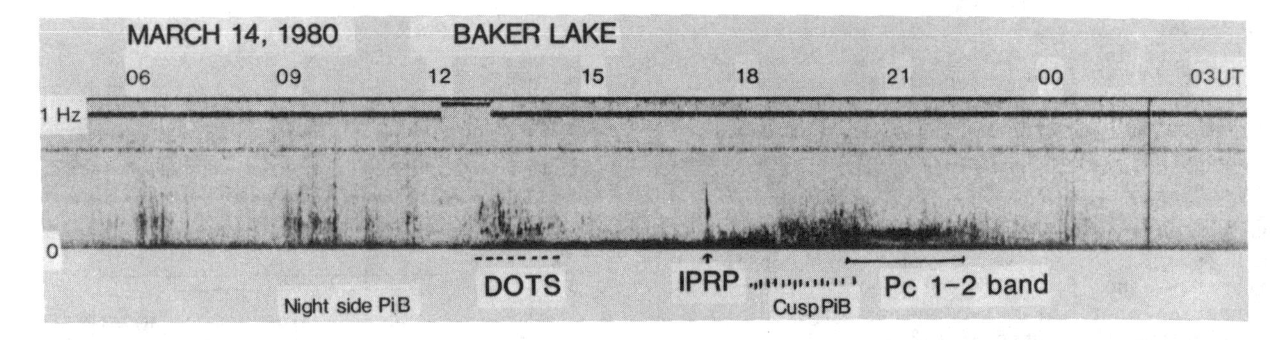

Figure 6. Dynamic spectrum of magnetic pulsations in the Pc 1-2 frequency range at Baker Lake station (74.6 invariant latitude) on March 14, 1980. Magnetic local noon is ~1930 UT (courtesy of K. Hayashi).

The average durations of IPRP and Pc 1b are 2-6 min and 1-4 min, respectively [Troitskaya, 1980]. The frequency range of IPRP events observed at Davis station in Antarctica is 0.1-1 Hz, and the start and finish frequencies of the events are 0.31 Hz and 0.18 Hz on average [Morris and Cole, 1985]. Pc 1b-type discrete emissions observed at Davis are seen in both the Pc 1 and the Pc 2 frequency ranges (16 % in the Pc 1 range and 84 % in the Pc 2 range). Therefore, Morris and Cole [1985] proposed the nomenclature IPCP (Intervals of Pulsations with Constant Period) instead of Pc 1b.

Pc 1-2 band emissions occur primarily around magnetic local noon and in local summer months [Troitskaya et al., 1980]. On the other hand, IPRP events occur primarily in the 6-13 MLT interval and in local winter, and Pc 1b (IPCP) events occur primarily in the 6-9 MLT interval and in local winter [Morris and Cole, 1985]. It has been pointed out that, statistically, both Pc 1-2 band events and IPRP events maximize in amplitude and frequency of occurrence under the projection of the polar cusp [Troitskaya et al. 1980; Bolshakova et al. 1980]. Hayashi et al. [1984] showed that the maximum amplitudes of dots (IPCP events) tend to occur at sub-cusp latitudes, while Morris and Cole [1985] gave evidence that IPCP events also occur in the dayside cusp region.

Since the band width of Pc 1-2 band, IPRP and Pc 1b (IPCP) is narrow, the ion cyclotron instability is a most possible generation mechanism for them. Bolshakova et al. [1980] suggested that pulsations in the Pc 1-2 band are generated mainly in the dayside magnetosheath on field lines, crossing the magnetopouse and entering the dayside cusp. Hayashi et al. [1984] demonstrated that dots (IPCP events) tend to occur associated with Pc 5 activity. This relationship suggests that Pc 5 hydromagnetic waves modulate the growth rate of ion cyclotron waves. However, there are few studies concerning the relationships between cusp-associated Pc 1-2 pulsations and Pc 3-5 pulsations. A recent paper by Arnoldy et al. [1988] shows evidence for discrete-type Pc 1-2 pulsations at the time of at least some FTE's. The relationships of these observations to IPRP and IPCP need to be further explored. The time scales of IPRP events are nearly equal to those of FTE's. Therefore, the frequency decrease in IPRP events might be related to the cusp dynamics associated with FTE's.

Ground Signatures of FTE's

As suggested by Glassmeier et al. [1984], Holzer and Reed [1975], Southwood [1987], Lanzerotti and Maclennan [1987], and Lanzerotti et al. [1987b], FTE's might be an important energy source for generating long-period magnetic pulsations in the dayside magnetosphere. In regions adjacent to the magnetopause, FTE's are relatively easily identified by speacecraft instuments using two general criteria : (1) the component of the magnetic field normal to the magnetopause must exhibit an approximately bimodal signature, and (2) the plasma must show enhancements of electron density and ion flow velocity within the bimodal magnetic field region [LaBelle et al., 1987]. In contrast, it is quite difficult to identify uniquely ground signatures of FTE's since the field lines from the FTE regions in the magnetosphere can not be readily traced to the ground. Therefore, there is presently no conclusive opinion on the ground signatures of FTE's. A number of authors, however, have reported theoretical and observational investigations of possible ground signatures of FTE's [Southwood, 1985, 1987; Lee, 1986; Lee et al., 1988; Goertz et al., 1985; Todd et al., 1986; Lanzerotti et al., 1986a, 1987a; Sandholt et al., 1986; Oguti et al., 1987; Glassmeier et al.,1987].

From STARE electron drift velocity data obtained during a magnetic storm period in which the magnetopause reached the vicinity of L=6.6, Goertz et al. [1985] found two events in which a sporadic and spatially isolated flow region moved westward and poleward across the convection boundary between sunward and antisunward flows.

The events occurred on time scales of a few minutes and in regions which ranged in size from 50 km to 300 km in the ionosphere. Goertz et al. [1985] interpreted the convection reversal boundary to correspond to the inner edge of the magnetospheric boundary layer. They then estimated that a typical magnitude for the magnetic flux transported across the boundary is 5×10^6 Wb and the flux transport rate is 2.5×10^4 Wb s^{-1} for each event. The estimated values are consistent with the values for FTE's observed on spacecraft, suggesting that these two events might be possible ground signatures of FTE's.

Another possible ground signature of FTE's has been presented by Sandholt et al. [1985, 1986] using optical aurora data from cusp-latitude stations. The dayside aurora is characterized by many discrete arcs forming a fan-shaped configuration with an apex at the polar cusp region of no discrete auroras [Snyder and Akasofu, 1976 ; Akasofu and Kan, 1980 ; Meng and Lundin, 1986]. The dynamical behavior of dayside auroras is characterized by short-lived poleward expansions superimposed on large-scale, gradual auroral oval dynamics. Using meridian-scanning photometer data and all-sky camera data from New Alesund (75.6 invariant latitude) and Longyearbyen (74.6 invariant latitude) on Svalbard, Sandholt et al. [1986] classified the dynamical behavior of dayside discrete auroras into two categories, small-scale and large-scale events, according to different space and time scales. Sandholt et al. [1986] suggested that the small-scale events have the optical signatures expected for FTE's, while the large-scale events have those expected for the quasi steady state reconnection process in the dayside magnetopause boundary layer.

A candidate for magnetic signatures of FTE's was proposed by Lanzerotti et al. [1986a]. They observed a number of spike-like magnetic variations with time scales of a few minutes in the South Pole data. Examples of the magnetic spikes are shown in Figure 7. In general, the vertical (B_V) component shows one-side spikes, while the north-south (B_H) and east-west (B_D) components show bimodal spikes. Therefore, Lanzerotti et al. [1986a] suggested that these magnetic spikes result from a moving ionospheric Hall current loop, and estimated amplitude of B_{Vo} at the surface of the earth beneath the center of the Hall current loop by the relation

$$B_{Vo} = cR\,\mu_o a^2 I_o / 4\pi h^3 \qquad (1)$$

where R (~ 2) is the ratio the Hall to Pedersen conductivity, μ_o is the permeability of free space, a is the radius of the flux tube in the ionosphere, I_o is the total field-aligned current, h is the ionospheric height, and c (~ 0.5) is constant depending upon the distribution of the field-aligned current. For h = a = 100 km, and $I_o = 10^5$ amps, $B_{Vo} = -100$ nT (i. e., $J_{11} > 0$ for $B_V < 0$).

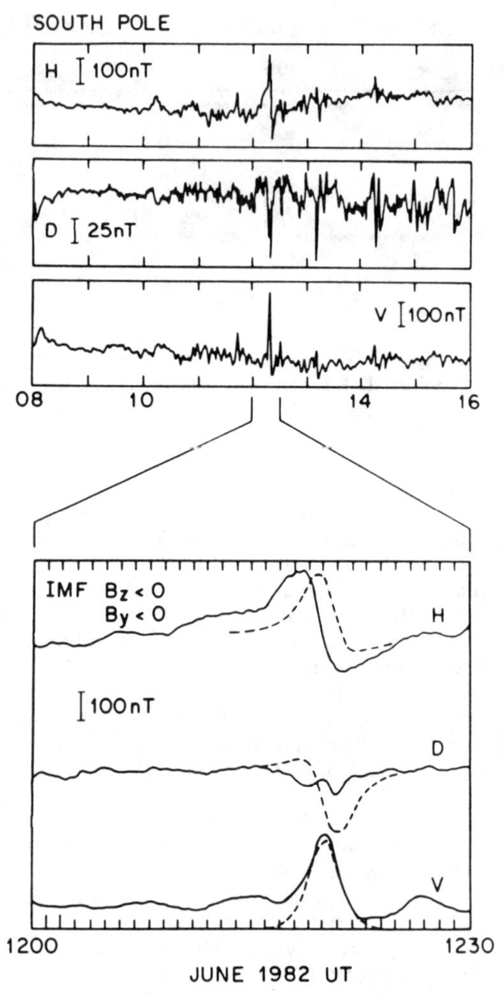

Figure 7. Example of spike-like magnetic variations observed at South Pole station. The upper panel shows the magnetic field data of H (north-south), D (east-west) and V (vertical) components for an eight-hour interval, while the lower panel shows higher resolution data for a thirty-minute interval. The magnetic noon of South Pole is ~ 16 UT. The dashed lines corresponds to magnetic variations estimated from a field-aligned current convected over South Pole (after Lanzerotti and Maclennan, 1987).

Magnetic field data acquired at the conjugate pair stations at the cusp latitudes South Pole/Frobisher Bay are significantly useful in order to examine the asymmetry of magnetic signatures between the northern and southern hemispheres. Using these data Lanzerotti et al. [1987a] and Trivedi et al.[1987] have demonstrated that spike-like magnetic variations occur simultaneously in the conjugate regions and that the directions of the changes in the vertical component are the same in both hemispheres. An example of spike-like variations observed at

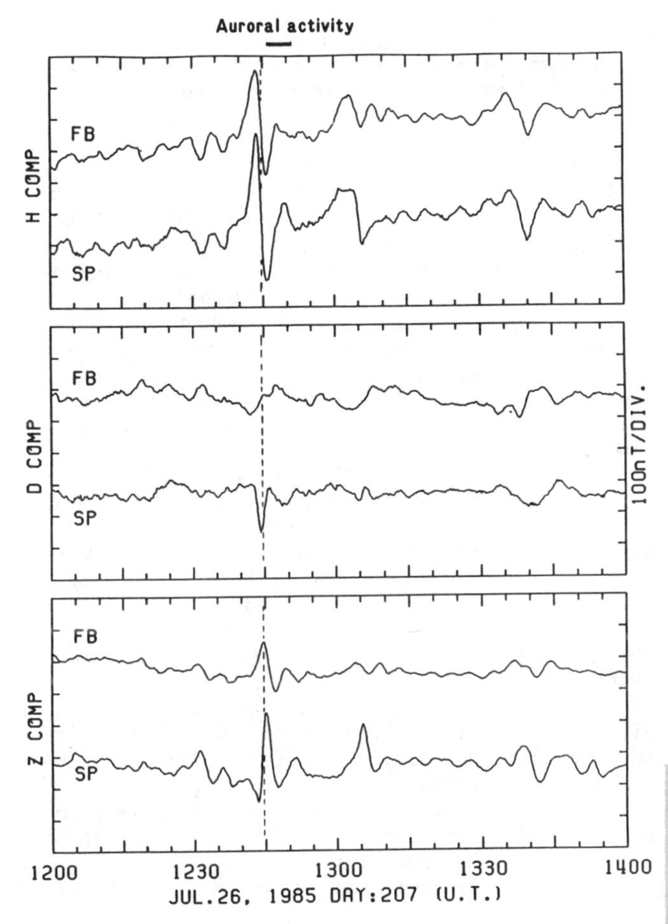

Figure 8. Example of spike-like magnetic variations simultaneously observed at the conjugate-pair stations South Pole/Frobisher Bay. The upward deflections of the traces are northward for the H component, eastward for the D component and upward (downward) for the Z component at South Pole (Frobisher Bay), respectively. The dashed vertical line is to guide the eye. A bar at the top indicates the time interval in which a sporadic auroral activity occurred (see Figure 9).

the South Pole-Frobisher Bay conjugate-pair stations is shown in Figure 8. This conjugate relationship suggests that the field-aligned currents associated with the Hall current loop are directed in the same direction in both hemispheres.

Recently Fukunishi et al. [1987] have examined the relationship between occurrences of spike-like magnetic variations and auroral activity. They have used all-sky camera data from South Pole station and magnetic field data from the South Pole-Frobisher Bay conjugate pair stations. Since South Pole station is located at 90° geographic latitude, auroras can be observed all day in winter. Therefore, this station is most suitable for studying optical signatures of FTE's. Fukunishi et al. [1987] found that spike-like magnetic field variations are always accompanied by a sporadic appearance of discrete auroras. Figure 9 shows all-sky photographs from South Pole station in the time interval 1232-1255 UT during which a spike-like event occurred (Figure 8). It is apparent that sporadic auroral activity which accompany westward and poleward motions occurred in the time interval of 1246-1251 UT. As indicated by the heavy bar at the top of Figure 8 (1246-1251 UT), it is also evident that the auroral activity occurred immediately after the peak in the vertical component intensity. The behavior of the magnetic variation suggests that a Hall current loop induced by a downward field-aligned current filament convected above both stations. The onset of the sporadic auroral activity after the current filament means that such auroral activity does not occur in the downward field-aligned current region identified from the direction of the vertical component.

Similar examples are shown in Figures 10 and 11. In these cases, damped-type Pc 5 pulsations appear to have been excited by magnetic spikes at 1230 UT and at 1335 UT. The damped-type Pc 5

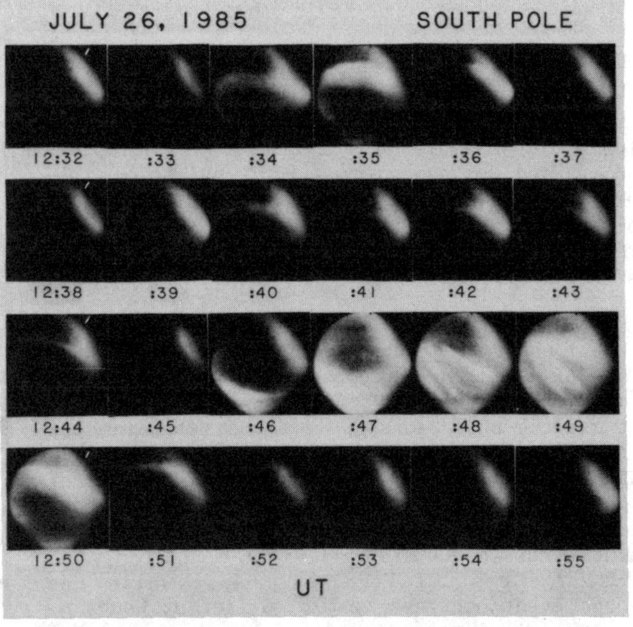

Figure 9. All-sky photographs taken at one minute intervals at South Pole station in the time interval 1232-1255 UT on July 26, 1985. A FTE-like magnetic spike occurred during this time interval, as shown in Figure 8. A white bar in the left-hand outside column of each row of photographs indicates the direction of the south invariant pole.

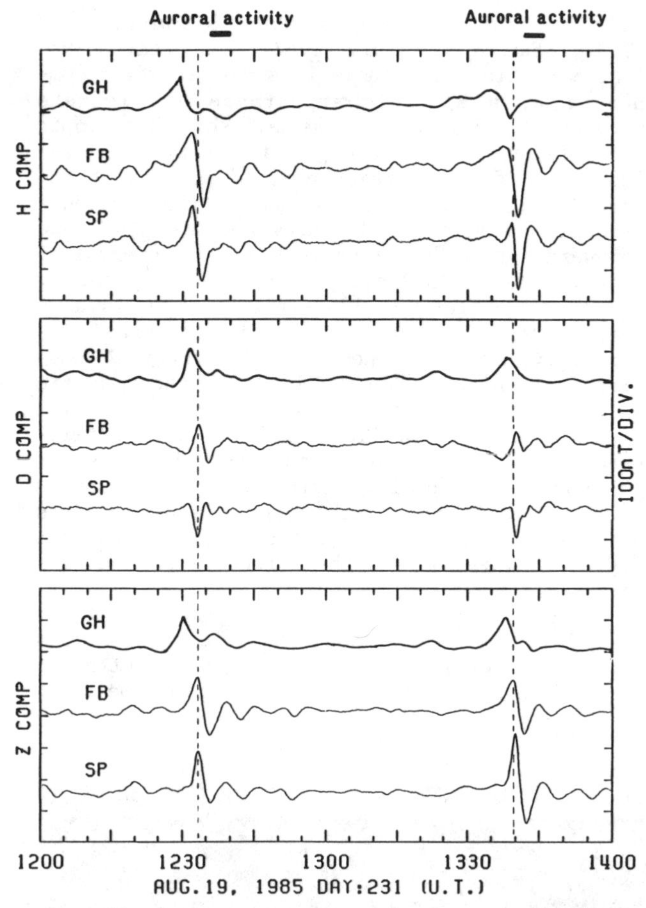

Figure 10. Example of FTE-like magnetic spikes accompanied by damped-type Pc 5 pulsations observed at the conjugate-pair stations South Pole/Frobisher Bay. The vertical dashed lines are to guide the eye. The notation is the same as that in Figure 8.

pulsations are characterized by in-phase oscillations of the H and Z components and out-of-phase oscillations of the D component, suggesting that the excited pulsations are odd mode standing oscillations of magnetic field lines. In Figure 10 magnetic field data digitized from the normal magnetogram at Godhavn (76.5 invariant latitude) are presented. The data show that the first event occurred ∿3 min earlier at Godhaven (∿ 900 km to the east of Frobisher Bay), while the second event occurred ∿2 min earlier at Godhavn. The magnetic local time of Godhavn is ∿ 1010 MLT in the first event, and ∿1120 MLT in the second event. If we assume that a field-aligned current filament was generated around magnetic noon, and convected with a constant speed to Frobisher Bay through Godhavn, the convection speed is estimated to be ∿ 5 km/sec for the first event. Lanzerotti et al.(1987a) have obtained a similar convection speed from a comparison between Frobisher Bay data and Sondre Stromfjord data. The all-sky photographs presented in Figure 11 demonstrate again that a sporadic auroral activity followed the occurrence of the first magnetic spike at 1230 UT. Similar auroral activity occurred following the second magnetic spike at 1235 UT.

As found by Trivedi et al. [1987] from an examination of data over the interval July-November 1985, and also as shown in Figures 8 and 10, the behavior of the vertical magnetic field perturbations imply that these perturbations were produced by symmetric field-aligned currents. Furthermore, the frequent occurrences of damped-type Pc 5 pulsations following flux transfer-type magnetic spikes suggest definitely that these pulsation events are generated on closed field lines.

Fukunishi et al. [1987] have provided further evidence for the occurrence region of magnetic spikes. Presented in Figure 12 are energy-time spectra and integrated fluxes of precipitating ions and electrons from the Defence Meteorological Satellite Program F7 satellite. The bottom panel in Figure 12 shows the distance between the DMSP F7 magnetic footprint and Frobisher Bay station. The magnetic footprint approached to within 110 km of Frobisher Bay station at 1336:25 UT. By fortunate coincidence, this time corresponds exactly to the onset time of the flux transfer-type event (see Figure 10). The DMSP particle data demonstrate that at this time Frobisher Bay station was located in a region characterized by stable precipitation of >1 keV electrons. The features of precipitating par-

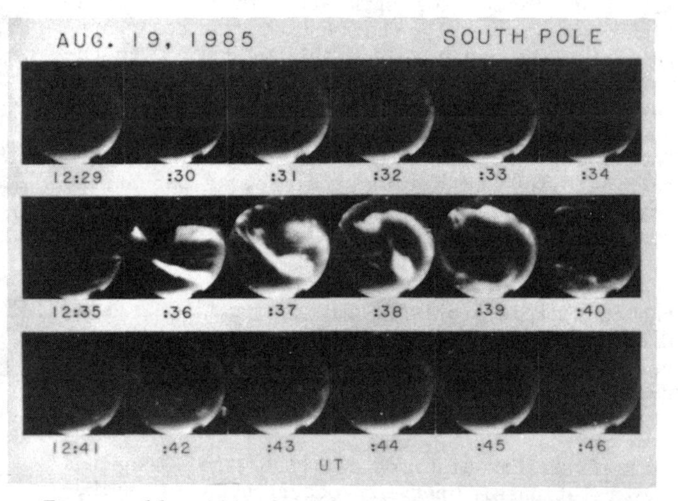

Figure 11. All-sky photographs taken at one minute intervals at South Pole station in the time interval of 1229-1246 UT on August 19, 1985. The bright area at the lower edge of each photograph is due to twilight. The direction of all-sky photograhs is the same as that in Figure 9.

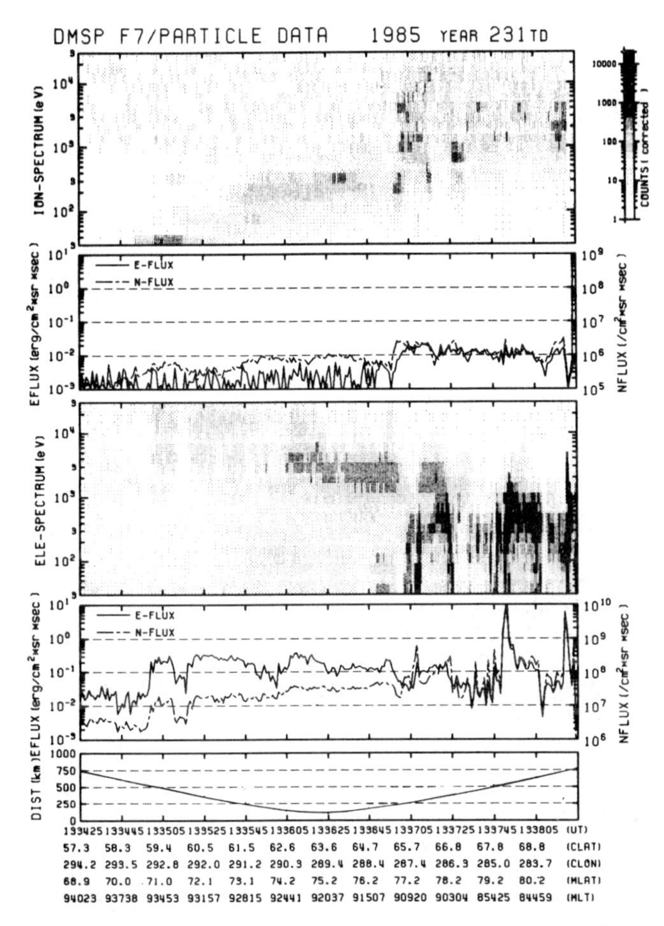

Figure 12. DMSP F7 particle data in the time interval 13h34m25s-13h38m05s UT on August 19, 1985. The top and third panels show energy-time spectra of precipitating ions and electrons, respectively. The second and fourth panels show total number fluxes and total energy fluxes for precipitating ions and electrons, respectively. The bottom panel shows the distance between the 110-km level magnetic footprint of the DMSP F7 satellite and Frobisher Bay. The universal time, geographic latitude and longitude and corrected geomagnetic latitude and magnetic local time of the DMSP magnetic footprint are given at the bottom.

ticles in the energy-time spectrum changed drastically at 77.2 MLAT (corrected magnetic latitude). The region poleward of 77.2 MLAT is characterized by intense, structured, low energy (<1 keV) electron precipitation. This region appears to correspond to the low-latitude boundary layer characterized by the injection of magnetosheath plasma. The region equatorward of 77.2 MLAT appears to correspond to the stably trapped particle region. The DMSP particle data suggests that the magnetic spikes accompanying damped-type Pc 5 pulsation events are observed in the closed field line region inside the low-latitude boundary layer.

Discussions and Summary

Energy Sources of Long-Period Magnetic Pulsations at Cusp Latitudes

The magnetic pulsation activity at cusp latitudes is characterized by occurrences of long-period irregular pulsations in the daytime hours. The spectral analyses have demonstrated that these irregular pulsations have broadband spectral structures, with spectral peaks in the Pc 3 and Pc 5 period bands. Further, the cross-spectrum analyses have clarified that the spectral peaks in the Pc 3 and Pc 5 period bands are highly correlated with those in the auroral zone. These results suggest strongly that the wave energies for exciting Pc 3 and Pc 5 pulsations at cusp latitudes are transferred to the auroral zone across the L shells.

Many researchers have considered upstream waves in the solar wind as an energy source of the Pc 3, 4 pulsations observed at middle and low latitudes. This idea was first discussed in some detail by Troitskaya et al. [1971]. These authors pointed out that the period of middle-latitude Pc 3, 4 pulsations is inversely proportional to the IMF magnitude ($T \sim 160/B$), and that a likely physical source for the Pc 3, 4 pulsations is a proton cyclotron resonant interaction in the solar wind. Simultaneous ground-satellite observations demonstrated that the occurrences and amplitudes of Pc 3, 4 pulsations increase as the cone angle between the IMF and the earth-sun line approaches 0° [Gul'yel'mi, 1974; Greenstadt and Olson, 1977; Wolfe et al., 1980; Russell et al., 1983; Odera, 1986]. Further Pc 3, 4 activity is often found to switch off or on corresponding to a sudden change in the IMF direction [Webb and Orr, 1976; Odera, 1986]. These close relationships between Pc 3, 4 pulsation activity and the magnitude and direction of the IMF confirm that upstream waves are important sources of Pc 3, 4 pulsations.

However, there is evidence that the Pc 3, 4 pulsation activity at middle and low latitudes can also depend upon the solar wind flow speed [Singer et al. 1977; Greenstadt and Olson, 1976; Odera, 1982; Wolfe et al, 1980]. This relation suggests that surface waves generated by the Kelvin-Helmholtz instability at the magnetopause are another energy source of the Pc 3, 4 pulsation activity, since this instability is more likely to be driven by larger solar wind velocities. Wolfe et al. [1980] have demonstrated that in the Pc 3, 4 period range the solar wind velocity and the cone angle of the IMF are of equal statistical significance in producing enhanced power. Thus, magnetic field pulsations in the Pc 3-4 range could be produced by the two separate sources.

The control of the solar wind and IMF parameters on Pc 3, 4 pulsation activity at cusp latitudes has been studied by Engebretson et al. [1986] and Wolfe et al. [1987]. Engebretson et al. [1986], analyzing several well-defined, discrete events, found that the frequency of their Pc 3 events is dependent on the magnitude of the IMF, and suggested that Pc 3 pulsations at cusp latitudes are upstream waves which penetrate to low altitudes through cusp field lines. On the other hand, from a statistical study, Wolfe et al. [1987] found the strong solar wind velocity effect and the absence of a significant IMF cone angle effect. Therefore, Wolfe et al. [1987] concluded that the Kelvin-Helmholtz instability at the dayside magnetopause is the dominant energy source for the magnetic pulsation activity in the Pc 3-4 period range at cusp latitudes.

The different conclusions appear to come from the different criteria for selection of the Pc 3, 4 pulsations. Engebretson et al. [1986] selected large-amplitude Pc 3 pulsations (up to 20 nT peak to peak) with packet-like structure and nearly monochromatic character, while Wolfe et al. [1987] used hourly power spectral density amplitudes integrated over two period bands, 20-45s and 45-150s, to quantify the magnetic activity in the Pc 3 and Pc 4 bands, respectively. In the analysis of Wolfe et al. [1987], broadband power enhancements appear to contribute significantly to the Pc 3 and Pc 4 powers.

Recently, Engebretson et al. [1987] have analyzed the magnetometer data from the AMPTE (Active Magnetospheric Particle Tracer Explorers) CCE (Charge Composition Explorer) satellite with an apogee at 18.7 Re and the IRM satellite with an apogee at 8.8 Re. Pc 3-4 magnetic pulsations (10- to 100-s period) were found to be occurring simultaneously at locations upstream of the earth's bow shock and inside the magnetosphere from L = 4 to 9. The observed dayside magnetospheric pulsations consisted of two categories : (1) harmonically structured, azimuthally polarized pulsations with period governed by local resonant conditions ; (2) more monochromatic compressional pulsations with periods identical to those observed simultaneously in the solar wind. The compressional pulsations were seldom observed far from the local noon region in which harmonically-structured pulsations were frequently observed. The occurrence of pulsations was controlled by the cone angle of the IMF, and the amplitude of harmonic pulsations increased when the solar wind velocity increased. These observations may suggest that compressional Pc 3 pulsations are upstream waves penetrated to the magnetosphere across the magnetopause, while the harmonically structured pulsations are produced partly by upstream waves and partly by surface waves at the magnetopause.

Southwood [1974] and Chen and Hasegawa [1974] showed that surface waves which penetrate deep into the magnetosphere as evanescent fast mode (compressional) waves can excite shear Alfven-mode standing oscillations of local magnetic field lines. However, Engebretson et al. [1987] have concluded that harmonically structured pulsations derive more power from upstream waves which have penetrated through the cusp region since compressional waves were seldom observed when harmonically-structured pulsations occurred. They have suggested that impulsive (or broadband) power entering the magnetosphere through the cusp region excite field lines into azimuthal resonances in the same way that a violinist generates vibrations by bowing the strings of the violin near the end. The cusp entry mechanism was also considered by Lanzerotti et al. [1981] and Bol'shakova and Troitskaya [1984].

Harmonically-structured Pc 3, 4 pulsations have frequently been observed at synchronous orbit in the dayside magnetosphere [Takahashi et al. 1984]. Further, multiple harmonics were found from ground observations at the Syowa-Husafell conjugate-pair stations in the auroral zone [Tonegawa and Fukunishi, 1984a ; Tonegawa et al., 1984b]. Tonegawa and Fukunishi [1984a] have showed that Pc 5 pulsations observed in the dayside auroral zone are fundamental oscillations of local resonant field lines, while Pc 3 pulsations are higher harmonics (usually third to sixth harmonics). They have also pointed out that Pc 3 and Pc 5 pulsations are apparently generated by different energy sources despite of their harmonic relation. It is likely that surface waves generated by the Kelvin-Helmholtz instability excite the fundamental oscillations of local resonant field lines in the Pc 5 range, while upstream waves generate higher harmonics in the Pc 3 period range. Lanzerotti et al. [1987b] have suggested that magnetic spikes at cusp latitude are another important energy source of Pc 5 pulsations. This subject is discussed further. The source regions of long-period magnetic pulsations discussed above are summerized in Figure 13.

Energy Souces of Pc 1-2 Magnetic Pulsations at Cusp Latitudes

Magnetic pulsation activity in the Pc 1-2 period range at cusp latitudes is characterized by continuous occurrences of Pc 1-2 band and sporadic occurrences of IPRP and Pc 1b (IPCP) events. Since the band width of these pulsations is narrow, the ion cyclotron instability is a most possible generation mechanism for them. It was found that the source region of periodic Pc 1 pulsations (pearl-type Pc 1's) follows the region corresponding to the average plasmapause locations, while the source region of non-periodic Pc 1 pulsations seems to correspond to the detached plasma region [Hayashi et al., 1984]. The observations suggest that Pc 1 pulsations are generated in the regions in which a mixture of hot ring current protons and dense

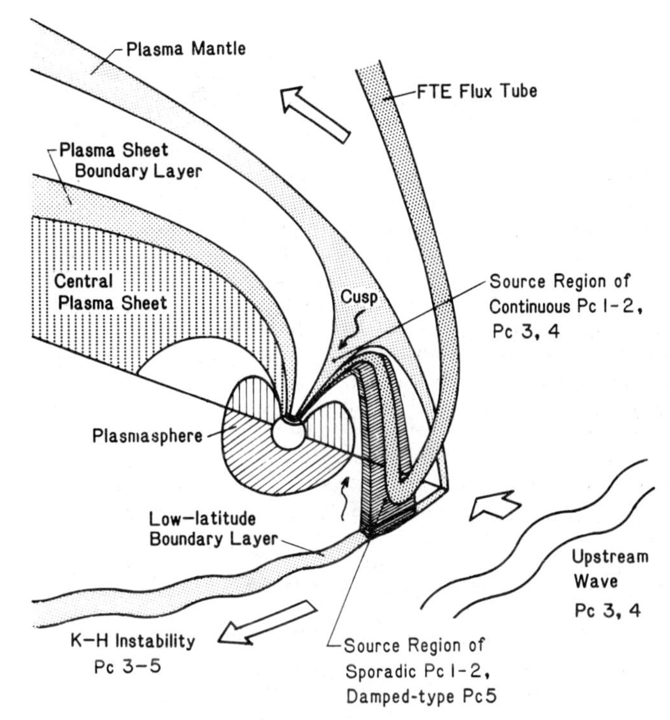

Figure 13. Schematic diagram showing source regions of various types of magnetic pulsations observed at cusp latitudes.

cold plasma occurs. In these regions the growth rate of the ion cyclotron instability is greatly enhanced due to a decrease in the resonant energy.

The effects of heavy ions such as He^+ and O^+ are also significantly important for the generation and propagation of ion cyclotron waves. The GEOS-1, -2 and ATS-6 data demonstrated clearly that Pc 1 emissions observed at geosynchronous orbit are well organized by He^+ and O^+ gyrofrequencies [Young et al., 1981; Frazer and McPherron, 1982; Gendrin, 1983]. From a detailed comparison between ground-based observations of Pc 1 waves and the GEOS-1 and -2 observations in the conjugate equatorial region, Perraut et al. [1984] suggested that the emissions below the helium gyrofrequency in the equatorial region are almost always recorded on the ground at the same frequency; in contrast, only 50% of the emissions generated above the helium gyrofrequency can reach the ground and the frequencies in space and on the ground are often different.

The coordinated satellite-ground conjugate observations in the cusp region have not been carried out. Therefore, there is little information on the source regions of Pc 1-2 pulsations observed at cusp latitudes. However, the source region of the Pc 1-2 band is predicted to be the high-altitude dayside cusp region, since this type of pulsation occurs continuosly around magnetic noon at cusp latitudes and the frequency is 0.1-0.4 Hz. The DE-1 plasma data showed that cold heavy ions (He^+, O^+) are supplied from the ionosphere into the dayside cusp region [Lockwood et al., 1985]. Therefore, the growth rate of the ion cyclotron instability could be enhanced in the high-altitude dayside cusp region by a mixture of hot protons supplied from the magnetosheath and cold atmospheric plasma supplied from the ionosphere. The observed summer maximum in occurrence of activity in the Pc 1-2 band suggests that cold plasma supplied from the ionosphere plays an important role in their generation.

On the other hand, sporadic Pc 1-2 events such as IPRP's and IPCP's occur primarily in local winter. If sporadic Pc 1-2 events occur in a spatially limited area, this seasonal variation would appear to be explainable by a propagation effect. The Pc 1-2 waves transmitted into the ionosphere cannot propagate horizontally through the ionospheric waveguide in summer since the ionosphere in the polar cap is sunlit all day; however, they can propagate in winter in the dayside ionosphere which has no solar radiation. Cole et al. [1982] suggested that IPRP events are generated in the dayside equatorial boundary layer and that the falling-tone structure with time scale of a few minutes is due to a decrease in the ion gyrofrequency as the generation region moves toward the dawn or dusk flanks of the boundary layer. The time scale of IPRP and the expected motion of the IPRP event source region are similar to those of FTE's. Therefore, IPRP events might be related to proton beams generated in the FTE tube.

IPCP events occur primarily in the morning hours at cusp latitudes and appear to be related to substorm activity. However, the generation mechanism of these events remains unsolved. A possible source region of IPCP's is the low latitude boundary layer since high-energy protons are likely to be produced by wave-particle interaction processes associated with field-aligned current flow in this region. The source regions of various types of Pc 1-2 pulsations discussed above are summarized in Figure 13.

Excitation of Standing Shear Alfven Waves by FTE's

A close relationship between magnetic spikes at cusp latitudes and Pc 5 pulsations in the auroral zone was noted first by Hirasawa [1970]. He showed that magnetic spikes at South Pole station are accompanied by damped-type Pc 5 pulsations at Byrd station (68.0 invariant latitude) located nearly on the same longitude in the auroral zone. The magnetometer data from the South Pole-Frobisher Bay conjugate pair have demonstrated conclusively that damped-type Pc 5 pulsations are excited by flux transfer-type spikes. Note that these events are different from Psc events, which are damped-type pulsations excited by storm sudden commencements (ssc's) or sudden impulses (si's). Psc events are observed from high latitudes to low latitudes on the earth's surface [Saito, 1969; Fukunishi, 1979; Nagano et al., 1987] and at geostationary orbit [Baumjohann et al., 1984; Nagano and Araki, 1985]. Ground signatures of ssc's and si's are well identified using the data from a global network of magnetometer stations. The flux transfer-type magnetic spikes observed at cusp stations had no magnetic signature at low-latitude stations. Lanzerotti et al. [1986a] have pointed out that these magnetic spikes in the dayside cusp region appear to be condidates for a ground signature of FTE's, as discussed in detail in the previous section. The DMSP particle data confirmed that damped-type Pc 5 pulsations following the magnetic spikes occurred on closed field lines.

The physical mechanisms for excitation of standing Alfven waves by FTE's have been proposed by Holzer and Reid [1975], Southwood [1987], and Lee et al. [1988]. In Southwood's model, part of the incident field-aligned current surge associated with a FTE is carried on the adjacent closed field lines and is then reflected by the ionosphere. If the reconnection has ceased by the time the surge returns to the magnetopause vicinity, the surge bounces back and forth on the closed tube, evolving into a damped standing Alfven wave. On the other hand, Lee [1986] pointed out from MHD simulations that magnetic islands (magnetic flux tubes) are repeatedly formed and convected out of the boundary. The sporadic formation of convecting magnetic islands deforms impulsively adjacent closed field lines. Therefore, it is also expected from Lee's model that damped-type standing Alfven waves will be excited on closed field lines associated with the formation region of magnetic islands [Lee et al., 1988].

In order to confirm the predictions discussed above, satellite-ground conjugate studies are important. Glassmeier et al. [1984] showed that an impulsive Pc 5 event occurred on the ground and on GEOS 2 at the geosynchronous orbit at the time when the ISEE satellite pair observed a FTE-like magnetic field disturbance at the magnetopause. Kokubun et al. [1987] found from a comparison between ground magnetometer data obtained from a Canadian chain of stations and the GOES magnetometer data that there is a close relationship between spike-like magnetic variations observed at cusp stations and impulsive Pc 5 pulsations at geosynchronous orbit.

It is important to study the relationships between FTE's and auroral activity. As shown by the examples in Figures 8-11, FTE-type magnetic spikes have been found to be accompanied by enhanced auroral activity. Discrete auroras appear to follow the passage of the downward field-aligned current region as inferred from the ground magnetic signatures. This fact suggests that FTE-type magnetic spikes are caused by pairs of downward and upward field-aligned currents moving from the dayside to the nightside, and that auroras are excited by electrons accelerated in the upward field-aligned current region. However, this suggestion needs further carefull examination using magnetic field and aurora data obtained from a network of stations at cusp latitudes. Additonal observational and theoretical works are required for future.

Acknowledgements. We are particularly grateful to Y. Tonegawa for help with data processing. We also would like to thank T. Ono, C.-I. Meng, WDC-C2 for Aurora and WDC-A for supplying us DMSP particle data. We thank D. Southwood, A. Nishida, and L. C. Lee for their valuable discussions and comments.

References

Akasofu, S-I., and J. R. Kan, Dayside and nightside auroral arc systems, Geophys. Res. Lett., 7, 753-756, 1980.

Arnoldy, R.L., M.J. Engebretson, and L.J. Cahill, Jr., Bursts of Pc 1-2 near the ionospheric footprint of the cusp and their relationship to flux transfer events, J. Geophys. Res., 93, 1007-1016, 1988.

Baumjohann, W., H. Junginger, G. Haerendel, and O.H. Bauer, Resonant Alfven waves excited by a sudden impulse, J. Geophys. Res., 89, 2765-2769, 1984.

Bol'shakova, O.V., and V. A. Troitskaya, The relation of the high latitude maximum of Pc 3 intensity to the dayside cusp, Geomag.Aeron., 24, 633-635, 1984.

Bolshakova, O.V., V.A. Troitskaya, and K.G. Ivanov, High-latitude Pc 1-2 geomagnetic pulsations and their connection with location of the dayside polar cusp, Planet. Space Sci., 28, 1-7, 1980.

Chen, L., and A. Hasegawa, A theory of long-period magnetic pulsation, 1. Steady state excitation of field line resonance, J. Geophys. Res., 79, 1024-1032, 1974.

Cole, K. D., R. J. Morris, E. T. Matveera, V. A. Troitskaya, and O. A. Pokhotelov. The relationship of the boundary layer of the magnetosphere to IPRP events, Planet. Space Sci. 30, 29-136, 1982.

Dungey, J.W., Interplanetary magnetic field and the auroral zone, Phys. Rev. Lett., 6, 47-48, 1961.

Engebretson, M., C.-I.Meng, R.L. Arnoldy, L.J. Chahill, Jr., Pc 3 pulsations observed near the south polar cusp, J. Geophys Res., 91, 8909-8918, 1986.

Engebretson, M.J., L.J. Zanetti, T. A. Potemra, W. Baumjohann, H. Luhr, and M.H. Acuna, Simultaneous observation of Pc 3-4 pulsations in the solar wind and in the earth's magnetosphere, J. Geophys. Res., 92, 10053-10062, 1987.

Fraser, B.J., and R.L. McPherron, Pc 1-2 magnetic pulsation spectra and heavy ion effects at synchronous orbit :ATS 6 results, J. Geophys. Res., 87, 4560-4566, 1982.

Fraser-Smith, A.C., ULF/Lower-ELF electromagnetic field measurements in the polar caps, Rev. Geophys. Space Phys., 20, 497-512, 1982.

Fu, Z.F., and L.C. Lee, Simulation of multiple X-line reconnection at the dayside magnetopause, Geophys. Res. Lett., 12, 291-295, 1985.

Fukunishi, H., Latitude dependence of power spectra of magnetic pulsations near L=4 excited by ssc's and si's, J. Geophys.Res., 84, 7191-7200, 1979.

Fukunishi, H., Pc 1-2 pulsations and related phenomena: Review,in Achievements of the International Magnetospheric Study (IMS), ESA Publications, SP-217, pp.437-448, 1984.

Fukunishi, H., Y. Tonegawa, and L.J. Lanzerotti, Auroras associated with FTE-type events recorded at South Pole-Frobisher Bay conjugate pair, in Abstract of the 82nd Fall Meeting of SGEPSS, Jpn., 1987.

Gendrin, R., Effects of heavy ions on microscopic plasma physics in the magnetosphere, in High Latitude Space Plasma Physics, edited by B. Hultqvist and T. Hagfors, pp.415-436, Plenum, New York, 1983.

Glassmeier, K.H., M. Hoenisch, and J. Untiedt, A new type of short period transient magnetic variations at high latitudes, preprint, 1987.

Glassmeier, K.H., M. Lester, W.A.C. Mier-Jedrzejowicz, C.A. Green, G. Rostoker, D. Orr, U. Wedeken, H. Junginer, and E. Amata, Pc 5 pulsations and their possible source mechanisms: A case study, J. Geophys., 55, 108-119, 1984.

Goertz, C. K., Nielsen, A. Korth, K.-H Glassmeier, C. Haldoupis, P. Hoeg, and D. Hayward, Observations of a possible ground signature of flux transfer events, J. Geophys. Res., 90, 4069-4078, 1985.

Greenstadt, E.W., and J.V. Olson, Pc 3,4 activity and interplanetary field orientation, J. Geophys. Res., 81, 5911-5920, 1976.

Greenstadt, E.W., and J.V. Olson, A contribution to ULF activity in the Pc 3-4 range correlated with IMF radial orientaition, J. Geophys. Res., 82, 4991-4996, 1977.

Greenstadt, E.W., M.M. Mellott, R.L. McPherron, C.T. Russell, H.J. Singer, and D.J. Knecht, Transfer of pulsation-related wave activity across the magnetopause: Observations of corresponding spectra by ISEE 1 and ISEE 2, Geophys. Res. Lett., 10, 659-662, 1983.

Gul'elmi, A.V., Diagnostic of the magneosphere and interplanetary medium by means of pulsations, Space Sci. Rev., 16, 331, 1974.

Haerendel, G., G. Paschmann, N. Sckopke, H. Rosenbauer, and P.C. Hedgecock, The frontside boundary layer of the magnetosphere and the problem of reconnection, J. Geophys. Res., 83, 3195-3216, 1978.

Hayashi, K., S. Kokubun, T. Oguti, T. Kitamura, O. Saka, and T. Watanabe, Substorm associated Pc 1 emission in the morning sub-cleft latitudes, in Achievements of the International Magnetospheric Study (IMS), ESA Publications, SP-217, pp. 603-607, 1984.

Heacock, R. R., Midday Pc 1-2 pulsations observed at a subcleft location, J. Geophys. Res., 79, 4239-4245, 1974.

Heacock, R. R., and J. K. Chao, Type Pi magnetic field pulsations at very high latitudes and their relation to plasma convection in the magnetosphere, J. Geophys. Res., 85, 1203-1213, 1980.

Heacock, R. R., and R. D. Hunsucker, Pc 1 - 2 magnetic field pulsations on dayside cleft field lines, Nature, 269, 313-314, 1977.

Hirasawa, T., Long-period geomagnetic pulsations (pc 5) with typical sinusoidal waveforms, Rep. Iono. Space Res. Jpn., 24, 66-79, 1970.

Holtzer, T.E., and G.C. Reid, The response of the day side magnetosphere-ionosphere system to time-varing field line reconnection at the magnetopause, 1. Theoretical model, J. Geophys. Res., 80, 2041-2049, 1975.

Kato, Y., Y. Tonegawa, and K. Tomonura, Dynamic spectral study of Pc 3-5 magnetic pulsations observed in the north polar cusp region, Mem. Nat. Inst. Polar Res. Spec. Issue, Jpn., 36, 58-63, 1985.

Kokubun, S., T. Yamamoto, H. Kawano, K. Hayashi, and T. Oguti, Impulsive and pulsive magnetic variations near the polar cusp region, in Abstract of the 82nd Fall Meeting of SGEPSS, Jpn., 1987.

LaBelle, J., R.A. Treumann, G. Haerendel, O.H. Bauer, G. Paschmann, W. Baumjohann, H. Luhr, R.R. Anderson, H.C. Koons, and R.H. Holzworth, AMPTE IRM observations of waves associated with flux transfer events in the magnetosphere, J. Geophys. Res., 92, 5827-5843, 1987.

Lanzerotti, L. J., and C. G. Maclennan, Hydromagnetic waves associated with possible flux transfer events, to be published in Astrophys. Space Sci., 1987.

Lanzerotti, L.J., L.V. Medford, C.G.Maclennan, A. Hasegawa, M.H. Acuna, and S.R. Dolce, Polarization characteriatics of hydromagnetic waves at low geomagnetic latitudes, J.Geophys. Res., 86, 5500-5506, 1981.

Lanzerotti, L. J., L. C. Lee, C. G. Maclennan, A. Wolfe, and L.V. Medford, Possible evidence of flux transfer events in the polar ionosphere, Geophys. Res. Lett., 13, 1089-1092, 1986a.

Lanzerotti, L.J., C.G. Maclennan, L.V. Medford, and D.L. Carpenter, Study of QP/GP event at very high latitudes, J. Geophys. Res., 91, 375-380, 1986b.

Lanzerotti, L.J., R. D. Hunsucker, D. Rice, L.C. Lee, A. Wolfe, C.G. Maclennan, and L. V. Medford, Ionospheric and ground-based response to field-aligned currents near the magnetospheric cusp regions, J. Geophys. Res., 92, 7739-7743, 1987a.

Lanzerotti, L.J., C.T. Russell, C.G. Maclennan, A. Wolfe, and L.V. Medford, Satellite and ground-based studies of possible ionospheric signatures of flux transfer events, in Abstracts of IAGA, IUGG XIX General Assembly, Vancouver, 1987b.

Lee, L.-C., Magnetic flux transfer at the earth's magnetopause, in Solar Wind-Magnetosphere Coupling, edited by Y. Kamide and J. A. Slavin, pp.297-314, Terra Sci. Pub, Co., Tokyo, 1986.

Lee, L.C., and J. V. Olson, Kelvin-Helmholtz instability and the variation of geomagnetic pulsation activity, Geophys. Res. Lett., 7, 777-780, 1980.

Lee, L.C., Y. Shi, and L.J. Lanzerotti, A mechanism for the generation of cusp-region hydromagnetic waves, submitted to Geophys. Res. Lett., 1988.

Lockwood, M., M.O. Chandler, J.L. Horwitz, J.H. Waite, Jr., T.E. Moore, and C.R. Chappell, The cleft ion fountain, J.Geophys. Res., 90, 9736-9748, 1985.

Meng, C.-I., and R. Lundin, Auroral morphology of the midday oval, J. Geophys. Res., 91, 1572-1584, 1986.

Miura, A., Anomalous transport by magnetohydrodynamic Kelvin-Helmholtz instabilities in the solar wind-magnetosphere interaction, J. Geophys. Res., 89, 801-818, 1984.

Morris, R. J., and K. D. Cole, Pc 1-2 discrete irregular daytime pulsation bursts at high latitudes, Planet. Space Sci., 33,53-67, 1985.

Nagano, H., and T. Araki, A statistical study on Psc 4 and Psc 5 observed by geostationary satellites, Planet. Space Sci.,33, 365-372, 1985.

Nagano, H., T. Araki, T. Iyemori, H. Fukunishi, N. Sato, and M. Ayukawa, Geomagnetic sudden commencements observed at the Syowa-Iceland conjugate stations, Mem. Nat. Inst. Polar Res. Jpn, Spec. Issue, No. 47, 78-91, 1987.

Odera, T.J., Solar wind controlled pulsations : A Review, Rev. Geophys. 24, 55-74, 1986.

Oguti, T., T. Yamamoto, K. Hayashi, S. Kokubun, A. Egeland, and J.A. Holtet, Dayside auroral activities and related magnetic impulses in the polar cusp region, submitted to J. Geomag. Geoelect., 1987.

Olson, J. V., ULF signature of the polar cusp, J. Geophys. Res., 91, 10055-10062, 1986.

Paschmann, G., G. Haerendel, I. Papamastorakis, N. Sckopke, S.J. Bame, J.T. Gosling, and C.T. Russell, Plasma and magnetic field characteristics of magnetic flux transfer events, J. Geophys. Res., 87, 2159-2168, 1982.

Perraut, S., R. Gendrin, A. Roux, and C. de Villedary, Ion cyclotron waves : Direct comparison between graund-based measurements and observations in the source region, J. Geophys. Res., 89, 195-202, 1984.

Rijnbeek, R.P., S.W.H. Cowley, D.J. Southwood, and C.T. Russell, A survey of dayside flux transfer events observed by ISEE 1 and 2 magnetometers, J. Geophys. Res., 89, 786-800, 1984.

Rostoker, G., J.C. Samson, and Y. Higuchi, Occurrence of Pc 4, 5 micropulsation activity at the polar cusp, J. Geophys. Res.,77, 4700-4706, 1972.

Russell, C.T., and R. C. Elphic, Intial ISEE magnetometer results: Magnetopause observations, Space Sci, Rev.,22, 681, 1978.

Russell, C. T., and R. C. Elphic, ISEE observations of flux transfer events at the dayside magnetopause, Geophys. Res. Lett., 6, 33-36, 1979.

Russell, C.T., and M.M. Hoppe, Upstream waves and particles, Space Sci. Rev., 34, 155-172, 1983.

Russell, C.T., J.G. Luhmann, T.J. Odera, and W.F. Stuart, The rate of occurrence of dayside Pc 3, 4 pulsations : The L-value dependence of IMF cone angle effect, Geophys. Res. Lett., 10, 663-666, 1983.

Saito, T., Geomagnetic pulsations, Space Sci. Rev., 10, 319-412, 1969.

Sandholt, P.E., A. Egeland, J.A. Holtet, B. Lybekk, K. Svenes, S. Asheim, and C.S. Deehr, Large- and small-scale dynamics of the polar cusp, J. Geophys. Res., 90, 4407-4414, 1985.

Sandholt, P. E., C. S. Deehr, A. Egeland, B. Lybekk, R. Viereck, and G. J. Romick, Signatures in the dayside aurora of plasma transfer from the magnetosheath, J. Geophys. Res., 91, 10063-10079, 1986.

Saunders, M. A., C.T. Russell, and N. Sckopke, Flux transfer events : scale size and interior structure, Geophys. Res. Lett., 11, 131-134, 1984.

Singer, H.J., C.T. Russell, M.G. Kilvelson, E.W. Greenstadt, and J.V. Olson, Evidence for the controle of Pc 3, 4 magnetic pulsations by the solar wind velocity, Geophys. Res. Lett., 4, 377-379, 1977.

Snyder, A. L., and S.-I. Akasofu, Auroral oval photographs from the DMSP 8531 and 10533 satellites, J. Geophys. Res., 81, 1799-1804, 1976.

Southwood, D.J., The hydromagnetic instability of the magnetospheric boundary, Planet. Space Sci., 16, 587, 1968.

Southwood, D.J., Some features of field line resonance in the magnetosphere, Planet. Space Sci., 22, 483-419, 1974.

Southwood, D. J., Theoretical aspects of ionosphere-magnetosphere-solar wind coupling, Adv. Space Res., 5, No.4, 7-14 1985.

Southwood, D. J., The ionospheric signature of flux transfer events, J. Geophys. Res., 92, 3207-3213, 1987.

Takahashi, K., R.L. McPherron, and W.J. Hughes, Multispacecraft observations of the harmonic structure of Pc 3-4 magnetic pulsations, J. Geophys. Res., 89, 6758-6774, 1984.

Todd, H., B.J.I. Bromage, S.W.H. Cowley, M. Lockwood, A.P. van Eyken, and D.M. Willis, EISCAT observations of bursts of rapid flow in the high latitude dayside ionosphere, Geophys. Res. Lett., 13, 909-912, 1986.

Tonegawa,Y., and H. Fukunishi, Harmonic structure of Pc 3-5 magnetic pulsations observed at the Syowa-Husafell conjugate pair, J. Geophys Res., 89, 6737-6748, 1984a.

Tonegawa, Y., H. Fukunishi, T. Hirasawa, R. L. McPherron, T. Sakurai, and Y. Kato, spectral characteristics of Pc 3 and Pc 4/5 magnetic pulsation bands observed near L=6, J. Geophys. Res., 89, 9720-9730, 1984b.

Tonegawa, Y., H. Fukunishi, L.J. Lanzerotti, C.G. Maclennan, L.V. Medford, and D.L. Carpenter, Studies of the energy source for hydromagnetic waves at auroral latitudes, Mem. Nat. Inst. Polar Res. Jpn., 38, 73-82, 1985.

Trivedi, N., L.J. Lanzerotti, A. Wolfe, C.G. Maclennan, and L.V. Medford, Search for cusp-latitude magnetic field signatures of possible flux transfer events, in Abstracts of IAGA, IUGG XIX General Assembly, Vancouver, 1987.

Troitskaya, V.A., T.A. Plyasova-Bakunina, and A.V. Gul'elmi, Relationship between Pc 2-4 pulsations and the interplanetary magnetic field, Dokl. Akad. Nauk SSSR, 197, 1312, 1971.

Troitskaya, V.A., O.V. Bolshakova, and V.P. Hessler, Polar micropulsations, Scientific Report. Geophysical Institute, Univ. of Alaska, 1972.

Troitskaya, V. A., O. V. Bolshakova, and E. T. Matveeva, Geomagnetic pulsation in the polar cap, J. Geomag. Geoelectr. 32, 309-324, 1980.

Vero, J., Experimental aspects of low-latitude pulsations-A review, J. Geophys., 60,106-119,1986.

Watanabe, Y., and T. Terasawa, On the excitation mechanism of the low frequency upstream waves, J. Geophys. Res., 84, 6623-6630, 1984.

Webb, D., and D. Orr, Geomagnetic pulsations (5-50 mHz) and the interplanetary magnetic field, J. Geophys. Res., 81, 5941-5947, 1976.

Wolfe,A.,and R.L. Kaufmann, MHD wave transmission and production near the magnetopause, J. Geophys. Res., 80, 1764-1775, 1975.

Wolfe, A., L.J. Lanzerotti, and C.G. Maclennan, Dependence of hydromagnetic energy on solar wind velocity and interplanetary magnetic field direction, J. Geophys. Res., 85, 114-118, 1980.

Wolfe, A., A. Meloni, L.J. Lanzerotti, C.G. Maclennan, J. Bamber, and Venkatesan, Dependence of hydromagnetic energy spectra near L=2 and L=3 on upstream solar wind parameters, J. Geophys. Res., 90, 5117-5131, 1985.

Wolfe, A., E. Kamen, L. J. Lanzerotti., C. G. Maclennan, I. F. Bamber and D. Venkatesan, ULF geomagnetic power at cusp latitudes in response to upstream solar wind conditions, J. Geophys. Res., 92, 168-173, 1987.

Young, D.T., S. Perraut, A. Roux, C. de Villedary, R. Gendrin, A. Korth, G. Kremser, and D. Jones, Wave particle interactions near Ω_{He^+} ovserved on GEOS 1 and 2, 1, Propagation of ion cyclotron waves in He^+-rich plasma, J. Geophys Res., 86, 6755-6772, 1981.

Yumoto, K., Long - period magnetic pulsations generated in the magnetospheric boundary layers, Planet. Space Sci., 32, 1205-1218, 1984.

Yumoto, K., Generation and propagation mechanisms of low-latitude magnetic pulsations - A review, J. Geophys. 60, 79-105, 1986.

CHARACTERISTICS OF THE MAGNETOHYDRODYNAMIC WAVES OBSERVED IN THE EARTH'S MAGNETOSPHERE AND ON THE GROUND

Masayuki Kuwashima and Shigeru Fujita

Kakioka Magnetic Observatory, 595 Kakioka Yasato-machi
Niihari-gun Ibaraki 315-01, Japan

Abstract. The recent research activities in the physics of the magnetohydrodynamic (MHD) waves in the earth's magnetosphere and on the ground will be reviewed focussing on Pc 3, Pc 4-5, Pi 2 and ssc-associated magnetic pulsations. MHD studies have been extensively developed due to improvement both in observational techniques and in theoretical works since IMS period (1976-1979) and succeeding years. However, there are still unresolved problems to be reserved to future studies. In the present paper, furture research directions will be also given.

1. Introduction

Since the first report by Stewart almost one hundred years ago, characteristics of the magnetohydrodynamic (MHD) waves have been studied under the name of magnetic pulsations, which are recognized as originating from plasma phenomena. Magnetic pulsations can also be thought as the lowest frequency waves that can be sustained in the magnetosheric plasma. The present review is devoted to some areas of current research interest in the physics of the MHD waves in the earth's magnetosphere and on the ground.

Magnetic pulsations observed on the earth's surface were suggested to be closely associated with MHD waves in the magnetosphere [Dungey, 1954], and the great impetus for the study of MHD waves came during the International Geophysical Year (IGY) in 1957-1958. During this program, many more magnetometer stations were established around the world in order to study the global occurrences of geomagnetic phenomena such as magnetic storms and magnetic pulsations. During IGY and succeeding years, the first global morphology of magnetic pulsations was established [Jacobs et al., 1964; Saito, 1969]. These early studies ultimately led to the classification of magnetic pulsations into categories according to the regularity and the period of variations. Magnetic pulsations having a rather well defined spectral peak are classified as Pc (continuous pulsations), while those involving a wide spectral range are classified as Pi (irregular pulsations). The observed range of the period of magnetic pulsations extends from fractions of a second to several-ten minutes.

The study of magnetic pulsations was extensively carried out during the International Magnetospheric Study program (IMS) in 1976 - 1979. During IMS and succeeding years, characteristics of MHD waves have become increasing clear due to improvements in observational techniques, which are concurrent magnetic observation by ground-based magnetometer chains from high- to low-latitudes at several local time meridians, multiple satellite observations in the magnetosphere as well as in the interplanetary field, and coordinated observations of various geophysical parameters, namely, electric field, thermal plasma density, energetic particles and others.

One of the most important findings is the existence of upstream MHD waves in the earth's foreshock region as a main source of Pc 3 magnetic pulsations observed on the earth's surface. Observational relationships between the solar wind parameters and magnetic field parameters clarified that the magnetosonic upstream waves in the Pc3-4 frequency range in the earth's foreshock region can be transmitted to the magnetosphere without significant changes in spectra, and can propagate across the ambient magnetic field into the inner magnetosphere coupling with various MHD oscillations [Greenstadt et al., 1983; Yumoto, 1985a, 1986; Yumoto et al., 1985a; Wolf et al., 1985].

The Kelvin-Helmholz type instability has also been found to be an important energy source of MHD waves. The early studies on MHD waves in the 1960s and 1970s revealed many aspects of MHD standing oscillation structures. Especially, the observational results based on the conjugate-pair stations on the ground suggested that many MHD signals are generated by the odd-mode MHD standing oscillations [Sugiura and Wilson, 1964; Kokubun and Nagata, 1965; Sakurai, 1970]. Observational results by the magnetometer chain on the ground and in the magnetosphere [Samson and Rostoker, 1972; Lanzerotti and Fukunishi, 1974; Hughes et al., 1978] revealed that Pc 4-5 at high-latitudes are caused by coupling of the MHD magnetosonic waves caused by the Kelvin-Helmholz instability near the dayside magnetopause and the Alfvén waves on the localized field lines in the magnetosphere [Southwood, 1974; Chen and Hasegawa, 1974a and b].

Satellite observations in the magnetosphere also revealed the existence of the compressional MHD waves in the Pc 4-5 frequency range as well as shear Alfvén waves [Barfield and McPherron, 1972; Kremmser et al., 1981; Kokubun, 1985; Takahashi and Higbie, 1986]. Those compressional waves are considered to be closely associated with the hot plasma in the ring current, where simple MHD theory is no longer adequate and microscopic kinetic processes must be incorporated. Compressional waves are concentrated near the equatorial plane in the magnetosphere and are rarely observed on the ground. One of the characteristics of those waves is association with the energetic particle modulations. Several kinds of generation mechanism for compressional MHD waves has been proposed theoretically, which are drift-mirror instability [Hasegawa, 1969], modified drift-mirror instability [Walker et al., 1982] and bounce resonance interaction [Southwood, 1973, 1976].

Large scale impulsive response of the magnetosphere also is an important energy source of MHD waves. It is more than twenty years ago that Saito and Matsushita [1967] analyzed the magnetic pulsations excited by the storm sudden commencement (ssc). Namely, various kinds of ssc-associated magnetic pulsations (Psc1- Psc5) are excited by ssc in the wide latitudinal region from high- to low- latitudes. The simultaneous appearance of ssc-associated magnetic pulsations with different eigen period has been confirmed by the observation by the dense magnetometer array and STARE observation [Poulter and Nielsen, 1982; Glassmeier et al., 1984]. According to these observations, the variation of ssc-associated magnetic pulsation period is essentially continuous along the latitudes suggesting the MHD toroidal oscillation of decoupled field shells.

Sudden reconfiguration of the nightside magnetosphere associated with the substorm expansion is another energy source of MHD waves, Pi2 magnetic pulsations for example. The main morphological characteristics of Pi2 were established during IMS period [Saito et al., 1976; Baumjohaan and Glassmeier, 1984; Yumoto, 1986]. It is generally believed that a main cause of Pi2 is a transient MHD signals associated with a sudden change of the structure of the nightside magnetosphere, which is caused by ionospheric channeling of the dawn-to-dusk tail current during the substorm expansion [Mallinkrodt and Carlson, 1978; Nishida ,1979; Sakurai and McPherron, 1983]. However, there are still unresolved problems in the Pi2 study, global Pi2 polarization behavior from high- to low- latitudes [Yumoto, 1986, 1987] for example.

It is commonly believed that oscillations with periods of about 1 second (Pc1) arise from processes associated with the ion cyclotron instability. In the early years, it was assumed that only protons and electrons are contained in the magnetospheric plasma. However, satellite observations in the magnetosphere revealed the important role of heavy ions (helium and oxygen) for the generation of Pc1 waves [Young et al., 1981; Fraser and McPherron, 1982]. The generation of Pc1 waves through wave-particle interaction in the ring current and their propagation along the field lines into the ionosphere as well as in the ducting layer in the ionosphere from high- to low-latitudes have been studied using realistic magnetosphere and ionosphere models [Kuwashima et al.,

1981; Webster and Fraser, 1985; Fujita, 1987].

Before reaching the ground to be detected as magnetic pulsations, MHD waves generated in the magnetosphere have to pass through the ionosphere and atmosphere. The role of the ionoshpere and atmosphere on the modification of the incident MHD waves has been extensively studied by many researchers [Nishida, 1964; Inoue, 1973; Hughes and Southwood, 1976a and b; Glassmeier, 1984; Fujita, 1987]. However, the experimental evidences of the ionospheric modulation effects are not so many because only limited number of data sets exists which can be used to examine ionospheric effects on MHD waves. Simultaneous observations of the MHD waves both above and below the ionosphere are needed to clarify the ionospheric effects in more detail.

The present paper will summarize above-mentioned current research topics on MHD waves in the earth's magnetosphere and on the ground. In Section 2, upstream waves in the earth's foreshock region and their transmission into and propagation through the magnetosphere will be discussed with relationships of Pc3 magnetic pulsations on the ground. In Section 3, characteristics of ssc-associated magnetic pulsations will be discussed. Instabilities with the hot plasma in the ring current in the magnetosphere will be discussed with the relationships of compressional Pc 4-5 waves in Section 4. Characteristics of Pi2 magnetic pulsations will be discussed in Section 5. Then, the role of the ionosphere on the modifications of MHD waves will be discussed in Section 6. In the last section, the results discussed in the present paper will be summarized and also future research directions on MHD waves will be discussed.

2. Upstream Waves as Sources of Pc 3

The region upstream from the earth's bow shock is filled with a variety of MHD waves [Fairfield, 1969; Russell et al., 1971; Tsurutani and Rodriguez, 1981; Russell and Hoppe, 1983]. Troitskaya et al. [1971], Greenstadt [1972] and Gul'elmi et al. [1973] originated the idea that Pc 3 magnetic pulsations observed on the earth's surface is closely related with the bow-shock associated wave phenomena which are associated with particle reflected or accelerated by the bow shock. According to his observational results, large amplitude magnetic fluctuations are observed in the upstream region that is connected to the bow shock by the interplanetary magnetic field (IMF). The fluctuations are intensified when the cone angle between IMF and the bow shock normal is small. The nature of the upstream waves has been clarified by many observations [Gosling et al., 1978; Paschmann et al., 1981; Hoppe et al., 1981; Russell and Hoppe, 1983]. At the leading edge of the foreshock region, waves with frequencies around 1 Hz are found in connection with the reflected ion beams. Further downstream, in the region populated by the intermediate and diffuse ions, the large amplitude MHD waves with period range of 30 seconds are observed. Those monochromatic sinusoidal waves are considered to be magnetosonic waves with rest-frame frequency 0.05 - 0.2 times the ion cyclotron frequency and wave length of about 1 Re . These waves are identified as magnetosonic right-handed waves excited by the ion cyclotron instability in the back streaming energetic particles, which are caused with reflection by the bow shock [Paschmann et al., 1980, 1981; Schwartz et al., 1983] or ion heating [Tidman and Krall, 1971; Edmiston et al., 1982].

There are many evidences that the MHD waves generated in the upsream region are convected through the bow shock and magnetosheath, and finally couple into the magnetosphere exciting Pc 3 magnetic pulsations [Yumoto, 1986; Odera, 1986]. Zhuang and Russell [1982] suggested that the fast mode of the upstream waves whose incident cone angle is less than the critical angle($\sim 20°$)can pass through and be amplified by the bow shock. A number of points suggests importance of the IMF direction for the generation and transmission of the upstream MHD waves. Small values of the cone angle is more favorable to the generation of the upstream MHD waves in the solar wind by bow-shock reflected ions [Fairfield, 1969; Kovner et al., 1976; Russell and Hoppe, 1983]. Small values of the cone angle also can facilitate significant wave transmission and amplification through the bow shock [McKenzie and Westphal, 1970] Also, small values of the cone angle can aid the convection of such waves through the magnetosheath to the dayside magnetopause [Greenstadt and Olson, 1976; Russell et al., 1983]. The cone angle effect also affects the generation of MHD waves at the magnetopause boundary by the Kelvin-

Helmholz instability [Lee and Olson, 1980].

The fast magnetosonic waves generated beyond the bow shock can propagate and convect through the magnetosheath, finally to reach the magnetopause. Both a tangential discontinuity model of the magnetopause [Verzariu, 1973; Wolf and Kaufmann, 1975; Greenstadt et al., 1983] and a rotational discontinuity one [Lee, 1982; Kwok and Lee, 1984] could allow such compressional wave to be transmitted through the magnetopause, from where the waves will propagate earthward through the outer magnetosphere. The transmission coefficient averaged over a hemispherical distribution of the incident MHD waves through the tangential discontinuity magnetopause is expected to be 1 - 2 %. Therefore, the energy input of MHD waves can contribute significantly to the energy budget of the magnetosphere [Verzariu, 1973]. Concerning to MHD transmission at a rotational discontinuity magnetopause, the transmission coefficient of 1 % is also expected theoretically [Lee, 1982]. In this case, the wave transmission is strongly dependent on the ratio of the normal component of the ambient magnetic field to the tangential one, solar wind velocity and the rotation angle of the magnetic field [Kwok and Lee, 1984].

Many theoretical and observational results have clarified that the magnetosonic upstream waves transmitted from outside the magnetosphere propagate across the ambient magnetic field into the inner magnetosphere. Therefore, it is considered that a main source of low-latitude Pc 3 pulsations is the magnetosonic upstream waves generated in the earth's foreshock region. From the comparison of power spectra of magnetic field data from ISEE 1 and 2 observed simultaneously on both sides of the magnetopause, Greenstadt et al. [1983] found that the MHD waves with similar frequencies were observed on the two sides of the magnetopause boundary. Ratios of the magnetic power in the magnetosphere to that in the magnetosheath in Pc 3-4 frequency range were 0.001 - 0.08, which are consistent with the theoretical results by Verzariu [1973] for tangential discontinuities and Lee [1982] for rotational discontinuities. The next question is how such transmitted waves propagate deep into the magnetosphere across the ambient magnetic field. The propagation mechanism of Pc 3 source waves in the deep magnetosphere has been studied extensively by Yumoto [1984, 1985a,b and 1986], Yumoto and Saito [1982, 1983] and Yumoto et al. [1984, 1985a,b and 1987]. They have clarified the propagation mechanisms of the Pc 3 source wave as shown in Fig. 1. According to their results, the magnetosonic upstream waves generated in the earth's foreshock region can propagate into the magnetosphere across the ambient magnetic field, and finally can couple with various MHD oscillations, which are surface waves at the plasmapause, trapped oscillations in the plasmasphere and eigen-oscillations of local field lines at low latitudes. Finally,the waves are observed as low-latitude Pc 3.

As illustrated in Fig. 1, Pc3-4 magnetic pulsations also are generated as surface waves excited by Kelvin-Helmholz type instability at the dayside high-latitude magnetopause. These surface waves can propagate effectively along the boundary layer (magnetopause), however, they have evanescent characteristics in the radial direction [Southwood, 1974; Chen and Hasegawa, 1974a; Miura, 1984; Yumoto et al., 1984]. Therefore, damping rate of typical Pc 3 surface waves in the radial direction is very large. According to the calculation by Yumoto [1986], the ampliude at L = 2.0 is seven orders of magnitude below that at the magnetopause. The existence of Pc 3 magnetic pulsations at very low latitude is difficult to explain by the surface waves at the magnetopause. Yumoto et al. [1984, 1985a] found that the correlation between the IMF parameters and the pulsation ones is higher at the low-latitude stations (Onagawa, Japan) than at the high-latitude (College, Alaska). That result suggests that Pc 3 magnetic pulsations at the high latitudes are originated both upstream waves and the other source, which is MHD waves generated as surface waves at the magnetopause.

3. Impulsive Responses of Magnetosphere as a Source of Psc

Dungey [1954] derived the axisymmetric toroidal MHD wave mode. This mode has an azimuthally (east-west) polarized magnetic field in the magnetosphere, and the MHD field line oscillations of adjacent L shells are decoupled with each other in a dipole magnetic field. A set of geomagnetic field shells can therefore be impulsively excited to "ring" simultaneously at their separate eigen periods. In such a condition, periods of magnetic pulsations vary continuously

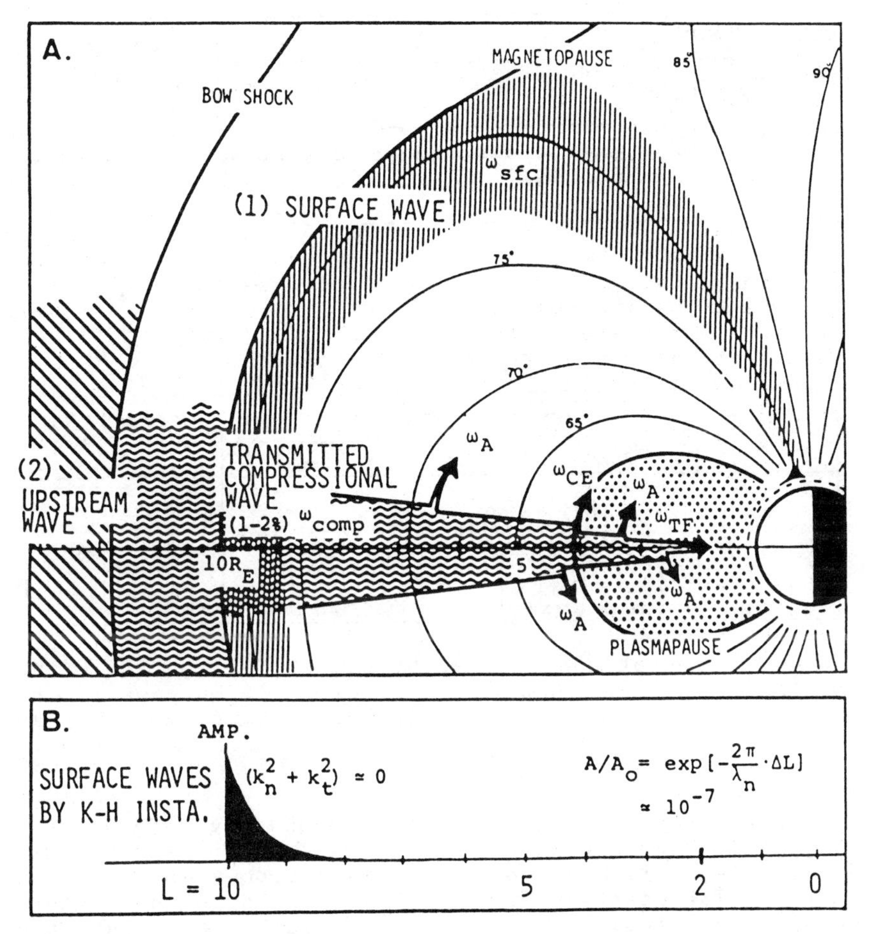

Fig. 1. Two possible generation mechanisms for Pc3 magnetic pulsation observed on the ground. One is surface waves in the Pc3 frequency range excited by the Kelvin-Helmholz type in stabilities in the day-side high-latitude boundary layer, which can transmit only into the high-latitude ionosphere. The other is upstream waves in the earth's foreshock region ,which can propagate into the magnetosphere across the field lines coupling with various MHD oscillations in the inner magnetosphere [Yumoto, 1986].

with latitude where a field line is anchored. For ringing toroidal mode resonance, the azimuthal wave number "m" is expected to be small in order to satisfy the axisymmetry condition. Coupling of magnetospheric impulse to ringing toroidal oscillations has enveloped theoretically by Chen and Hasegawa [1974b] and called "non-collective mode". The origin of ringing toroidal oscillations is expected to be impulsive response of the magnetosphere to shocks in the solar wind (ssc or si).

Existence of decoupled oscillations of individual magnetic field shells has firstly been suggested by Saito and Matsushita [1967], who studied occurrence characteristics of MHD waves associated with ssc as shown in Fig. 2. Magnetic pulsations associated with ssc are generally named as Psc. Psc is further classified into Psc 1 - Psc 6, according to the corresponding period of Pc's. As shown in Fig. 2, Psc 5, Psc 4 and Psc 2-3 were observed with damped-type wave forms at the high-(Big Delta), middle- (Fredericksburg) and low-(Onagawa) latitudes, respectively. Existence of decoupled oscillation associated with ssc has been confirmed experimentally by the result of STARE(Scandinavian Twin Auroral Rader Experiment) and SMA (Scandinavian Magnetometer Array) observations. According to the observation by STARE [Poulter and

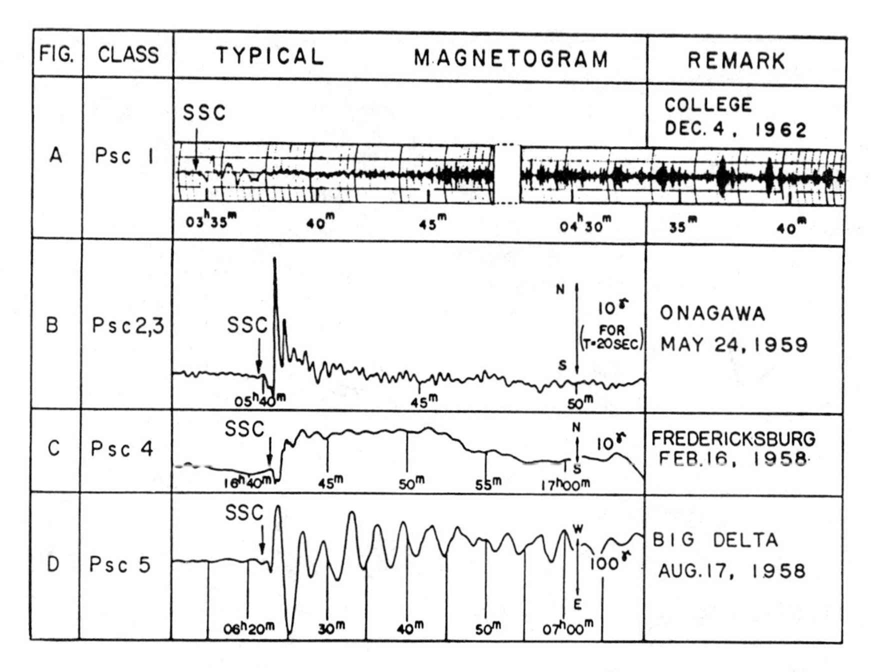

Fig. 2. Magnetic pulsations excited by ssc at College (65.0°in geomagnetic latitude, 258.7°in geomagnetic longitude), Onagawa (28.8°, 208.5°), Fredericksburg(49.5°, 350.4°) and Big Delta (64.4°, 259.8°) [Saito, 1969].

Nielsen, 1982], the Psc 5 period variation is essentially continuous along the latitudes, suggesting the oscillation of decoupled field shells. The observed Psc 5 period variation is also consistent with toroidal eigen periods calculated for the dipole field model [Allan and Knox, 1982]. The general behavior of these variable period event is consistent with geomagnetic field shells "ringing" independently at their own resonant periods after an impulsive stimulus, ssc for example. The axisymmetry condition to excite ringing toroidal mode resonances has also been observed at SMA magnetometer observation. Glassmeier et al. [1984] showed that Psc 5 period varies along the latitude, while it is almost constant along the longitude, satisfying the axisymmetry condition for ringing toroidal mode resonance.

The Psc waves in the magnetosphere have been studied by the observations at geosynchronous altitude [Nagano and Araki, 1984, 1985, 1986; Kuwashima and Fukunishi, 1985; Kuwashima et al, 1985]. According to their results, the amplitude of compressional Psc wave has a tendency to be proportional to the magnitude of ssc, suggesting the compressional Psc wave (fast mode MHD wave) has a direct relationship with disturbances at the magnetopause generated by the interplanetary shock. Those compressional mode will propagate towards the earth across the field shells in the magnetosphere. For the azimuthal oscillation of Psc waves at geosynchronous altitude, Kuwashima and Fukunishi [1985] found that direction of the initial movement is eastward in the morning sector and westward in the afternoon sector as shown in Fig. 3. When the dynamic pressure of the solar wind is enhanced suddenly, magnetic field lines are swept back tailward causing the counterclockwise polarization in the plane perpendicular to the ambient magnetic field in the morning sector and vice versa in the afternoon sector [Nagano and Araki, 1985]. The distribution of the initial movement of Psc waves summarized in Fig. 3 is consistent with the distribution predicted from the above mentioned polarization behavior. Those results indicate that the bending of field lines in the magnetosphere is more intensified associated with ssc causing perturbation in the plane perpendicular to the ambient magnetic field direction. That perturbation will propagate along the field lines with Alfvén velocity, and arrive at the auroral ionosphere. The Alfvén signal will be reflected there, and then propagate again along the field lines causing

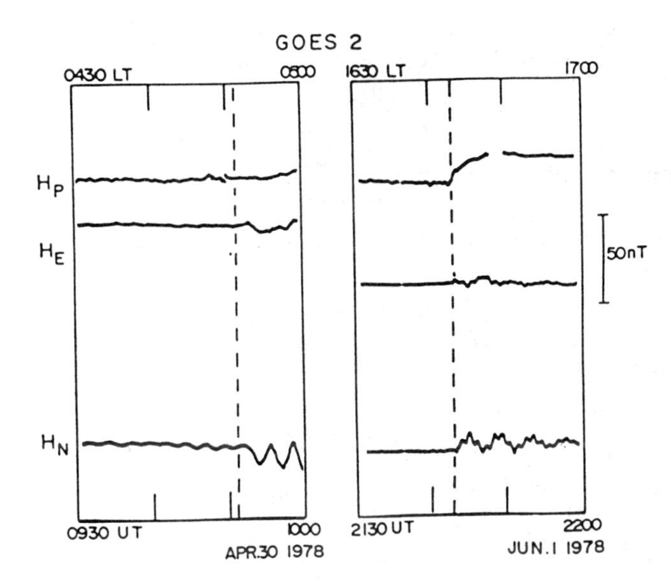

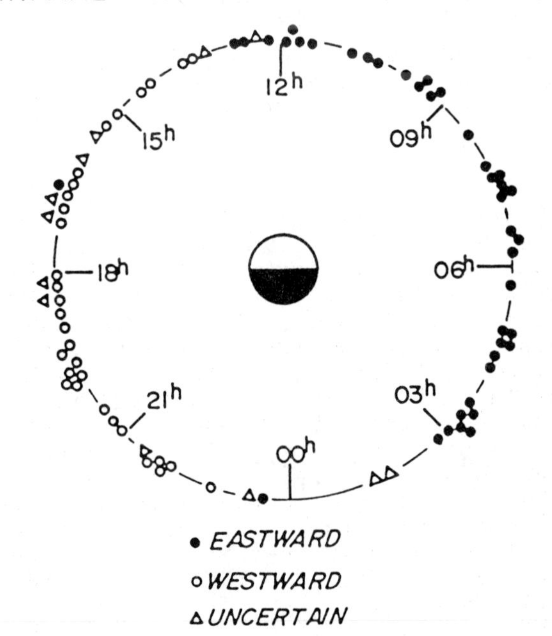

Fig. 3. Distribution of the direction of the initial movement of ssc-associated magnetic pulsations in the azimuthal (east-west) component at geosynchronous altitude in the manetosphere. The direction of the initial movement is eastward in the morningside, while it is westward in the afternoonside.

the MHD standing oscillations. The damped type waveform of Psc oscillations will be related also with ionospheric Joule dissipation as will be discussed later.

4. Instabilities in Hot Plasma as a Source of Compressional Pc5

Radially polarized Pc 5 magnetic pulsations with a significant compressional component have been found in the magnetosphere from the early afternoon to midnight sectors [Barfield and McPherron, 1972; Barfield et al., 1972]. Those compressional waves are localized near the equatorial plane in the magnetosphere and are scarcely observed on the ground. According to the results by Barfield et al. [1972], these compressional Pc5 waves observed during magnetically disturbed conditions as shown in Fig. 4. So, Barfield et al. [1972] firstly called

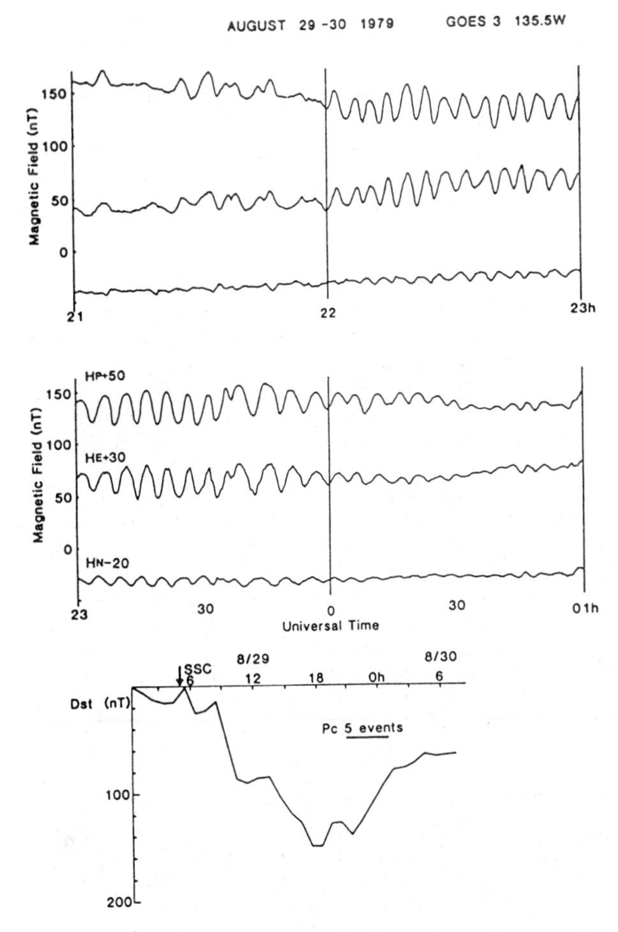

Fig. 4. Compressional Pc5 pulsations at geosynchronous altitude (GOES 3) observed during the recovery phase of the magnetic storm [Kokubun, 1985]. HP is parallel to the earth's rotational axis, HE is radially earthward and HN is azimuthally westward.

those phenomena as storm time Pc 5 pulsations. Further observational results show that those Pc 5 events have very large azimuthal wave number, m = 30 - 50 [Hughes et al., 1978; Allan et al., 1982]. Storm time Pc 5 events show close relationships with energetic particle oscillations [Kremser et al., 1981; Higbie et al., 1982; Walker et al., 1982; Wedeken et al., 1984; Takahashi et al., 1985a and b; Takahashi and Higbie, 1986]. The classical MHD theories are no longer adequate to describe these energetic particle associated phenomena. Microscopic kinetic processes must be considered in the finite β plasma ($\beta \sim 1$), where β is the ratio of plasma pressure to magnetic field pressure. In the situation of energetic particle associated phenomena , an important driving mechanism is the coupling of unstable energetic particles to generate guided Alfvén -like waves.

Relationships between instabilities in the ring current region and excitation of MHD waves have been studied theoretically by many researchers [Southwood et al., 1969; Hasegawa, 1969, 1971, 1975; Southwood, 1973, 1976, 1977; Lanzerotti and Hasegawa, 1975; Tamao, 1978, 1984; Lin and Cheng, 1984; Patel et al., 1983; Walker et al., 1982; Southwood and Kivelson, 1981, 1982; Kivelson and Southwood, 1983, 1985]. One of the plasma instabilities is the drift mirror instability, in which the classical mirror instability is modified by gradient in the magnetic field and the hot plasma density. The unstable particle distribution can generate compressional waves which couple to guided Alfvén-like waves. For the wave of this type, azimuthal wave number "m" is expected to be large, and strong compressional component of the wave magnetic field should occur near the equatorial plane in the magnetosphere. Many researchers suggested that the compressional waves observed in the magnetosphere (storm time Pc 5) might be caused by the coupling of Alfvén waves and unstable drift mirror waves. Lanzerotti and Hasegawa [1975] showed that the drift mirror instability will be excited when the Alfvén velocity is smaller than the drift speed of the energetic particles associated with diamagnetic or gradient-curvature drift. Walker et al. [1982] modified the model by Hasegawa [1969] to interpret the observational results at GEOS, including the coupling between guided poloidal standing Alfvén waves along the field lines and particle driven drift mirror waves. Coupling between the drift mirror waves and standing Alfvén waves has been discussed both with the fundamental mode [Walker et al., 1982] and with the second harmonics [Southwood, 1973, 1976]. Southwood [1977] showed that localized compressional MHD waves with the second harmonics result from a field line resonance in a hot inhomogeneous plasma.

The compressional Pc 5 waves are characterized by modulation in energetic particle fluxes in the energy range of several tens to several hundreds KeV. According to the results by Takahashi et al. [1985a and b], the time lag of the onset of the particle mudulation at the two geosynchronous satellites showed the wave velocity and the azimuthal wave number to be 8 Km/s westward and -60, respectively. Based on the results at GEOS 2, Kremser et al. [1981] showed that periodic oscillations of the magnetic field at geosynchronous altitude were accompanbied by anti-phase oscillations in the energetic proton flux. Oscillations in the electron flux could be either in-phase or anti-phase relationship with the magnetic field pulsations. Kremser et al. [1981] and Takahashi et al. [1987] suggested that the drift mirror instability proposed by Hasegawa [1969] was a good candidate of an excitation mechanism for the compressional Pc 5. However, Allan et al. [1982] pointed out that several properties of the Pc 5 waves observed with the STARE are not consistent with existing theories. According to the results by Wedeken et al. [1984], the observational results of ground-satellite coordinated study show only partial agreement with the prediction of the drift mirror instability theory. Takahashi and Higbie [1986] found that the particle modulation with the compressional Pc 5 waves could be explained by an antisymmetric standing wave structure. This result implies that the compressional Pc 5 waves do not have the fundamental standing wave structure. Southwood [1973, 1976] argued that in the ring current the second harmonic standing waves are likely to be excited through bounce-resonant interaction of the ring current particles.

Barfield and McPherron [1978] suggested that the compressional Pc 5 in the afternoon sector often occurs associated with the substorm expansion. The time delay of the storm time Pc 5 onset behind the substorm expansion at geosynchronous altitude compares well with the speed of westward propagation of about 50 KeV protons with gradient curvature drift [Kokubun, 1985]. It is concluded that the source of the short-duration cmpressional Pc 5 is the substorm associated newly injected protons of energy about 50 KeV, which drift westward with the speed of the gradient curvature drift.

5. Reconfiguration of the Nightside Magnetosphere as a Source of Pi 2

Pi 2 magnetic pulsations are interpreted as transient MHD signals associated with the sudden reconfiguration in the structure of the nightside magnetosphere during the substorm expansion [Saito, 1969; Orr, 1973; Lanzerotti and Fukunishi, 1974; Fukunishi, 1975; Saito et al., 1976; Kuwashima and Saito, 1981; Samson and Rostoker, 1983; Sakurai and McPherron, 1983; Baumjohann and Glassmeier, 1984; Yumoto, 1986; Kuwashima, 1988]. In association with the onset of a substorm expansion, Pi 2 waves with damped-type waveforms are simultaneously observed over a wide latitudinal range from the auroral region through low latitudes. According to this characteristic, Pi 2 has been used as one of the most sensitive indicator for the onset of the substorm expansion.

The sudden reconfiguration of the nightside magnetosphere is expected to be caused by a short-circuiting of the cross-tail current to the auroral ionosphere via field-aligned currents [Clauer and McPherron, 1974; Sakurai and Mcpherron, 1983]. In that sense, the Pi 2 waves associated with the sudden formation of field-aligned currents play an important role in the dynamic coupling between the magnetosphere and the ionosphere. Fig. 5 shows the geometry of the current wedge in the magnetosphere together with the associated changes in the magnetic field [Clauer and McPherron, 1974; Baumjohann and Glassmeier, 1984]. The upward field-aligned current in the premidnight sector causes an eastward deflection of the magnetic field, while the downward field-aligned current in the postmidnight sector causes a westward magnetic field deflection at geosynchronous altitude as shown in the figure. In addition to the transverse east-west magnetic field deflections, a compressional magnetic field perturbation will be also excited at geosynchronous altitude, because the sudden disappearance of part of the dawn to -dusk directed cross-tail currents leads to a magnetic field perturbation similar to that of the appearance of a dusk- to -dawn directed current flow

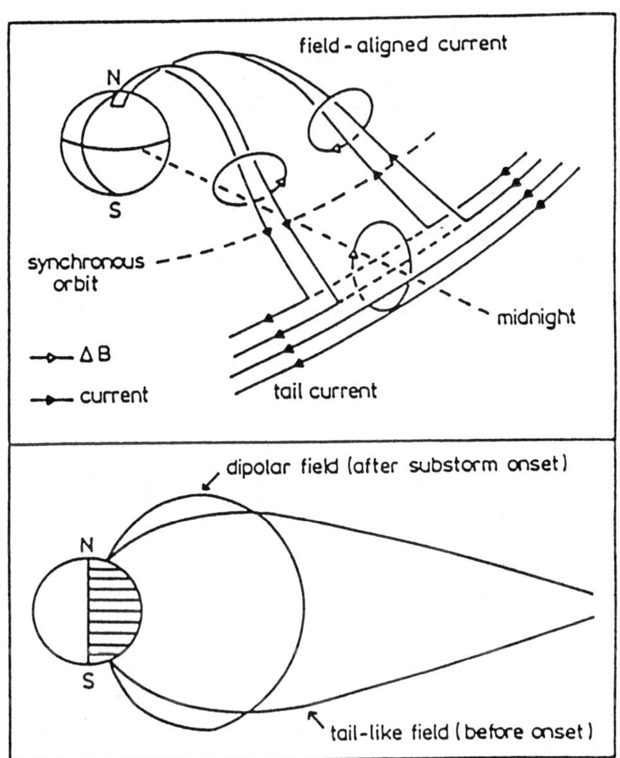

Fig. 5. Upper panel; The substorm current wedge and associated magnetic field changes during ionospheric channeling of the dawn-to-dusk tail current. Lower panel; Reconfiguration of the nightside magnetic field associated with the disruption of the dawn-to-dusk tail current at the onset of the substorm expansion [Baumjohann and Glassmeier, 1984].

which has a northward (field aligned) component on the equatorial plane in the earthward of the current disruption region. This is observed as the dipolarization of the inner nightside magnetosphere.

The above- mentioned predictions have been confirmed experimentally by the observation of the geosynchronous satellite. According to the observational results by Sakurai and McPherron [1983], the initial deflection in the azimuthal component (D- component) of the Pi 2 pulsation is in the same direction as that due to the field aligned current (eastward in the premidnight sector and westward in the postmidnight sector). These results suggest close relationships between the Pi 2 generation and the reconfiguration of the nightside magnetosphere associated with a short circuiting of the cross-tail current at the onset of the substorm expansion. The transverse signal in the azimuthal (east-west) component could propagate along the field lines with Alfvén velocity. That Alfvén signal will arrive at the auroral ionosphere, and effectively be reflected there causing the MHD resonant oscillations. Nishida [1979] theoretically showed that the transient east-west electric field generated from ionospheric channeling of the dawn-to-dusk tail current during the substorm expansion[Aggson and Heppner, 1977] propagates along the field line to the auroral ionosphere. At the reflection at the ionosphere, the polarity of the incident electric field is reversed, and the reflected electric field propagates along the field line to the opposite auroral ionosphere forming MHD standing oscillation. The excited oscillation is a damping one and the period is expected to be equal to the bounce period of the Alfvén waves propagating along the field line anchored both at the northern and southern ionospheres. Kan et al. [1982] suggested that the transient response to a step function voltage source in the magnetosphere can also produce an overshoot damped type MHD standing oscillation, resembling to the Pi 2 signature. The relationship between the Alfvén waves and the magnetosphere-ionosphere channeling through the field aligned current has further been developed to two-dimensional models [Lysak and Dum, 1983; Sun and Kan, 1985]. The damped oscillatory nature of Pi 2 pulsations observed on the ground could be produced by the transient ionospheric response and energy dissipation through Joule heating [Newton et al., 1978; Glassmeier et al., 1984].

The MHD standing oscillation model of Pi 2 has been experimentally established by the results of the conjugate observations on the ground and in the magnetosphere. Fig. 6 shows the wave-phase difference of Pi 2 at a conjugate pair in the auroral region, Syowa Station in Antarctica and Husafell in Iceland. The figure shows that the H (north -south) component of Pi 2 wave oscillates almost in-phase (with wave-phase difference around 0°), while the D (east -west) component oscillates almost anti-phase (with wave-phase difference around 180°) at the conjugate pair. The large scattering of the D-component shown in Fig. 6 is the result of smaller wave amplitude of that component compared with that of the H-component. Considering the 90° rotation of the polarization axis

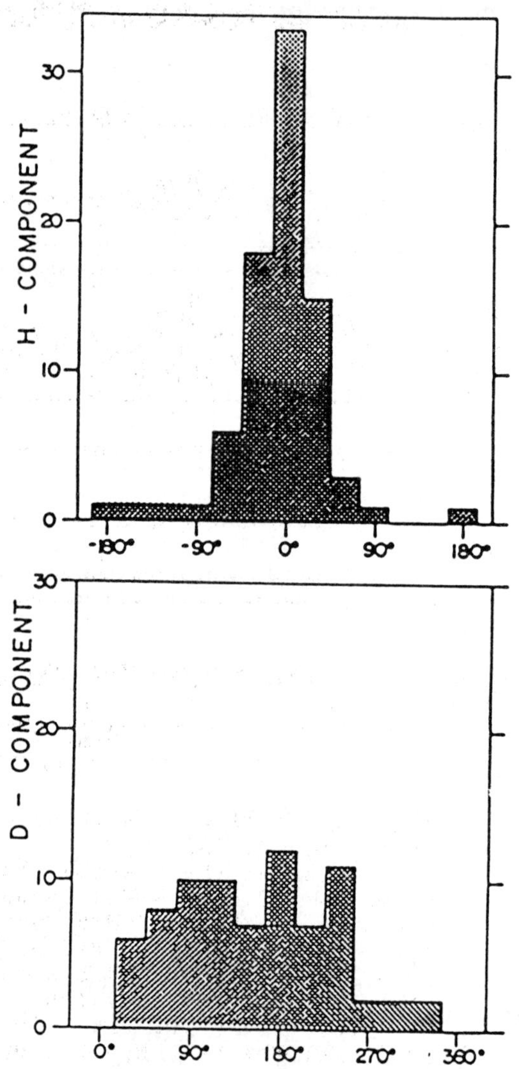

Fig. 6. Wave-phase relationships of Pi2 magnetic pulsations at the conjugate pairs in the auroral region, Syowa Station (SYO), Antarctica and Husafell (HUS), Iceland. The H- component shows a dominant peak around 0° showing in-phase relationship, while the D-component shows a broad maximum around 180° showing anti-phase relationship.

through the ionosphere, dominant Pi 2 oscillations in the D (azimuthal) component is expected in the magnetosphere. These results suggest that Pi 2 is caused by the MHD standing oscillation of the field lines. The oscillation is expected to be the fundamental-mode

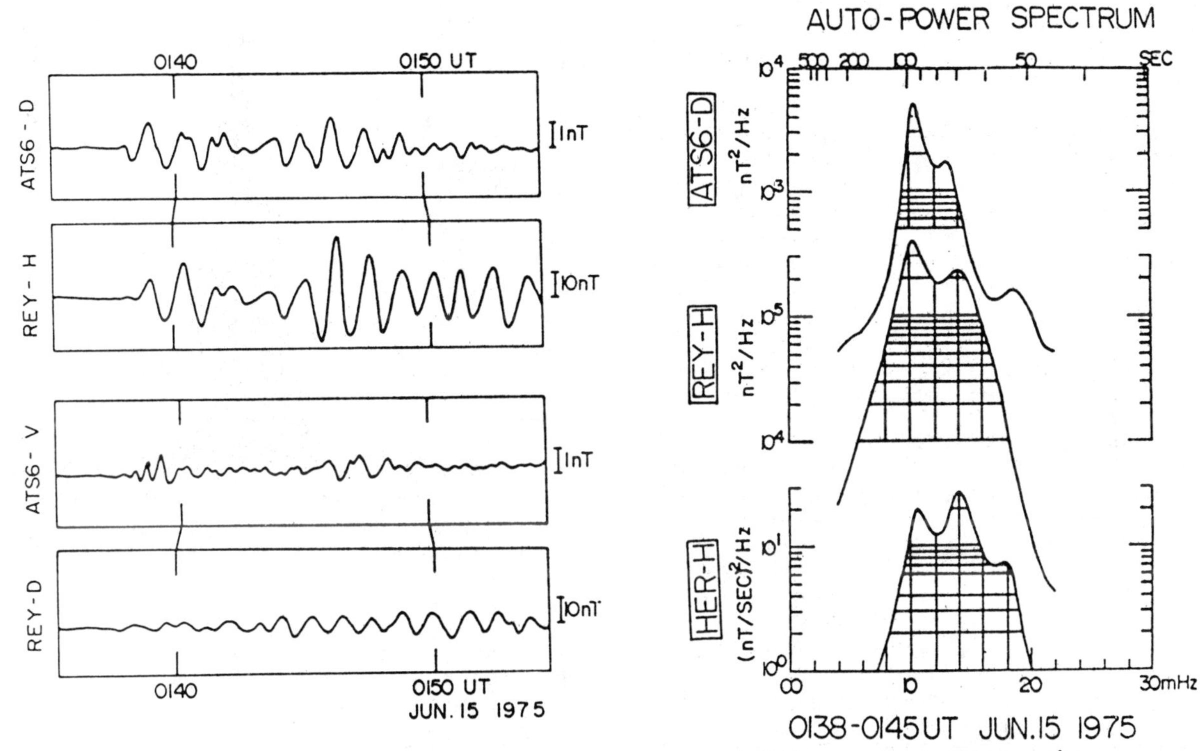

Fig. 7. Leftside; Simultaneous appearance of Pi2 event at geosynchronous altitude (ATS 6) in the magnetosphere and at the conjugate ground based station (Reykjavik). The Pi2 oscillations are dominant in the azimuthal (D:east-west) component in the magnetosphere, while they are dominant in the H- (north-south) component on the ground. Rightside; Auto-power spectra of Pi2 at geosynchronous altitude (ATS 6), conjugate ground-based station (Reykjavik) and the low-latitude station located along almost the same meridian (Hermanus).The spectra were calculated for the azimuthal (D) component in the magnetosphere, while for the H- component on the ground.

because the behavior of the observed Pi 2 period is consistent with the calculated fundamental period of the MHD standing oscillation [Kuwashima and Saito, 1981]. That prediction has been confirmed by the direct comparison of Pi 2 waves between at the geosynchronous altitude in the magnetosphere and on the ground as shown in Fig. 7, in which two successive Pi 2 events with pure transverse waves (dominant oscillations in the D-component) were observed at geosynchronous altitude (ATS 6) at about 0139 and 0145 UT on June 15,1975. At that time, ATS 6 was located 4° above the geomagnetic equatorial plane with the geomagnetic longitude of 70°, which was very close to the meridian of Reykjavik in Iceland. Pi 2 waves were observed both at geosynchronous altitude in the magnetosphere (ATS 6) and on the conjugate ground (Reykjavik) with similar wave forms as shown in the figure. Azimuthal oscillations (D- component) are more dominant than radial oscillations (V- component) in the magnetosphere, while the H- component is more dominant than in the D- component on the ground. Considering the 90° rotation of the polarization axis through the ionosphere, the results shown in Fig. 7 are consistent with the MHD standing oscillation model as a cause of Pi 2, because the azimuthal oscillations correspond to the shear Alfvén waves. The small wave amplitude (~1 nT) at the geosynchronous altitude compared with that on the ground (~20nT) also suggests the fundamental-mode of MHD standing oscillations as a cause of Pi 2 [Cummings et al., 1969].

Compressional waves will be also excited in the source region of Pi 2 as indicated in Fig. 5. Those compressional wave could propagate across the field line in the magnetosphere [Sakurai and McPherron, 1983; Singer et al., 1983], then be coupled with the surface waves on the plasmapause causing the secondary Pi 2 there [Lanzerotti and Fukunishi, 1974; Fukunishi , 1975; Lester et al., 1983, 1984]. Saito et al. [1976] and Kuwashima and Saito [1981] found the latitudinal dependence of the Pi 2 amplitude , which shows the primary maximum in the auroral oval, while the secondary maximum near the plasmapause. The spectra of Pi 2 waves shown in Fig. 7 show two spectral components, where the lower frequency component (~11mHz) is more dominant at the high latitude, while the higher frequency one (~14mHz) is more dominant at the lower latitude. Those results also suggest the existence of two source regions of Pi 2, one is at the auroral oval and the other is at the plasmapause.

Though Pi 2 waves are simultaneously observed over a wide latitudinal range from the high- to low- latitudes, their polarization characteristics have not been interpreted systematically. Complicated polarization patterns of high-latitude Pi 2 pulsations [Kuwashima, 1978; Pashin et al., 1982; Samson and Harrold, 1983] can not be explained by the simple MHD standing oscillations localized at the auroral oval. Samson and Rostoker [1983] have considered a current system where an oscillating field aligned current expands both eastward and westward in the ionosphere to explain observed Pi 2 polarization structures. Pashin et al. [1982] showed that essential characteristics in the auroral break-up region can be simulated by considering the transient response of the localized upward field-aligned currents. The observed complicated Pi 2 polarization structure could be explained by the westward movement of the upward field-aligned current.

Pi 2 polarization behavior at midlatitudes is rather simple compared with that at high latitudes. The predominant left-handed polarization is observed at mid-latitudes (56°-43° in geomagnetic latitudes) associated with the predominant westward propagation of the Pi 2 wave independent of occurrence local time [Rostoker, 1967; Mier-Jedrzejowicz and Southwood, 1979; Samson and Harrold, 1983; Lester et al., 1984]. However, there are very limited number of theoretical works in understanding the predominant left- handed polarization at midlatitudes. Southwood and Hughes [1985] proposed a superposition of two circularlly polarized waves propagating azimuthally in opposite directions, namely a left- handed polarized westward propagating wave and a right-handed polarized east- ward propagating wave with amplitude half of the westward propagating one. However, their theoretical assumption of exsistence of two kinds of circulary polarized waves has not been confirmed yet experimentally. So, the problem with the polarization behavior is remained to the future studies.

Pi 2 polarization structure at low-latitude has not yet been conclusively understood theoretically. Many researchers indicated that the low-latitude Pi 2 shows right-handed polarization in the premidnight sector, while shows left-handed polarization in the postmidnight sector [Kato et al., 1956; Sakurai, 1970; Sutcliffe , 1981]. These polarization behaviors at low-latitude Pi2 are not consistent with those at the higher latitudes ,and cannot be understood by using the present theoretical models. It is noteworthy that the apparent azimuthal wave numbers of Pi 2 have a latitudinal dependence, m = 3-20 at high latitude, m= 2-4 at mid- and low-latitude, and m ~1 at very low-latitudes [Yumoto, 1986]. Any existing theory can not explain those global characteristics of Pi 2 wave. Further studies are needed to explain the wave characteristics of Pi 2 from high- to low-latitudes.

6. Ionosphere and Its Modulation on MHD Waves

Before reaching the ground to be detected as magnetic pulsations, MHD waves generated in the magnetosphere have to pass through the ionosphere. Because of the finite electric field in the ionosphere, the Pedersen currents are induced, and those current shield the incident magnetic field from the ground. The signal below the ionosphere is due to the ionospheric Hall currents. Such an ionospheric effect on MHD waves has been extensively studied theoretically by Nishida [1964], Inoue [1973] and Hughes and Southwood [1976a and b] and others. One of the important ionospheric effects on the MHD wave modulation is rotation of the major axis of the polarization ellipse between the magnetosphere and on the ground. For the Alfvén mode waves, the incident Alfvén waves propagating along the field lines from the magnetosphere in the northern hemisphere are rotated through 90° counterclockwise direction when viewed toward the ground. In the southern hemisphere, the Alfvén waves are rotated through 90° clockwise direction in the ionosphere. Therefore, if the satellite is located above the equator, namely in the northern hemisphere, the azimuthal component (D-component) at the satellite will show in-phase relationship with the north-south component (H- component) on the ground, in the case of the fundamental- mode MHD standing oscillation as Pc 5 waves. The first experimental confirmation with the ionospheric effect on MHD waves was done by the observation of OGO 5 satellite [Kokubun et al., 1976]. Fig. 8 shows simultaneous observations of a Pc 5 event at a conjugate pair of the high-latitude stations (Great Whale River and Byrd Station) and OGO 5 satellite which was located in the neighborhood of the field lines that pass the conjugate ground stations. The satellite was located above the equatorial plane at magnetic latitude of about 10°, namely OGO 5 satellite was located in the northern hemisphere. On the ground, the oscillations were dominantly in the north-south (H) component as compared with the east-west (D) component. At the conjugate station, the H- component showed in-phase relationship, while the D-component seemed to be in the anti-phase relationship. In the magnetosphere (OGO 5), the oscillation is the most dominant in the azimuthal component (By, east-west). The azimuthal component in the magnetosphere showed a clear in- phase relationship with the H-component on the cojugate ground stations, corresponding to the 90° rotation of the polarization ellipse through counterclockwise direction. These conjugate relationships correspond to the odd-mode MHD standing oscillations. Similar rotation effect can be also seen in the simultaneous Pi 2 observation between the magnetosphere and the cunjugate ground station as shown in Fig. 7. STARE results also confirmed the existence of the rotation of the polarization ellipse between above and below the ionosphere [Walker et al., 1979].

Relationships of the ionosphere and its MHD modulation effects have been studied further theoretically by Glassmeier [1984] and Fujita [1987].

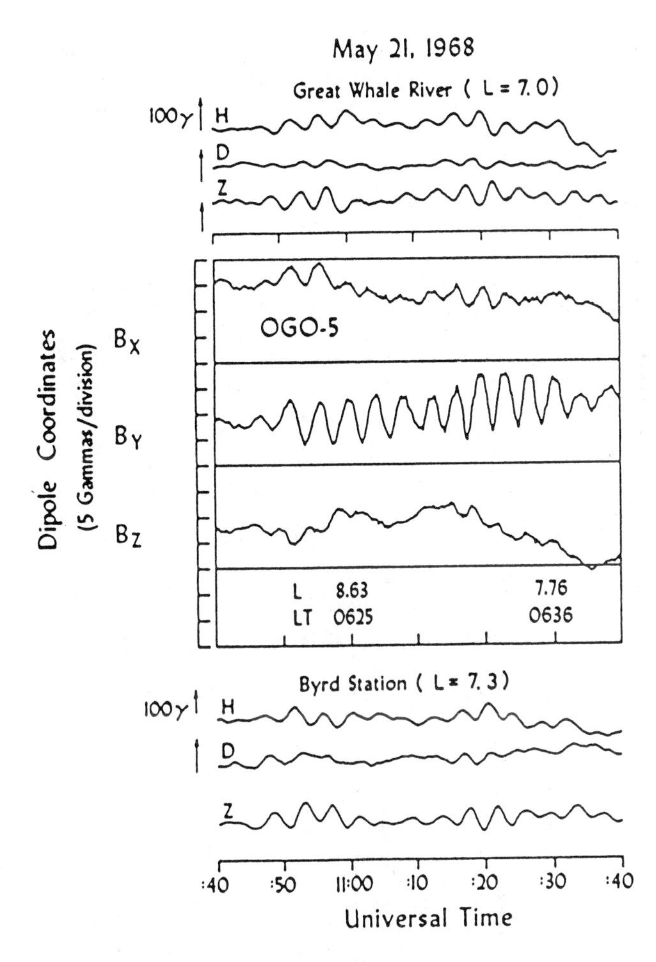

Fig. 8. Ground-satellite correlations of Pc5 magnetic pulsations on May 21,1968. OGO 5 was inbound at a magnetic latitude of 10°-13° at 0620-0646 UT. The conjugate pair of stations, Great Whale River (L=7.0) and Byrd (L=7.2), were situated in the region of 0520-0620 LT. On the ground conjugate pair stations, the H- and D- components show in-phase and anti-phase relationship indicating the odd-mode MHD standing oscillations. The east-west (By) component is dominant above the equatorial plane (OGO 5) in the magnetosphere, and shows in-phase relationship to the H-component oscillations on the conjugate ground stations [Kokubun et al., 1976].

Glassmeier [1984] has studied the ionospheric effects including the condition of the non-uniform electric conductivity, while Fujita [1987] has studied the ionospheric response for the localized incident Pc 1 waves, using 5-layer model shown in Fig. 9. The 5-layer model includes magnetosphere, ducting layer,

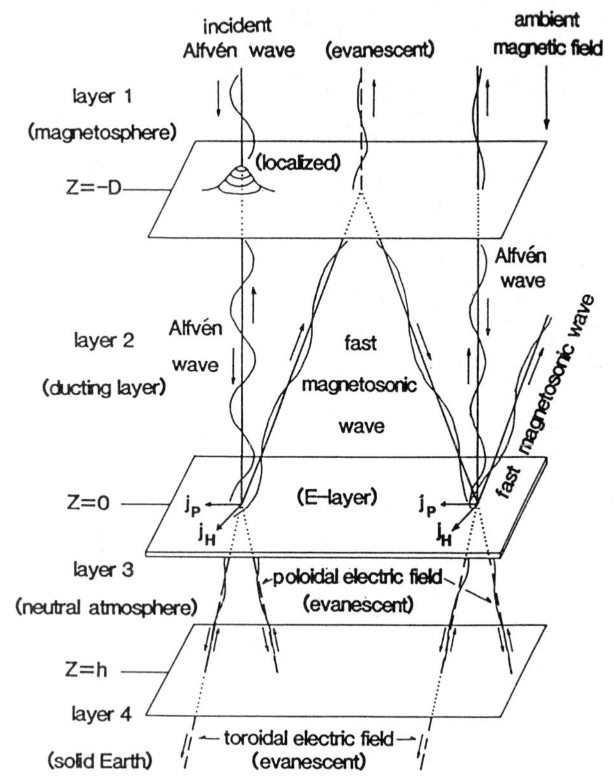

Fig. 9. Schematical illustration of 5-layer model which includes the magnetosphere, ducting layer, ionospheric E-layer, neutral atmosphere and solid earth.

ionospheric E-layer, neutral atmosphere and solid earth. Considering the observational result in the Pc 1 source region [Hayashi et al., 1981; Inhester et al., 1984], localized injection of a shear Alfvén Pc 1 waves is adoptted and ambient magnetic field lines are assumed to be vertical downward. The incident Alfvén wave is converted to a fast magnetosonic wave through the ionospheric Hall current in the ionospheric E-layer, while the Pedersen current generates a reflected Alfvén wave propagating along the ambient magnetic field lines. The Pedersen current also generates a downward evanescent magnetic toroidal type perturbation in the neutral atmosphere. The toroidal mode has negligibly small magnetic field intensity in the neutral atmosphere, while up-going Alfvén wave is partly reflected at a transition altitude where a sharp increase of the Alfvén speed takes place, and the other part is transmitted into the magnetosphere. The Hall current generates up-going fast magnetosonic wave spreading in all directions in the ducting layer, and a downward evanescent magnetic poloidal mode type perturbation in the neutral atmosphere. The up-going fast magnetosonic wave is partly reflected at the transition altitude and partly transmitted into the magnetosphere. The downward poloidal mode is partly reflected back and the other part is transmitted as downward evanescent poloidal mode in the solid earth. The poloidal mode has finite magnetic field intensity in the neutral atmosphere.

The calculated results are summarized schematically in Fig. 10, where the top

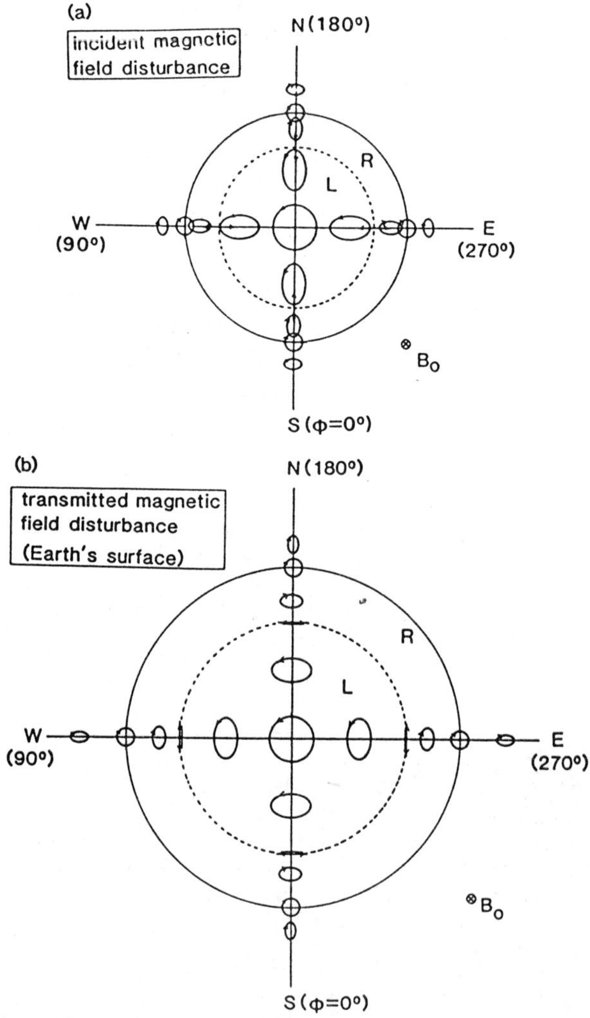

Fig. 10. Schematical diagrams illustrating spatial distributions of polarization patterns of horizontal magnetic field vector (top; the incident wave above the ionosphere, bottom; the trasmitted wave on the ground). Broken and solid circles are demarcations of the polarization sense and the major-axis direction.

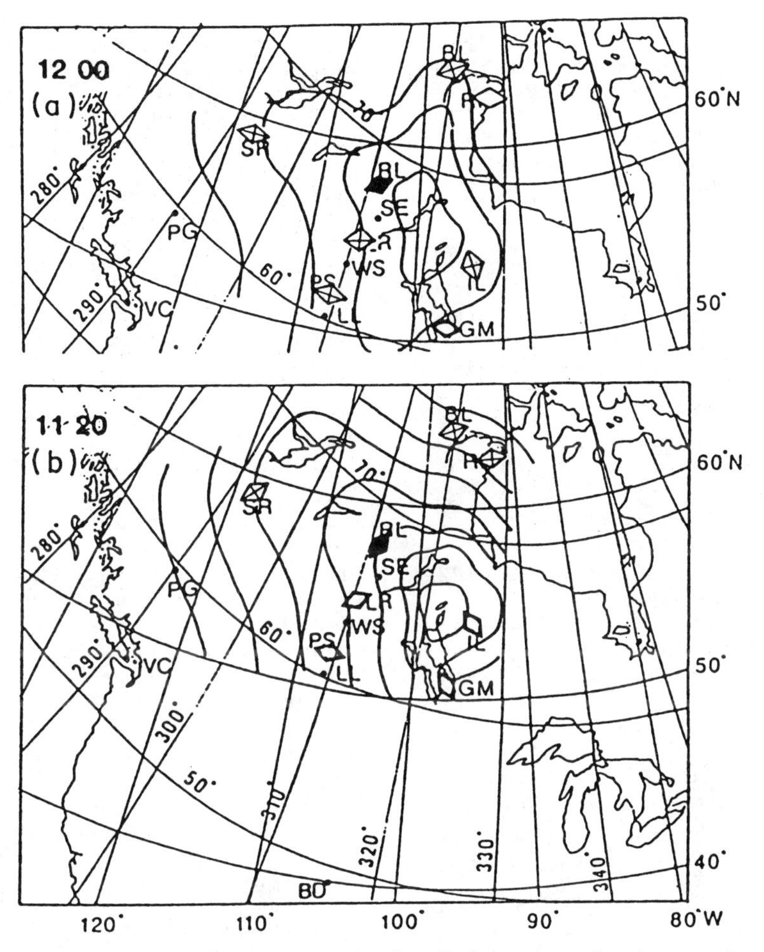

Fig. 11. Spatial distributions of the Pc1 magnetic intensity and its polarization. Each curve indicates 5dB contour. Each symbol indicates orientation of the major axis of polarization ellipse and sense of polarization, LH (Left-handed), RH (Right-handed) or linear according to the inside mark, open,solid or cross,respectively [Hayashi et al., 1981].

shows spatial distributions of polarization patterns for the incident wave, and the bottom shows the patterns for the transmitted waves on the ground. The polarization ellipse is rotated through 90°, while the polarization sense is conserved by the tranmission through the ionospheric E-layer. On the ground, left-handed polarization is dominant near the injection center of Pc 1 waves, and oscillation is dominantly in the azimutal direction. In the middle region, there is a right-handed elliptically polarized disturbance with a major axis in the azimuthal direction. In the outer region, a right-handed polarized magnetic field disturbance is observed and the major axis is pointing to the center of the injection area. Webster and Fraser [1975] had much success in determination of the Pc 1 source region by using directions of the major axis of the polarization ellipse. The calculated results shown in Fig. 10 support the method of Fraser [1975]. The distribution shown in Fig. 10 is also consistent with the observational results obtained around the Pc 1 source region shown in Fig. 11. The figure shows that Pc 1 waves are localized in the source region and left-handed polarization is dominant in the central region of the Pc 1 injection, while right-handed polarization becomes dominant in the surrounding region.

7. Summary and Future Research Direction

The present paper has discussed several aspects of current research interest in the physics of the MHD waves in the earth's magnetosphere and on the ground. The study of MHD waves has progressed due to improvements in observational techniques, which are concurrent magnetic observations by ground based magnetometer chains from high- to low-latitude, multiple satellite observations in the magnetosphere as well as in the earth's foreshock region, simultaneous observations of various kinds of phenomena (electric field, energetic particles and others). According to those observations, characteristics of MHD waves have been extensively clarified. However, there are still many unresolved problems that remains for future studies.

On the basis of observational relationships of the solar wind parameters with magnetic pulsation parameters, it is established that the magnetosonic upstream waves in the Pc 3 frequency range in the earth's bow shock region are transmitted to the magnetosphere without significant changes in spectra ,and are the main source of low- latitude Pc 3 magnetic pulsations at L= 1.5 - 3.0. These transmitted compressional waves can propagate across the ambient magnetic field lines into the inner plasmasphere, and then can couple with various MHD oscillations, which are surface eigen oscillation at the plasmapause, trapped oscillations of the fast magnetosonic waves in the Alfvén trough, and fundamental or higher- harmonics standing oscillation of localized field lines. Pc 3 magnetic pulsations observed at middle- and low- latitude would be a superposition of these MHD oscillations in the plasmasphere. Further coordinated study is needed to clarify which modes of various MHD waves dominate in the plasmasphere. Pc 3 magnetic pulsations at very low latitudes (less than 22° in geomagnetic latitude) are not yet sufficiently clarified. Kuwashima et al. [1979] pointed out that the diurnal variation of very low- latitude Pc 3 polarizations at Chichijima (17.1° in geomagnetic latitude and 208.9° in geomagnetic longitude) is different from that of low-latitude Pc 3 polarizations at Memambetsu(34.0°, 208.4°). At very low-latitudes, the field lines are almost entirely in the ionosphere, where MHD standing oscillations can never appear.Recently, Yumoto [1987] has proposed that Pc 3 magnetic pulsations at very low-latitudes might be caused by an ionospheric Pedersen eddy current induced by the inductive electric field of the compressional Pc 3 source waves. More extensive observations at low- and very-low- latitudes should be further operated to clarify how very low-latitude Pc 3 can be excited.

Satellite observation revealed existence of the compressional Pc 5 magnetic pulsations in the magnetosphere. Compressional Pc 5 waves are thought to be caused by instabilities of hot plasma in the ring current. Kremser et al. [1981] suggested that the drift mirror instability proposed by Hasegawa [1969] was a good candidate for an excitation mechanism of compressional Pc 5 waves. The modification of Hasegawa's basic idea by Walker et al. [1982], including guided poloidal standing Alfvén waves along the field line, was also applied to explain the GEOS 2 observations. On the other hand, Takahashi et al. [1986] suggested that second harmonic standing waves excited through bounce-resonant interaction of ring current particles [Southwood, 1976] are main cause of compressional Pc 5 waves. More works are needed to understand the nature of the compressional Pc 5 waves and their relationships to the energetic particle modulations.

The existence of the MHD oscillation of decoupled field shells is revealed by the observation of ssc-associated magnetic pulsations. This mode has an azimuthally (east-west) polarized magnetic field in the magnetosphere, and the oscillation of adjacent L shell is decoupled in a axisymmetry condition where the azimuthal wave number is very small. STARE results showed that the period variation of the ssc-associated magnetic pulsation is essentially continuous along the occurrence latitude, suggesting the oscillation of decoupled field shells. The general behavior of those variable period events is consistent with geomagnetic field shells "ringing" independently at their own resonant periods associated with the impulsive disturbance (ssc). Concurrent observations from high- to low- latitudes on the ground as well as in the magnetosphere are needed to clarify those ringing effects in more detail.

Pi 2 magnetic pulsations associated with the onset of the substorm expansion have been widely studied by a number of researchers. Pi 2 is generally interpreted as a transient magnetic signal associated with a sudden reconfiguration of the nightside magnetosphere during the substorm expansion associated with channeling to the auroral ionospheres of the dawn-to-dusk tail current. However, the behavior of Pi 2 magnetic pulsations has not yet been sufficiently understood as pointed out by Yumoto [1986, 1987]. Pi 2 waves are simultaneously observed over the wide latitudinal range from the high- to low- latitudes with similar waveforms [Saito et al., 1976; Kuwashima and Saito, 1981]. However, their wave characteristics are not consistent over the wide latitudinal range. The Pi 2 azimuthal wave numbers are m = 3 - 20 at high-latitudes [Lester et al., 1983, 1984; Samson and Harrold, 1985], m = 2 -4 at middle latitude [Meir-Jedrzejowicz and Southwood, 1979; Lester et al., 1983, 1984], and $m < 2$ at low latitudes. The Pi 2 polarization behavior at high latitude could be explained by the relationship between the substorm associated upward field-aligned current and the auroral ionospheric current system [Pashin et al., 1982; Samson and Rostoker, 1983]. Left -handed Pi 2 polarization behavior at middle latitude could be explained by the superposition of two waves with different amplitude, travelling in opposite direction [Southwood and Hughes, 1985]. However, any existing theory can not explain the Pi 2 polarization behavior observed at low-latitudes [Kato et al., 1956; Sakurai, 1970; Sutcliffe, 1981]. Further coordinated simultaneous observations at separate stations from high- to low-latitudes are needed at several local time meridians to clarify the behavior of Pi 2 polarization. Occurrence and wave characteristics of daytime Pi 2 magnetic pulsations [Yumoto, 1987] are also unresolved problems.

Before reaching the ground to be observed as magnetic pulsations, MHD waves generated in the magnetopsphere have to pass through the ionosphere. A highly conducting ionosphere does reflect most of the incident MHD wave energy back along the field line. However, the electric field is finite in the ionosphere and Pedersen currents flow there. Those Pedersen currents in the ionosphere shield the incident magnetic field from the ground. The signal below the ionosphere is due to the ionospheric Hall currents. As the result, the polarization vector is rotated through 90° between the magnetosphere and the ground. However, the 90°rotation of the polarization ellipse of MHD waves through the ionosphere have been confirmed only for several cases because of quite small existing sets of data which can be used to examine possible ionospheric effect on MHD waves. Simultaneous observations of MHD waves both above and below the ionosphere are needed to clarify the ionospheric effects in more detail.

Finally, the importance of the earth induction effect should be emphasized. MHD waves originated in the magnetosphere and the ionosphere could produce induction currents in the solid earth. Those currents are called as Geomagnetic Induced Current (GIC). Recently, GIC has been remarked as a cause of various kinds of disturbances and disruptions for cable communication system and long pipe lines. GIC must be also included on the study of MHD waves to establish a complete MHD plasma physics, which are associated with interplanetary space, earth's magnetosphere, ionosphere, neutral atmosphere and the solid earth.

Acknowledgement. The authors wish to express their sincere gratitude to Prof. H. Oya of Tohoku University for his encouragement in carrying out the present study as well as preparing the present manuscript. The authors especially are indebted to Prof. T. Saito and Dr. K. Yumoto of Tohoku Univerity for their very valuable discussions and suggestions. Thanks are also due to Dr. R. Murakami, the director of the Kakioka Magnetic Observatory, for his support to the present study. The authors would like to express their appreciations to Prof. A. Nishida of Institute of Space and Astrophysical Science and Prof. D. Southwood of Imperial College for their appropriate criticism with the present manuscript.

References

Aggson, T. L., and J. P. Heppner, Observation of large transient magnetospheric electric fields, J.Geophys.Res., 82, 5155, 1977.

Allan, W., and E. M. Poulter, The spatial structure of different ULF pulsation types: A review of STARE rader results, Rev. Geophys. Space Phys. 22, 85, 1984.

Allan, W., and E. M. Poulter, The spatial structure of different ULF pulsation types: A review of STARE rader results, Pc 5 pulsation with large azimuthal wave number, J.Geophys. Res., 87, 6163, 1982.

Araki, T., and J. H. Allen, Latitudinal reversal of polarization of the geomagnetic sudden commencement, J. Geophys., Res. 87, 5207, 1982.

Barnes, A., Theory of generation of bow-shock associated hydromagnetic waves in the upstream interplanetary medium, Cosmic. Electrodyn., I, 90, 1970.

Barfield, J. N., and R. L. McPherron, Statistical characteristics of storm-associated Pc 5 micropulsations

observed at the synchronous equatorial orbit, J. Geophys. Res., 77, 4720, 1972.
Barfield, J. N., and R. L. McPherron, Stormtime Pc 5 magnetic pulsations observed at synchronous orbit and their correlation with the partial ring current, J. Geophys. Res., 83, 7393, 1978.
Barfield, J. N., and C. S. Lin, Remote determination of the outer radial limit of stormtime Pc 5 wave occurrence using geosynchronous satellite, Geophys. Res. Lett., 10, 671, 1983.
Barfield, J. N., et al., Storm-associated Pc 5 micropulsation events observed at the synchronous equatorial orbit, J. Geophys. Res., 77, 143, 1972.
Baumjohann, W., and K. H. Glassmeier, The transient response mechanism and Pi 2 pulsations at substorm onset : Review and outlook, Planet. Space Sci., 32, 1361, 1984.
Chen, L., and A. Hasegawa, A theory of long-period magnetic pulsations, 1. Steady state excitation of field line resonance, J. Geophys. Res., 79, 1024, 1974a.
Chen. L., and A. Hasegawa, A theory of long-period magnetic pulsations, 2. Impulse excitation of surface eigenmode J.Geophys. Res., 79, 1033, 1974b.
Clauer, C. R. and R. L. McPherron, Mapping of localtime universaltime development of magnetospheric substorms using mid-latitude magnetic observations, J.Geophys. Res., 79, 2811, 1974.
Cummings, W. D., et al., Standing Alfven waves at the synchronous orbit, J. Geophys. Res., 74, 778, 1969.
Cummings, W. D., et al., Measurements of the Poynting vector of standing hydromagnetic waves at geosynchronous orbit, J. Geophys. Res., 83, 697, 1978.
Dungey, J. W., Electromagnetics of the outer atmosphere, The Pennsylvania State Univ. Sci. Rep., 69, 229, 1954.
Edmiston, J. P., et al., Escape of heated ions upstream of quasi- parallel shocks Geophys. Res. Lett., 9, 531, 1982.
Fairfield, D. H., Bow shock associated waves observed in the far upstream interplanetary medium, J. Geophys. Res. 74, 3541, 1969.
Fairfield, D. H., and K. W. Behannon, Bow shock and magnetosheath waves at Mercury, J. Geophys. Res., 81, 3897, 1976.
Fraser, B. J., Ionospheric duct propagation and Pc 1 pulsation sources, J. Geophys. Res., 80, 2790, 1975.
Fraser, B. J., and R. L. McPherron, Pc 1-2 magnetic pulsation spectra and heavy ion effects at synchronous orbit: ATS 6 results, J. Geophys. Res., 87, 4560, 1982.
Fredricks, R. W., A model for generation of bow-shock associated upstream waves, J. Geophys. Res., 80, 7, 1975.
Fujita, S., Duct propagation of a short-period hydromagnetic wave based on the international reference ionosphere model, Planet. Space Sci., 35, 91, 1987.
Fukunishi, H., Polarization changes of geomagnetic Pi 2 pulsations associated with the plasmapause, J. Geophys. Res., 80, 98, 1975.
Fukunishi, H., Latitudinal dependence of power spectra of magnetic pulsations near L = 4 excited by ssc's and si's, J.Geophys. Res., 84, 7191, 1979.
Glassmeier, K. H., On the influence of ionospheres with non-uniform conducti-ty distribution on hydromagnetic waves, J. Geophys. Res., 54, 125, 1984.
Glassmeier, K. H., et al., Ionospheric joule dissipation as a damping mechanism for high latitudes ULF pulsations: Observational evidence, Planet. Space Sci., 32, 1463, 1984.
Gosling, J. T., et al., Observations of two distinct populations of bow-shock ions, Geophys. Res. Lett., 5, 957, 1978.
Gray, S. P., et al., The electromagnetic ion beam instability upstream of the Earth's bow shock, J.Geophys. Res., 86, 6691, 1981.
Green, C. A., and R. A. Hamilton, An ionospheric effect on the conjugate relationship of Pi 2 magnetic pulsations, J. Atmos. Terr. Phys., 43, 1133, 1981.
Green, C. A., et al.,The relationship between the strength of the IMF and the frequency of the magnetic pulsations on the ground and in the solar wind, Planet. Space Sci., 31, 559, 1983.
Greenstadt, E. W., Field-determined oscillations in the magnetosheath as possible source of medium-period daytime micropulsations, Proceedings of Conference on Solar Terrestrial Relations, 515, Univ. of Calgary, April 1972.
Greenstadt, E. W., and T. V. Olson, Pc 3,4 activity and interplanetary field orientation, J. Geophys. Res., 81, 5911, 1976.
Greenstadt, E. W., et al., Transfer of pulsation-related wave activity across the magnetopause: Observation of corresponding spectra by ISEE 1 and ISEE 2, Geophys. Res. Lett., 10, 659, 1983.

Gul'elmi, A. V., et al., Relationship between the period of geomagnetic pulsations Pc 3,4 and parameters of the interplanetary medium at the earth's orbit, Geomag. Aeron. 13, 331 , 1973.

Hasegawa, A., Drift mirror instability in the magnetosphere, Phys. Fluids, 12, 2642, 1969.

Hasegawa, A., Drift-wave instabilities of a compressional mode in a high-plasma, Phys, Rev. Lett., 27, 11, 1971.

Hasegawa, A., Plasma instabilities and non-linear effects, Springer-Verlag, New York, 1975.

Hayashi, K., et al., The extent of Pc 1 source region in high latitudes, Can. J. Phys., 59, 1097, 1981.

Higbie, P. R., et al., The global Pc 5 event of November 14-15 1979, J. Geophys. Res., 87, 2337, 1982.

Hoppe, M. M., and C. T. Russell, Particle acceleration at planetary bow shock waves, Nature, 295, 41, 1982.

Hoppe, M. M., and C. T. Russell, Plasma rest frame frequencies and polarizations of the low-frequency upstream waves: ISEE1 and 2 observations, J. Geophys. Res., 88, 2021, 1983.

Hoppe, M. M., et al., Upstream hydromagnetic waves and their association with backstreaming ion pulsations:ISEE 1 and 2 observations, J. Geophys. Res., 86, 4471, 1981.

Hughes, W. J., The effect of the atmosphere and ionosphere on long period magnetospheric micropulsations, Planet. Space Sci., 22, 1157, 1974.

Hughes, W. J., and D. J. Southwood, The screening of micropulsation signals by the atmosphere and ionosphere, J. Geophys. Res., 81, 3234, 1976a.

Hughes, W. J., and D. J. Southwood, An illustration of modification of geomagnetic pulsation structure by the ionosphere, J. Geophys. Res., 81, 3241, 1976b.

Hughes, W. J., et al., Geomagnetic pulsations observed simultaneously on three geostationary satellites, J. Geophys. Res., 83, 1109, 1978.

Inhester, B., et al., Ground-satellite coordinated study of the April 5,1979 events: Observation of O cyclotron waves, J. Geophys. Res., 55, 134, 1984.

Inoue, Y., Wave polarization of geomagnetic pulsation observed in high latitudes on the earth's surface, J. Geophys. Res., 78, 2959, 1973.

Jacobs, J. A., et al., Classification of geomagnetic micropulsations, J. Geophys. Res., 69, 180, 1964.

Kan, J. R., A transient response model of Pi 2 pulsations, J. Geophys. Res., 87, 7483, 1982.

Kato, Y., et al., Investigation on the magnetic disturbance by the induction magnetograph, Part 5, On the rapid pulsation psc, Sci. Rept. Tohoku Univ. Ser.5 Geophys., 7, 136, 1956.

Kennel, C. F., et al., Nonlocal plasma turbulence associated with interplanetary shocks, J.Geophys. Res., 87, 17, 1982.

Kivelson, M. G., Instability phenomena in detached plasma regions, J. Atmos. Terr. Phys., 38, 1115, 1976.

Kivelson, M. G., and D. J. Southwood, Charged particle behavior in low-frequency geomagnetic pulsations, 3.Spin phase dependence, J. Geophys. Res., 88, 174, 1983.

Kivelson, M. G., and D. J. Southwood, Charged particle behavior in low-frequency geomagnetic pulsations, 4. Compressional waves, J. Geophys. Res., 90, 1486, 1985.

Kokubun, S., Observations of Pc pulsations in the magnetopshere: Satellite-ground correlation, J. Geomag. Geoelectr. 32, 39, 1980.

Kokubun, S., Statistical characteristics of Pc 5 waves at geostationary orbit, J. Geomag. Electr., 37, 759, 1985.

Kokubun, S., and T. Nagata, Geomagnetic pulsation Pc 5 in and near the auroral zones, Rept. Ionos. Space Res. Japan, 19, 158, 1965.

Kokubun, S., et al., Ogo 5 observations of Pc 5 waves: Ground- magnetosphere correlations, J. Geophys. Res., 81, 5141, 1976.

Kovner, M. S., et al., On the generation of low-frequency waves in the solar wind in the front of the bow shock, Planet. Space Sci., 24, 261, 1976.

Kremser, G., et al., Observations of quasi-periodic flux variations of energetic ions and electrons associated with Pc 5 geomagnetic pulsations, J.Geophys. Res., 86, 3345, 1981.

Kuwashima, M., Long period geomagnetic pulsations associated with storm sudden commencements (Psc 5), Mem. Kakioka Mag. Obs., 15, 31, 1972.

Kuwashima, M,, Some characters of substorm associated geomagnetic phenomena in the southern auroral region, Mem. Kakioka Mag. Obsl., 16, 1, 1975.

Kuwashima, M., Wave characteristics of magnetic Pi 2 pulsations in the auroral region : Spectral and polarization studies, Mem. Natl. Inst. Polar Res., 15, 1, 1978.

Kuwashima, M., Wave characteristics of magnetic Pi 2 pulsations in the magnetosphere and on the ground, Mem Kakioka Mag. Obs., 20, 1, 1983.

Kuwashima, M., Conjugate relationships of magnetic Pi 2 pulsation in the auroral region, Mem. Kakioka Mag. Obs., 23, 1, 1988 .

Kuwashima, M., and T. Saito, Spectral characteristics of magnetic Pi 2 pulsations in the auroral region and lower latitudes, J. Geophys. Res. 86, 4686, 1981.

Kuwashima, M., and H. Fukunishi, Local time asymmetries of the ssc-associated hydromagnetic variations at the geosynchronous altitude, Planet. Space Sci., 33, 711, 1985.

Kuwashima, M., et al., On the geomagnetic pulsation Pc (Part3), Spectral and polarization characteristics of middle- and low-latitude Pc 3, Mem Kakioka Mag. Obs., 18, 1, 1979.

Kuwashima, M., et al., Comparative study of magnetic Pc 1 pulsations between low latitudes and high latitudes: Statistical study, Mem. Natl. Inst. Polar Resl., 18, 101, 1981.

Kuwashima, M., et al., SSC associated magnetic variations at the geos ynchronous altitude, J. Atmos. Terr. Physl , 47, 451, 1985.

Kwok, Y. C., and L. C. Lee, Transmission of magnetohydrodynamic waves through the rotational discountinuity at the earth's magnetosphere, J. Geophys. Res. 89, 10697, 1984.

Lanzerotti, L. J., and H. Fukunishi, Modes of magnetohydrodynamic waves in the magnetosphere, Rev. Geophys. Space Phys., 12, 724, 1974.

Lanzerotti, L. J., and H. Fukunishi, Relationships of the characteristics of magnetohydrodynamic waves to plasma density gradients in the vicinity of the plasmapause, J. Geophys. Res., 80, 4627, 1975.

Lanzerotti, L. J., and A. Hasegawa, High plasma instabilities and storm time geomagnetic pulsations, J. Geophys. Res., 80, 1019, 1975.

Lee, M. A., Coupled hydromagnetic wave excitation and ion acceleration upstream of the earth's bow shock, J. Geophys. Res., 87, 5063, 1982.

Lester, M., et al., Polarization patterns of Pi 2 magnetic pulsations and the substorm current wedge, J. Geophys. Res., 88, 7958, 1983.

Lester, M., et al., Longitudinal structure in Pi 2 pulsations and the substorm current wedge, J. Geophys. Res., 89, 5489, 1984.

Lin, C. S., and C. Z. Cheng, Tail field effects on drift mirror instability, J.Geophys. Res., 89, 10771, 1984.

Lysak, R. L., and C. T. Dum, Dynamics of magnetosphere- ionosphere coupling including turbulent transport, J. Geophys. Res., 88, 365, 1983.

Mallinckrodt, A. J., and C. W. Carlson, Relations between transverse electric fields and field-aligned currents, J. Geophys. Res., 83, 1426, 1978.

Maltsev, Y. P., et al., Pi 2 pulsations as a result of evolution of an Alfven impulse originating in the ionosphere during a brightening of aurora, Planet. Space Sci., 22, 1519, 1974.

McKenzie, J. F., and K. O. Westphal, Interaction of hydromagnetic waves with hydromagnetic shocks, Phys. Fluids, 13, 630, 1970.

Mier-Jedrzejowicz, W. A. C., and D. J. Southwood, The east-west structure of mid-latitude geomagnetic pulsation in the 8 - 25 mHz band, Planet. Space Sci., 27, 617, 1979.

Miura, A., Anomalous transport by magnetohydrodynamic Kelvin- Helmholz instabilities in the solar wind magnetosphere interaction, J. Geophys. Res., 89, 801, 1984.

Nagano, H., and T. Araki, Long-duration Pc 5 pulsations observed by geostationary satellites, Geophys. Res. Lett. 10, 908, 1983.

Nagano, H., and T. Araki, Polarization of geomagnetic sudden commencements observed by geostationary satellites, J. Geophys. Res., 89, 11018, 1984.

Nagano, H., and T. Araki, A statistical study on Psc 4 and Psc 5 observed by geostationary satellites, Planet. Space Sci., 33, 365, 1985.

Nagano, H., and T. Araki, Seasonal variation of amplitude of geomagnetic sudden commencements near midnight at geostationary orbit, Planet. Space Sci. 34, 205, 1986.

Newton, R. S., et al., Damping of geomagnetic pulsations by the iono-sphere, Planet. Space Sci., 26, 201, 1978.

Nishida, A., Ionospheric screening effect and storm sudden commencement, J. Geophys. Res., 69, 1861, 1964.

Nishida, A., Possible origin of transient dusk-to-dawn electric field in the magnetosphere, J. Geophys. Res., 84, 3409, 1979.

Odera, T. J., Solar wind controlled pulsations: A review, Rev. Geophys., 24, 55, 1986.

Orr, D., Magnetic pulsations within the magnetosphere: A review, J. Atoms. Terr. Phys., 35, 1, 1973.

Paschmann, G., et al., Energization of solar wind ions by reflection from the Earth's bow shock, J. Geophys. Res. 85, 4689, 1980.

Paschmann, G., et al., Characteristics of reflected and diffuse ions upstream from the Earth's bow shock, J. Geophys. Res., 86, 4355, 1981.

Pashin, A. B., et al., Pi 2 magnetic pulsations, auroral break-ups and the substorm current wedge: A case study, J.Geophys., 51, 223, 1982.

Patel, V. L., et al., Drift wave model for geomagnetic pulsations in a high plasma, J. Geophys. Res., 88, 5677, 1983.

Poulter, E. M., and E. Nielsen, The hydromagnetic oscillation of individual shells of the geomagnetic field, J. Geophys. Res., 87, 10432, 1982.

Rostoker, G., The frequency spectrum of Pi 2 micropulsation activity and its relationship to planetary magnetic activity, J. Geophys. Res., 72, 2032, 1967.

Rostoker, G., and J. C. Samson, Pc micropulsations with discrete latitude-dependent frequencies, J. Geophys. Res., 77, 6249, 1972.

Rostoker, G., and J. C. Samson, Polarization characteristics of Pi 2 pulsations and implications for their source mechanisms: Location of source regions with respect to the auroral electrojets, Planet. Space Sci., 29, 225, 1981.

Russell, C. T., and N. M. Hoppe, Upstream waves and particles, Space Sci. Rev., 34, 155, 1983.

Russell, C. T., et al., Ogo 5 observations of upstream waves in the interplanetary medium: Discrete wave packets, J. Geophys. Res., 76, 845, 1971.

Russell, C. T., et al., The rate of occurrence of dayside Pc 3,4 pulsations: The L-value dependence of the IMF cone angle effect, Geophys. Res. Lett., 10, 663, 1983.

Russell, C. T., et al., Upstream waves simultaneously observed by ISEE and UKS, J.Geophys. Res., 92, 7354, 1987.

Saito, T., Geomagnetic pulsations, Space Sci. Rev., 10, 319, 1969.

Saito, T., and S. Matsushita, Geomagnetic pulsations associated with sudden commencements and sudden impulses, Planet. Space Sci., 15, 573, 1967.

Saito, T., et al., Mechanism of association between Pi 2 pulsation and magnetospheric substorm, J.Atmos. Terr. Phys., 38, 1265, 1976.

Saito, T., et al., Substorm and Pi 2 observed simultaneously at circum northern Pacific and Alaskanchain stations, Paper presented at the IAGA meeting, Edinburgh, 1981.

Sakurai, T., Polarization characteristics of geomagnetic Pi 2 micropulsations, Sci. Rep. Tohoku Univ., Ser.5 Geophys., 20, 107, 1970.

Sakurai, T., and R. L. McPherron, Satellite observations of Pi 2 activity at synchronous orbit, J. Geophys. Res., 88, 7015, 1983.

Samson, J. C., Pi 2 pulsations: High latitude results, Planet. Space Sci., 30, 1239, 1982.

Samson, J. C., and G. Rostoker, Latitude dependent characteristics of hgh latitude Pc 4 and Pc 5 micropulsations, J. Geophys. Res., 77, 6133, 1972.

Samson, J. C., and B. G. Harro;d, Maps of the polarizations of high latitude Pi 2's, J. Geophys. Res., 88, 5736, 1983.

Samson, J. C., and B. G. Harro;d, Characteristic time constants and velocities of high-latitude Pi 2's, J. Geophys. Res., 90, 12173, 1985.

Samson, J. C., and G. Rostoker, Polarization characteristics of Pi 2 pulsations and implications for their source mechanisms: Influence of the westward travelling surge, Planet. Space Sci., 31, 435, 1983.

Schwartz, S. J., et al., Ion upstream of the earth's bow shock, A theoretical comparison of alternative source populations, J. Geophys. Res., 88, 2039, 1983.

Singer, H. J., et al., Evidence for the control of Pc 3,4 magnetic pulsations by the solar wind velocity, Geophys. Res., Lett., 4, 377, 1977.

Singer, H. J., et al., The localization of Pi 2 pulsations, Ground-Satellite observations, J. Geophys. Res., 88, 7029, 1983.

Southwood, D. J., The behaviour of ULF waves and particles in the magnetosphere, Planet. Space Sci., 21, 53, 1973.

Southwood, D. J., Recent studies in micropulsation theory, Space Sci.Rev., 16, 413, 1974.

Southwood, D. J., A general approach to low frequency instability in the ring current plasma.J.Geophys. Res., 81, 3340, 1976.

Southwood, D. J., Localized compressional hydromagnetic waves in the magnetospheric ring current, Planet. Space Sci., 25, 549, 1977.

Southwood, D. J., and M. G. Kivelson, Charged particle behavior in low-frequency geomagnetic pulsations, 1.Transverse waves, J. Geophys. Res., 86, 5643, 1981.

Southwood, D. J., and M. G. Kivelson, Charged particle behaviour in low-frequency geomagnetic pulsations, 2.Graphical approach, J. Geophys. Res., 87, 1707, 1982.

Southwood, D. J., and W. J. Hughes, Concerning the structure of Pi 2 pulsations, J. Geophys. Res., 90, 386, 1985.

Southwood, D. J., et al., Bounce resonant interaction between pulsations and trapped particles, Planet. Space Sci., 17, 349, 1969.

Sugiura, M., and C. R. Wilson, Oscillation of the geomagnetic field lines and associated magnetic perturbations at conjugate points, J. Geophys. Res., 69, 1211, 1964.

Sun, W., and J. R. Kan, A transient response theory of Pi 2 pulsations, J. Geophys. Res., 90, 4395, 1985.

Sutcliffe, P. R., Ellipticity variations in Pi 2 pulsations at low latitudes, Geophys. Res. Lett., 8, 91, 1981.

Takahashi, K., and R. L. McPherron, Standing hydromagnetic oscillations in the magnetosphere, Planet. Space Sci., 32, 1343, 1984.

Takahashi, K., and P. R. Higbie, Antisymmetric standing wave structure associated with the compressional Pc 5 pulsation of November 14, 1979, J. Geophys. Res., 91, 11163, 1986.

Takahashi, K., et al., Azimuthal propagation and frequency characteristic of compressional Pc 5 waves observed at geostationary orbit, J. Geophys. Res., 90, 1473, 1985a.

Takahashi, K., et al., Energetic electron flux pulsations observed at geostationary orbit: Relation to magnetic pulsations, J. Geophys. Res., 90, 8308, 1985b.

Takahashi, K., et al., An eastward propagating compressional Pc 5 waves observed by AMPTE/CCE in the midnight sector, J. Geophys. Res.,92,13472,1987.

Tamao, T., Coupling modes of hydromagnetic oscillations in non-ufiform finite pressure plasma, Two-fluids model, Planet. Space Sci., 26, 1141, 1978.

Tamao, T., Interaction of energetic particles with HM-waves in the magnetosphere, Planet. Space Sci., 32, 1371, 1984.

Tidman, D. A., and N. A. Krall, Shock waves in collisionless plasmas, Wiley-Interscience, New York, 1971.

Tonegawa, Y., Compressional Pc 4 pulsations observed at synchronous orbit, Mem. Natl. Inst. Polar Res., 22, 17, 1982.

Tonegawa, Y., and H. Fukunishi, Harmonic structure of Pc 3-5 magnetic pulsations observed at the Syowa-Husaffel conjugate pair, J. Geophys. Res., 89, 6737, 1984.

Tonegawa, Y., et al., Spectral characteristics of Pc 3 and Pc 4/5 magnetic pulsation band observed near L = 6, J. Geophys. Res., 89, 9720, 1984.

Tsurutani, B. T.,and P. Rodriguez, Upstream waves and particles: An overview of ISEE results, J. Geophys. Res., 86, 4317, 1981.

Troitskaya, V. A., et al., Relationships between Pc 2-4 pulsations and the interplanetary magnetic field, Dokl., Akad., Nauk. SSSR, 197, 1312, 1971.

Verzariu, P., Reflection and refraction of hydromagnetic waves at the magnetopause, Planet. Space Sci., 21, 2213, 1973.

Walker, A. D. M., et al., STARE auroral rader observations of Pc 5 geomagnetic pulsations, J. Geophys. Res., 80, 3373, 1979.

Walker, A. D. M., et al., STARE and GEOS 2 observations of a stormtime Pc 5 ULF pulsation, J. Geophys. Res., 87, 9135, 1982.

Watanabe, Y., and T. Terasawa, On the excitation mechanism of the low frequency upstream waves, J. Geophys. Res., 89, 6623, 1984.

Webb, D., and D. Orr, Geomagnetic pulsations (5 - 50 mHz) and the interplanetary magnetic field, J. Geophys. Res., 81, 5941, 1976.

Webb, D., et al., Hydromagnetic wave observations at large longitudinal separations, J.Geophys.Res., 82, 3329, 1977.

Webster, D. J., and B. J. Fraser, Source regions of low-latitude Pc 1 pulsations and their relationship to the plasmapause, Planet. Space Sci., 33, 777, 1985.

Wedeken, U., et al., Ground-satellite coordinated study of the April 5, 1979 event: Flux variations of energetic particles and associated magnetic pulsations,J.Geophys.Res.,55,120,1984.

Wolfe, A., and R. L. Kaufmann, MHD wave transmission and production near the magnetopause, J. Geophys. Res., 80, 1764, 1975.

Wolfe, A., et al., Dependence of hydromagnetic energy spectra on solar wind velocity and interplanetery magnetic field direction, J. Geophys. Res., 85, 114, 1980.

Wolfe, A., et al., Dependence of hydromagnetic energy spectra near L = 2 and L = 3 on upstream solar wind parameters, J. Geophys. Res., 90, 5117, 1985.

Young, D. T., et al., Wave particle interactions near He observed on GEOS-1 and -2, 1.Propagation of ion cyclotron waves in a He rich plasma, J. Geophys. Res., 86, 6755, 1981.

Yumoto, K., Low-frequency upstream waves as a probable source of low-latitude Pc 3-4 magnetic pulsations, Planet. Space Sci., 33, 239, 1985a.

Yumoto, K ., Characteristics of localized resonance coupling oscillations of the slow magnetosonic wave in a non-uniform plasma, Planet. Space Sci., 33, 1029, 1985b.

Yumoto, K., Generation and propagation mechanisms of low latitude magnetic pulsations, A review, J. Geophys., 60, 79, 1986.

Yumoto, K., Characteristics of daytime bay and Pi 2 magnetic variations: A case study, Planet. Space Sci., 35, 799, 1987.

Yumoto, K., and T. Saito, Nonlinear resonance theory of Pc 3 magnetic pulsation, J.Geophys. Res., 87, 5159, 1982.

Yumoto, K., and T. Saito, Relation of compressional HM waves at GOES 2 to low-latitude Pc 3 magnetic pulsations, J. Geophys. Res., 88, 10041, 1983.

Yumoto, K.,et al., Relationships between the IMF magnitude and Pc 3 magnetic pulsations in the magnetosphere, J. Geophys. Res., 89, 9731, 1984.

Yumoto, K., et al., Low-latitude Pc 3 magnetic pulsations observed at conjugate stations (L 1.5), J. Geophys. Res., 90, 12201, 1985a.

Yumoto, K., et al., Propagation mechanism of daytime Pc 3-4 pulsations observed at synchronous orbit and multiple ground-basd stations, J. Geophys. Res., 90, 6439, 1985b.

Yumoto, K., et al., Pc 3 magnetic pulsations observed at low latitudes: A possible model, Mem. Natl. Inst. Polar Res. 47, 139, 1987.

Zhuang, H. C., and C. T. Russell, Interaction of small-amplitude fluctuation with a strong magnetohydrodynamic shock, Pyhs.Fluids, 25, 748, 1982.

OBSERVATIONS OF NON-THERMAL RADIATION FROM PLANETS

M.L. Kaiser

Goddard Space Flight Center, Greenbelt, MD, USA 20771

Abstract. Observations of the four known non-thermal radio planets are reviewed. Several of the properties are quite similar, suggesting that the same processes take place at all four planets.

Introduction

A radio astronomical instrument in interplanetary space can detect only two of the many magnetoionic wave modes generated within a distant planetary magnetosphere, namely, the left-hand ordinary mode and the right-hand extraordinary mode. Since the solar wind electron plasma frequency is higher than the electron gyrofrequency throughout the region from Mercury to Neptune, both of these modes can be detected provided that they are always at frequencies higher than the solar wind electron plasma frequency along the ray path between the source and the observing spacecraft. This minimum frequency restriction is a strong function of the heliocentric distance of the planet. In the case of Mercury, for example, the solar wind plasma frequency is high enough to significantly hinder or even prevent detection of Mercurian radio emissions. At the other extreme, the solar wind in the vicinity of the outer planets poses little or no barrier for radio astronomical observations because the solar wind plasma frequency is less than 1-2 kHz.

Correct determination of the mode of the emission by a space radio telescope is of prime importance in order to understand the conditions that exist in the source region where spacecraft rarely, if ever, visit. However, only a subset of the radio astronomy instruments flown in space to date have had the capability to directly determine the mode of the emission. Table 1 shows the radio astronomical instruments that have been flown to date, the planet or planets which they have observed, and the actual **observations** they have made (as opposed to **deduced** properties). The measurable quantities are shown in terms of the Stokes parameters and the direction of arrival of the emission, either one dimensional (1D) as in the case of simple spin modulation, or two dimensional (2D) as in the case of lunar occultation or orthogonal dipoles. The Stokes parameter I is equivalent to total intensity, V is effectively the sense of circular polarization (left or right hand), and Q and U measure the degree of linear polarization [see Kraus, 1986]. Most of the instruments flown have made observations (i.e. published observations) only of the earth even

TABLE 1. Instruments that have made Observations of Non-thermal Planetary Radio Emissions

Spacecraft	Planet	Stokes' param. I	Q	U	V	Direction 1D	2D
ALOUETTE-2	E	X					
ELECTRON-2,4	E	X					
OGO-1	E	X					
PROGNOZ-10	E	X					
RAE-1	EJ	X					
EXOS-B,C	E	X				X	
GEOS 1,2	E	X				X	
HAWKEYE	E	X				X	
IMP-6,8	E	X				X	
ISEE-1,2	E	X				X	
ISIS-1,2	E	X				X	
VIKING	E	X				X	
AMPTE	E	X				X	
RAE-2	EJ	X					X
VOYAGER-1,2	EJSUN	X			X		
DE-1	E	X			X	X	
ISEE-3	E	X	x	x	X		X
GALILEO	(V)EJ	X					
ISTP-W,P,G	E(JS)	X	X	X	X		X
ULYSSES	EJS	X	X	X	X		X

V=Venus
E=Earth
J=Jupiter
S=Saturn
U=Uraunus
N=Neptune

Published in 1989 by American Geophysical Union.

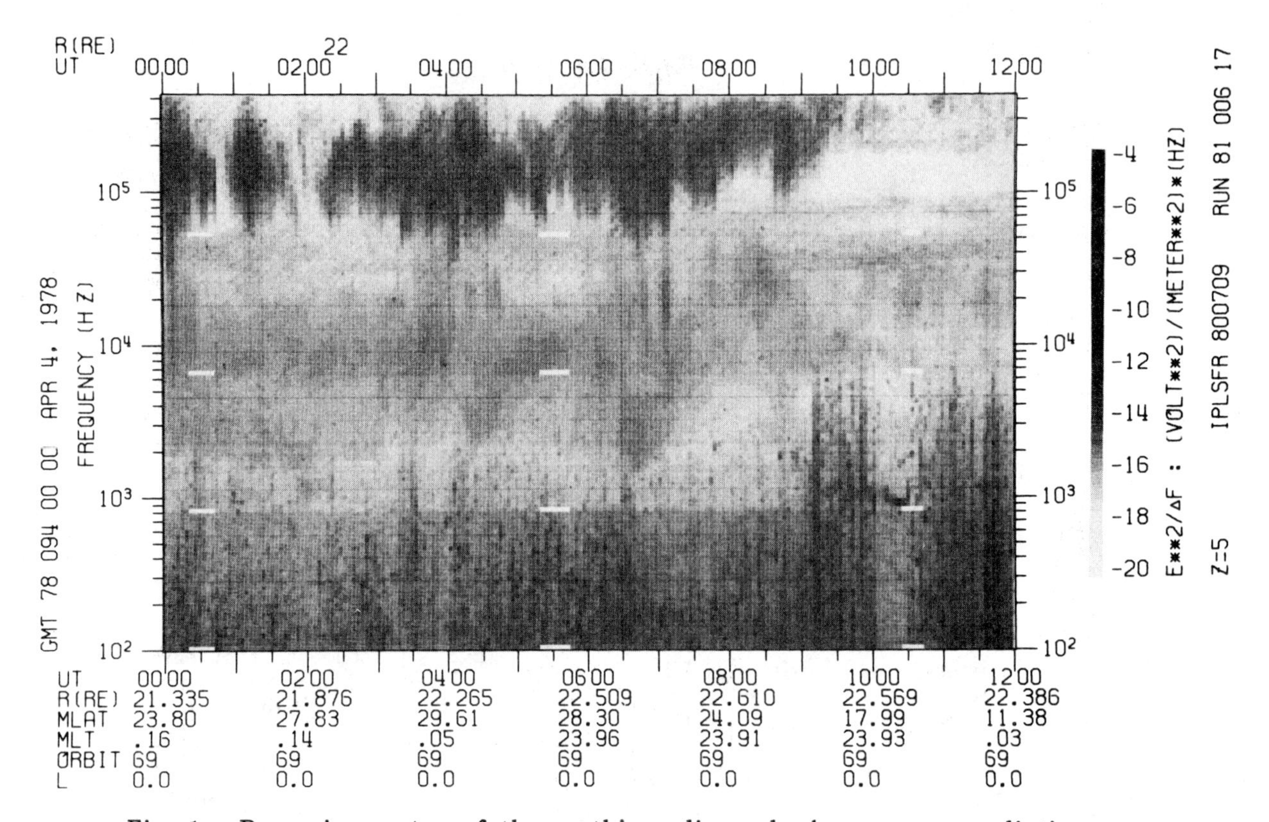

Fig. 1. Dynamic spectra of the earth's radio and plasma wave radiation as observed by the ISEE-1 spacecraft [after Gurnett and Anderson, 1981]. AKR is the dark band in the upper portion of the panel. The AKR emission extends from approximately the local electron gyrofrequency to several hundred kHz, with peak flux at approximately 200 kHz.

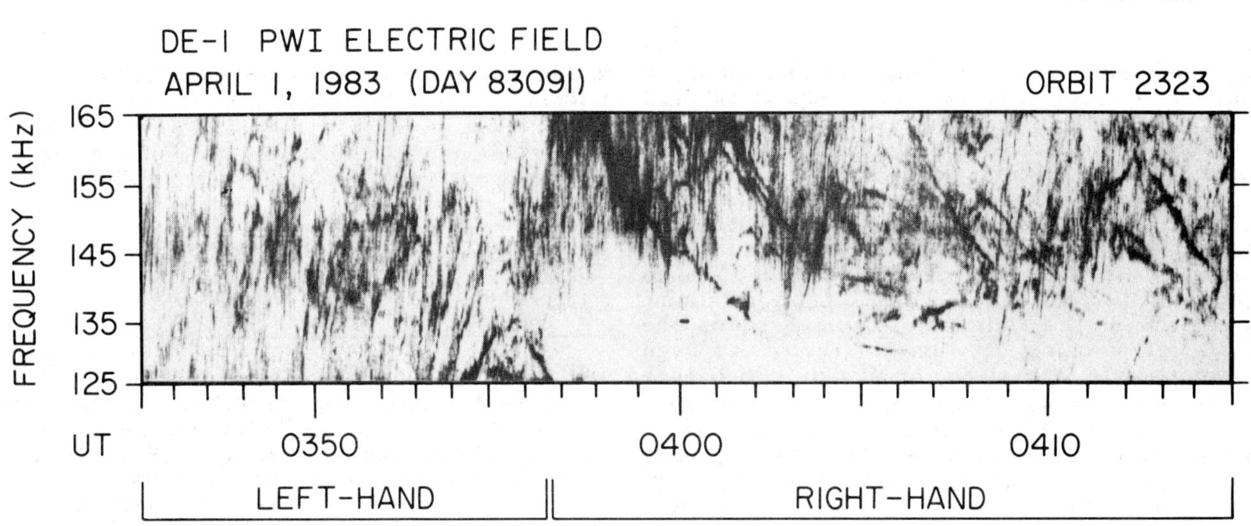

Fig. 2. High time resolution dynamic spectra from the DE-1 spacecraft [after Benson et al., 1988]. The emission consists of numerous narrowband tones rising and falling with time.

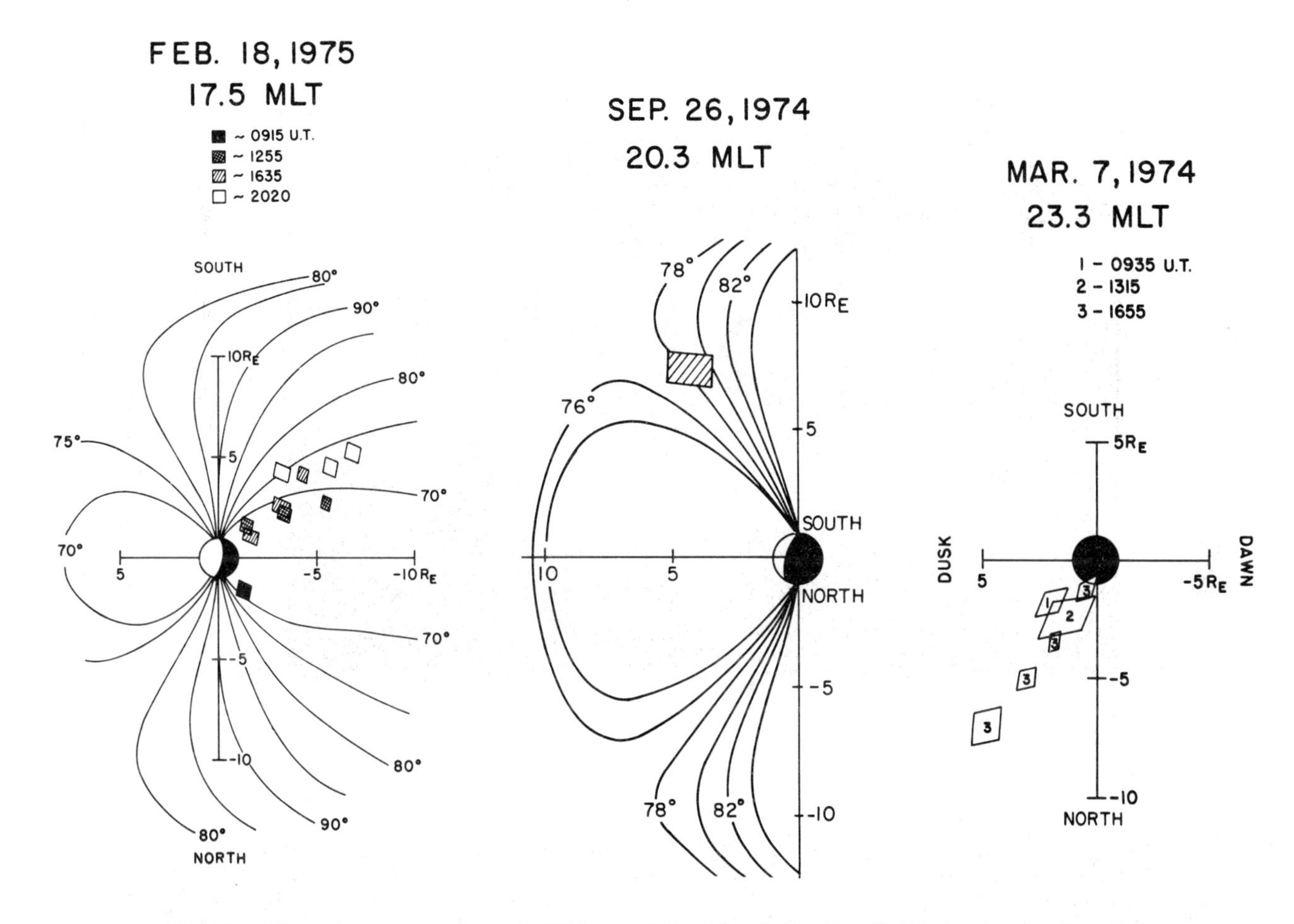

Fig. 3. The source location of AKR as determined by the RAE-2 spacecraft using the lunar occultation technique [after Alexander and Kaiser, 1976]. Each parallelogram represents the source location of an individual occultation projected onto the plane of the sky.

though the radio emissions from both Jupiter and Saturn could have been detected. Almost all of the data on the outer planets has been collected by the Voyager Planetary Radio Astronomy (PRA) and Plasma Wave Science (PWS) experiments. Also shown in the table are some of the currently funded future missions equipped with radio astronomical telescopes.

To date, four planets are known to be non-thermal radio sources in the astronomical sense, that is, sources which can be detected over interplanetary distances. These planets are the earth, Jupiter, Saturn, and Uranus. It is expected that the 1989 Voyager-2 encounter with Neptune will increase this number to five. Mercury probably generates radio noise near the electron gyrofrequency (≲20 kHz) in its magnetosphere [Kaiser, 1977], but the solar wind electron plasma frequency near Mercury is approximately 100 kHz, so the presumed Mercurian emissions do not escape its magnetosphere.

Radio emissions from the earth, Jupiter and Saturn were reviewed by Kaiser and Desch [1984]. Although some duplication of material is unavoidable, the intent of this present review is to concentrate on developments in planetary radio astronomy occurring since the time of that earlier review.

Observations, Planet by Planet

Earth

The earth has three sources of radio emission, the powerful auroral kilometric radiation (AKR), the much weaker "continuum" or myriametric radiation, and the little-known hectometer wavelength component. Although AKR was first discovered by Benediktov et al. [1965], Gurnett [1974] is generally credited with first realizing that the earth's auroral regions were a powerful source of naturally occurring radio noise. He showed that the earth emitted as much as one billion watts at kilometer wavelengths and that these bursts of

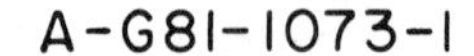

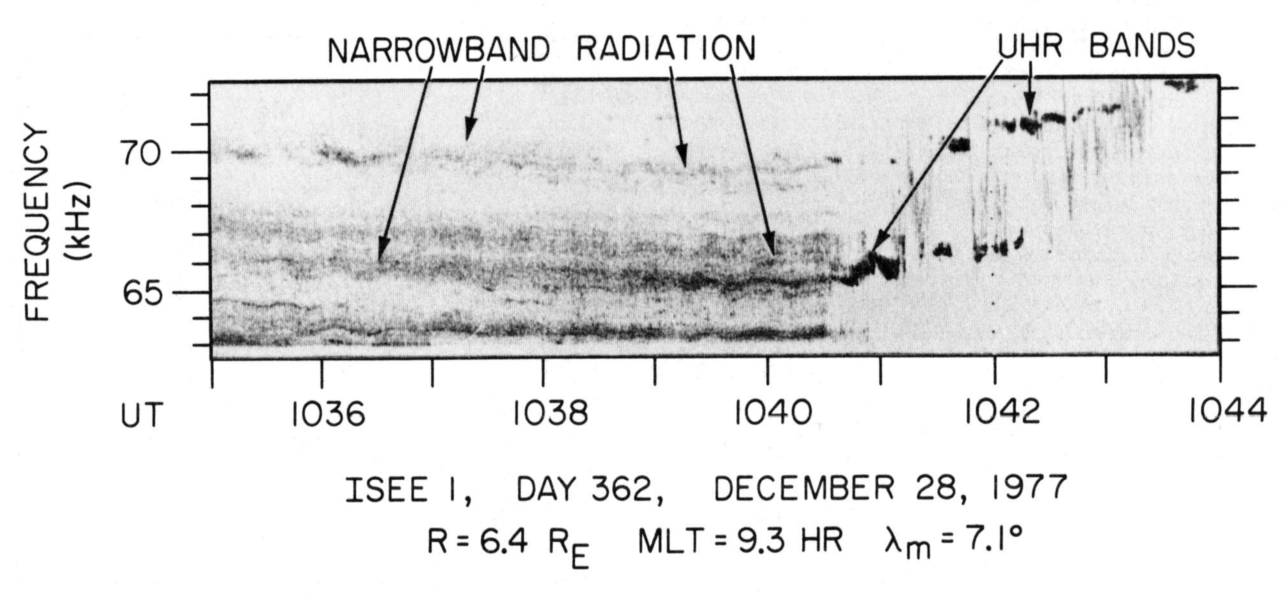

Fig. 4. Dynamic spectra of terrestrial myriametric radiation [after Kurth, 1982]. The narrow discrete tones are clearly seen to be related to the upper hybrid electrostatic emission which is enhanced at the n+1/2 harmonics of the gyrofrequency.

radio emission were strongly correlated with the appearance of discrete auroral arcs. Since that initial work, literally dozens of papers have appeared in the literature concerning the observation and analysis of the earth as a radio source. It is beyond the scope of this paper to review all of this body of literature.

AKR falls in the band from about 50 kHz to as high as 700 kHz with peak power occurring near 250-300 kHz [see Kaiser and Alexander, 1977]. Figure 1 is a dynamic spectra of radio emission and plasma waves taken by the ISEE-1 spacecraft. Increasing intensity is coded as shown at the right. AKR is the intense band near the top of the figure. The peak intensity occurs near 200 kHz, and the emission extends to as high as 400 kHz. The low frequency extent of AKR extends to the local electron gyrofrequency. This apparent cutoff is consistent with AKR being predominantly X-mode [e.g. Kaiser et al., 1978; Shawhan and Gurnett, 1982], although there does appear to be convincing evidence for weaker O-mode AKR [e.g. see Oya and Morioka, 1983; Hashimoto, 1984; Mellott et al., 1984].

At higher time resolution (Figure 2), AKR appears to consist of a myriad of narrow band tones drifting both upward and downward with time. Recent ISEE-1 and 2 interferometric observations of these narrow tones [Baumback et al., 1986; Baumback and Calvert, 1987] have shown that there are occasions when they are less than 5 Hz in bandwidth, and that they generally arise from a very small (<20 km) source.

The AKR emission occurrence pattern is fixed relative to the sun and the earth's magnetic field such that the emission maximizes near 22 hours local time and 70° dipole latitude at altitudes corresponding to the electron gyrofrequency (typically 2-4 R_E) [see Gallagher and Gurnett, 1979]. Figure 3 shows two dimensional AKR source locations as determined by Alexander and Kaiser [1976] using lunar occultations of the source region as viewed by the RAE-2 spacecraft from several different orientations. The source locations clearly cluster around the magnetic extensions of the nightside auroral regions although some AKR appears to emanate from the dayside cusp region, which is thought to be due to scattering from high density regions in the magnetosheath [Alexander et al., 1979].

Terrestrial "continuum" emission was also first described by Gurnett [1975]. It exists in two forms, a series of discrete "lines", and a smooth "continuum". This emission is observed most readily at frequencies below 100 kHz (hence, myriametric), with peak flux occurring at the lowest frequencies which usually correspond to the local electron plasma frequency. Figure 4 is a wide band dynamic spectra showing this component in its discrete form arises from the conversion of odd half harmonics of the electron gyrofrequency into an ordinary electromagnetic mode wave. These narrow tones propagate throughout the earth's magnetosphere. At frequencies below the local solar wind plasma frequency, they are trapped, i.e. they reflect at the magnetopause so that the conglomerate spectrum appears smooth and is called continuum emission. At frequencies above the solar wind plasma frequency, the narrow tones like those in Figure 4 are not reflected and retain their original character. Recent work by Jones et al. [1987] has shown convincingly that the discrete lines arise from the earth's plasmapause near the equatorial plane, in a region where the electron density gradient is nearly perpendicular to the magnetic field as predicted by Jones' linear mode conversion theory (see Figure 5).

Terrestrial emissions in the 2-5.5 MHz range have also been reported by James et al. [1974] and Oya et al. [1985] have reported emissions in this same general range up to the foF2 peak. This emission is apparently O-mode emission emitted from the very highest regions of the ionosphere and may be the result of conversion of upper hybrid mode emission to escaping O-mode emission.

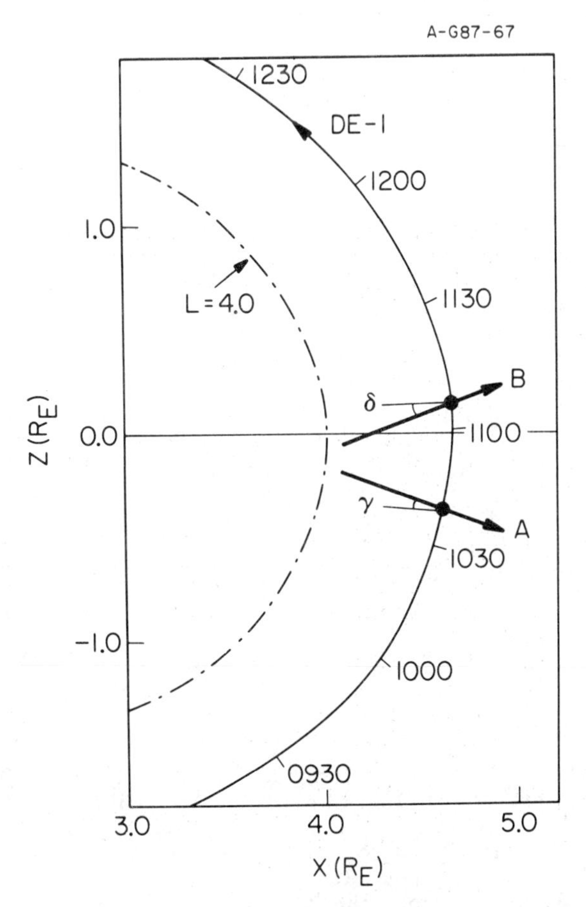

Fig. 5. Direction of arrival information from DE-1 reveals that the source of myriametric radiation is just inside the plasmapause near the magnetic equator [after Jones et al., 1987]. The emission appears in two directed beams at each frequency, one above and one below the magnetic equator.

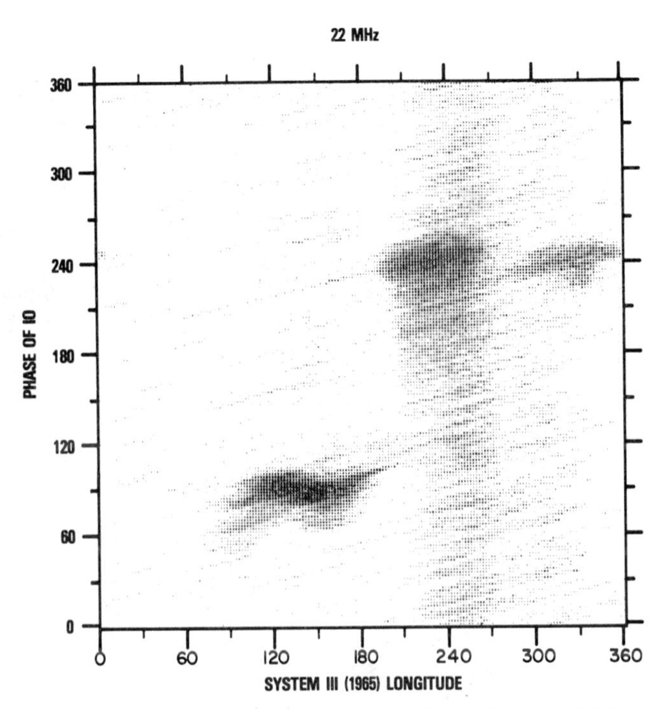

Fig. 6. Occurrence probability of Jovian 22 MHz emission as a function of the departure of Io from superior geocentric conjunction (Io phase) and central meridian longitude [after Thieman, 1979].

Jupiter

Jupiter is the most complex planet in the radio spectrum [see extensive review by Carr et al., 1983], having many totally unrelated spectral components. These components include synchrotron emission generated by relativistic particles trapped in the Jovian Van Allen belts, AKR-like emissions associated with either the satellite Io or its circumjovian plasma torus, low-frequency narrow-band emissions generated in the Io plasma torus, bursty kilometer-wavelength emissions that come from either the Io plasma torus or from the Jovian auroral ionosphere, and earth-like continuum emission.

Jupiter's radio emissions above about 10-15 MHz (called DAM for decametric wavelength emissions) have been studied by ground-based radio telescopes since their initial discovery in the mid-1950s [Burke and Franklin, 1955]. The curious periodic occurrence of these emissions were the first suggestion that Io was an unusual satellite [Bigg, 1964]. Figure 6 shows the occurrence probability of 22 MHz radio emissions displayed as a function of the central meridian longitude on Jupiter and the departure of Io from geocentric superior conjunction. The radio emissions fall into two categories, those that are organized in both the Io frame of reference and the Jovian longitude system (the Io-controlled emissions) and those that are

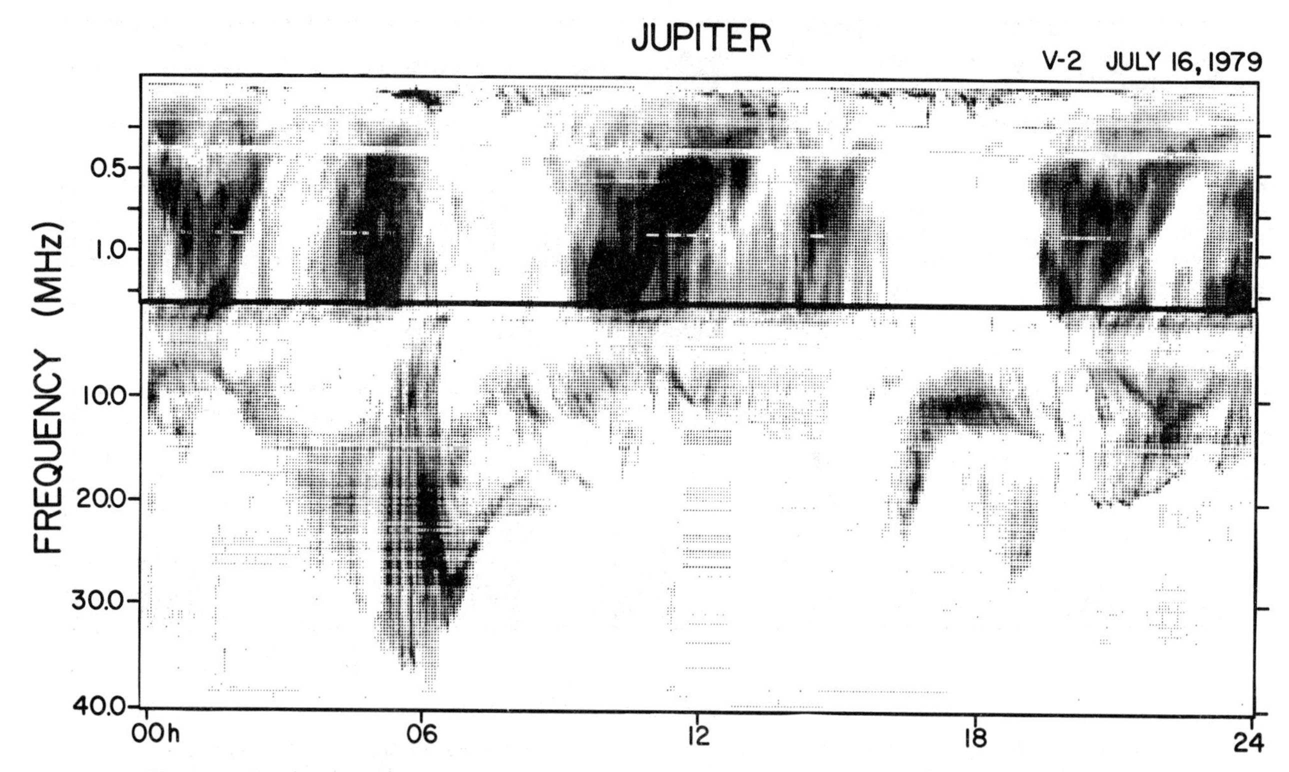

Fig. 7. Dynamic spectra of Jupiter covering the frequency range from 1 kHz to 40 MHz taken by the Voyager PRA instrument [after Kaiser and Desch, 1984]. At frequencies above 1-2 MHz, the emission appears to be resolved into a series of "arc-like" structures in the frequency-time plane.

only localized in the Jovian longitude system (the Io-independent emissions). Approximately 10% of the DAM emission appears in the form of rapidly drifting narrow band bursts of millisecond duration known as S-bursts.

Figure 7 shows a Voyager PRA dynamic spectra covering the band from the top-most Jovian emission frequency (40 MHz) down to 1 kHz. The prominent feature in the frequency range of 10 to 40 MHz corresponds to one of the Io controlled components, specifically, the one centered at 90° Io phase and 90° to 180° system III in Figure 6. The emission appears to be resolved into a series of arc-like structures in the frequency-time plane (great arcs). Similar arc-like structures, but with quite different curvatures, are also found in the region between 1 and 10 MHz (lesser arcs) [see Leblanc, 1981; Goldstein and Thieman, 1981]. The range between a few hundred kHz and 1 to 2 MHz (called HOM for hectometer wavelength emissions) shows of a very confused spectral signature consisting of the overlap of the low frequency extensions of the lesser and great arcs, a separate component of unknown origin [Lecacheux, et al., 1980], and the high frequency extent of one of the Voyager discovered very low frequency components.

The very lowest PRA frequencies are shown best in Figure 8. There exist two spectral components both with maximum power near 100 kHz, but with quite different properties. The top panel of Figure 8 shows an impulsive emission called bKOM (for broadband kilometric wavelength emissions) which consists of events that have the longest duration at low frequencies and successively shorter duration at higher frequencies [see Leblanc and Daigne, 1985], extending as high as 1 Mhz or above. These bKOM events are most prominent when the North magnetic dipole tip of Jupiter nods toward the spacecraft. In the bottom panel is a very smooth component called nKOM (for narrowband kilometric wavelength emissions) which consists of an emission centered near 100 kHz, but with a bandwidth of only a few tens of kHz. In contrast to bKOM, the nKOM emission clearly does not repeat at the Jovian rotation period, but slips behind by 3 to 5% [Kaiser and Desch, 1980; Daigne and Leblanc, 1987].

Figure 9, from the Voyager PWS instrument, shows the Jovian analog of terrestrial myriametric emission. This emission component has been observed at great distances from Jupiter [Kurth et al., 1983] due to waveguide-like properties of the long Jovian magnetotail.

Figure 10 shows the current understanding of

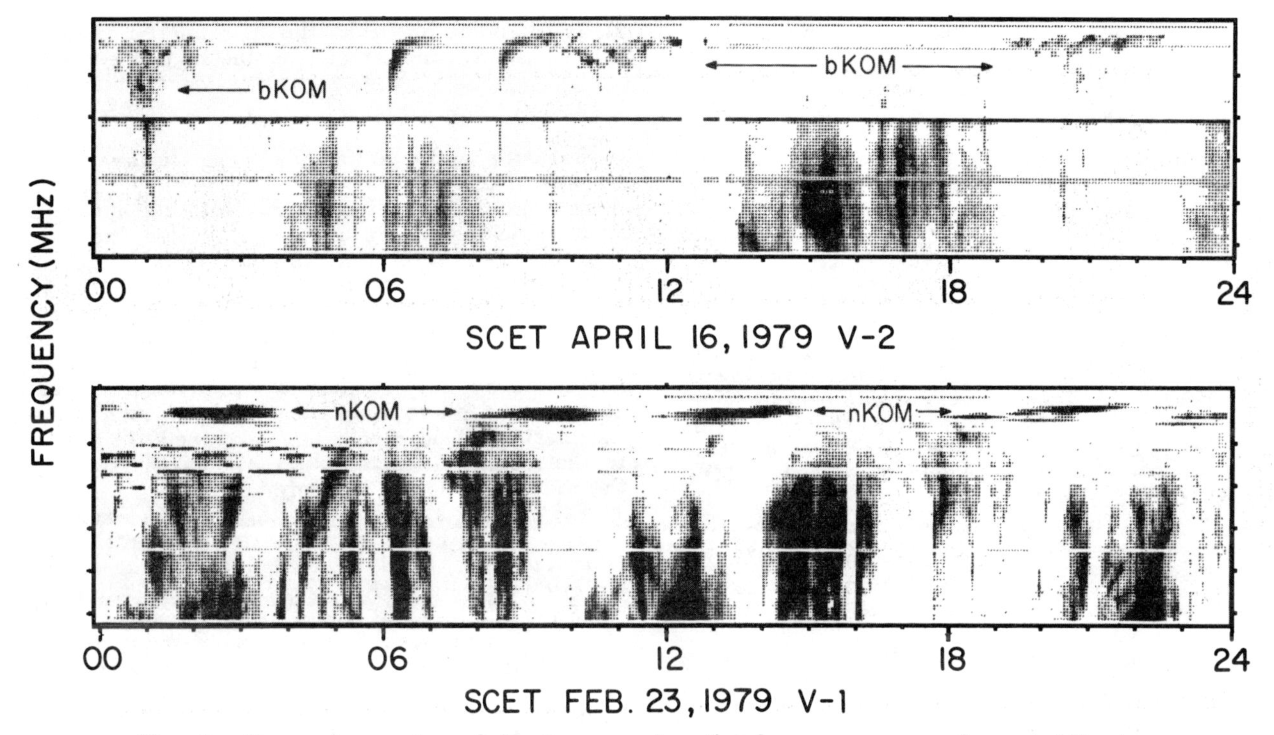

Fig. 8. Dynamic spectra of Jupiter covering the frequency range from 1 kHz to 1326 kHz [after Kaiser and Desch, 1984]. The upper panel shows the so-called bKOM which has a high occurrence probability at 100 kHz, but falls with increasing frequency, thus forming a tappered appearance. In contrast, the lower panel shows nKOM which has its peak flux also near 100 kHz, but is only a few tens of kHz in bandwidth and drifts relative to Jovian rotation by approximately 3%.

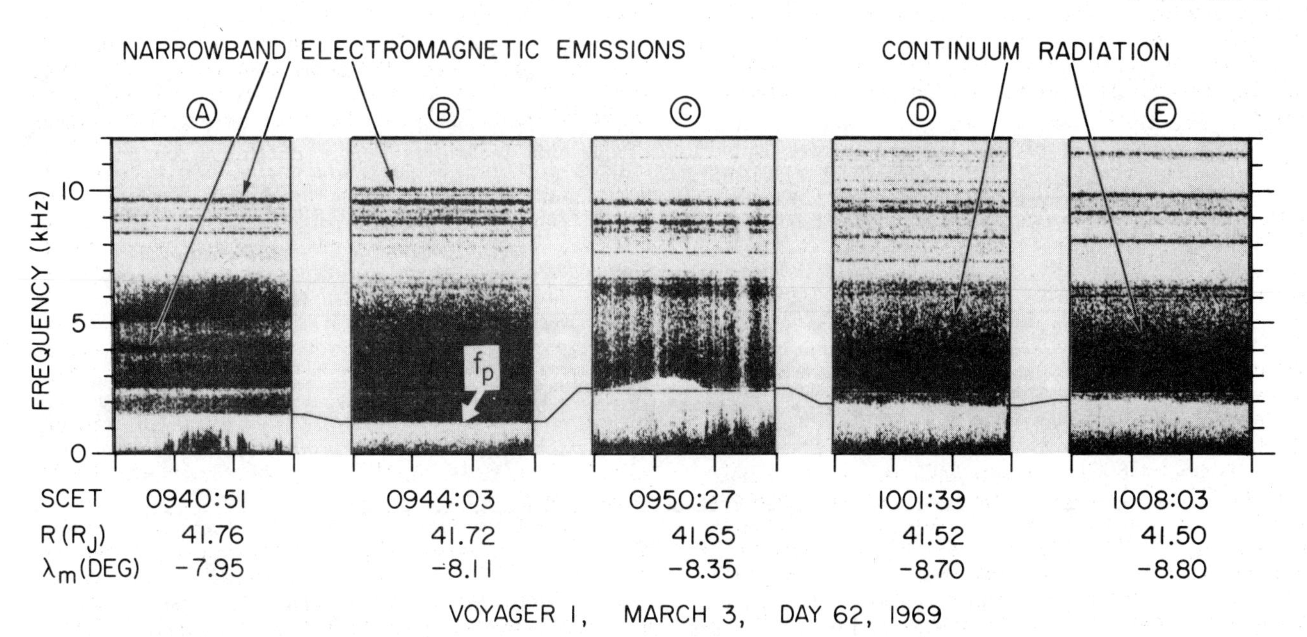

Fig. 9. Dynamic spectra taken by the Voyager PWS instrument showing the Jovian analog of terrestrial myriametric radiation [after Gurnett et al., 1983].

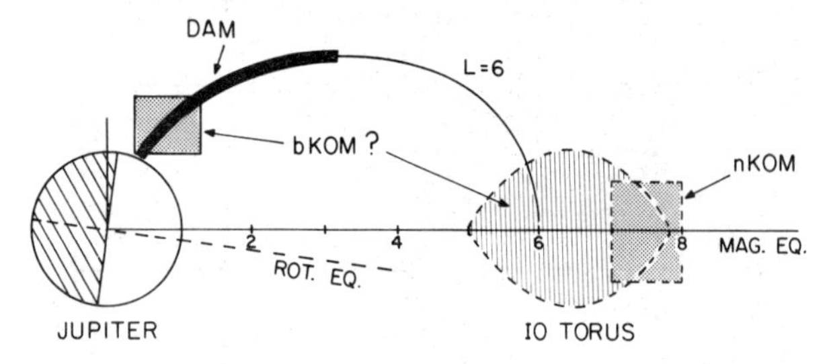

Fig. 10. The deduced source locations of several of Jupiter's radio components [after Kaiser and Desch, 1984]. The DAM occurs along a magnetic field line threading through either Io or the Io plasma torus at altitudes corresponding to the electron gyrofrequency (i.e. X-mode cutoff). Both bKOM and nKOM are believed to arise from the Io plasma torus, although the possibility exists that bKOM could originate at low altitudes above the auroral zones.

some of the source locations of the various radio components. The Io controlled DAM is believed to be related to the magnetic flux tube linking Io to the polar ionosphere [see review by Carr et al, 1983]. The emission is believed to be X-mode and originates at altitudes on this flux tube corresponding to the electron gyrofrequency. The two kilometric components, bKOM and nKOM, may both originate in the Io torus [Jones and Leblanc, 1987; Jones, 1987], although it is still possible that bKOM could have its source above the polar ionosphere. Both of these components are believed to be O-mode emission. To date, no explanation exists for the origin of lesser arcs and for HOM, only regions of exclusion have been shown [Lecacheux, 1980]. The Jovian analog to the terrestrial myriametric radiation may originate from the conversion of electrostatic upper hybrid waves at the magnetosheath [Leblanc et al, 1986], far beyond the boundaries of Figure 10. Kurth et al. [1981] point out that the myriametric radiation is very similar in character to the nKOM emission.

Saturn

Saturn's radio emissions were discovered by Voyager [Kaiser et al, 1980], although they are present in the data from earth orbiting spacecraft such as ISEE-C. Saturn is more like the earth than like Jupiter in that there appears to be only an extraordinary mode AKR-like source in the auroral regions and some weak continuum emission. However, Saturn's radio emission is not without interest. Its auroral emissions are strongly modulated by the planetary rotation, an effect seen at all other planets. However, in Saturn's case, the magnetic field dipole axis shows no tilt relative to the rotation [Ness et al., 1981], so the source of the radio modulation is a mystery.

Figure 11 shows a Voyager PRA dynamic spectra of 24 hours of data taken approximately 100 planetary radii from Saturn. The Saturn emissions (called SKR for Saturnian kilometric radiation) fall in the band from about 3 kHz to 1200 kHz with peak flux occurring near 300 kHz [see review by Kaiser et al., 1985]. In this figure, three episodes of emission separated by the planetary rotation period of 10h 39.4m [Desch and Kaiser, 1981] can be seen. The SKR emission falls somewhere between Jupiter and the earth in terms of the repeatability of features in the dynamic spectra, Jupiter being highly repeatable. The SKR was observed to be exclusively right hand polarized from Saturn's northern hemisphere and left hand from Saturn's southern hemisphere which led to the conclusion that SKR is X-mode.

During the two Voyager encounters with Saturn, the SKR source region was occulted by the planet allowing workers to deduce the probable origin of the emission [Kaiser and Desch, 1982; Lecacheux and Genova, 1983]. Figure 12 shows the footprint of the two SKR sources (north and south) derived from this technique. The northern footprint is a small region in the auroral zone where the SKR somehow maximizes when that set of field lines is

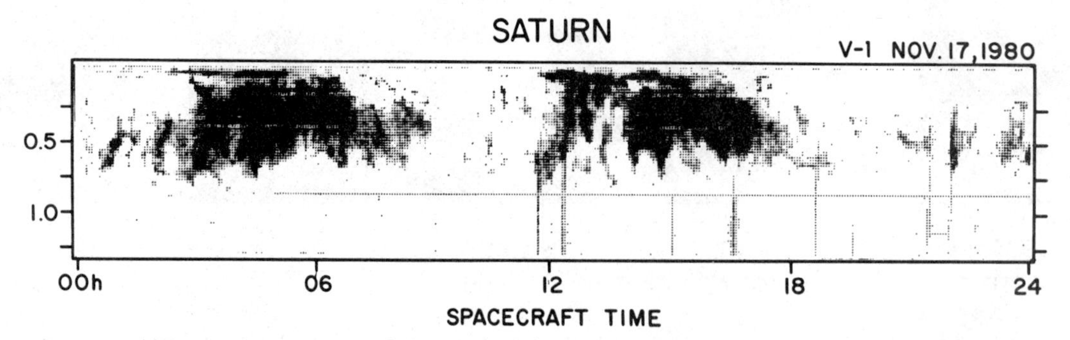

Fig. 11. Dynamic spectra of Saturn's kilometric radiation taken by the Voyager PRA instrument [after Kaiser and Desch, 1984]. The emission tends to recur with a 10h 40m period even though the dipole magnetic field is not tilted relative to the rotation axis.

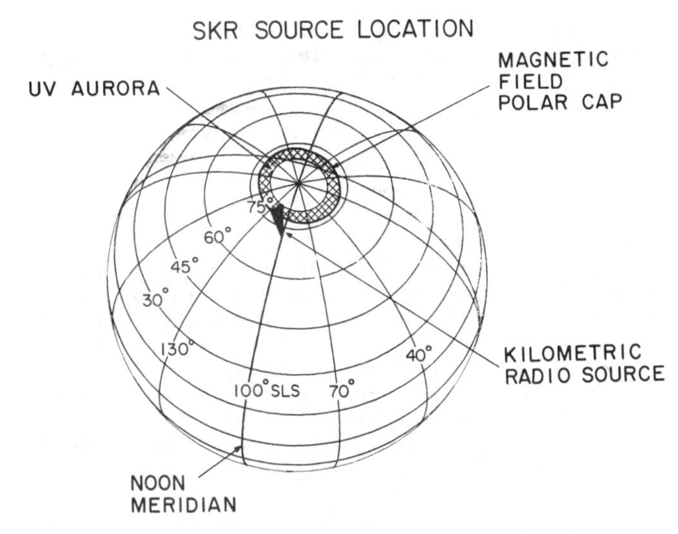

Fig. 12. Deduced source location of SKR has its northern hemisphere footprint near the noon meridian and in the auroral zone [after Kaiser and Desch, 1982]. The southern hemisphere source is not as well defined, but may also be coincident with the auroral zone.

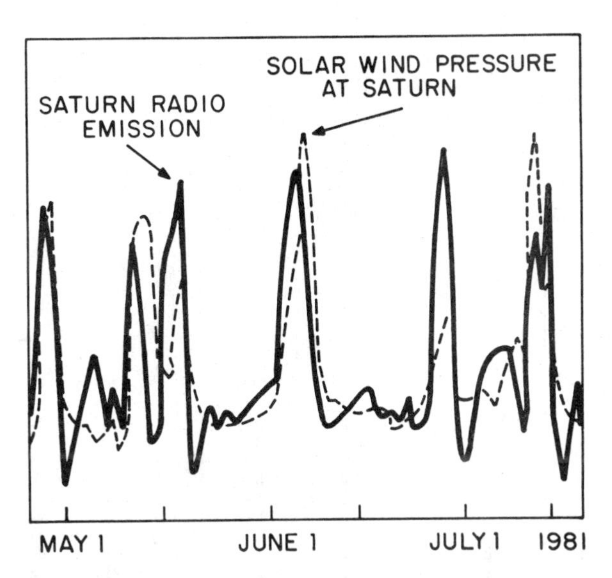

Fig. 13. SKR energy over a period of several months plotted along solar wind ram pressure [after Desch and Rucker, 1983]. The correlation between these two quantities is better at Saturn than for any other planet.

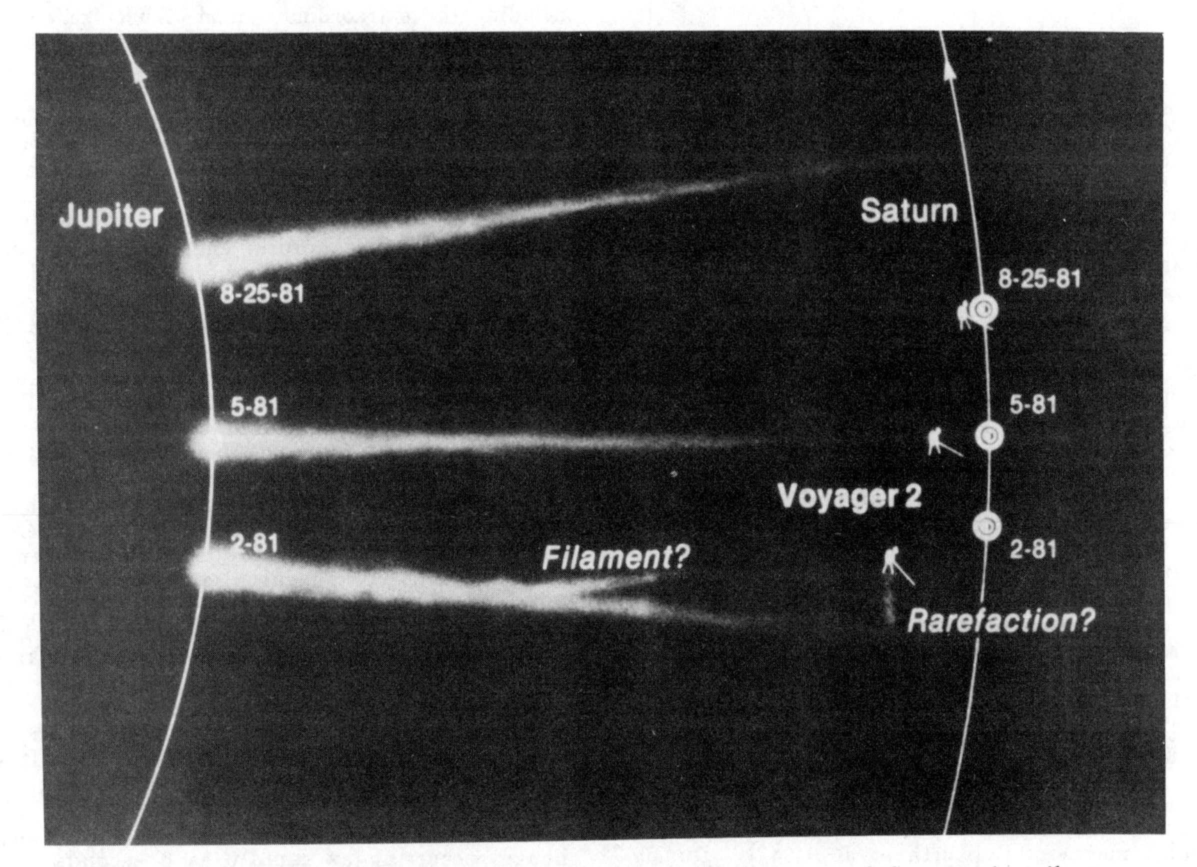

Fig. 14. Schematic of the possible orientation of Jupiter's extended magnetic tail at the time of the Voyager-2 Saturn encounter [after Scarf, 1979].

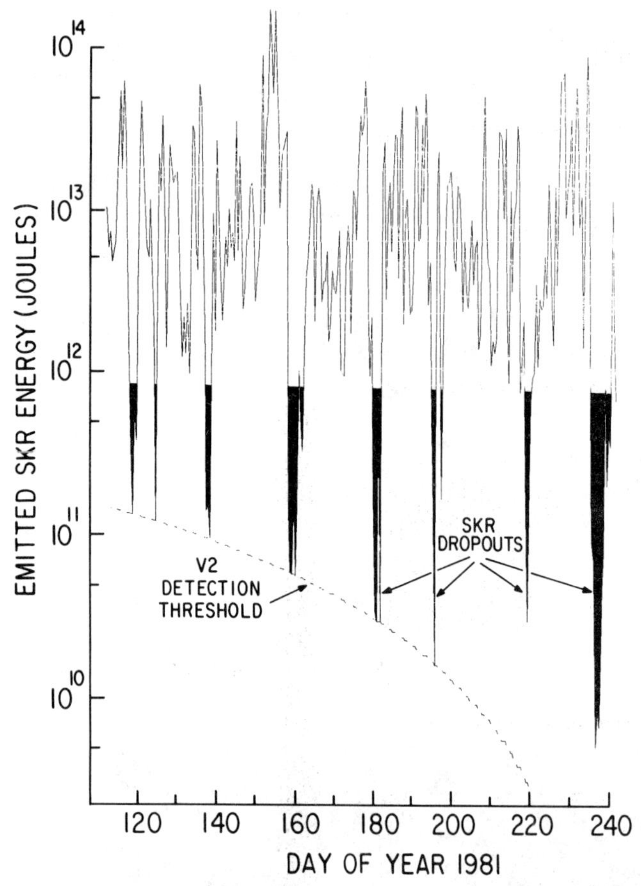

Fig. 15. SKR energy as a function of time for the period near the Voyager-2 encounter shows several dropouts which correspond to intervals when Saturn was immersed in Jupiter's magnetic tail [after Desch, 1983].

on or near the noon meridian. This same range of longitudes (100° to 115°) is also strongly related to the appearance of spokes in the rings of Saturn [Porco and Danielson, 1982].

In addition to the strong, but unexplained rotation modulation, Saturn's auroral emissions are a function of the solar wind pressure, sometimes totally disappearing during quiet solar wind conditions. Figure 13 shows the correlation between the observed SKR energy and the solar wind ram pressure [Desch and Rucker, 1983]. Similar effects are also seen for terrestrial AKR [Gallagher and D'Angelo, 1981] and Jupiter's HOM [Desch and Barrow, 1984]. Dramatic evidence of the influence of the solar wind on SKR occurred during the Voyager-2 encounter with Saturn in 1981. Figure 14 shows the fortuitous radial alignment of Jupiter and Saturn. This allowed Saturn to be immersed, at times, in Jupiter's extended magnetotail [Kurth et al., 1983]. During these intervals, the solar wind was totally excluded from the Saturn environment. Figure 15 shows the SKR energy as a function of time. The heavily shaded regions show episodes when SKR energy dropped to below the PRA detection threshold. These episodes corresponded almost exactly with intervals when Saturn was immersed in Jupiter's tail [Desch, 1983].

Uranus

The most recent addition to the list of radio planets is Uranus, whose emissions were measured by the Voyager-2 PRA and PWS instruments during the January, 1986 fly-by [Warwick et al., 1986; Gurnett et al., 1986]. Although its emissions are intrinsically weak, they are nearly as complex as Jupiter's, caused probably by the highly unusual orientation of the planet and its extremely tilted magnetic field relative to the sun [Ness et al., 1986].

Figure 16 shows a series of dynamic spectra centered approximately on the time of Voyager-2 closest approach (C.A.) to Uranus. The extreme asymmetry in the radio signature between the inbound trajectory above the Uranian dayside and the outbound trajectory above the Uranian nightside is apparent. Uranus appears to have at least a half dozen separate spectral components including the extraordinary mode AKR-like emission observed at all planets (numbers 4 and 5), several narrow bandwidth components of unknown origin (numbers 1 and 2, plus a spectral component near 5 kHz not shown here), and ordinary mode emission (number 3) associated with the current epoch dayside magnetic dipole tip [see Leblanc et al., 1986; Desch and Kaiser, 1986]. This latter component may be magnetically linked to the dominant AKR-like auroral emissions which occur primarily over the night hemisphere and may represent, for the first time, direct observations of emission arising from conjugate points on a field line.

The dominant emissions from Uranus (4 and 5) had a very repeatable pattern as shown in Figure 17. The emission consisted of two components, both in the band between about 50 and 850 kHz, but with quite different temporal properties. The component labelled "4" in Figure 16 has a relatively smooth profile where large intensity changes occur over several minutes or tens of minutes. This component also exhibits a striking emission dropout at frequencies above about 300 kHz. The repeatability of this dropout led Desch et al. [1986] to determine the previously unknown rotation period of Uranus, 17.24 hours. This dropout occurs when the spacecraft is near maximum magnetic latitude on each rotation. At times during each rotation when the spacecraft was near the magnetic equator, the other nightside component ("5" in Figure 16) was visible. This component is very bursty with large intensity changes occurring as rapidly as 6 seconds. This overall pattern of smooth emission at high latitudes, with the embedded dropout, followed by

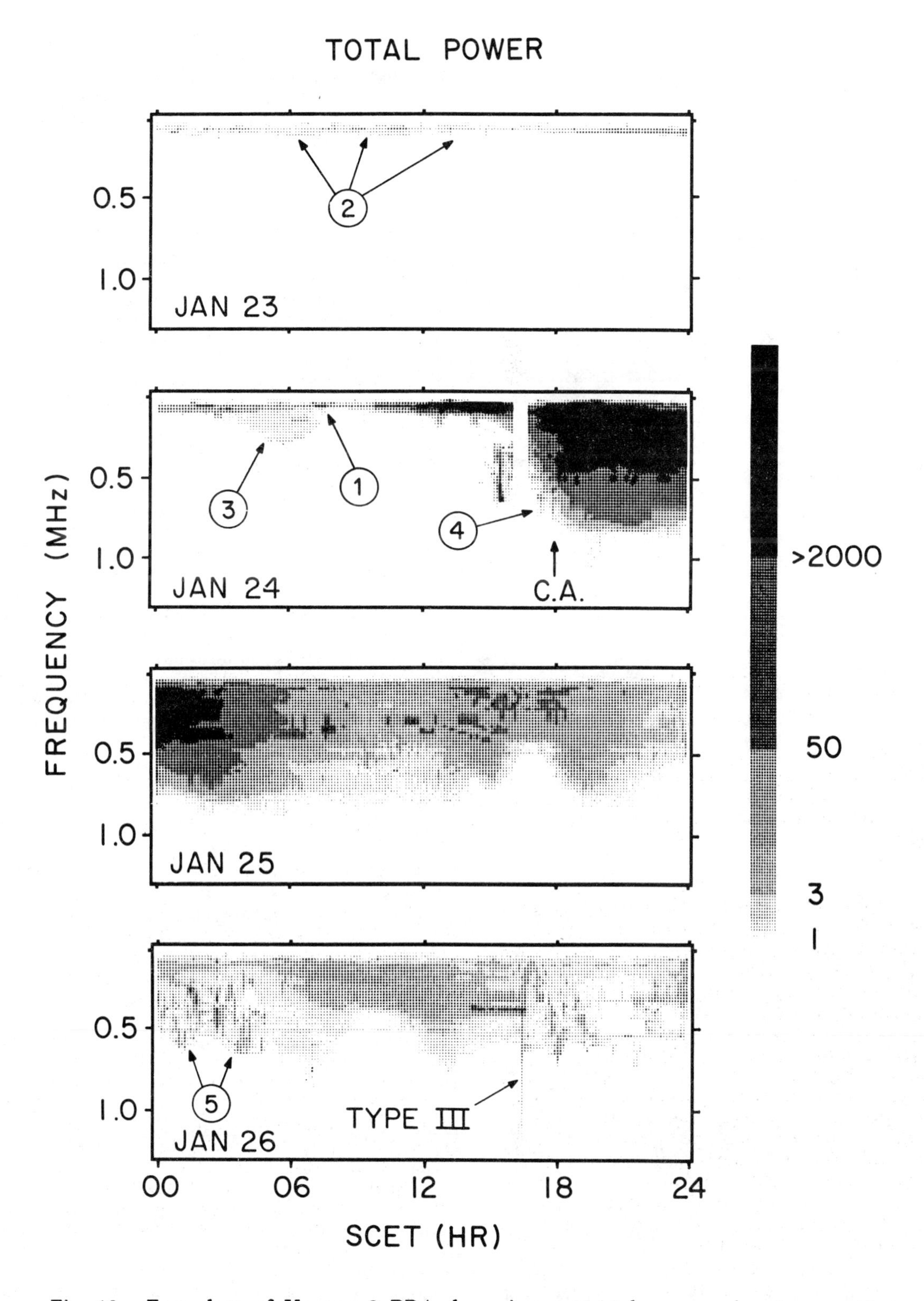

Fig. 16. Four days of Voyager-2 PRA dynamic spectra taken near closest approach (C.A.) to Uranus [after Warwick et al., 1986]. Five separate radio components are visible and a sixth component was detected by the Voyager PWS experiment.

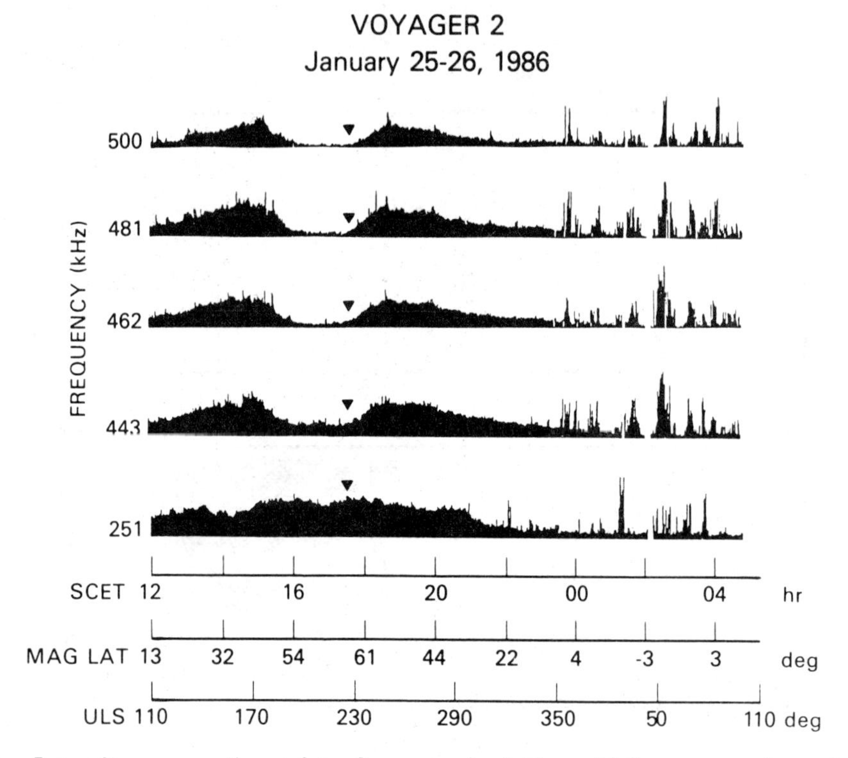

Fig. 17. Intensity versus time plots for several of the mid-frequency channels from the Voyager-2 PRA experiment [after Warwick et al., 1986]. These data, taken from the night hemisphere, show the two components labelled 4 and 5 in Figure 16. Component 4 has a distinct intensity dropout near the time the spacecraft reaches maximum magnetic latitude on each planetary rotation.

bursty emissions at low latitudes repeats on virtually every rotation of the planet.

Several workers have attempted to determine the source location of the smooth nightside emission by analyzing the dropout feature [Kaiser et al., 1987; Zarka and Lecacheux, 1987; Lecacheux and Ortega-Molina, 1987]. All of these workers agree that the dropout is most likely the result of a hollow radiation beam oriented antiparallel to the magnetic field near Uranus South magnetic dipole tip. Figure 18 shows a schematic of the possible source location and radiation beam pattern for emission at 600 kHz. The emission is extraordinary mode and is emitted near the electron gyrofrequency, possibly on closed field lines (i.e. equatorward of the polar cap). Desch and Kaiser [1987] point out that the component labelled "3" in Figure 16 could possibly arise from the conjugate footprint of the set of active field lines.

Kurth et al. [1986] report the detection of very low frequency (3 to 10 kHz) narrowband tones during the Voyager-2 outbound trajectory. While these tones are somewhat reminiscent of escaping continuum observed at the earth, Jupiter and Saturn, they show considerable structure at time scales as rapid as 1 second, suggesting a very different type of generation mechanism.

Discussion

Table 2 summarizes the basic observable parameters for the four radio planets and indicates the current best estimate of the actual source locations. The values for myriametric radiation are for that portion of the emission able to propagate into the solar wind. In terms of absolute power, Jupiter appears to be the strongest radio planet in the solar system, followed by Saturn, the earth, and Uranus, although there is evidence that Uranus can at times exceed the earth in radio power [Brown, 1976]. Figure 19 shows the median spectrum observed by the Voyager PRA instrument for each of the radio planets. Although these medians probably accurately reflect the relative signal strength of each planet, it must be remembered that the Voyager measurements were made from a very limited range of local times and magnetic latitudes, i.e beaming is probably very important. The detailed comparison of radiated power is not possible at this time. The corresponding values for myriametric radiation when normalized to the 1 A.U. distance scale, are all below the lower left edge of the figure.

From the polarization measurements of Table 2 and the deduced source locations, it is possible to

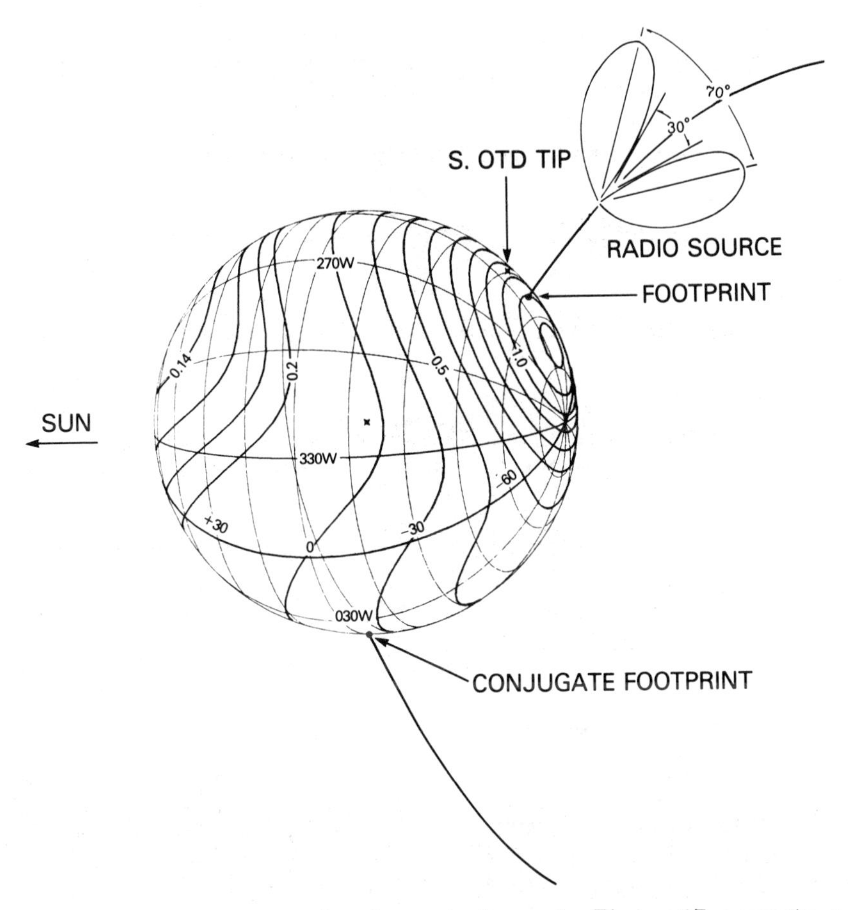

Fig. 18. Analysis of the intensity dropout shown in Figure 17 suggests a source region for component 4 at gyro frequency altitudes on a small set of field lines emanating from near the Uranian South dipole tip [after Kaiser et al., 1987]. This small set of field lines may be closed field lines with their conjugate footprint near the present-epoch terminator near the equator. Radio component number 3 from Figure 16 may arise from the conjugate end of the active nightside field line.

determine the emission mode. The current state of knowledge is shown in Figure 20. The dominant source at each planet appears to be AKR-like auroral emission in the X-mode. O-mode emissions, which are observed at most planets, tend to be weaker by a substantial amount. The emission mechanism believed to be responsible for the strong X-mode (and some of the related O-mode) emission is the Doppler shifted cyclotron maser mechanism as formulated by Wu and Lee [1979] and described by Lee [this volume]. Additional mechanisms involving mode conversion are discussed by Jones [see review in this volume] to explain some of the O-mode emissions.

Even though there is a spread of several orders of magnitude in the absolute radio power, there is a remarkably constant scaling factor relating the total solar wind input power into each planetary system and the AKR-like auroral emissions. Figure 21 shows the "radiometric Bodes' law" of Desch and Kaiser [1984] where solar wind input in watts is plotted against observed total radio power in watts. The earth's AKR, Jovian HOM, Saturnian SKR, and Uranian nightside emissions all fall on a straight line indicating a constant "efficiency" factor of 5 parts per million. With this scaling "law" and predictions of a 0.5 Gauss Neptunian magnetic field [Curtis and Ness, 1986], the Voyager-2 spacecraft could possibly detect Neptunian AKR-like emissions during the early summer of 1989 (closest approach Aug 25, 1989), assuming they are not preferentially emitted toward the night hemisphere or high latitudes.

The Future of Planetary Radio Astronomy

Although the Challenger disaster in January of 1986 caused long delays in the space program, planetary radio astronomy continues to have a

TABLE 2. Summary of Planetary Radio Observations

Property	Earth	Jupiter	Saturn	Uranus
frequency range (kHz)	50-700 (A) 30-100 (M)	20-1000 (b) 50- 175 (n) 500-40000(D) 5-50 (M)	3-1200 (S) 1-10 (M)	20-150 (1,2) 100-850 (4,5) 100-250 (3)
average power (MW)	30 (A) ? (M)	400 (b) 50 (n) 6000 (D) ? (M)	1000 (S) ? (M)	4 (1,2) 10 (4,5) <1 (3)
polarization sense in magnetic north	RH/LH (A) LH (M)	RH (b) LH (n) RH/LH (D) LH (M)	RH (S) LH (M)	LH (1,2) LH* (4,5) LH (3)
source location	discrete aurora	(b) torus ? (n) torus (D) IFT	cusp	(1,2) N mag pole ? (4,5) S mag pole (3) N mag pole
emission mode	R-X/L-O(A) L-O (M)	L-O (b) L-O (n) R-X (D) L-O (M)	R-X (S) L-O (M)	L-O (1,2) ? R-X (4,5) L-O (3)
	A=AKR	b=bKOM n=nKOM D=DAM	S=SKR	(1-5) as in Figure 16 * = South only

M=myriametric (all planets)

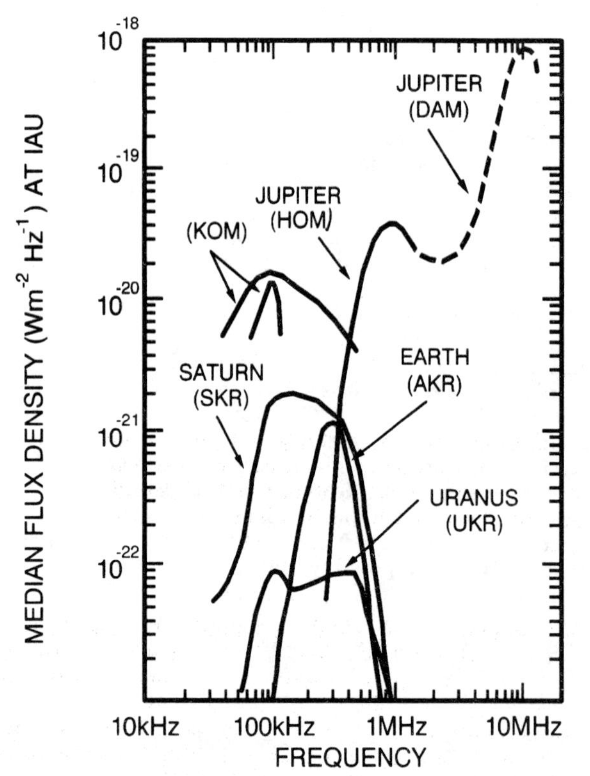

Fig. 19. The median flux density observed by the Voyager PRA instrument from a distance of 1 A.U. from each planet.

bright future. The Voyager detection and measurement of Neptunian radio emissions will occur soon and the current Shuttle launch schedule calls for a Galileo launch in 1989 and a Ulysses launch in 1990. Galileo will arrive at Jupiter in the mid-1990s and will make measurements below 5 MHz. Unlike Voyager, its orbit will allow complete coverage of Jupiter in local time. Ulysses will fly by Jupiter in late 1991 and make measurements of the direction of arrival and all four Stokes' parameters at frequencies below 1 MHz. Beginning in the mid-1990s, a series of earth-orbiting spacecraft in the International Solar Terrestrial Program will be capable of making complete wave and particle measurements of AKR and "continuum". Finally, planning is well underway for a Saturn orbiter mission named Cassini. This spacecraft, which could arrive at Saturn just after the turn of the century, could make measurements of SKR from essentially all local times and all latitudes over its projected four year lifetime.

Of course, much remains to be done with the available data bases. A number of outstanding problems remain unanswered. For example, what is the source of the strong rotation modulation of SKR? Can the two kilometer wavelength components of Jupiter be used to remotely monitor conditions in the Io plasma torus? Study of the recently acquired Uranus data has just begun. A major problem in planetary radio astronomy is finding available interested people to work on the vast amount of data.

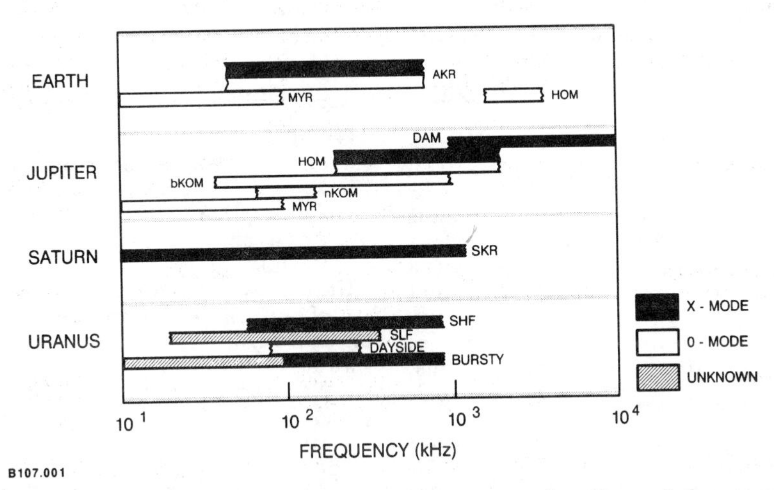

Fig. 20. Emission mode deduced for each planet as a function of frequency. The strong auroral zone emissions (AKR, DAM, SKR, and UKR) all appear to be predominantly X-mode. Some of the other emissions are also probably X-mode, but there are a number of O-mode components, particularly at Jupiter and Uranus.

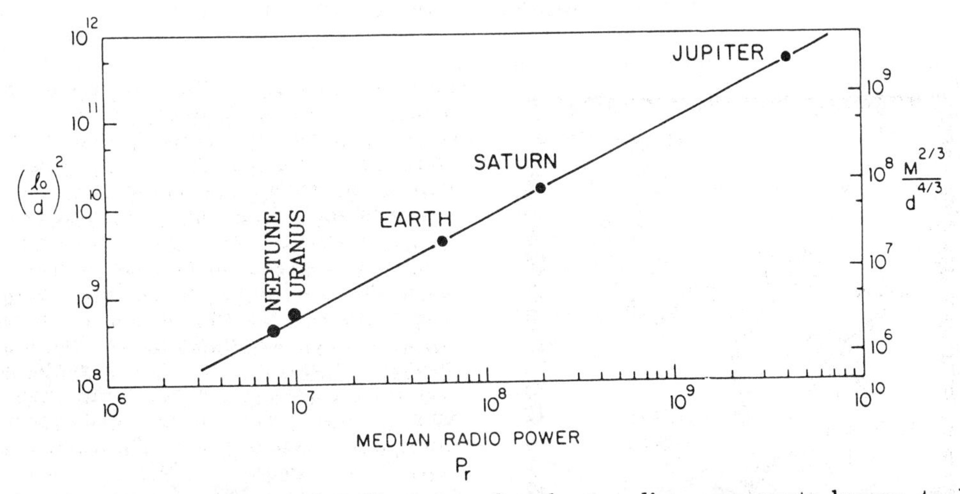

Fig. 21. The total emitted radio power for those radio components known to be associated with the solar wind ram pressure is in a nearly constant ratio of 5 parts per million to the total solar wind input power into the magnetosphere [after Desch and Kaiser, 1984]. Estimates for Neptune indicate that it will have about the same strength as Uranus.

References

Alexander, J.K. and M.L. Kaiser, Terrestrial kilometric radiation 1. Spatial structure studies, J. Geophys. Res., 81, 5948, 1976.

Alexander, J.K., M.L. Kaiser, and P. Rodriguez, Scattering of terrestrial kilometric radiation at very high altitudes, J. Geophys. Res., 84, 2619, 1979.

Baumback, M.M., D.A. Gurnett, W. Calvert and S.D. Shawhan, Satellite interferometric measurements of auroral kilometric radiation, Geophys. Res. Lett., 13, 1105, 1986.

Baumback, M.M. and W. Calvert, The minimum bandwidths of auroral kilometric radiation, Geophys. Res. Lett., 14, 119, 1987.

Benediktov, E.A., G.G. Gemantsev, Yu. A. Sazonov, and A.F. Tarasov, Preliminary results of measurements of the intensity of distributed extraterrestrial radio frequency emission at 725 and 1525 kHz frequency by the satellite Electron-2, Kosm. Issled., 3, 614, 1965.

Benson, R.F., M. Mellott, D.A. Gurnett, and R.L. Huff, Ordinary mode auroral kilometric radiation fine structure observed by DE-1, J. Geophys. Res., in press, 1988.

Bigg, E.K., Influence of the satellite IO on Jupiter's decametric emission, Nature, 203, 1008, 1964.

Brown, L.W., Possible radio emission from Uranus at 0.5 MHz, Astrophys. J., 207, L209, 1976.

Burke, B.F. and K.L. Franklin, Observations of a variable radio source associated with the planet Jupiter, J. Geophys. Res., 60, 213, 1955.

Carr, T.D., M.D. Desch and J.K. Alexander, Phenomenology of magnetospheric radio emissions, in Physics of the Jovian Magnetosphere, ed. A.J. Dessler, Cambridge, 1983.

Curtis, S.A. and N.F. Ness, Magnetostrophic balance in planetary dynamos: Predictions for Neptune's magnetosphere, J. Geophys. Res., 91, 11003, 1986.

Daigne, G. and Y. Leblanc, Narrowband Jovian kilometric radiation: Occurrence, polarization, and rotation period, J. Geophys. Res., 91, 7961, 1987.

Desch, M.D. and M.L. Kaiser, Voyager measurement of the rotation period of Saturn's magnetic field, Geophys. Res. Lett., 8, 253, 1981.

Desch, M.D., Radio emission signature of Saturn immersions in Jupiter's magnetic tail, J. Geophys. Res., 88, 6904, 1983.

Desch, M.D. and H.O. Rucker, The relationship between Saturnian kilometric radiation and the solar wind, J. Geophys. Res., 88, 8999, 1983.

Desch, M.D. and C.H. Barrow, Direct evidence for solar wind control of Jupiter's hectometer-wavelength radio emission, J. Geophys. Res., 89, 6819, 1984.

Desch, M.D. and M.L. Kaiser, Predictions for Uranus from a radiometric Bode's law, Nature, 310, 5980, 1984.

Desch, M.D., J.E.P. Connerney, and M.L. Kaiser, The rotation period of Uranus, Nature, 322, 42, 1986.

Desch, M.D. and M.L. Kaiser, Ordinary mode emission from Uranus, J. Geophys. Res., 92, 15211, 1987.

Gallagher, D.L. and D.A. Gurnett, Auroral kilometric radiation: time averaged source location, J. Geophys. Res., 84, 6501, 1979.

Gallagher, D.L. and N. D'Angelo, Correlations between solar wind parameters and auroral kilometric radiation intensity, Geophys. Res. Lett., 8, 1087, 1981.

Goldstein, M.L. and J.R. Thieman, The formation of arcs in the dynamic spectra of Jovian decametric bursts, J. Geophys. Res., 86, 8569, 1981.

Gurnett, D.A., The earth as a radio source: Terrestrial kilometric radiation, J. Geophys. Res., 79, 4227, 1974.

Gurnett, D.A., The earth as a radio source: The nonthermal continuum, J. Geophys. Res., 80, 2751, 1975.

Gurnett, D.A. and R.R. Anderson, The kilometric radio emission spectrum: relationship to auroral acceleration processes, AGU Geophysical Monograph 25, Physics of Auroral Arc Formation, ed. by S.-I. Akasofu and J.R. Kan, 1981.

Gurnett, D.A., W.S. Kurth, and F.L. Scarf, Narrowband electromagnetic emissions from Jupiter's magnetosphere, Nature, 302, 385, 1983.

Gurnett, D.A., W.S. Kurth, F.L. Scarf, and R.L. Poynter, First plasma wave observations at Uranus, Science, 233, 106, 1986.

Hashimoto, K., Reconciliation of propagation modes of auroral kilometric radiation, J. Geophys. Res., 89, 7459, 1984.

James, H.G., E.L. Hagg, and D.L.P. Strange, Narrowband radio noise in the topside ionosphere, AGARD Conf. Proc., AGARD-CP-138, 24-1, 1974.

Jones, D., Io plasma torus and the source of Jovian narrow-band kilometric radiation, Nature, 327, 492, 1987.

Jones, D. and Y. Leblanc, Source of broadband Jovian kilometric radiation, Annales Geophys., 5A, 29, 1987.

Jones, D., W. Calvert, D.A. Gurnett, and R.L. Huff, Observed beaming of terrestrial myriametric radiation, Nature, 328, 391, 1987.

Kaiser, M.L., A low frequency radio survey of the planets with RAE-2, J. Geophys. Res., 82, 1256, 1977.

Kaiser, M.L. and J.K. Alexander, Terrestrial kilometric radiation 3. Average spectral properties, J. Geophys. Res., 82, 3273, 1977.

Kaiser, M.L., J.K. Alexander, A.C. Riddle, J.B. Pearce and J.W. Warwick, Direct measurement by Voyagers 1 and 2 of the polarization of terrestrial kilometric radiation, Geophys. Res. Lett., 5, 857, 1978.

Kaiser, M.L. and M.D. Desch, Narrowband Jovian kilometric radiation: A new radio component, J. Geophys. Res., 85, 4248, 1980.

Kaiser, M.L., M.D. Desch, Voyager detection of nonthermal radio emission from Saturn, Science, 209, 1238, 1980.

Kaiser, M.L. and M.D. Desch, Saturnian kilometric radiation: Source locations, J. Geophys. Res., 87, 4555, 1982.

Kaiser, M.L. and M.D. Desch, Radio emissions from the planets Earth, Jupiter, and Saturn, Rev. Geophys. & Spa. Sci., 22, 373, 1984.

Kaiser, M.L., M.D. Desch, W.S. Kurth, A. Lecacheux, F. Genova, B.M. Pedersen, and D.R. Evans, Saturn as a radio source, in Saturn, ed. by T. Gehrels and M. Matthews, U. Arizona Press, Tucson, 1984.

Kaiser, M.L., M.D. Desch, and S.A. Curtis, The sources of Uranus' nightside radio emissions, J. Geophys. Res., 92, 15169, 1987.

Kraus, J.D., Radio Astronomy (2nd edition), Cygnus -Quasar Books, Powell, Ohio, 1986.

Kurth, W.S., D.A. Gurnett, and R.R. Anderson,

Escaping nonthermal continuum radiation, J. Geophys. Res., 86, 5519, 1981.
Kurth, W.S., J.D. Sullivan, D.A. Gurnett, F.L. Scarf, H.S. Bridge, and E.C. Sittler,Jr., Observations of Jupiter's distant magnetotail and wake, J. Geophys. Res., 87, 10373, 1983.
Kurth, W.S., Detailed observations of the source of terrestrial narrowband electromagnetic radiation, Geophys. Res. Lett., 9, 1341, 1982.
Kurth, W.S., D.A. Gurnett, and F.L. Scarf, Sporadic narrowband radio emissions from Uranus, J. Geophys. Res., 91, 11958, 1986.
Leblanc, Y., On the arc structure of the DAM Jupiter emissions, J. Geophys. Res., 86, 8546, 1981.
Leblanc, Y. and G. Daigne, The broadband Jovian kilometric radiation: new results on polarization and beaming, J. Geophys. Res., 90, 12073, 1985.
Leblanc, Y., D. Jones, and H.O. Rucker, Jovian 1.2 kHz nonthermal continuum radiation, J. Geophys. Res., 91, 9995, 1986.
Leblanc, Y., M. Aubier, A. Ortega-Molina, and A. Lecacheux, Overview of the Uranian radio emissions: Polarization and constraints on source locations, J. Geophys. Res., 92, 15125, 1987.
Lecacheux, A., B. Moller-Pedersen, A.C. Riddle, J.B. Pearce, A. Boischot and J.W. Warwick, Some spectral characteristics of the hectometric Jovian emission, J. Geophys. Res., 85, 6877, 1980.
Lecacheux, A., Ray tracing in the Io plasma torus: Application to the PRA observations during Voyager-1's closest approach, J. Geophys. Res., 86, 8523, 1981.
Lecacheux, A. and F. Genova, Source localization of Saturn kilometric radiation, J. Geophys. Res., 88, 8993, 1983.
Lecacheux, A. and A. Ortega-Molina, Polarization and localization of the Uranian radiosources, J. Geophys. Res., 92, 15148, 1987.
Mellott, M.M., W. Calvert, R.L. Huff, and D.A. Gurnett, DE-1 observations of ordinary mode and extraordinary mode auroral kilometric radiation, Geophys. Res. Lett., 11, 1188, 1984.
Ness, N.F., M.H. Acuna, R.P. Lepping, J.E.P. Connerney, K.W. Behannon, L.F. Burlaga, and F.M. Neubauer, Magnetic field studies by Voyager-1: Preliminary results at Saturn, Science, 212, 211, 1981.
Ness, N.F., M.H. Acuna, K.WE. Behannon, L.F. Burlaga, J.E.P. Connerney, R.P. Lepping, and F.M. Neubauer, Magnetic fields at Uranus, Science, 233, 85, 1986.
Oya, H. and A. Morioka, Observational evidence of Z and L-O mode waves as the origin of auroral kilometric radiation from the Jikiken (EXOS-B) satellite, J. Geophys. Res., 88, 6189, 1983.
Oya, H., A. Morioka, and T. Obara, Leaked AKR and terrestrial hectometric radiations discovered by the plasma wave and planetary plasma sounder experiments on board the Ohzora (EXOS-C) satellite, J. Geomag. Geoelectr., 37, 237, 1985.
Porco, C.C. and G.E. Danielson, The periodic variation of spokes in Saturn's rings, Astron. J., 87, 826, 1982.
Scarf, F.L., Possible traversals of Jupiter's distant magnetic tail by Voyager and by Saturn, J. Geophys. Res., 84, 4422, 1979.
Shawhan, S.D. and D.A. Gurnett, Polarization measurements of auroral kilometric radiation by Dynamics Explorer-1, Geophys. Res. Lett., 9, 913, 1982.
Thieman, J.R., A catalog of Jovian decameter radio observations from 1957-1978, Tech. Memo TM-80308, NASA, Goddard Space Flight Center, Greenbelt, MD, 1979.
Warwick, J.W., D.R. Evans, J.R. Romig, C.B. Sawyer, M.D. Desch, M.L. Kaiser, J.K. Alexander, T.D. Carr, D.H. Staelin, S. Gulkis, R.L. Poynter, M. Aubier, A. Boischot, Y. Leblanc, A. Lecacheux, B.M. Pedersen, and P. Zarka, Voyager-2 radio observations of Uranus, Science, 233, 102, 1986.
Wu, C.S. and L.C. Lee, A theory of terrestrial kilometric radiation, Astrophys. J., 230, 621, 1979.
Zarka, P. and A. Lecacheux, Beaming of Uranian kilometric radio emission and inferred source location, J. Geophys. Res., 92, 15177, 1987.

THEORIES OF NON-THERMAL RADIATIONS FROM PLANETS

L. C. Lee

Geophysical Institute and Department of Physics, University of Alaska, Fairbanks, AK 99775-0800

Abstract. The Earth, Jupiter, Saturn, and Uranus are observed to emit intense radio waves, which include auroral kilometric radiation (AKR), Jovian decametric radiation (DAM), Jovian kilometric radiation (KOM), Saturnian kilometric radiation (SKR), Uranian kilometric radiation (UKR), and terrestrial, Jovian, Saturnian, and Uranian continuum radiation. A number of theories have been proposed in the past years to explain the non-thermal radio emissions from these planets. A brief review of various radio emission mechanism is given in the paper.

The mechanisms of non-thermal emissions from planets can be divided into two classes: (A) direct emission mechanisms and (B) mode-conversion mechanisms. Each class can be subdivided into linear and nonlinear processes. In the direct emission mechanisms, the electromagnetic X-mode and O-mode waves are directly generated by plasma instabilities and escape from the source region. In the mode-conversion mechanisms, electrostatic waves are first produced and then converted into electromagnetic waves. In this review, we emphasize the cyclotron maser mechanism (a direct emission process) and the linear mode-conversion process.

The electron cyclotron maser process, which requires a population inversion in the electron velocity distribution, provides a very efficient means for the generation of radio emissions. The process appears to be the favored mechanism for generating auroral kilometric radiation and its analogues, Jovian decametric radiation, Saturnian kilometric radiation, and Uranian kilometric radiation. In particular, the cyclotron maser process can explain the X-mode, O-mode, Z-mode, and second harmonic X-mode emissions observed above the auroral ionosphere. On the other hand, the linear mode-conversion of upper-hybrid waves to O-mode waves is the favored mechanism for the generation of the non-thermal continuum radiation observed in the magnetosphere of these planets.

1. Introduction

The Earth, Jupiter, Saturn, and Uranus are observed to emit intense and coherent radio waves. These waves include the auroral kilometric radiation (AKR) emitted from the Earth [Gurnett, 1974; Kurth et al., 1975; Kaiser and Stone, 1975; Alexander and Kaiser, 1976; Benson and Calvert, 1979; Shawhan and Gurnett, 1982; Oya and Morioka, 1983; Gurnett et al., 1983; Mellott et al., 1984, 1986], the Jovian decametric radiation (DAM) and the Jovian kilometric radiation (KOM) from the Jupiter [Burke and Franklin, 1955; Bigg, 1964; Carr and Gulkis, 1969; Kurth et al., 1979, 1980; Boischot et al., 1981; Carr et al., 1983; Calvert, 1983], the Saturnian kilometric radiation (SKR) [Warwick et al., 1981; and Kaiser et al., 1984], and the Uranian kilometric radiation (UKR) [Warwick et al., 1986; Leblanc et al., 1987; Evans et al., 1987; Kaiser et al., 1987; Curtis et al., 1987; Buti and Lakhina, 1988]. In addition, the non-thermal continuum (NTC) radiations are also observed in the magnetospheres of the Earth, of the Jupiter, of the Saturn, and of the Uranus [Gurnett and Shaw, 1973; Gurnett et al., 1981, 1986; Kurth et al., 1982; Jones, 1985]. A detailed review of observations of planetary non-thermal radiation is referred to Kaiser [1988; in this volume]. The readers are also referred to several review papers by Grabbe [1981], Barbosa [1982], Lee [1983], Goldstein and Goertz [1983], and Wu [1985].

Many theories have been proposed in the past years to explain the non-thermal radio emissions from planets. The proposed mechanisms of non-thermal emission from planets can be divided into two classes: (A) direct emission mechanisms and (B) mode-conversion mechanisms. Each class can be subdivided into linear and nonlinear processes. In the direct emission mechanisms, electromagnetic (EM) waves are generated directly by plasma instabilities and escape from the source region. In the mode-conversion mechanisms, electrostatic (ES) waves are produced first and then converted into electromagnetic waves that can escape from the source region. The escapable electromagnetic waves consist of the ordinary mode (O-mode) and the fast extraordinary mode (X-mode) waves. A brief summary of various mechanisms is given in Table 1.

For both direct and indirect mechanisms, a free energy from the energetic particle distribution is needed to generate the electrostatic or electromagnetic waves. The free energy is available if the particle velocity distribution function is non-Maxwellian, e.g., a distribution with an anisotropic temperature ($T_{\parallel} \neq T_{\perp}$), with a beam, or with a loss-cone. In Section 2, we will discuss the free energy for each wave mode. Among the direct emission mechanisms, the electron cyclotron maser process proposed by Wu and Lee [1979] appears to be the favored mechanism for the generation of auroral kilometric radiation, Jovian decametric radiation, and Saturnian kilometric radiation. The mechanism is reviewed in Section 3. Recently, Borovsky [1988] suggested that the gyrophase-bunched sheet beams of electrons which emanate from double layer may also directly generate auroral kilometric radiation. In Section 4, we discuss the linear mode-conversion process proposed by Oya [1971, 1974], Benson

TABLE 1. Mechanisms for Nonthermal Emission

	Direct Emission	Indirect Mode-Conversion
Linear Process	Melrose [1976] Wu and Lee [1979] Borovsky [1988]	Oya [1971, 1974] Benson [1975] Jones [1976, 1977] Okuda et al., [1982] Ashour-Abdalla and Okuda [1984]
Nonlinear Process	Palmadesso et al., [1976] Grabbe et al., [1980]	Galeev and Krasnoselkikh [1976] Barbosa [1976] Roux and Pellat [1979] Istomin [1980] Ben-Ari [1980] Melrose [1981] Goldstein et al., [1983] Buti and Lakhina [1985] Chian and Alves [1988]

[1975], Jones [1976], and Okuda et al. [1982]. In section 5, we discuss the nonlinear direct emission mechanisms and the nonlinear mode-conversion processes.

2. Free Energy for ES and EM Waves

For the direct process, a free energy source for the electromagnetic O-mode and X-mode waves is needed. For the indirect process, the free energy is needed for generating non-escapable Langmuir waves, upper-hybrid (UH) waves, slow-extraordinary mode (Z-mode) waves, or whistler (W-mode) waves. The growth rate of the generated wave γ_n from contribution through the n-th harmonic resonance in the magnetized plasma can be written approximately as,

$$\gamma_n \simeq \text{constant} \int G\delta(\omega - \frac{n\Omega_e}{\gamma} - k_{\parallel}v_{\parallel})dv \qquad (1)$$

where

$$G \equiv (\frac{\omega}{k_{\parallel}}\frac{\partial f}{\partial v_{\perp}}) - (v_{\parallel}\frac{\partial f}{\partial v_{\perp}}) + (v_{\perp}\frac{\partial f}{\partial v_{\parallel}}) \equiv G_1 + G_2 + G_3 \qquad (2)$$

In (1) and (2), ω is the wave frequency, $k_{\parallel}$ is the component of the wave vector **k** parallel to B_0, Ω_e is the electron cyclotron frequency, $\gamma = (1 - v^2/c^2)^{-1/2}$, v is the particle velocity, and f is the particle distribution function. In (2), we let G_1, G_2, and G_3 denote, respectively, the three terms of G.

2.1. Free Energy for the O-Mode and X-Mode Waves

For the O-mode and X-modes, the wave index of refraction $N < 1$. Hence, $\omega/k_{\parallel} > \omega/k = c/N > c \gg v_{th}$, where v_{th} is the thermal speed for the weakly relativistic electrons with energy $\leq$ 50kev. It is noted that the X-mode and O-mode waves cannot directly interact with electrons through the Landau resonance ($n = 0$).

2.1.1. Bi-Maxwellian distribution. For a bi-Maxwellian distribution,

$$f \sim \exp(-\frac{v_{\perp}^2}{\alpha_{\perp}^2} - \frac{v_{\parallel}^2}{\alpha_{\parallel}^2}) \qquad (3)$$

where $\alpha_{\perp}$ and $\alpha_{\parallel}$ are, respectively, the perpendicular and parallel thermal speeds. Since $(\omega/k_{\parallel} - v_{\parallel}) > 0$ and $\partial f/\partial v_{\perp} < 0$ for $v_{\perp} \geq 0$, we have

$$G_1 + G_2 = (\frac{\omega}{k_{\parallel}} - v_{\parallel})\frac{\partial f}{\partial v_{\perp}} < 0. \qquad (4)$$

Since $\partial f/\partial v_{\parallel} < 0$ for $v_{\parallel} > 0$, we have

$$G_3 = v_{\perp}\frac{\partial f}{\partial v_{\parallel}} < 0. \qquad (5)$$

and

$$G = G_1 + G_2 + G_3 < 0. \qquad (6)$$

Therefore, the growth rate is always negative. This indicates that the O-mode and X-mode waves cannot be generated by electrons with a bi-Maxwellian distribution.

2.1.2. Distribution with an electron beam. For electrons with a beam distribution, we may have a region with $\partial f/\partial v_{\parallel} > 0$. Thus we have

$$G_3 = v_{\perp}\frac{\partial f}{\partial v_{\parallel}} > 0 \qquad (7)$$

in the region associated with the beam. However, for the first harmonic wave,

$$G_1 + G_2 = (\frac{\omega}{k_{\parallel}} - v_{\parallel})\frac{\partial}{\partial v_{\perp}} \sim c\frac{\partial}{\partial v_{\perp}} < 0. \qquad (8)$$

In order to obtain a positive growth rate ($\gamma_n > 0$), G_3 must be greater than $|G_1 + G_2|$ and hence a very large temperature anisotropy is needed. Melrose [1976] found that a temperature anisotropy with $T_{\perp}/T_{\parallel} > 50$ is necessary for the electron beam to generate AKR or DAM. However, such a large value of $T_{\perp}/T_{\parallel}$ is not observed along the auroral field lines. Therefore, it is unlikely that the O-mode or X-mode emission can be generated directly by the electron beams.

2.1.3. Loss-cone distribution (population inversion). Since $\omega/k_{\|} > c \gg v_{th}$ for the X-mode and O-mode waves, the term G_1 in (2) dominates. Therefore, Eq. (2) can be written as

$$G \simeq G_1 \simeq c\frac{\partial f}{\partial v_{\perp}} > 0 \tag{9}$$

for a distribution with a population inversion ($\partial f/\partial v_{\perp} > 0$). Hence the growth rate can be greater than zero. Therefore, electrons with a loss-cone distribution may directly generate the O-mode and X-mode waves. It will be illustrated in Section 3 that a relativistic resonance condition is also needed in order to generate the electromagnetic waves. The population inversion and the relativistic resonance condition constitute of the core of the electron cyclotron maser mechanism proposed by Wu and Lee [1979].

2.2. Free Energy for Langmuir, Upper-Hybrid, Whistler, and Z-Mode Waves

For the Langmuir, upper-hybrid, Z-mode, or the whistler waves, we have $\omega/k_{\|} \ll c$.

2.2.1. Bi-Maxwellian distribution ($T_{\perp} \neq T_{\|}$). For the whistler mode, we have [e.g., Kennel and Petschek, 1966; Cuperman, 1981]

$$G \sim [-\omega + (1 - \frac{T_{\perp}}{T_{\|}})(\omega - \Omega_e)]f \tag{10}$$

Hence

$$G > 0, \qquad \text{if} \quad \frac{T_{\perp}}{T_{\|}} - 1 > \frac{\omega/\Omega_e}{1 - \omega/\Omega_e}. \tag{11}$$

Therefore, the bi-Maxwellian distribution can lead to the generation of electron whistler waves.

2.2.2. Beam distribution. The electron Langmuir or the upper-hybrid waves can be generated by the electron beam through Landau resonance ($n = 0$). For $n = 0$, $\omega = k_{\|}v_{\|}$, and hence

$$\gamma_{n=0} \sim G = G_3 = v_{\perp}\frac{\partial f}{\partial v_{\|}} > 0. \tag{12}$$

The presence of the upper-hybrid waves is essential for the generation of the non-thermal continuum radiation through mode-conversion [Gurnett and Frank, 1976; Jones, 1976; Kurth et al., 1981].

2.2.3. Loss-cone distribution. The loss-cone distribution can also lead to the generation of Z-mode wave and W-mode wave [Lee and Wu, 1980; Wu et al., 1982; Omidi et al., 1984; Omidi and Wu, 1985; Kennel et al., 1987]. For the generation of the whistler waves, the loss-cone distribution can be shown to be equivalent to a temperature anisotropy.

In summary, the Langmuir waves, the upper-hybrid waves, the Z-mode waves, and the whistler waves can be generated from various sources of free energy. However, the EM waves (O-mode, X-mode) can only be generated by a population inversion, e.g., a loss-cone distribution.

3. The Cyclotron Maser Mechanism

Auroral kilometric radiations (AKR) are intense radio emissions with frequency $\sim 100 - 600$kHz observed above the auroral ionosphere during geomagnetic substorms [Gurnett, 1974; Kurth et al., 1975; Kaiser and Stone, 1975; Alexander and Kaiser, 1976; Benson and Calvert, 1979; Shawhan and Gurnett, 1982; Oya and Morioka, 1983; Gurnett et al., 1983; Mellott et al., 1984, 1986]. The observed characteristics of AKR are: (1) the radiation is generated mostly in the X-mode, although the O-mode and Z-mode radiations are also observed, (2) the generation of AKR is closely related with the precipitating inverted-V electrons with energy greater than 1 keV, (3) the radiation occurs within a range of local electron density depletion where the ratio of the electron plasma frequency (ω_{pe}) to the electron cyclotron frequency (Ω_e) is less than 0.3, and (4) the source region is located along the auroral field lines at a geocentric distance between 1.5 and $3R_E$.

Inspired by observations of auroral kilometric radiation, Wu and Lee [1979] proposed the cyclotron maser mechanism to explain the generation of AKR. They demonstrated that a direct amplification of the X-mode and O-mode radiation is possible when a population of superthermal electrons possess a loss-cone distribution function (also see the discussions in Section 2). In the past years, various aspects of the cyclotron maser process operating along the auroral field lines have been further discussed by Lee and Wu [1980], Lee et al. [1980], Wu et al. [1982], Omidi and Gurnett [1982], Melrose et al. [1982], Calvert [1982], Dusenbery and Lyons [1982], Hewitt et al. [1982], Wong et al. [1982], Le Quéau et al. [1984], Omidi et al. [1984], Pritchett [1984, 1986], Pritchett and Strangeway [1985], and Benson and Wong [1987]. In addition, Curtis [1986] applied the cyclotron maser mechanism to the Saturnian kilometric radiation.

A comprehensive review of the kinetic cyclotron maser process has been given by Wu [1985]. In the following, we present a brief review of the theoretical as well as simulation study of the cyclotron maser mechanism. In the cyclotron maser theory of auroral kilometric radiation proposed by Wu and Lee [1979], the radiation is emitted by energetic auroral electrons (1–10 keV) with loss-cone velocity distribution. The AKR source region is further found to be in the region of low electron density, in which the low-energy electrons are depleted due to the presence of the parallel electric field.

3.1. Growth Rate

For simplicity we assume that the AKR source region consists of two electron components: a cold background electrons and a population of energetic electrons with a loss-cone distribution. Furthermore, the density of the cold background electrons is assumed to be higher than that of the energetic electrons such that the dispersion relation of the radiation is dictated by the cold background electrons. The growth rate ω_i of the extraordinary mode (X-mode) radiation can then be written approximately as [Wu and Lee, 1979].

$$\omega_i \simeq \frac{\pi^2}{4}\omega_{pe}^2\frac{n_e}{n_0}\int d^3u u_{\perp}\frac{\partial F_e}{\partial u_{\perp}}\delta(\omega_r - \frac{\Omega_e}{\gamma} - k_{\|}u_{\|}/\gamma) \tag{13}$$

where $\mathbf{u} = \gamma\mathbf{v}$ is the relativistic velocity of electrons, ω_{pe} is the electron plasma frequency, n_0 is the background electron

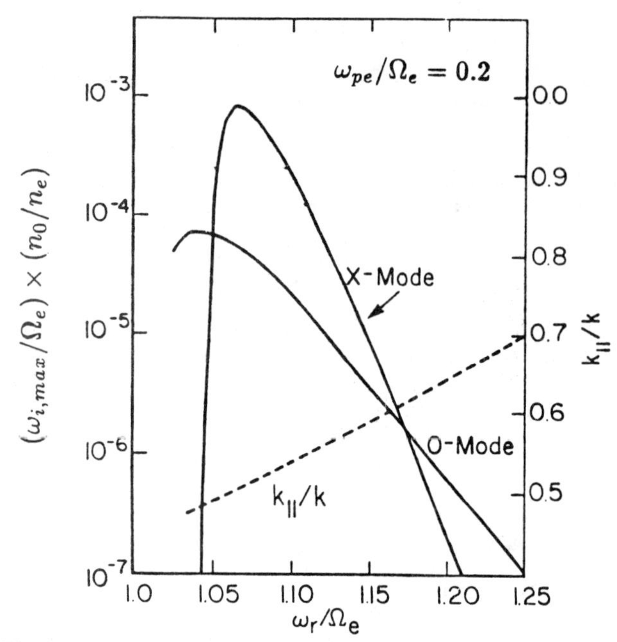

Fig. 1. The maximum growth rate $\omega_{i,max}$ and the corresponding value of $k_{\parallel}/k$ as functions of ω_r/Ω_e [Lee et al., 1980].

number density, n_e is the energetic electron density, and F_e is the velocity distribution of the energetic electrons. Note that the relativistic factor can also be written as $\gamma = (1 + u^2/c^2)^{1/2}$.

3.2. Relativistic Resonance Condition

The relativistic cyclotron resonance condition can be obtained from the delta function in Eq. (13) as

$$(1 + \frac{u^2}{c^2})^{1/2}\omega_r - \Omega_e - N\cos\theta\omega_r\frac{u_{\parallel}}{c} = 0 \qquad (14)$$

where $N = kc/\omega_r$ is the refractive index, ω_r is the real part of frequency ω, and $\cos\theta = k_{\parallel}/k$. The resonance condition is an ellipse in the $(u_{\parallel} - u_{\perp})$ space of the electron relativistic velocity if $N\cos\theta < 1$. This ellipse may be expressed as [Wu, 1985]

$$\frac{u_{\perp}^2}{A^2} + \frac{(u_{\parallel} - u_0)^2}{B^2} = 1 \qquad (15)$$

where

$$\frac{u_0}{c} = \frac{\Omega_e}{\omega_r}\frac{N\cos\theta}{1 - N^2\cos^2\theta},$$

$$\frac{A^2}{c^2} = \frac{\Omega_e^2}{\omega_r^2}\frac{1}{1 - N^2\cos^2\theta} - 1, \qquad (16)$$

and

$$\frac{B^2}{c^2} = \frac{A^2}{c^2}\frac{1}{1 - N^2\cos^2\theta}.$$

For comparison, the ordinary nonrelativistic resonance condition $\omega_r - \Omega_e = k_{\parallel}v_{\parallel}$ is a straight line parallel to the $v_{\perp}$-axis. It turns out that the relativistic effect in the resonance condition plays a crucial role in the generation of AKR. Without the relativistic effect, the electromagnetic wave will be damped, instead of being amplified, even if the energetic electrons have a loss-cone distribution.

The above discussion explains why in the AKR theory by Wu and Lee [1979], even for electrons with energies of several keV, the relativistic effect is essential in the generation of radio emission.

Figure 1 shows the maximum growth rate $\omega_{i,max}$ and the corresponding value of $k_{\parallel}/k$ as functions of the normalized frequency ω_r/Ω_e. The frequency ratio ω_{pe}/Ω_e is chosen to be 0.2. The energetic electrons are assumed to be 6 keV and with a loss-cone distribution. The peak growth rate of the fast extraordinary mode (X-mode) is much higher than that of the ordinary mode (O-mode). Note that the growth rate of the X-mode radiation peaks at frequency just above the electron cyclotron frequency Ω_e.

Figure 2 shows the growth rate as a function of ω_{pe}/Ω_e for the same 6 keV electrons. It is seen that the growth rate peaks at $\omega_{pe}/\Omega_e \simeq 0.2$ and becomes extremely small for $\omega_{pe}/\Omega_e > 0.3$. The above results are consistent with the fact that strong AKR is mainly in the X-mode and is observed only for $\omega_{pe}/\Omega_e \leq 0.2$ [Benson and Calvert, 1979]. The plasma region with $\omega_{pe}/\Omega_e \leq 0.2$ corresponds to the region with depleted electron density, which is caused by the field-aligned potential drop associated with the inverted-V acceleration region [e.g., Lee et al., 1980]. It is interesting to note that for $\omega_{pe}/\Omega_e > 0.25$, the O-mode has a higher growth rate than the X-mode.

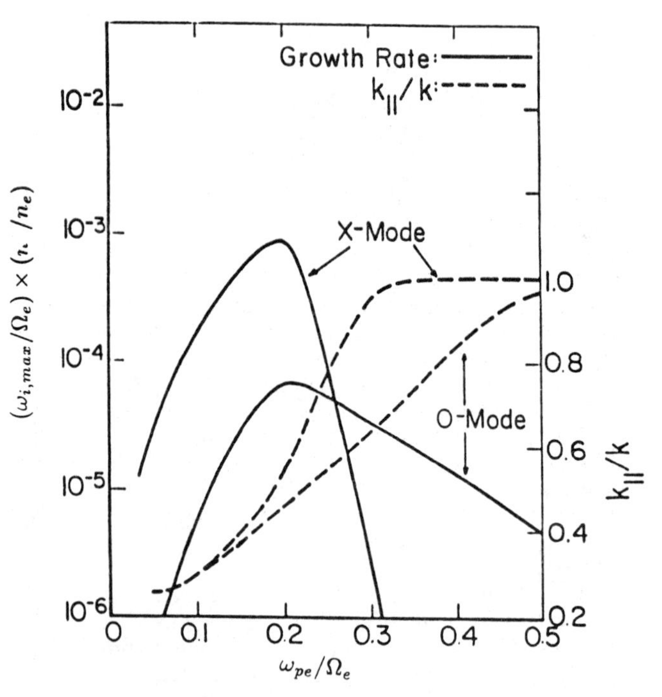

Fig. 2. The maximum growth rate $\omega_{i,max}$ as a function of the frequency ratio ω_{pe}/Ω_e [Lee et al., 1980].

3.3. Second Harmonic Radiation

The cyclotron maser mechanism can also generate waves at the higher harmonics [Lee et al., 1980; Wu and Qiu, 1983; Melrose et al., 1984; Mellott et al., 1986; Benson and Wong, 1987]. The above studies predicted the presence of harmonic radiation, but with small growth rates in the usual AKR source region where $\omega_{pe}/\Omega_e \leq 0.2$. However, the second harmonic radiation can have a high growth rate in the region of relatively high plasma density ($\omega_{pe}/\Omega_e \geq 0.3$). It is found that the second harmonic X-mode radiation is also more important than the second harmonic O-mode radiation.

3.4. Computer Simulation

Computer simulations of the cyclotron maser process have been carried out by Wagner et al. [1983, 1984], Pritchett and Strangeway [1985], and Pritchett [1986]. The results obtained by the above authors will be briefly summarized in this section.

In Wagner et al., [1983, 1984], a relativistic electromagnetic particle simulation code with one dimension in space and three dimensions in velocity is used to study the generation of auroral kilometric radiation. In their simulation the relativistic particle velocity $\mathbf{u}$ and the relativistic momentum $\mathbf{p}$ are used, which are related to the particle velocity $\mathbf{v}$ by $\mathbf{u} = \mathbf{p}/m_e = \gamma\mathbf{v}$. Two electron populations are used in the simulation of auroral kilometric radiation: (1) the hot (5-20 keV) electrons with a loss cone distribution, which corresponds to the upgoing energetic electrons observed along the auroral field lines during substorms and (2) the cold (20-500 eV) electrons with a Maxwellian distribution, which corresponds to the cold electrons of ionospheric origin.

The simulation results for the case with a frequency ratio $\omega_{pe}/\Omega_e = 0.2$ are shown in Figures 3-4. In Figure 3, the hot electrons initially possess a double loss-cone velocity distribution, with a loss-cone angle $\alpha = 45^o$. The wave propagation angle θ is 45° with respect to the magnetic field.

The initial particle distribution is shown in Figure 3*a*, in which the particle parallel momentum $p_\parallel$ is plotted against perpendicular momentum $p_\perp$. The double loss-cone feature in the hot species is evident. The cold Maxwellian component is the dense group of particles near the origin. Figure 3*b* shows the particle distribution in $p_\parallel - p_\perp$ space at the end of the run, $t = 250\omega_{pe}^{-1}$. Figure 3*b* shows that some hot electrons have diffused into the loss-cone, which is due to the turbulent scattering by the amplified X-mode radiation. Wave scattering of particles is very efficient for those particles in cyclotron resonance with the amplified waves. The resonance condition is given in Eq. (14), which is an ellipse in the $p_\parallel - p_\perp$ space. In Figure 3*b* the three most important modes and their corresponding resonant ellipse are plotted. These modes are ($m = 23$, $\omega = 5.5\omega_{pe}$), ($m = 24$, $\omega = 5.65\omega_{pe}$), and ($m = 25$, $\omega = 5.8\omega_{pe}$). The wave number k is related to the mode number m by $k = 2\pi m/L$, where L is the simulated length.

Figure 4 shows the relative power of the X-, O-, and Z-modes as a function of wave frequency. A large narrow peak of the X-mode occurs at mode 24. The X-mode peak has 5 times more power than the most significant O-mode peak at $m = 17$. The Z-mode was observed at very low power levels, with 45 times less power than the X-mode peak. The efficiency of energy conversion from the energetic particles to

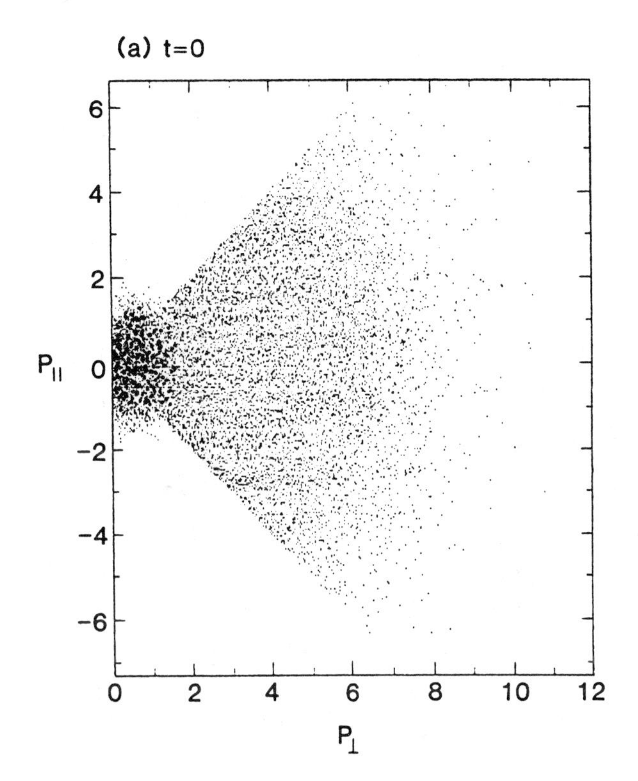

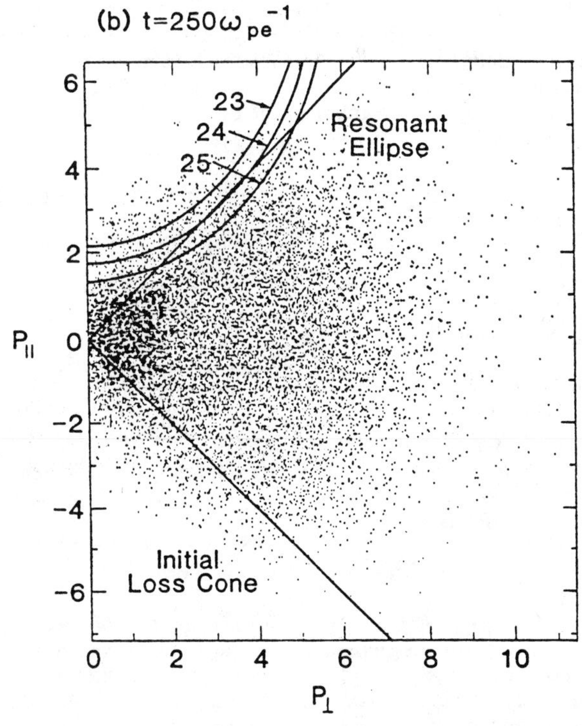

Fig. 3. (*a*) Initial distribution of the hot and cold electrons in the $p_\parallel$-$p_\perp$ space. (*b*) The electron distribution at $t = 250\omega_{pe}^{-1}$. The three resonant ellipses are also plotted [Wagner et al., 1984].

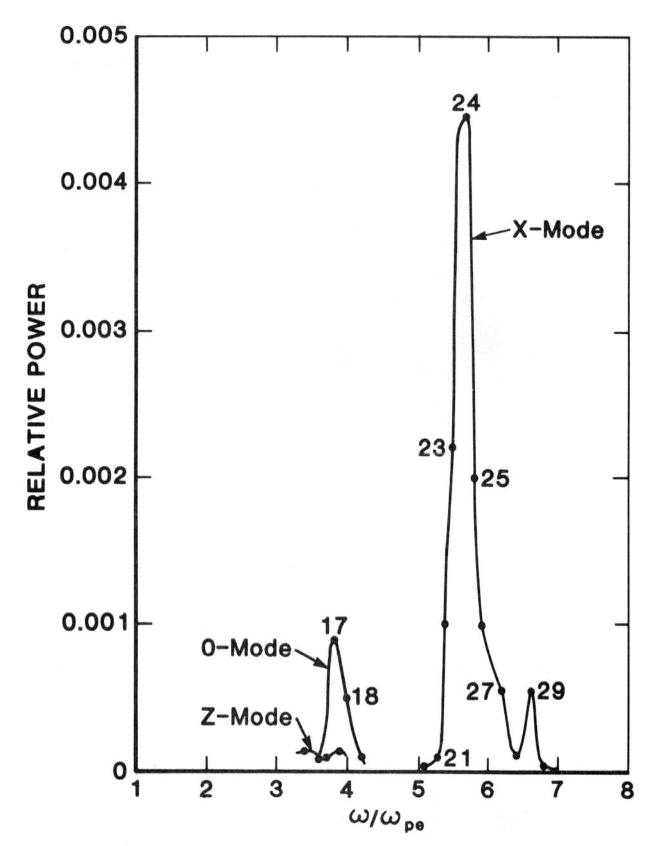

Fig. 4. Relative powers of the X-, O- and Z-mode radiations as a function of wave frequency ω. The numbers near the dots indicate the mode numbers of the radiation [Wagner et al., 1984].

radiation is about 1%, with most of the power in the X-mode radiation branch.

Pritchett and Strangeway [1985] carried out a two-dimensional simulation of the cyclotron maser process for the auroral kilometric radiation. They found that the peak in the radiation is very close to 90° to the ambient magnetic field, with an offset θ_γ a few degrees in the direction of the loss cone. The transverse electric energy density $|E_T(k_x, k_y)|^2$ as a function of the wave numbers k_x and k_y for a two-dimensional simulation is shown in Figure 5. The efficiency of energy conversion can be higher in the two-dimensional simulation [Pritchett, 1986].

4. Linear Mode-Conversion

As discussed in Section 2, various sources of free energy can lead to the generation of Langmuir, upper-hybrid, Z-mode, or whistler modes. These waves cannot propagate freely in space. However, as they propagate through an inhomogeneous medium, electromagnetic waves can be produced through the linear mode-conversion process. The mode-conversion process has been discussed by Stix [1965], Oya [1971, 1974], Jones [1976, 1977], Benson [1975], Okuda et al., [1982], and Ashour-Abdalla and Okuda [1984]. Oya [1971, 1974] was the first to apply the mode-conversion process to explain the planetary radio emission.

The methods for the study of mode-conversion process can be divided into two categories: (a) mode-conversion at a sharp boundary, (b) mode-conversion through a slowly varying medium.

4.1. Mode-Conversion Through a Sharp Boundary

As the Z-mode wave impinges on a sharp boundary, the O-mode and X-mode waves may be generated either as transmitted waves or as reflected waves [Oya, 1974]. This process is similar to the wave reflection and transmission through a shock, a tangential discontinuity, or a rotational discontinuity [McKenzie and Westphal, 1970; Lee, 1982; Kwok and Lee, 1984]. The process is illustrated in Fig. 6, which is obtained from Oya [1974]. As shown in the figure, for an incident slow-extraordinary mode (Z-mode), two transmitted waves and two reflected waves are produced. The wave amplitude of each reflected or transmitted wave is determined by matching the boundary conditions required by Maxwell equations. Oya [1974] found that it is possible to obtain a 1% – 50% efficient for the conversion of ES mode to EM mode at the sharp boundary. However, the sharp boundary required for an efficient mode-conversion may not exist in the planetary magnetospheres. Fig. 7 shows schematically the generation of Jovian radio emission through the linear mode-conversion process. First, on path I, the longitudinal plasma waves are converted into extraordinary upper-hybrid mode, which propagates toward the Jovian plasmasphere on path II. The incidence of this upper-hybrid mode onto the Jovian plasmasphere generates mainly a reflected O-mode which then escapes into the interplanetary space on path III. The transmitted extraordinary mode waves taking path IV and propagating toward the planetary surface are, however, not able to escape into interplanetary space. Therefore, the mode-conversion would lead to a dominant O-mode radiation. However, the observed AKR or DAM are mostly in the X-mode radiation. Nevertheless, the linear mode-conversion process may contribute to the observed O-mode component of AKR or DAM.

4.2. Mode-Conversion Through a slowly varying medium

The terrestrial nonthermal continuum radiation observed in the Earth's magnetosphere consists of an escaping component with frequency $\sim 30-100$kHz and a trapped component with frequency $5-20$kHz [Brown, 1973; Gurnett and Shaw, 1973]. Gurnett and Frank [1976] suggested that the continuum radiation is generated from the electrostatic instabilities near the upper hybrid resonance frequency. Jones [1976, 1985] also suggested that the mode-conversion of upper-hybrid waves to the O-mode radio waves leads to the generation of continuum radiation. Kurth et al. [1981] provided a strong observational evidence for this scenario.

Okuda et al. [1982] calculated the conversion efficiency of upper-hybrid (UH) waves to the O-mode radio waves by using cold plasma theory in a one-dimensional inhomogeneous plasma. Figure 8 is a sketch of their model in which the plasma density $n(x)$ varies linearly with x and the magnetic field B_0 is uniform and in the z direction. In the figure, $Q_x(x) = (\omega^2/c^2)[1 + \omega_{pe}^2(\omega_{pe}^2 - \omega^2)/\omega^2(\omega^2 - \omega_{uh}^2)]$, and $Q_0(x) = (\omega^2/c^2)(1 - \omega_{pe}^2/\omega^2)$. x_R and x_L are locations

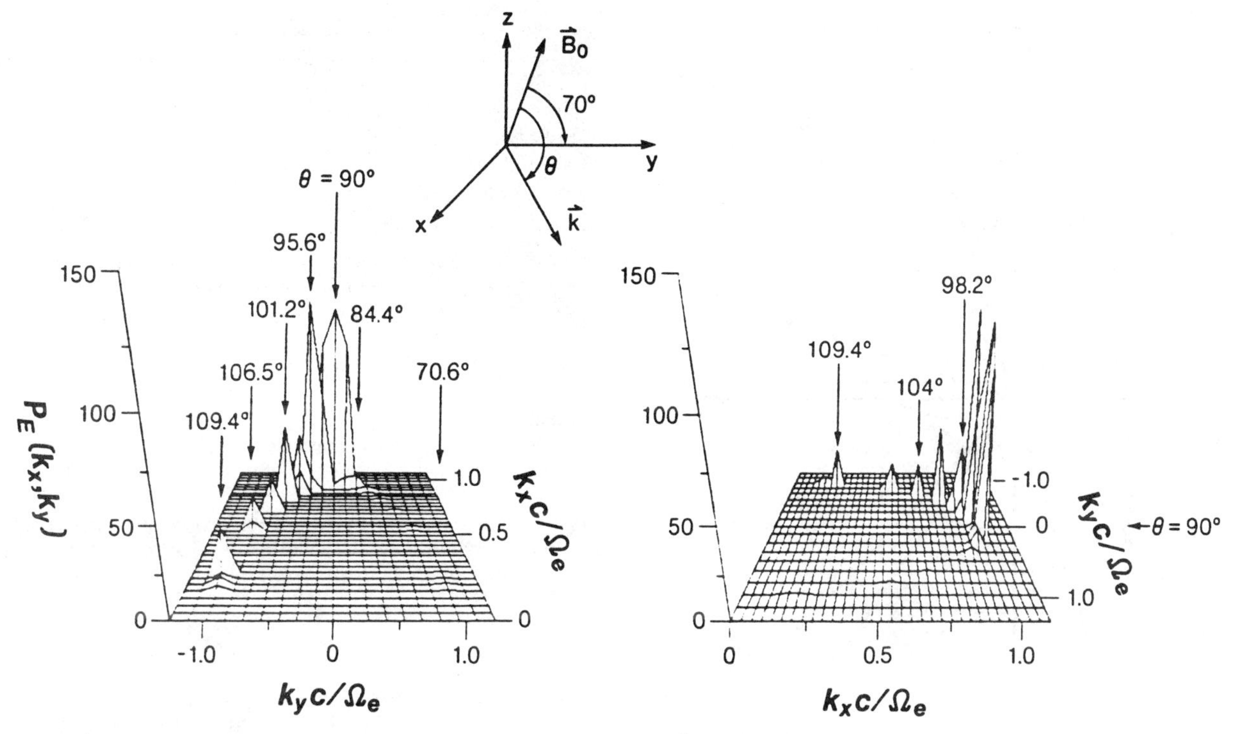

Fig. 5. The transverse electric energy density $|E_T(k_x, k_y)|^2$ for a two-dimensional simulation with 1.5 R_E auroral parameters [Pritchett and Strangeway, 1985].

where the right- and left-hand cutoff occur, respectively, for the extraordinary mode propagation. In the absence of collisions, $Q_x(x)$ goes to infinity at $x = x_h$. The propagating solutions of the corresponding plasma wave equations exist only for $x < x_R$ and $x_h < x < x_L$ while evanescent elsewhere.

By assuming a nearly perpendicular propagating condition and a source of current sheet at $x = x_s$ modeling the plasma instabilities which generate the upper-hybrid mode waves, they were able to solve the one-dimensional linear equations of cold plasma. The O-mode wave that is generated by the electrostatic fluctuations at the upper hybrid layer can propagate toward the low-density region, where the original extraordinary mode cannot reach. They obtained a ratio of the amplitude of the electrostatic field at the resonance to that of the O-mode electromagnetic radiation just beyond the source in the range of 10^2 to 10^3. The polarization of the electromagnetic wave which can escape from the source region is found to be predominantly O-mode as observed.

Recently, Budden and Jones [1987] studied the conversion of UH waves to O-mode and X-mode waves at the Earth's plasmapause and magnetopause. They found that ray theory predicts only O-mode production, whereas full wave theory predicts that both O-mode and X-mode waves are produced, their relative intensities depending on the plasma parameters.

5. Nonlinear Processes

Nonlinear processes consist of two types: (a) direct conversion of particle energy into the electromagnetic waves and (b) indirect mode conversion process. In the indirect process, the non-radiative waves are generated first by the free energy associated with particle distribution and the electromagnetic waves are generated through the nonlinear mode conversion process.

5.1. Direct Nonlinear Process

The free electron laser (FEL) can be considered as a direct nonlinear process, in which the electron beam energy is directly converted into the EM wave energy in the presence of a magnetic wiggler. Palmadesso et al., [1976] proposed that in the presence of a plasma density wiggler caused by electrostatic ion cyclotron (EIC) waves, electron beam energy can be fed into the electromagnetic waves. Both the X-mode and O-mode waves can be generated through this process. However, only the O-mode waves can propagate to the free space. These O-mode waves may explain the observed weak component of O-mode waves associated with AKR. Grabbe et al., [1980] modified the mechanism and found that only X-mode can be generated by the auroral electron beam. They also found that $\omega_{pe} < 0.2\Omega_e$ is required for this direct nonlinear process to operate along the auroral field lines. This condition is consistent with the Isis-1 observations [Benson and Calvert, 1979].

The direct nonlinear process is relatively an efficient process since electron beam energy can be directly converted into the electromagnetic wave energy. However, the existence of a coherent low-frequency wave to produce the plasma density wiggler is required for this mechanism to operate.

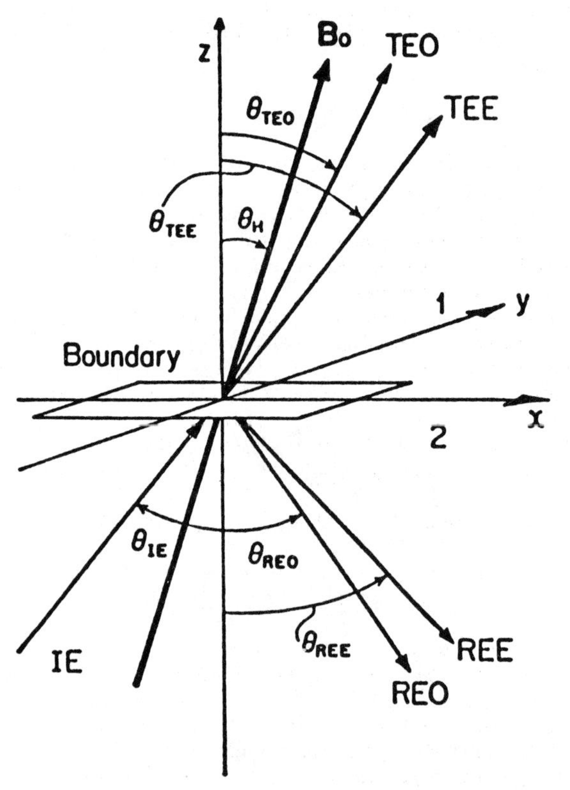

Fig. 6. Sharp boundary between the media 1 and 2 with Cartesian coordinate (x, y, z) whose z-axis coincides with the normal direction of the boundary. The extraordinary mode incident wave IE is split into the reflected ordinary mode wave REO, the reflected extraordinary mode wave REE, the transmitted ordinary mode wave TEO and the transmitted extraordinary mode wave TEE. B_0 is the magnetic field [Oya, 1974].

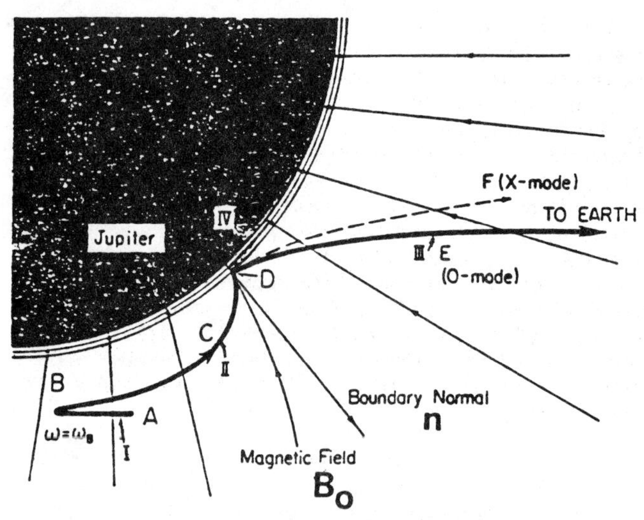

Fig. 7. Schematic illustration of the propagation paths during the plasma wave conversion into the electromagnetic waves in the Jovian plasmasphere with respect to the magnetic field B_0 [Oya, 1974].

5.2. Indirect Nonlinear Conversion

The indirect nonlinear process consists of two types (1) soliton radiation, and (2) nonlinear wave-wave interaction.

5.2.1. Soliton radiation: Langmuir soliton. Galeev and Krasnoselkikh [1976] suggested that the Langmuir soliton may lead to the emission of electromagnetic waves. In their model, a beam of auroral electrons can generate Langmuir turbulence in the magnetosphere. The ponderomotive force associated with intense Langmuir waves can then generate modulational instability, which produces density cavities (cavitons) that trap the waves. The resulting Langmuir solitons undergo collapse and radiate electromagnetic waves at second harmonic of electron plasma frequency ($\omega \simeq 2\omega_{pe}$).

5.2.2. Soliton radiation: electrostatic electron cyclotron soliton. Istomin et al. [1978] suggested that energetic beam of auroral electrons can excite electrostatic electron cyclotron waves at frequencies near harmonics of electron cyclotron frequency. These unstable waves can grow nonlinearly to form three-dimensional solitons. In contrast to ordinary wave packets that spread out in space due to the dispersive effect, these solitons are very effective in accumulating large amount of energy in a localized region of space through the formation of magnetic cavity. The electric field in these solitons can reach levels high enough to generate radio emission at the second harmonics $\omega \simeq 2\Omega_e$. This mechanism can be relevant to the generation of the second harmonic AKR.

5.2.3. Radiation associated with whistler soliton. Buti and Lakhina [1985] suggested that the nonlinear interaction between whistler solitons and the upper hybrid waves may lead to the generation of electromagnetic radiation.

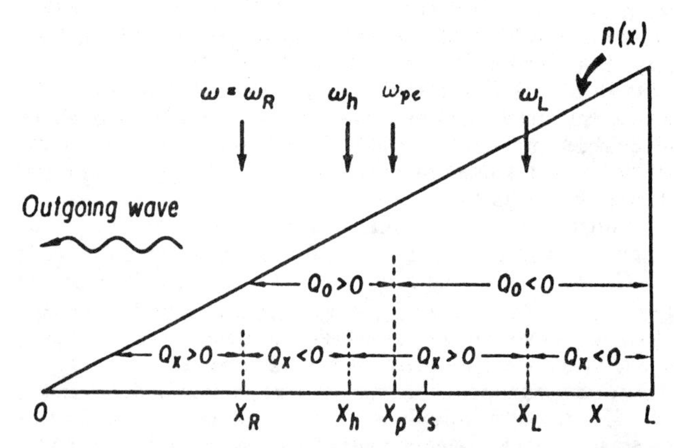

Fig. 8. A sketch of the mode-conversion model. Plasma density varies linearly with x from $x = 0$ to $x = L$ while the magnetic field B_0 is uniform, satisfying $\overline{\omega}_{pe}^2 \gg \Omega_e^2$, where $\overline{\omega}_{pe}$ is the average plasma frequency. A current source $J_y(x) = J_s\delta(x - x_s)$ is located between x_h and x_L to generate the X-mode. Boundary conditions for electromagnetic radiations are such that they are evanescent for $x \gg x_L$ and outgoing for $x \ll x_R$ [Okuda et al., 1982].

5.2.4. Wave-wave interaction. Most nonlinear wave-wave interaction models that generate radio emission at the fundamental frequency of the pump wave fall into the category of electromagnetic decay or fusion instability in which a high-frequency electrostatic pump wave, generated by a beam-plasma instability, interacts with another low-frequency electrostatic wave, leading to radio emissions near the fundamental pump frequency. Alternatively, two electrostatic waves of the same kind can interact with each other to emit radiation at the second harmonic of the pump waves.

For example, Barbosa [1976] studied the generation of AKR by nonlinear coupling of two electrostatic waves having $\omega \simeq \omega_{uh}$ (upper hybrid frequency). In this study, the electrostatic waves are assumed to be incoherent and the random phase approximation is used. Symbolically, this process can be denoted by

$$\omega_{uh} + \omega_{uh} \rightarrow \text{O-mode or X-mode}, \quad \text{at } \omega \simeq 2\omega_{uh}. \quad (17)$$

However, the mode-conversion efficiency in the incoherent case is low.

Roux and Pellat [1979] considered the coherent three-wave process involving waves having $\omega \simeq \omega_{uh}$ or ω_{lh}, where ω_{lh} is the lower hybrid frequency. Four possible combinations may occur:

$$\omega_{uh} + \omega_{lh} \rightarrow \text{O-mode}, \quad \text{at } \omega \simeq \omega_{uh}, \quad (18)$$

$$\omega_{uh} + \omega_{lh} \rightarrow \text{X-mode}, \quad \text{at } \omega \simeq \omega_{uh}, \quad (19)$$

$$\omega_{uh} + \omega_{uh} \rightarrow \text{O-mode}, \quad \text{at } \omega \simeq 2\omega_{uh}, \quad (20)$$

$$\omega_{uh} + \omega_{uh} \rightarrow \text{X-mode}, \quad \text{at } \omega \simeq 2\omega_{uh}, \quad (21)$$

Roux and Pellat [1979] suggested that the most intense radiation for the auroral kilometric radiation should be the X-mode at $\omega \simeq 2\omega_{uh}$. However, observations of AKR indicate that the fundamental X-mode is the dominant radiation. This coherent three-wave process may be responsible for the second harmonic AKR. Goldstein et al. [1983] examined in detail the process in (19) and suggested that this process may be responsible for the Jupiter decametric radiation.

Melrose [1981] suggested that the coalescence of the upper hybrid waves and the low-frequency ion-cyclotron waves may lead to the generation of nonthermal radio continuum in the terrestrial and Jovian magnetospheres. This process may be denoted by

$$\omega_{uh} + \Omega_i \rightarrow \text{O-mode, at } \omega \simeq \omega_{uh} \quad (22)$$

However, Barbosa [1982] pointed out that there is no published evidence for the simultaneous presence of both the upper-hybrid and ion-cyclotron waves in the source regions.

Three-wave process has the disadvantage of being convective. Thus the waves excited can propagate out of the unstable region before reaching sufficiently large amplitude. This difficulty is removed by the electromagnetic oscillating two-stream instability (EOTSI), which is a four wave process in which two counter-streaming electrostatic pump waves excite two EM waves by means of localized density perturbations. EOTSI is an absolute instability, and hence the wave can grow indefinitely in the localized region until some nonlinear effects saturate the instability. An example of this mechanism using Langmuir wave as the pumps has been applied to explain radio emission upstream of planetary bow shocks [Chian, 1987] and type III radio bursts [Chian and Alves, 1988].

6. Summary

The various theories of the nonthermal radiation from planets are briefly reviewed. It is demonstrated that the only free-energy available for the direct linear emission process is the electron loss-cone distribution. The electron energy is directly converted to the X-mode or O-mode through the cyclotron maser process. On the other hand, the Langmuir waves, upper-hybrid waves, Z-mode waves, or whistler waves can be linearly excited by various energy sources, which includes electron beam, temperature anisotropy, and loss-cone distribution. However, these waves have to be converted into the X-mode or O-mode electromagnetic waves through the linear or nonlinear mode-conversion process.

Finally, it is pointed out that the nonlinear direct process proposed by Palmadesso et al., [1976] and Grabbe et al., [1980] is of particular interest. This process is similar to the mechanism for free electron laser and should be further examined for cases with the presence of various other wave modes.

Acknowledgments. This work is supported by Department of Energy grant DE-FG06-86ER13530 and National Science Foundation grant ATM85-21115 to the University of Alaska. The author would like to thank Professor A. C.-L. Chian at INPE, Brazil for the helpful discussion on the nonlinear emission processes. The author also likes to thank D. Q. Ding, Y. Shi, Z. F. Fu, and A. L. La Belle-Hamer for their help in preparing this paper.

References

Alexander, J. K. and M. L. Kaiser, Terrestrial kilometric radiation: 1. Spatial structure studies, *J. Geophys. Res., 81*, 5948, 1976.

Ashour-Abdalla, M. and H. Okuda, Generation of ordinary mode electromagnetic radiation near the upper hybrid frequence in the magnetosphere, *J. Geophys. Res., 89*, 9125, 1984.

Barbosa, D. D., Electrostatic mode coupling at $2\omega_{uh}$: A generation mechanism for AKR, Ph.D thesis, Univ. of Calif ., Los Angeles, 1976.

Barbosa, D. D., Low-level VLF and LF radio emissions observed at the Earth and Jupiter, *Rev. Geophys. Space Phys., 20*, 316, 1982.

Ben-Ari, M., The decametric radiation of Jupiter: A nonlinear mechanism, Ph.D. Thesis, Tel-Aviv Univ., Ramat-Aviv, Israel, 1980.

Benson, R. F., Source mechanism for terrestrial kilometric radiation, *Geophys. Res. Lett., 2*, 52, 1975.

Benson, R. F. and W. Calvert, ISIS 1 observations at the source of auroral kilometric radiation, *Geophys. Res. Lett., 6*, 479, 1979.

Benson, R. F. and H. K. Wong, Low-altitude ISIS 1 observations of auroral radio emission s and their significance to the cyclotron maser instability, *J. Geophys. Res., 92*, 1218, 1987.

Bigg, E. K., Influence of the satellite Io on Jupiter's decametric emission, *Nature, 203*, 1008, 1964.

Boischot, A., A. Lecacheux, M. L. Kaiser, M. D. Desch, J. K. Alexander, and J. W. Warwick, Radio Jupiter after Voyager: An overview of the planetary radio astronomy observations, *J. Geophys. Res., 86*, 8213, 1981.

Borovsky, J., Production of auroral kilometric radiation by gyrophase-bunched double-layer-emitted electrons: Antennae in the magnetospheric current regions, *J. Geophys. Res., 93*, 5727, 1988.

Brown, L. W., The galactic radio spectrum between 130kHz and 2600kHz, *Astrophys. J., 180*, 359, 1973.

Budden, K. G. and D. Jones, Conversion of electrostatic upper hybrid emissions to electromagnetic O- amd X-mode waves in the Earth's magnetosphere, *Annales Geophysicae, 5A*, 21, 1987.

Buti, B. and G. S. Lakhina, Coherent generation mechanism for auroral kilometric radiation, *J. Geophys. Res., 90*, 2785, 1985.

Buti, B. and G. S. Lakhina, Source of bursty emissions from Uranus, *Geophys. Res. Lett., 15*, 1149, 1988.

Burke, B. F. and K. L. Franklin, Observations of a variable radio source associated with the planet Jupiter, *J. Geophys. Res., 60*, 213, 1955.

Calvert, W., The source location of certain Jovian decametric radio emissions, *J. Geophys. Res., 88*, 6165, 1983.

Carr, T. D. and S. Gulkis, The magnetosphere of Jupiter, *Ann. Rev. Astron. Astrophys., 7*, 577, 1969.

Carr, T. D., M. D. Desch, and J. K. Alexander, Phenomenology of magnetospheric radio emissions, in *Physics of the Jovian Magnetosphere*, edited by A. J. Dessler, Cambridge University Press, 1983.

Chian, A. C.-L., Nonlinear generation of the fp radiation upstream of planetary bow shock, in *Proc. Chapman Conference on Plasma Waves and Instabilities in Magnetospheres and at Comets*, edited by H. Oya and B. T. Tsurutani, 1987.

Chian, A. C.-L. and M. Alves, Nonlinear generation of the fundamental radiation of interplanetary type III radio burst, *Astrophys. J., 330*, 277, 1988.

Cuperman, S., Electromagnetic kinetic instabilities in multicomponent space plasmas: Theoretical predictions and computer simulation experiments, *Rev. Geophys., 19*, 307, 1981.

Curtis, S. A., R. P. Lepping, and E. C. Sittler, Jr., The centrifugal flute instability and the generation of Saturnian kilometric radiation, *J. Geophys. Res., 91*, 10989, 1986.

Curtis, S. A., M. D. Desch, and M. L. Kaiser, The radiation belt origin of Uranus' nightside radio emission, *J. Geophys. Res., 92*, 15199, 1987.

Dusenbery, P. B. and L. R. Lyons, General concepts on the generation of auroral kilometric radiation, *J. Geophys. Res., 87*, 7467, 1982.

Evans, D. R., J. H. Romig, and J. W. Warwick, Bursty radio emissions from Uranus, *J. Geophys. Res., 92*, 15206, 1987.

Galeev, A. A. and V. Krasnoselkikh, Strong Langmuir turbulence in the earth's magnetosphere as a source of kilometric radio wave emission, *JETP Lett.*, Engl. Transl., 24, 515, 1976.

Goldstein, M. L., R. R. Sharma, M. Ben-Ari, A. Eviatar, and K. Papadopoulos, A theory of Jovian decametric radiation, *J. Geophys. Res., 88*, 792, 1983.

Goldstein, M. L. and C. K. Goertz, Theories of radio emissions and plasma waves, in *Physics of the Jovian Magnetosphere*, edited by A. J. Dessler, Cambridge University Press, Cambridge, 1983.

Grabbe, G. L., Auroral kilometric radiation: A theoretical review, *Rev. Geophys. Space Phys., 19*, 627, 1981.

Grabbe, G. L., K. Papadopoulos, and P. J. Palmadesso, A coherent nonlinear theory of auroral kilometric radiation: 1. Steady state model, *J. Geophys. Res., 85*, 3337, 1980.

Gurnett, D. A., The earth as a radio source: Terrestrial kilometric radiation, *J. Geophys. Res., 79*, 4227, 1974.

Gurnett, D. A. and R. R. Shaw, Electromagnetic radiation trapped in the magnetosphere above the plasma frequence, *J. Geophys. Res., 78*, 8136, 1973.

Gurnett, D. A., and L. F. Frank, Continuum radiation associated with low-energy electrons in the outer radiation zone, *J. Geophys. Res., 81*, 3875, 1976.

Gurnett, D. A., F. L. Scarf, W. S. Kurth, R. R. Shaw, and R. L. Poynter, Determination of Jupiter's electron density profile from plasma wave observations, *J. Geophys. Res., 86*, 8199, 1981.

Gurnett, D. A., S. D. Shawhan, and R. R. Shaw, Auroral hiss, Z mode radiation, and auroral kilometric radiation in the polar magnetosphere: DE 1 observations, *J. Geophys. Res., 88*, 329, 1983.

Gurnett, D. A., W. S. Kurth, F. L. Scarf, R. L. Poynter, First plasma wave observation at Uranus, *Science, 233*, 106, 1986.

Hewitt, R. G., D. B. Melrose, and K. G. Rönnmark, The loss-cone driven electron-cyclotron maser, *Aust. J. Phys., 35*, 447, 1982.

Istomin, Ya. N., V. I. Petviashvily, and O. A. Pokhotelov, Terrestrial radio emission in the kilometric range by cyclotron solitons, *Sov. J. Plasma Phys., 4*, 76, 1978.

Jones, D., Source of terrestrial non-thermal radiation, *Nature, 260*, 225, 1976.

Jones, D., Mode-coupling of Z-mode waves as a source of terrestrial kilometric and Jovian decametric radiation, *Astron. Astrophys., 55*, 245, 1977.

Jones, D., Non thermal continuum radiation at the radio planets, in *Planetary radio Emissions*, edited by H. O. Rucker and S. J. Bauer, Österreichische Akademie der Wissenschaften, Wien, 1985.

Kaiser, M. L., Observations of non-thermal radiation from planets, in this proceeding, 1988.

Kaiser, M. L. and R. G. Stone, Earth as an intense planetary radio source: similarities to Jupiter and Saturn, *Science, 198*, 285, 1975.

Kaiser, M. L., M. D. Desch, W. S. Kurth, A. Lecacheux, F. Genova, and B. M. Pedersen, in *Saturn*, edited by T. Gehrels and M. Mathews, University of Arizona Press, Tucson, 1984.

Kaiser, M. L., M. D. Desch, and S. A. Curtis, The sources of Uranus' dominant nightside radio emissions, *J. Geophys. Res., 92*, 15169, 1987.

Kennel, C. F. and H. E. Petschek, Limit on stably trapped particles fluxes, *J. Geophys. Res., 71*, 1, 1966.

Kennel, C. F., R. F. Chen, S. L. Moses, W. S. Kurth, F. V. Coroniti, F. L. Scarf, and F. F. Chen, Z mode radiation in Jupiter's magnetosphere, *J. Geophys. Res., 92*, 9978, 1987.

Kurth, W. S., M. M. Baumback, and D. A. Gurnett, Direction finding measurements of auroral kilometric radiation, *J. Geophys. Res., 80*, 2764, 1975.

Kurth, W. S., D. D. Barbosa, F. L. Scarf, D. A. Gurnett, and R. L. Poynter, Low frequence radio emissions from Jupiter: Jovian kilometric radiation, *Geophys. Res. Lett., 6*, 747, 1979.

Kurth, W. S., D. A. Gurnett, and F. L. Scarf, Spatial and temporal studies of Jovian kilometric radiation, *Geophys. Res. Lett., 7*, 61, 1980.

Kurth, W. S., D. A. Gurnett, and R. R. Anderson, Escaping nonthermal continuum radiation, *J. Geophys. Res., 86*, 5519, 1981.

Kurth, W. S., F. L. Scarf, J. D. Sullivan, and D. A. Gurnett, Detection of nonthermal continuum radiation in Saturn's magnetosphere, *Geophys. Res. Lett., 9*, 889, 1982.

Kwok, Y. C. and L. C. Lee, Transmission of magnetohydrodynamic waves through the rotational discontinuity at the Earth's magnetopause, *J. Geophys. Res., 89*, 10697, 1984.

Leblanc, Y., M. G. Aubier, A. Ortega-Molina, and A. Lecacheux, Overview of the Uranian radio emission: Polarization and constraints on source locations, *J. Geophys. Res., 92*, 15125, 1987.

Lee, L. C., Transmission of Alfvén waves through the rotational discontinuity at magnetopause, *Planet. Space Sci., 30*, 1127, 1982.

Lee, L. C., Theory and simulation of auroral kilometric radiation, in *Proc. Conf. Achievements of the IMS*, 505, 1983.

Lee, L. C. and C. S. Wu, Amplification of radiation near cyclotron frequence due to electron population inversion, *Phys. Fluids, 23*, 1348, 1980.

Lee, L. C., J. R. Kan, and C. S. Wu, Generation of auroral kilometric radiation and the structure of auroral acceleration region, *Planet. Space Sci., 28*, 703, 1980.

Le Quéau, D., R. Pellat, and A. Roux, Direct generation of the auroral kilometric radiation by the maser synchrotron instability: physical mechanism and parametric study, *J. Geophys. Res., 89*, 2831, 1984.

McKenzie, J. F. and K. O. Westphal, Interaction of hydromagnetic waves with hydromagnetic shocks, *Phys. Fluids, 13*, 630, 1970.

Mellott, M. M., W. Calvert, R. L. Huff, and D. A. Gurnett: DE-1 observations of ordinary mode and extraordinary mode auroral kilometric radiation, *Geophys. Res. Lett., 11*, 1188, 1984.

Mellott, M. M., R. L. Huff, and D. A. Gurnett, DE-1 observations of harmonic kilometric radiation, *J. Geophys. Res., 91*, 13732, 1986.

Melrose, D. B., An interpretation of Jupiter's decametric radiation and the terrestrial kilometric radiation as direct amplified gyroemission, *Astrophys. J., 207*, 651, 1976.

Melrose, D. B., A theory for the nonthermal radio continua in the terrestrial and Jovian magnetosphere, *J. Geophys. Res., 86*, 30, 1981.

Melrose, D. B., K. G. Rönnmark, and R. G. Hewitt, Terrestrial kilometric radiation: The cyclotron theory, *J. Geophys. Res., 87*, 5140, 1982.

Melrose, D. B., R. G. Hewitt, and G. A. Dulk, Electron-cyclotron maser emission: Relative growth and damping rates for different modes and harmonics, *J. Geophys. Res., 89*, 897, 1984.

Okuda, H., M. Ashour-Abdalla, M. S. Chance, and W. S. Kurth, Generation of nonthermal continuum radiation in the magnetosphere, *J. Geophys. Res., 87*, 10457, 1982.

Omidi, N. and D. A. Gurnett, Growth rate calculations of auroral kilometric radiation using the relativistic resonance condition, *J. Geophys. Res., 87*, 2377, 1982.

Omidi, N. and C. S. Wu, The effect of background plasma density on the growth of ordinary and Z mode emissions in the auroral zone, *J. Geophys. Res., 90*, 6641, 1985.

Omidi, N., C.. S. Wu, and D. A. Gurnett, Generation of auroral kilometric and Z-mode radiation by the cyclotron maser mechanism, *J. Geophys. Res., 89*, 883, 1984.

Oya, H., Conversion of electrostatic waves into electromagnetic waves: Numerical calculation of the dispersion relation for all wavelengths, *Radio Sci., 6*, 1131, 1971.

Oya, H., Origin of Jovian decametric emission: Conversion from the electron cyclotron plasma wave to the ordinary mode electromagnetic wave, *Planet. Space Sci., 22*, 687, 1974.

Oya, H., and A. Morioka, Observational evidence of Z and L-O mode waves as the origin of auroral kilometric radiation from the Jikiken satellite, *J. Geophys. Res., 88*, 6189, 1983.

Palmadesso, P., T. P. Coffey, S. L. Ossakow, and K. Papadopoulos, Generation of terrestrial kilometric radiation by a beam-driven electromagnetic instability, *J. Geophys. Res., 81*, 1762, 1976.

Pritchett, P. L., Relativistic dispersion, the cyclotron maser instability, and auroral kilometric radiation, *J. Geophys. Res., 89*, 8957, 1984.

Pritchett, P. L., Cyclotron maser radiation from a source structure localized perpendicular to the ambient magnetic field, *J. Geophys. Res., 91*, 13569, 1986.

Pritchett, P. L. and R. J. Strangeway, A simulation study of kilometric radiation generation along an auroral field line, *J. Geophys. Res., 90*, 9650, 1985.

Roux, A. and R. Pellat, Coherent generation of the auroral kilometric radiation by nonlinear beating between electrostatic waves, *J. Geophys. Res., 84*, 5189, 1979.

Shawhan, S. D., and D. A. Gurnett, Polarization measurements of auroral kilometric radiation by Dynamics Explorer 1, *Geophys. Res. Lett., 9*, 913, 1982.

Stix, T. H., Radiation and absorption via mode conversion in an inhomogeneous collision-free plasma, *Phys. Rev. Lett., 15*, 878, 1965.

Wagner, J. S., L. C. Lee, C. S. Wu, and T. Tajima, Computer simulation of auroral kilometric radiation, *Geophys. Res. Lett., 6*, 483, 1983.

Wagner, J. S., L. C. Lee, C. S. Wu, and T. Tajima, A simulation study of the loss cone driven cyclotron maser applied to auroral kilometric radiation, *Radio Sci., 19*, 509, 1984.

Warwick, J. W., J. B. Pearce, D. R. Evans, T. D. Carr, J. J. Schauble, J. K. Alexander, M. L. Kaiser, M. D. Desch, M. Pedersen, A. Lecacheux, G. Daigne, A. Boischot, and C. H. Barrow, Planetary radio astronomy observations from Voyager 1 near Saturn, *Science, 212*, 239, 1981.

Warwick, J. W., D. R. Evans, J. H. Romig, C. B. Sawyer, M. D. Desch, M. L. Kaiser, J. K. Alexander, T. D. Carr, D. H. Staelin, S. Gulkis, R. L. Poynter, M. Aubier, A. Boischot, Y. Leblanc, A. Lecacheux, B. M. Pedersen, P. Zarka, Voyager 2 radio observations of Uranus, *Science, 233*, 102, 1986.

Wong, H. K., C. S. Wu, F. J. Ke, R. S. Schneider, and L. F. Ziebell, Electromagnetic cyclotron-loss-cone instability associated with weakly relativistic electrons, *J. Plasma Phys., 28*, 503, 1982.

Wu, C. S., Kinetic cyclotron and synchrotron maser instabilities: Radio emission processes by direct amplification of radiation, *Space Sci. Rev., 41*, 215, 1985.

Wu, C. S. and L. C. Lee, A theory of the terrestrial kilometric radiation, *Astrophys. J., 230*, 621, 1979.

Wu, C. S. and X. M. Qiu, Emissions of second harmonic auroral kilometric radiation, *J. Geophys. Res., 88*, 10072, 1983.

Wu, C. S., H. K. Wong, D. J. Gorney, and L. C. Lee, Generation of the auroral kilometric radiation, *J. Geophys. Res., 87*, 4476, 1982.